# Cilia, Mucus, and Mucociliary Interactions

# Cilia, Mucus, and Mucociliary Interactions

edited by

Gerald L. Baum
*Israel Lung Association: Tel Aviv*
*Tel Aviv, Israel*

Zvi Priel
*Ben-Gurion University of the Negev*
*Beer-Sheva, Israel*

Yehudah Roth
*Edith Wolfson Medical Center*
*Holon, Israel*

Nadav Liron
*Technion–Israel Institute of Technology*
*Hafia, Israel*

Ervin Ostfeld
*Hillel Jaffe Medical Center*
*Hadera, Israel*

MARCEL DEKKER, INC. NEW YORK • BASEL • HONG KONG

**Library of Congress Cataloging-in-Publication Data**
Cilia, mucus, and mucociliary interactions/edited by Gerald L. Baum, Zvi Priel, Yehudah Roth, Nadav Liron, and Ervin Ostfeld
p. cm.
Includes bibliographical references and index
ISBN 0-8427-0138-0
1. Mucociliary system—Physiology. 2. Mucociliary system—Pathophysiology. I. Baum, Gerald L.
QP215.C54 1998
612.2—dc21

97-44126
CIP

The publisher offers discounts on this book when ordered in bulk quantities. For more information, write to Special Sales/Professional Marketing at the address below.

This book is printed on acid-free paper.



MARCEL DEKKER, INC.
270 Madison Avenue, New York, New York 10016
*http://www.dekker.com*

Current printing (last digit):
10 9 8 7 6 5 4 3 2 1

**PRINTED IN THE UNITED STATES OF AMERICA**

# Preface

An international congress dealing with cilia, mucus, and mucociliary interactions was held February 23–27, 1997, in Jerusalem, Israel, to honor Professor Alex Silberberg. Professor Silberberg's death at the age of 70 in 1993 ended a highly productive career and saddened friends, colleagues, and former students around the world. This book is a collection of manuscripts representing some of the work presented at the congress and reflects both the basic science and clinical aspects, in the broadest sense, of the general subjects of the congress.

Professor Silberberg was the Director in the Polymer Department of the Weizmann Institute of Science at the time of his death and in that capacity had occasion to act as teacher and mentor of many of the authors and lecturers whose work appears within this book. It is the opinion of all who have been involved with the organization of the congress and the publication of this volume that the high quality of the work included in this book would have pleased Alex and stimulated him to pursue the subjects and learn even more about them.

For the first time, basic science aspects of mucociliary interactions and the components of the system are being presented along with the latest clinical applications of this information, giving an integrated view that had not previously appeared. Physicians dealing with the upper airways, including the nose and sinuses and the tracheobronchial tree, as well as those dealing with any area of the body where mucus and cilia are crucial, will find useful information. The unique quality of the meeting upon which this book is based was the possibility of communication between clinicians and basic scientists, and these proceedings give the reader an opportunity to participate in this unusual situation.

In particular, the basic science papers include details of intercellular communication, genetics of mucus-producing cells, molecular biology of dynein, ciliogenesis, ciliary coordination, modeling of mucociliary function, polymer biophysics of mucus, and many other topics. In the clinical arena, papers deal with mucociliary clearance in the upper airways including the nose and sinuses, the relationship between airway ion transport and mucociliary clearance, laboratory studies of mucus, airway bacteria and mucus production, medical modulation of mucus, diagnosis of primary and secondary mucociliary function, and pharmacological regulation of mucociliary activity.

It is hoped that this book will be of value to those interested in this general field and will stimulate work that will further clarify the many mysteries that challenge us all.

*Gerald L. Baum*
*Zvi Priel*
*Yehudah Roth*
*Nadav Liron*
*Ervin Ostfeld*

# About the Editors

**GERALD L. BAUM** is the Medical Director of the Israel Lung Association, Tel Aviv, Israel, as well as a Professor Emeritus at the Sackler School of Medicine, Tel Aviv University, Israel. The author, coauthor, editor, or coeditor of eight books and 180 journal papers and book chapters, he is a Fellow of the American College of Chest Physicians and the American College of Physicians, and a member of the American Thoracic Society and the Israel Medical Association, among others. Dr. Baum received the M.D. degree (1947) from the University of Wisconsin—Madison.

**ZVI PRIEL** is a Professor in the Department of Chemistry, Ben-Gurion University of the Negev, Beer-Sheva, Israel. The author or coauthor of more than 70 refereed articles in scientific journals dealing with biomotility, signal transduction, and allied subjects. Dr. Priel received the Ph.D. degree (1970) from the Weizmann Institute of Science, Rehovot, Israel.

**YEHUDAH ROTH** is the Director of the Institute for Nose and Sinus Therapy and Clinical Investigations, Department of Otolaryngology-Head and Neck Surgery, The Edith Wolfson Medical Center, Holon, Israel. Dr. Roth received the M.D. degree (1982) from the Sackler School of Medicine, Tel Aviv University, Israel.

**NADAV LIRON** is the Vice President for Academic Affairs as well as a Professor in the Department of Mathematics, Technion–Israel Institute of Technology, Haifa. The author or coauthor of over 50 publications in the areas of biomathematics, and fluid mechanics in biology and applied mathematics, Dr. Liron received the Ph.D. degree (1969) in applied mathematics from the Weizmann Institute of Science, Rehovot, Israel.

**ERVIN OSTFELD** is the Director of the Department of Otorhinolaryngology, Hillel Yaffe Hospital, Hadera, Israel. A member of the Israel Medical Association and the Israel Society for Otolaryngology, Dr. Ostfeld received the M.D. degree (1958) from Institutul Medico-Farmaceutic, Cluj, Romania.

# Acknowledgments

The editors would like to recognize the following individuals and groups for participation and sponsorship of the Congress on Cilia, Mucus, and Mucociliary Interactions, that took place February 23–27, 1997, in Jerusalem, Israel. The editors of this book comprised the Executive Committee of the Congress.

Honorary President: Ephraim Katchalski-Katzir; Organizing Committee (Israel): Simon Godfrey, Daniel Katznelson, Yaacov Klein, Miriam Lurie, Zvi Merom, Eliora Ron, Elvan Tabatchnik, and Tzila Zwas. Organizing Committee (International): Australia: Jonathan Rutland; Canada: Malcolm King, Michael Newhouse, and Arnold Noyek; France: Edith Puchelle; Italy: Desiderio Passali; Japan: Yasuo Sakakura and Keiichi Takahashi; The Netherlands: E. H. Huizing; Sweden: Sven Lindberg; South Africa: Charles Feldman; United Kingdom: Peter Cole and Michael A. Sleigh; United States: Carol Basbaum, Charles D. Bluestone, J. J. Blum, Richard C. Boucher, Ellen Roter Dirksen, I. R. Gibbons, Mitchell Litt, Thomas McCaffrey, Jay A. Nadel, Lynne M. Reid, Stephen Rennard, Burce K. Rubin, Peter Satir, Pedro Verdugo, and Adam Wanner.

For financial support, the editors would like to thank: Bayer plc, Pharmaceutical Division, United Kingdom; Ben Gurion University of the Negev, Israel; Canada-International Scientific Exchange Program (CISEPO); Department of Materials and Interfaces, Weizmann Institute of Science, Israel; El Al Israel Airlines; Genentech, Inc., United States; Glaxo Wellcome plc, United Kingdom; Government of Israel, Ministry of Science and Development, and Ministry of Tourism; Helen and Milton A. Kemmelman Center for Biomedical Structure and Assembly, Weizmann Institute of Science, Israel; Maurice and Gabriela Goldscheger Conference Foundation, Weizmann Institute of Science, Israel; Ministry of Foreign Affairs, France, and the French Embassy in Israel; Technion–Israel Institute of Technology; Teijin Ltd., Japan; Tel Aviv University, Israel; and Yad Hanadiv-Keren Rothschild, Israel.

For academic sponsorship, the editors would like to thank: American College of Chest Physicians; Ben Gurion University of the Negev, Israel; Canada-International Scientific Exchange Program (CISEPO); Canadian Society of Otolaryngology-Head and Neck Surgery; European Respiratory Society; European Rhinologic Society; Hebrew University, Jerusalem, Israel; Institute for Nose and Sinus Therapy and Clinical Investigation, Wolfson Medical Center, Israel; Technion–Israel Institute of Technology; Israel Society of Otolaryngology-Head and Neck Surgery; Israel Society of Pulmonology; Tel Aviv University, Israel; and Weizman Institute of Science, Israel.

Thanks are also due to the Kenes-Organizers of Congresses and Tour Operators Ltd., Tel Aviv, for organizing the congress.

# Contents

# Contributors

**Lubna H. Abdullah, Ph.D.** Department of Pulmonary Medicine, The University of North Carolina at Chapel Hill, Chapel Hill, North Carolina

**William M. Abraham, M.D.** Division of Pulmonary and Critical Care Medicine, University of Miami School of Medicine, Miami, Florida

**Kenneth B. Adler, Ph.D.** Department of Anatomy, Physical Sciences and Radiology, College of Veterinary Medicine, North Carolina State University, Raleigh, North Carolina

**Cuneyt M. Alper, M.D.** Department of Pediatric Otolaryngology, Children's Hospital of Pittsburgh, and University of Pittsburgh School of Medicine, Pittsburgh, Pennsylvania

**Ronald Anderson, Ph.D.** Department of Immunology, School of Pathology, University of Pretoria, Pretoria, South Africa

**Sandra D. Anderson, Ph.D., D.Sc.** Department of Respiratory Medicine, Royal Prince Alfred Hospital, Sydney, New South Wales, Australia

**Kaya L. Andrews, Ph.D.** Laboratory of Pulmonary Pathobiology, National Institute of Environmental Health Sciences, National Institutes of Health, Research Triangle Park, North Carolina

**Dale L. Bailey, Ph.D.** MRC Cyclotron Unit, Hammersmith Hospital, London, United Kingdom

**S. J. Banner, M.Sc.** Department of Child Health, Ninewells Hospital and Medical School, and University of Dundee, Dundee, United Kingdom

**Carol B. Basbaum, Ph.D.** Department of Anatomy, Cardiovascular Research Institute, School of Medicine, University of California, San Francisco, San Francisco, California

**George Bautovich, Ph.D., M.B., B.S., B.Sc.(Med), F.R.A.C.P., D.D.U.** Department of Nuclear Medicine, Royal Prince Alfred Hospital, Sydney, New South Wales, Australia

**William D. Bennett, Ph.D.** Department of Medicine, Center for Environmental Medicine, The University of North Carolina at Chapel Hill, Chapel Hill, North Carolina

**John R. Blake, Ph.D.** School of Mathematics and Statistics, The University of Birmingham, Birmingham, United Kingdom

**Charles D. Bluestone, M.D.** Department of Pediatric Otolaryngology, Children's Hospital of Pittsburgh, and University of Pittsburgh School of Medicine, Pittsburgh, Pennsylvania

**Jacob J. Blum, Ph.D.** Department of Cell Biology, Duke University Medical Center, Durham, North Carolina

**György Miklós Böhm, M.D., Ph.D.** Department of Pathology, University of São Paulo School of Medicine, São Paulo, Brazil

**Richard C. Boucher, M.D.** Department of Medicine, The University of North Carolina at Chapel Hill, Chapel Hill, North Carolina

**Alex Braiman** Department of Chemistry, Ben-Gurion University of the Negev, Beer-Sheva, Israel

**Peter Bye, M.B., B.S.(Syd), F.R.A.C.P., Ph.D.** Respiratory Investigation Unit, Royal Prince Alfred Hospital, Sydney, New South Wales, Australia

**Birgitta Carlén, Ph.D.** Department of Pathology and Cytology, University Hospital, Lund, Sweden

**Anders Cervin, M.D.** Department of Otorhinolaryngology-Head and Neck Surgery, University Hospital, Lund, Sweden

**W. M. Chan, M.D.** Host Defence Unit, Imperial College of Science, Technology and Medicine, National Heart and Lung Institute, London, United Kingdom

**Nonghoon Choe, Ph.D.*** Department of Anatomy, Physical Sciences and Radiology, College of Veterinary Medicine, North Carolina State University, Raleigh, North Carolina

* *Current affiliation:* Department of Pathology, Uniformed Services University of the Health Sciences, Department of Defense, Bethesda, Maryland

**C. Cobain, B.Sc.** Department of Child Health, Ninewells Hospital and Medical School, and University of Dundee, Dundee, United Kingdom

**Peter J. Cole, M.D., F.R.C.P.** Host Defence Unit, Imperial College of Science, Techology and Medicine, National Heart and Lung Institute, and Royal Brompton Hospital, London, United Kingdom

**Catarina Cramnert, M.T.** Department of Pathology and Cytology, University Hospital, Lund, Sweden

**C. William Davis, M.D.** Department of Physiology, The University of North Carolina at Chapel Hill, Chapel Hill, North Carolina

**Evangelia Daviskas, M.Biomed.E., Ph.D.** Department of Respiratory Medicine, Royal Prince Alfred Hospital, Sydney, New South Wales, Australia

**Robbert U. de Iongh, M.Sc., Ph.D.** Ciliary Studies Laboratory, Respiratory Unit, Concord Hospital, Concord, and Department of Anatomy and Histology, The University of Sydney, Sydney, New South Wales, Australia

**D. P. Dhillon, M.D.** Respiratory Medicine, Kings Cross Hospital, Dundee, United Kingdom

**Scott R. Diehl, Ph.D.** National Institute of Dental Research, National Institutes of Health, Bethesda, Maryland

**Robert H. Dillon, Ph.D.*** Department of Mathematics, Tulane University, New Orleans, Louisiana

**Ellen Roter Dirksen, Ph.D.** Department of Neurobiology, School of Medicine, University of California, Los Angeles, Los Angeles, California

**Davorin Djanić, M.D.** Ear, Nose and Throat Division, General Hospital Slavonski Brod, Slavonski Brod, Croatia

**Lisa J. Fauci, Ph.D.** Department of Mathematics, Tulane University, New Orleans, Louisiana

**Charles Feldman, M.D., Ph.D.** Johannesburg Hospital, and Department of Medicine, University of the Witwatersrand, Parktown, Johannesburg, South Africa

**Wei Feng, M.D.** Pulmonary Research Group, University of Alberta, Edmonton, Alberta, Canada

* *Current affiliation:* Department of Pure and Applied Mathematics, Washington State University, Pullman, Washington

**Eliane Festa, M.D., Ph.D.** Pulmonary Research Group, University of Alberta, Edmonton, Alberta, Canada

**Bernard M. Fischer, D.V.M., Ph.D.*** Department of Anatomy, Physical Sciences and Radiology, College of Veterinary Medicine, North Carolina State University, Raleigh, North Carolina

**Dan M. Fliss, M.D.** Head and Neck Surgical Oncology Unit, Soroka Medical Center, and Ben-Gurion University of the Negev, Beer-Sheva, Israel

**Diana Freas-Lutz, M.S.** National Institute of Dental Research, National Institutes of Health, Bethesda, Maryland

**John Freeman, M.D.** Pediatric Research Institute, St. Louis University School of Medicine, St. Louis, Missouri

**Saul Frenkiel, M.D., F.R.C.S.(C)** Department of Otolaryngology, Sir Mortimer B. Davis–Jewish General Hospital, and McGill University, Montreal, Quebec, Canada

**Larisa Gheber, Ph.D.** Department of Biology, The Johns Hopkins University, Baltimore, Maryland

**I. R. Gibbons, Ph.D.** Pacific Biomedical Research Center, University of Hawaii, Honolulu, Hawaii

**Chantal Giguère, M.D.** Department of Otolaryngology, Sir Mortimer B. Davis–Jewish General Hospital, and McGill University, Montreal, Quebec, Canada

**Elizabeth Gillanders, B.A.** National Institute of Dental Research, National Institutes of Health, Bethesda, Maryland

**T. Gray** Laboratory of Pulmonary Pathobiology, National Institute of Environmental Health Sciences, National Institutes of Health, Research Triangle Park, North Carolina

**Shay Gueron, Ph.D.** Department of Mathematics, Technion–Israel Institute of Technology, Haifa, Israel

**K. Guzman, Ph.D.** Laboratory of Pulmonary Pathobiology, National Institute of Environmental Health Sciences, National Institutes of Health, Research Triangle Park, North Carolina

* *Current affiliation:* Division of Pulmonary Diseases, Department of Pediatrics, Duke University Medical Center, Durham, North Carolina

**Roderick L. Hall, Ph.D.** Research Department, Bayer plc, Slough, United Kingdom

**Toshikazu Hamasaki, Ph.D.** Department of Anatomy and Structural Biology, Albert Einstein College of Medicine, Bronx, New York

**Qutayba Hamid, M.D.** McGill Nasal Research Group, Meakins–Christie Laboratories, and McGill University, Montreal, Quebec, Canada

**John Hamilton, F.R.C.S.** Department of Otolaryngology, University of Liverpool, Liverpool, United Kingdom

**Daniel J. Hassett, M.D.** Department of Molecular Genetics, Biochemistry and Microbiology, University of Cinncinati, Cinncinati, Ohio

**Ariane Hemming, B.Sc.** Respiratory Investigation Unit, Royal Prince Alfred Hospital, Sydney, New South Wales, Australia

**Anne-Laure Herard, Ph.D.** Unité INSERM, CHU Maison Blanche, Reims, France

**Keisuke Hirata, M.D.*** Department of Otorhinolaryngology, Mie University School of Medicine, Mie, Japan

**Samuel B. Ho, M.D.** Department of Medicine, University of Minnesota, and Veterans Affairs Medical Center, Minneapolis, Minnesota

**Michael Holwill, B.Sc., Ph.D.** Department of Physics, King's College, London, United Kingdom, and Albert Einstein College of Medicine, Bronx, New York

**Gordon Honeyman, B.Sc.(Hons), M.Sc.‡** Department of Child Health, Ninewells Hospital and Medical School, and University of Dundee, Dundee, United Kingdom

**Toyoharu Isawa, M.D., Ph.D.** Director, Yokohama Higashi National Hospital, Yokohama, Japan

**Jolanta T. Jackowski, M.D.†** Division of Pulmonary and Critical Care Medicine, University of Miami School of Medicine, Miami, Florida

**Hirofumi Kai, M.D.** Department of Pharmacological Sciences, Kumamoto University, Kumamoto, Japan

---

† Deceased

*Current affiliations:*

* Department of Otorhinolaryngology, Yamamoto Hospital, Kuwana, Japan

‡ Department of Biochemistry, University of Durham, Durham, United Kingdom

**Edward Earl Kassel, D.D.S., M.D., F.R.C.P.C.** Department of Medical Imaging, Mount Sinai Hospital, and University of Toronto, Toronto, Ontario, Canada

**P. J. Kemp, Ph.D.** Department of Child Health, Ninewells Hospital and Medical School, and University of Dundee, Dundee, United Kingdom

**Malcolm King, Ph.D.** Pulmonary Research Group, University of Alberta, Edmonton, Alberta, Canada

**Chikako Kishioka, M.D.*** Department of Pediatrics, Pediatric Research Institute, St. Louis University School of Medicine, St. Louis, Missouri

**Michael R. Knowles, M.D.** Department of Medicine, The University of North Carolina at Chapel Hill, Chapel Hill, North Carolina

**Hirokatsu Kobayashi, M.D.†** Department of Applied Physics and Chemistry, University of Electro-Communications, Tokyo, Japan

**Kenji Kobayashi, Eng.M.** Division of Pulmonary and Critical Care Medicine, University of Miami School of Medicine, Miami, Florida

**Ja Seok Koo, Ph.D.** Laboratory of Pulmonary Pathobiology, National Institute of Environmental Health Sciences, National Institutes of Health, Research Triangle Park, North Carolina

**Alon Korngreen, Ph.D.** Department of Chemistry, Ben-Gurion University of the Negev, Beer-Sheva, Israel

**Hyung-Hwan Lee, Eng.M.‡** Department of Applied Physics and Chemistry, University of Electro-Communications, Tokyo, Japan

**Konstantin Levit-Gurevich, M.Sc.** Department of Mathematics, Technion–Israel Institute of Technology, Haifa, Israel

**Chengming Li, M.D.** Department of Anatomy, Physical Sciences, and Radiology, College of Veterinary Medicine, North Carolina State University, Raleigh, North Carolina

**Hongfei Li, M.D.** Department of Anatomy, Physical Sciences, and Radiology, College of Veterinary Medicine, North Carolina State University, Raleigh, North Carolina

---

*Current affiliations:*
* Department of Otolaryngology, Mie University, Mie, Japan
† Tokyo Electron, Tokyo, Japan
‡ Daewoo Electronics Co., Ltd., Seoul, Korea

**Jian-Dong Li, M.D., Ph.D.** Department of Anatomy, Cardiovascular Research Institute, School of Medicine, University of California, San Francisco, San Francisco, California

**Sven Lindberg, M.D., Ph.D.** Department of Otorhinolaryngology-Head and Neck Surgery, University Hospital, Lund, Sweden

**Nadav Liron, Ph.D.** Department of Mathematics, Technion–Israel Institute of Technology, Haifa, Israel

**Mariangela Macchione, M.D.** Department of Pathology, University of São Paulo School of Medicine, São Paulo, Brazil

**Yuichi Majima, M.D.** Department of Otorhinolaryngology, Mie University School of Medicine, Mie, Japan

**Lindsay Marshall, B.Sc.(Hons), M.Sc.** Department of Child Health, Ninewells Hospital and Medical School, and University of Dundee, Dundee, United Kingdom

**Anil Mehta, M.Sc., M.R.C.P.** Department of Child Health, Ninewells Hospital and Medical School, and University of Dundee, Dundee, United Kingdom

**Damir Miličić, M.D.** Ear, Nose and Throat Division, General Hospital Slavonski Brod, Slavonski Brod, Croatia

**Takeshi Miyata, Ph.D.** Department of Pharmacological Sciences, Kumamoto University, Kumamoto, Japan

**Ranko Mladina, M.D., Ph.D.** Otorhinolaryngology Department, University Hospital Salata, Zagreb, Croatia

**Lucy Morgan, B.Med.** Ciliary Studies Laboratory, Respiratory Unit, Concord Hospital, Concord, New South Wales, Australia

**R. Muimo, Ph.D.** Department of Child Health, Ninewells Hospital and Medical School, and University of Dundee, Dundee, United Kingdom

**Tadashi Nakamura, Ph.D.** Department of Applied Physics and Chemistry, University of Electro-Communications, Tokyo, Japan

**Paul Nettesheim, M.D., D.M.S.** Laboratory of Pulmonary Pathobiology, National Institute of Environmental Health Sciences, National Institutes of Health, Research Triangle Park, North Carolina

**Michael T. Newhouse, M.D., M.Sc., F.R.C.P.(C), F.A.C.P., F.C.C.P.** McMaster Faculty of Health Sciences, McMaster University, and Barnett Medical Aerosol Research Laboratory, St. Joseph's Hospital, Hamilton, Ontario, Canada

**B. Ben Newton, B.Sc.(Hons)** Research Department, Bayer plc, Slough, United Kingdom

**Jason H. Nielson, M.D.** Department of Anatomy and Structural Biology, Albert Einstein College of Medicine, Bronx, New York

**Satoru Noguchi, Eng.M.*** Department of Information Network Science, University of Electro-Communications, Tokyo, Japan

**Peadar G. Noone, M.D., M.R.C.P.I., F.C.C.P.** Department of Medicine, The University of North Carolina at Chapel Hill, Chapel Hill, North Carolina

**Arnold M. Noyek, M.D., F.R.C.S.(C.)** Departments of Otolaryngology and Medical Imaging, Mount Sinai Hospital, and University of Toronto, Toronto, Ontario, Canada

**Lawrence E. Ostrowski, Ph.D.** Laboratory of Pulmonary Pathobiology, National Institute of Environmental Health Sciences, National Institutes of Health, Research Triangle Park, North Carolina

**Jacek Pawlik, M.D.** Pediatric Division, Institute of Tuberculosis and Lung Diseases, Rabka, Poland

**Andrzej Pogorzelski, M.D.** Pediatric Division, Institute of Tuberculosis and Lung Diseases, Rabka, Poland

**Pravin D. Potdar, Ph.D.†** Laboratory of Pulmonary Pathobiology, National Institute of Environmental Health Sciences, National Institutes of Health, Research Triangle Park, North Carolina

**Zvi Priel, Ph.D.** Department of Chemistry, Ben-Gurion University of the Negev, Beer-Sheva, Israel

**Edith Puchelle, Ph.D.** Unité INSERM, CHU Maison Blanche, Reims, France

**Jeff A. Regnis, Ph.D.** Respiratory Investigation Unit, Royal Prince Alfred Hospital, Sydney, New South Wales, Australia

**Stephen Rennard, M.D.** Department of Pulmonary and Critical Care Medicine, University of Nebraska Medical Center, Omaha, Nebraska

---

*Current affiliations*:
* ASM Japan, Tokyo, Japan
† Tata Memorial Centre, Cancer Research Institute, Parel, Bombay, India

**Michael Robinson, D.V.M.** Respiratory Investigation Unit, Royal Prince Alfred Hospital, Sydney, New South Wales, Australia

**Lori G. Rochelle, Ph.D.†** Department of Anatomy, Physical Sciences and Radiology, College of Veterinary Medicine, North Carolina State University, Raleigh, North Carolina

**Yehudah Roth, M.D.** Department of Otolaryngology-Head and Neck Surgery, The Edith Wolfson Medical Center, Holon, Israel

**Hans-Dieter Rott, M.D.** Institute of Human Genetics, University of Erlangen-Nuremberg, Erlangen, Germany

**Bruce K. Rubin, M.D., F.R.C.P.(C), F.C.C.P.*** Department of Pediatrics, St. Louis University School of Medicine, and Cardinal Glennon Children's Hospital, St. Louis, Missouri

**Thomas Runer, M.D.** Department of Otorhinolaryngology-Head and Neck Surgery, University Hospital, Lund, Sweden

**Ewa Rutkiewicz** Institute of Human Genetics, Polish Academy of Sciences, Poznan, Poland

**Jonathan Rutland, B.Sc., M.B., B.S., F.R.A.C.P., F.C.C.P.** Ciliary Studies Laboratory, Respiratory Unit, Concord Hospital, Concord, New South Wales, Australia

**Andrew Rutman, C.Biol., M.I.Biol.** Host Defence Unit, Imperial College of Science, Technology and Medicine, National Heart and Lung Institute, and Royal Brompton Hospital, London, United Kingdom

**Yasuo Sakakura, M.D., Ph.D.** Department of Otorhinolaryngology, Mie University School of Medicine, Mie, Japan

**Matthias Salathé, M.D.** Division of Pulmonary and Critical Care Medicine, University of Miami School of Medicine, Miami, Florida

**Paulo Hilário N. Saldiva, M.D., Ph.D.** Department of Pathology, University of São Paulo School of Medicine, São Paulo, Brazil

**Peter Satir, Ph.D.** Department of Anatomy and Structural Biology, Albert Einstein College of Medicine, Bronx, New York

---

*Current affiliations*:

†Department of Cystic Fibrosis/Pulmonary Research, The University of North Carolina at Chapel Hill, Chapel Hill, North Carolina

*Department of Pediatrics, Bowman Gray School of Medicine, Winston-Salem, North Carolina

**Tomo-Oki Satoh, M.D.‡** Department of Applied Physics and Chemistry, University of Electro-Communications, Tokyo, Japan

**Zsuzsanna Seybold, M.D.** Division of Pulmonary and Critical Care Medicine, University of Miami School of Medicine, Miami, Florida

**Michael A. Sleigh, Ph.D., D.Sc.** Biodiversity and Ecology Division, School of Biological Sciences, University of Southampton, Southampton, United Kingdom

**Cui-e Sun, B.S.** National Institute for Dental Research, National Institutes of Health, Bethesda, Maryland

**Kazuhiko Takeuchi, M.D.** Department of Otorhinolaryngology, Mie University School of Medicine, Mie, Japan

**Robert P. Tomkiewicz, M.D.*** Pediatric Research Institute, St. Louis University School of Medicine, St. Louis, Missouri

**Elina Toskala, M.D., Ph.D.†** Department of Otorhinolaryngology-Head and Neck Surgery, Tampere University Hospital, Tampere, Finland

**Peter van Asperen, M.D., F.R.A.C.P.** Department of Pediatric Medicine, Royal Alexandria Hospital for Children, Sydney, New South Wales, Australia

**Pedro Verdugo, M.D.** Center for Bioengineering, University of Washington, Seattle, Washington

**Shengbiao Wang, M.D.** National Institute for Dental Research, National Institutes of Health, Bethesda, Maryland

**Adam Wanner, M.D.** Division of Pulmonary and Critical Care Medicine, University of Miami School of Medicine, Miami, Florida

**Karen A. Waters, M.B., B.S., Ph.D., F.R.A.C.P.** Department of Respiratory Medicine, Royal Alexandria Hospital for Children, Sydney, New South Wales, Australia

**Peter John Wills, M.B., Ch.B., Ph.D.** Host Defence Unit, Imperial College School of Medicine, National Heart and Lung Institute, London, United Kingdom

---

*Current affiliations*:
‡Yamanouchi Pharmaceutical Company, Tsukuba, Japan
*Private practice
†Department of Pediatrics, Bowman-Gray School of Medicine, Winston-Salem, North Carolina

**Robert Wilson, M.D., F.R.C.P.** Host Defence Unit, Imperial College of Science, Technology and Medicine, National Heart Institute, and Royal Brompton Hospital, London, United Kingdom

**John H. Winter, M.D., F.R.C.P.(Ed), F.R.C.P.(Glas)** Department of Respiratory Medicine, Kings Cross Hospital, Dundee, United Kingdom

**Michał Witt, M.D., Ph.D.** Institute of Human Genetics, Polish Academy of Sciences, Poznań, Poland

**Ian James Witterick, M.D., F.R.C.S.(C)** Department of Otolaryngology, Mount Sinai Hospital, and University of Toronto, Toronto, Ontario, Canada

**George Wortzman, M.D.** Department of Medical Imaging, Mount Sinai Hospital, and University of Toronto, Toronto, Ontario, Canada

**Erin Daniel Wright, M.D.** Department of Otolaryngology, Meakins–Christie Laboratories, and McGill University, Montreal, Quebec, Canada

**David T. Wright, M.D.** Department of Anatomy, Physical Sciences and Radiology, College of Veterinary Medicine, North Carolina State University, Raleigh, North Carolina

**Donovan B. Yeates, Ph.D.** Department of Medicine, University of Illinois at Chicago, and VA Chicago Health Care System, Chicago, Illinois

**J.-Y. Yoon, M.D.** Laboratory of Pulmonary Pathobiology, National Institute of Environmental Health Sciences, National Institutes of Health, Research Triangle Park, North Carolina

**Jerzy Żebrak, M.D.** Pediatric Division, Institute of Tuberculosis and Lung Disease, Rabka, Poland

**Jean-Marie Zahm, Ph.D.** Unité INSERM, CHU Maison Blanche, Reims, France

**Irwin Ziment, M.D., F.R.C.P.** Department of Medicine, Olive View-UCLA Medical Center, and University of California, Los Angeles, School of Medicine, Los Angeles, California

# In Memory of Professor Alex Silberberg 1923–1993

*The following dedication was presented at the opening of the Congress by Professor Ephraim Katchalski-Katzir, former President of the State of Israel and presently Professor Emeritus of the Weizmann Institute of Science.*

Thank you for inviting me to address this International Congress on Cilia, Mucus, and Mucociliary Interactions. The congress is being held in memory of the late Professor Alex Silberberg, and I am happy to have the opportunity to share with you some of my memories of this outstanding scientist, who was also my good friend. Alex was one of the pillars of science in modern Israel. The topics to be discussed at this meeting cover a field in which he was particularly interested, and it is therefore most fitting to commemorate him by holding an international conference on this subject. It also gives me much pleasure to welcome you to Jerusalem whose name means "the city of peace." Your presence here demonstrates once again that scientists have a common language and a common desire to build a better world where differences among peoples are respected and freedom, democracy, and human dignity are the guiding principles of life.

I first met Alex Silberberg in 1954, when he came to Israel from Basel, where he had been working as an assistant to Professor Werner Kuhn, the distinguished theoretical polymer chemist. Soon after Alex's arrival, we became friends, and over the years I came to value him as an excellent scientist and an outstanding human being.

Alex was born in Vienna in 1923. In 1935, when the storm clouds were gathering over Europe, his family moved to South Africa. At the University of Witwatersrand in Johannesburg, he received his degree in chemical engineering—a training that proved most useful in his later scientific career. In 1944, he joined the South African army, where he served for two years. In 1946, having chosen to specialize in the rapidly developing field of high-molecular-weight materials, he went to Switzerland to study at the University of Basel under Professor Werner Kuhn, a pioneer of modern science and one of its formost theoreticians. Alex received his Ph.D. summa cum laude in 1952, and soon afterward was taken on as a special assistant to Professor Kuhn. It was at this time that he met my late brother Aharon, who had gone to Switzerland on a private visit and had used the opportunity to meet with Professor Kuhn. It was at this time that he met my late brother Aharon, who had gone to Switzerland on a private visit and had used the opportunity to meet with Professor Kuhn. Aharon was most favorably impressed with the young scientist, and invited him to join his newly established Department of Polymer Science at the Weizmann Institute of Science in Rehovot. Alex became a full professor in that department in 1970 and remained at the Institute all his life.

Thanks to his background and general education, Alex was blessed with charm, thoroughness, and modesty, and the Jewish thirst for learning and knowledge. For me, as a young Israeli scientist who had hardly ever ventured abroad, Alex was the epitome of the modern, urbane, young scholar who had received a first-class education at a great European university. Those who were fortunate enough to work with him learned firsthand how to carry out precise, well-planned experiments and to become immersed in the most intricate and challenging theoretical and practical problems.

Starting as a classical polymer chemist, Alex soon bcame intrigued by the physical behavior of solid and liquid materials, which led him to specialize in the various branches of rheology. In due course, he became fascinated by the achievements of modern biology and therefore directed his future reserarch efforts mainly towards modern biorheology. He became Editor-in-Chief of the jounral *Biorheology* and was acquainted with and sought after by rheologists and biorheologists all over the world. He was invited to visit and teach at the great centers of learning, including the Mellon Institute in Pittsburgh, the University of California at San Diego, the Oersted Institute at Copenhagen, the California Institute of Technology, King's College in London, and the College of France. His colleagues used to call him Santa Claus because of his practice of generously giving away some of his original ideas in the most disinterested way. There are many who built their careers on the development of hypotheses propounded by Alex, and he himself was always happy to see his own ideas bear fruit in the work of others. He was awarded honorary degrees from many learned societies and received the Poiseuille Gold Medal of the International Society of Biorheology. Yet, for all his international recognition and standing, he remained unassuming and charming to the last, ever considerate of his friends, students, and colleagues here in Israel and all over the world.

Alex started his scientific career as a polymer scientist, intersted in the physical and chemical characteristics of high-molecular-weight materials. His thesis on the interfacial tension and phase separators in two polymer solvent systems, submitted in Basel in 1952, proved that even at that early stage, Alex was ready to tackle complex physical and chemical problems. He continued to work on problems concerning the gelation of aqueous solutions of polymethacrylic acid, turning from there to the analysis of non-Newtonian behavior of dilute suspensions of macroscopic spheres in a capillary viscometer. Finally, while still working as a polymer physicist, he produced a remarkable theoretical and experimental analysis of the adsorption of flexible macromolecules. I was particularly interested in this area, and everything I know about the adsorption of synthetic and biopolymers by various synthetic and natural surfaces I owe to Alex.

While Alex was working as a polymer physical chemist at the Weizmann Institute, groundbreaking work was being done on biological systems in the various Institute departments of life sciences. No wonder, therefore, Alex was influenced by his many friends to direct his attention to problems of biorheology. In 1968, he analyzed hemorheology theory and in the same year predicted the role of mucus in ciliary transport. This was followed by an analysis of the viscoelastic properties of whole blood, studies of the microcirculation, and work on tissue fluid exchange and transport through deformable, swellable, porous systems.

A good part of Alex's time was devoted to the study of mucus. He characterized epithelial mucus and predicted its function in clearance of ciliary propulsion. He also applied all of the available experimental and theoretical tools to the study of the rehology of mucus, mucociliary interaction, and ciliary activity. In this work he took as his starting point the well-known postulate that mucus is the essential mechanical coupler between the ciliary beat and the epithelial transport of dust and other particles.

At the Congress of Biotechnology held in Tokyo in 1986, Alex outlined the scope and aims of modern biorehology. This is what he had to say: "We all, as practicing biorheologists, regardless of whether our activities and contributions concern theoretical experimental, or applied pursuits, will have to exercise the mission of disseminating biorheology as the important link missing thus far in most life sciences towards their advancement. . . . The rheological properties of biological materials and their cellular and molecular level interpretations are thus not studied purely as material attributes. They are inextricably linked to the transport and motility requirements and thus to the molecular, structural, and thermodynamic description of the biological system.''

I believe that Alex was right. We classical biochemists and molecular biologists have been preoccupied with studying the individual, separate, well-classified components of the cell. But Alex understood, as we too are now beginning to understand, that the cell represents a complex biological system whose function, behavior, and responses can be comprehended only by application of complex network theories. We also now recognize the need to understand flows, stress and strain, elasticity, and viscosity in biological systems. It is therefore especially pleasing to note that this meeting will address several topics of interest to molecular biologists. The subject of this congress in Jerusalem is without doubt what Alex would himself have chosen, and I feel sure that the lectures, discussions, and plans for future activities will be inspired by his spirit.

*Ephraim Katchalski-Katzir*

# 1

# Molecular Biology of Axonemal Dynein and Its Role in Ciliary Motility

**I. R. Gibbons**
*University of Hawaii, Honolulu, Hawaii*

## I. INTRODUCTION

The oscillatory beating motion of epithelial cilia is generated through the activity of the motor protein, dynein, located in the ciliary axonemes [1,2]. The dynein forms two rows of projections, usually termed the outer and inner arms, located along one side of each of the 9 doublet microtubules of the axoneme where they form crossbridges with the adjacent microtubules. In beating cilia, these dynein crossbridges repeatedly detach and reattach to successive tubulin subunits on the adjacent microtubule of the axoneme in a cyclic stepping process driven by the dephosphorylation of ATP. The localized sliding displacement of microtubules produced by this activity of the dynein arms is converted into transverse bending moments by other resistive structures and results in the formation and propagation of the ciliary bending waves. The axonemal dyneins that are responsible for ciliary motility were first discovered and characterized about 30 years ago [3]. They are closely related to cytoplasmic dynein, that was discovered much more recently, and is involved in diverse intracellular activities including intracellular transport, nuclear migration, and the orientation of the cell spindle at mitosis [4]. Both axonemal and cytoplasmic dyneins are characterized by possession of exceptionally large heavy-chain subunits of molecular weight > 500,000 Da, by their sensitivity to inhibition by vanadate, and by their function as minus-end directed motors in assays of motility on microtubules in vitro.

Over the past few years, several laboratories have explored the application of molecular genetics to dynein. This work has shown that there are genes encoding approximately 14 distinct heavy-chain isotypes of axonemal dynein present in most eukaryotic organisms that possess cilia or flagella, whereas the heavy chain of cytoplasmic dynein appears always to be encoded by a single gene [5]. Deduced amino acid sequences have become available for the complete heavy-chain subunits of several of these isoforms in organisms ranging from sea urchin to *Paramecium* and *Chlamydomonas*, as well as for many of their associated smaller subunits and for the subunits of cytoplasmic dynein. Since the important functional regions of an enzyme are usually well conserved through evolution, the comparison of these sequences has helped to clarify the functional structure of the individual dynein, as well as define the relationships between different members of the dynein family.

## II. FUNCTIONAL PROPERTIES OF THE β HEAVY CHAIN OF AXONEMAL DYNEIN

Because of the exceptionally high molecular weight of dynein and the rather small amounts of it present in most tissues, the choice of organism and tissue must usually be governed by the need to obtain sufficient molar quantities for study. One of the most available forms of axonemal dynein is from the sea urchin. The outer arm dynein from sea urchin sperm flagella can be prepared conveniently in quantities up to about 10 mg, corresponding to 5 nanomoles, which is sufficient for many procedures in enzymology and protein chemistry. The dynein from this source has the additional advantage that its α and β heavy-chain subunits can be isolated easily, with the β heavy chain being obtained in a form that retains its functional motor activity when assayed with microtubules in vitro [6]. The study of this β heavy chain by a variety of biochemical and molecular biological techniques has enabled the position of many of its major functional sites to be identified, as indicated on the linear map in Figure 1.

When the amino acid sequence of the β heavy chain was first deduced from the nucleotide sequence of the gene in 1991, one of the major surprises was that it possessed four copies of the consensus P-loop motif (GXXXXGKT) for a nucleotide-binding site [7] spaced nearly equidistant in the midregion of the heavy chain [8,9].

These putative binding sites have been designated P1, P2, P3, and P4, in order from the N- to C-terminus on the chain (Figure 1). The positions within the complete sequence of the heavy chain provide an important clue to the function of these P-loops. In particular, the P1 site, at

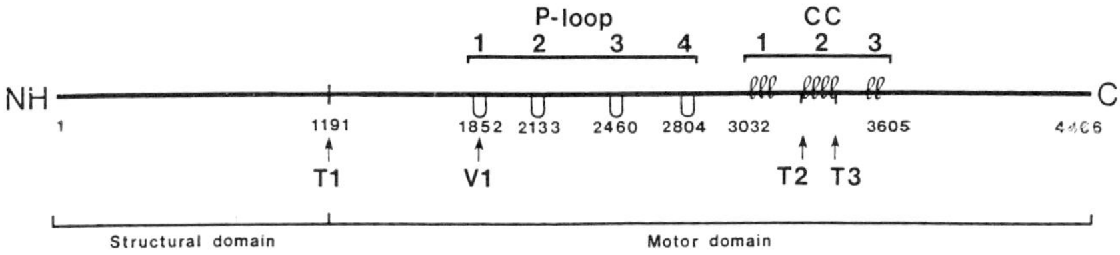


**Figure 1** Linear map of the β heavy chain of axonemal dynein in sea urchin showing the positions of the P-loops P1, P2, P3, and P4 that mark putative ATP-binding sites, the sites of rapid tryptic cleavage, T1, T2, T3; and the site of vanadate-mediated photocleavage V1. Also indicated are the locations of predicted α-helical coiled-coil CC1, CC2, and CC3. The numbers indicate position in the amino acid sequence of the complete heavy chain. (From Reference 5.)

position 1852–1859 of the 4466 total residues in the β heavy chain, appears most likely to engage in the cyclic binding and hydrolysis of ATP that underlies mechano-chemical energy transduction, for it is located close to the V1 site. Photolytic scission of the heavy chain occurs at this site upon irradiation in the presence of vanadate, a photosensitizing inhibitor of dynein ATPase [10].

This identification of the hydrolytic ATP-binding site is supported by the detailed levels of amino acid conservation within the four binding sites of different dynein isotypes. The P-loop motif of the P1 site is absolutely conserved, consisting of the same eight residues—Gly-Pro-Ala-Gly-Thr-Gly-Lys-Thr—in all known axonemal and cytoplasmic dyneins, whereas the sequences within the P2, P3, and P4 sites are less stringently conserved [11]. Thus, only the region of the P1 site has the extremely high level of conservation anticipated for the substrate-binding site in a family of motor proteins.

Although the detailed amino acid sequences around P2, P3, and P4 sites are less stringently conserved that that around the P1 site, the nucleotide-binding motif is recognizably present in all dynein heavy chains whose sequence is known [5]. Moreover, all four sites may be competent to bind ATP: direct assay of ATP binding has shown that the isolated α and β heavy chains of axonemal dynein each bind approximately 4 moles of ATP per mole of heavy chain with association constants of $10^5$ to $10^6$ $M^{+1}$ [12]. The lower level of conservation in the P2, P3, and P4 sites would be consistent with the binding of ATP at these sites having a less critical function, perhaps of a regulatory nature. Although the details remain obscure, the retention of these sites through evolution in organisms from algae to mammals strongly suggests that they serve some significant function in both axonemal and cytoplasmic dyneins.

More detailed comparison of the conservation within the P2, P3, and P4 sites shows that their sequence remains closer to the canonical P-loop motif, GXXXXGK[T,S,G], in cytoplasmic dynein than it does in axonemal dyneins [5]. For example, the canonical form is present unmodified in all four P-loops of cytoplasmic dyneins, whereas the variant form GVGGSGKQ is common in the P4 loop of many axonemal heavy chains, and still more varied forms including GVGGSGRH and GRSGVGRR occur in individual instances. It is particularly difficult to rationalize a strict requirement for multiple ATP-binding sites in cytoplasmic dynein, which has a seemingly simpler job than axonemal dynein yet possesses more stringently conserved P-loop motifs.

Although the sequence of dynein heavy chains shows no evidence of a long α-helical coiled-coil domain, such as forms the extended tails of myosin and of kinesin, it does contain three short regions of predicted helix—CC1, CC2, CC3—that possess a heptad hydrophobic repeat (Figure 1). The positions and spacings of these are well conserved in all axonemal and cytoplasmic dynein heavy chains, suggesting that they are of functional significance and may serve as interfaces between the heavy chains and regulatory light chain subunits [13]. This hypothesis has been elegantly supported by the mapping of two independent suppressor mutations that modify the regulation of axonemal dynein-arm activity to two short deletions in this region of the β heavy chain. These deletions, located adjacent to each other in the coiled-helix region CC1 shorten the "approximately" 13 nm length of this predicted helix by 1.0–1.5 nm [14]. While the regulatory subunit(s) that bind directly to this region have not been identified, a potential candidate is provided by the dynein-regulatory complex that has been identified in the region close to where the radial spoke structures of the axoneme attach close to the base of the dynein arms. An association of regulatory subunits with such an α helix located downstream of the ATP-binding region in dynein would be reminiscent of the positioning of the regulatory light chains in myosin [15].

## III. FUNCTIONAL IMPLICATIONS OF DYNEIN HEAVY CHAIN PHYLOGENY

When aligned over their full lengths, the amino acid sequences of axonemal and cytoplasmic heavy chains are about 25% identical, but this is quite nonuniform over their lengths. In the N-terminal third of the heavy chain, the level of identity is only ~18%, whereas it averages ~30% in the mid and C-terminal domains and runs as high as 47–52% in the 120 residue region around the P1 loop [11,16–18]. Even though the N-terminal region is relatively divergent, comparison with alignments of randomly shuffled sequences indicates that the probability of the level of matching found in the N-terminal region having occurred by chance is less than $10^{-7}$. This establishes that the heavy chains of axonemal and cytoplasmic dyneins evolved directly from a common ancestral gene of about the present size. The family of dynein motors thus appears more homogeneous than the families of kinesins and of myosins, in which a common ancestral motor domain appears to have been spliced onto different tail domains over the course of evolution [19–21].

The nearly equidistant spacing of the four P-loop motifs, located about 300 residues apart in the mid region of the dynein heavy chain, suggests that one step in the early evolution of the dynein heavy chain involved successive duplications of a single gene that encoded hypothetical ancestral protodynein of $M_r$ ~35,000 with a single ATP-binding site. Evidence of gene duplication can be seen most clearly in cytoplasmic dyneins, where patches of highly conserved sequence located approximately 50, 120, and 200 residues downstream of the P1 and P3 loops can be identically aligned with 22 of 42 amino acids [5,22]. Traces of the same patches appear in similar locations downstream of the P2 and P4 loops but are harder to discern. Axonemal dyneins show the same uniform spacing of the P-loops as cytoplasmic dynein, but evidence of conserved amino acid identity among their P-loop regions is less clear. Nevertheless, the transition from protodynein to dynein appears to have occurred long before the divergence of axonemal and cytoplasmic dyneins. The available data favor a transition of protodynein to dynein involving two duplicative events, with the first being $2 \times P1 \rightarrow P1 + P2$ and the second being $2 \times (P1 + P2) \rightarrow (P1 + P2) + (P3 + P4)$, although other lineages are possible.

Some other aspects of the separate evolution of the cytoplasmic and axonemal heavy chains are apparent. In currently studied organisms, cytoplasmic dynein has a simpler, apparently more primitive, structure than axonemal dynein. For example, cytoplasmic dynein is a homodimer [17,23], with its two heavy chains encoded by a single well-conserved gene that has been found in organisms from protozoa and fungi to mammals. Axonemal dyneins, on the other hand, are heterodimers or heterotrimers, formed by different combinations of the heavy chains that are encoded by a multigene family with 10–15 members. The greater diversity of the axonemal forms may have arisen during the early evolution of the 9 + 2 axonemal structure, for this would likely have been a competitive stage when fewer constraints permitted their relatively rapid divergence. During this period the members of the axonemal dynein subfamily presumably separated into the small number of groups that subsequently became frozen and now appear as universal as the 9 + 2 axonemal structure itself [11].

The phylogenetic tree (Figure 2) for the motor domains of heavy chains shows three major groupings: cytoplasmic dyneins and axonemal dyneins, which diverge from the two ends of an extended main trunk, and a third group represented by the single gene DHCIB diverging separately from the trunk at a position nearly equidistant between the cytoplasmic and axonemal forms. The subfamily of cytoplasmic dyneins forms a single cluster on the left. The order of divergence within this subfamily shows

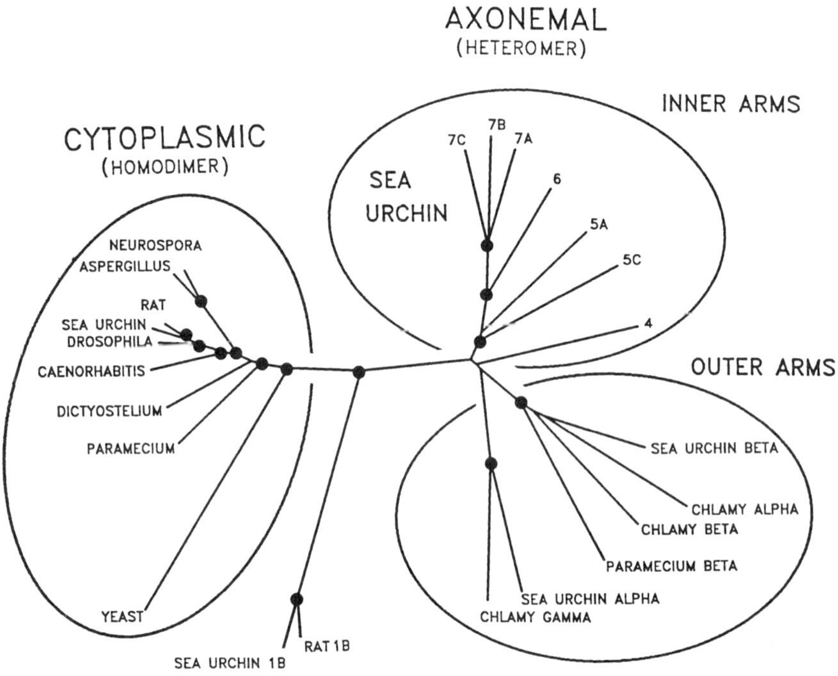


**Figure 2** Phylogenetic tree showing the sequence relatedness of dynein heavy chains. The ellipses indicate the subfamilies of cytoplasmic dynein, outer-arm axonemal dynein, and putative inner-arm axonemal dynein. Nodes with a bootstrap probability of 95% or better are indicated with an enlarged black dot. Sequences are from various authors and are deposited in the GenBank database. (Modified from Reference 5.)

the yeast *Saccharomyces cerevisiae* branching first, followed by the protozoan *Paramecium*, the slime mold *Dictyostellium*, the ascomycetes *Emericella* and *Neurospora*, the nematode *Caenorhabditis*, the arthropod *Drosophila*, the echinoderm *Tripneustes*, and the mammal rat. This order agrees well with that reported in previous sequence comparisons of other proteins [24].

The subfamily of axonemal dyneins separates as two groups, the outer-arm dyneins and the inner-arm dyneins, from an imperfectly resolved multifurcation at the other end of the main trunk. Within the group of outer arms, the branch containing the *Chlamydomonas* $\gamma$ chain and the probable sea urchin $\alpha$ chain (also known as DHC3A) separates rapidly from the branch containing the $\beta$ chains of sea urchin (DHC2), *Chlamydomonas* and *Paramecium*, as well as the $\alpha$ chain of *Chlamydomonas*. These groupings indicate that the *Chlamydomonas* $\gamma$ chain [25] is the homolog of the sea urchin $\alpha$ chain. The odd chain out in the trimeric outer arms of *Chlamydomonas* appear to be the $\alpha$ chain, which possibly arose by gene duplication of the $\beta$ chain at a relatively late stage and has now evolved to have a distinct function [2].

The group of seven heavy-chain genes identified only by molecular genetics (DHC4, DHC5A, DHC5C, DHC6, DHC7A, DHC7B, DHC7C) has been tentatively assigned as axonemal inner arm dyneins [11]. Gene knockout has identified DHC4 as encoding one of the heavy chains in the I1 form of the inner arm [26], but positive evidence linking the other gene products with particular inner arm heavy chains [2,27] is not yet available. This uncertainty notwithstanding, the group of inner arm heavy chains appears to have diverged from the main trunk almost concurrent with the group of outer arm heavy chains. Thus, to the extent that the rate of the molecular clock has remained constant, the divergence of axonemal and cytoplasmic dyneins appears to have occurred about twice as far back in evolutionary time as the divergence of *Paramecium* from the line leading to higher animals.

The third subfamily of dynein heavy chains contains only the DHC1B isoform, which forms a long branch separating from the main trunk nearly half way between the branch point of yeast cytoplasmic dynein and the multifurcation where the axonemal forms diverge. The deduced amino acid sequence of DHC1B is quite divergent, but lies somewhat closer to that of the cytoplasmic dyneins than to any of the axonemal dyneins. This DHC1B isoform is expressed in widely different organisms, for in addition to sea urchin embryos it has been cloned from the mRNA of human liver [28], as well as identified in the amoeba-flagellate *Naegleria gruberi* [B. H. Gibbons, unpublished data]. However, it does not appear to occur in any organism that does not possess a cilium or flagellum at any stage of its life cycle, such as yeast or Dictyostelium. The only direct evidence regarding the function of DHC1B is that its transcript becomes upregulated in sea urchin embryos that are actively regenerating new cilia [11]. Whether it plays a general cytoplasmic role in most eukaryotes or serves some more specialized role involved with ciliary growth or function remains an enigma.

## IV. FUTURE QUESTIONS

Our present understanding of axonemal dynein poses several questions of significant general interest to be addressed in the future.

### A. Regulation of the Sliding Activity of Dynein

The sliding activity of dynein arms on the axonemal microtubules must be subject to appropriate spatial and temporal regulation for it to yield the asymmetrical oscillatory beating motion that is characteristic of cilia. To what extent is this regulation intrinsic to the large and complex structure of dynein itself and to what extent is it due to extrinsic factors, such as stress-dependent changes in the microtubule lattice or transient phosphorylation of other axonemal structures [29]?

### B. Dynein Size

Why is dynein so big? Its known functions in intracellular transport and microtubule sliding scarcely explain why its conserved motor domain is so much larger than those of kinesin and myosin, which fulfill similar roles. One possibility is that a large contact surface on the microtubule might be required for dynein's capability of rotating the microtubule [30]. This rotation could provide a eukaryotic rotational motor and underlie the asymmetrical migration of the centrosomes around the nucleus that leads to development of chiral asymmetry in cellular structures [31–34].

### C. Evolution of Cytoplasmic Dynein

Does the relatively primitive homodimeric structure of cytoplasmic dynein indicate that it evolved first and later split off the axonemal dyneins that then became freer to evolve into more specialized and complex heteromers? This would be supported by the finding that the motor domains of axonemal and cytoplasmic dyneins have evolved at similar rates within recent evolutionary time [5]. If true, it would suggest that the role of centrosomes in cell division predates the evolution of basal bodies and the flagellar axoneme, rather than the converse as often supposed [35].

### D. Relationship of Dynein Within All Molecular Motors

What is the evolutionary relationship of dynein within the hyperfamily of all molecular motors in which kinesin and myosin appear to be distant members relatable by similarity in protein folding [36]? The four evenly spaced sites for binding nucleotide in the mid region of the heavy chain may indicate that the dynein motor domain evolved through double duplication of an ancestral gene that encoded a hypothetical "protodynein" motor [5] similar in size to the present kinesin motor domain.

## ACKNOWLEDGMENTS

I thank Barbara Gibbons for helpful discussion and Cheryl Phillipson for excellent technical assistance in sequencing sea urchin isoforms and for preparation of illustrations. This work has been supported in part by grant GM30401 from the National Institute of General Medical Sciences.

## REFERENCES

1. Gibbons IR. Cilia and flagella of eukaryotes. J Cell Biol 1981; 91 (3 Pt 2): 107s–124s.
2. Mitchell DR. Cell and molecular biology of flagellar dyneins. Int Rev Cytol 1994; 155:141–180.
3. Gibbons IR, Rowe AJ. Dynein: a protein with adenosine triphosphatase activity from cilia. Science (Washington DC) 1965; 149:424–426.
4. Holzbaur ELF, Vallee RB. Dyneins: molecular structure and cellular function. Annu Rev Cell Biol 1994; 10:339–372.
5. Gibbons IR. The dynein family of motor proteins: present status and future questions. Cell Motil Cytoskel 1995; 32:136–144.
6. Sale WS, Fox LA Isolated β-heavy chain subunit of dynein translocates microtubules in vitro. J Cell Biol 1988; 107:1793–1797.
7. Saraste M, Sibbald PR, Wittinghofer A. The P-loop–a common motif in ATP–and GTP–binding proteins. TIBS 1990; 15:430–434.
8. Gibbons IR, Gibbons BH, Mocz G, Asai DJ. Multiple nucleotide binding sites in the sequence of dynein β heavy chain. Nature 1991; 352:640–643.
9. Ogawa K. Four ATP–binding sites in the mid–region of the β heavy chain of dynein. Nature 1991; 352:643–645.
10. Gibbons IR, Lee-Eiford A, Mocz G, Phillipson CA, Tang W-JY, Gibbons BH. Photosensitized cleavage of dynein heavy chains: cleavage at the "V1

site" by irradiation at 365 nm in the presence of ATP and vanadate. J Biol Chem 1987; 262:2780–2786.
11. Gibbons BH, Asai DJ, Tang W-JY, Hays TS, Gibbons IR. Phylogeny and expression of axonemal and dynein genes in sea urchins. Mol Biol Cell 1994; 5:57–70.
12. Mocz G, Gibbons IR. Phase partition analysis of nucleotide binding to axonemal dynein. Biochemistry 1996; 35:9204–9211.
13. Mitchell DM, Brown KS. Sequence analysis of the *Chlamydomonas* α and β dynein heavy chain genes. J Cell Sci 1994; 107:635–644.
14. Porter ME, Knott JA, Gardner LC, Mitchell DR, Dutcher SK. Mutations in the SUP-PF-1 locus of *Chlamydomonas reinhardtii* identify a regulatory domain in the β–dynein heavy chain. J Cell Biol 1994; 126:1495–1507.
15. Rayment I, Rypniewski WR, Schmidt-Base K, Smith R, Tomchick DR, Benning MM, Winkelmann DA, Wesenberg G, Holden HM. Three-dimensional structure of myosin-subfragment 1: a molecular motor. Science 1993; 261: 50–58.
16. Koonce MP, Grissom PM, McIntosh JR. Dynein from *Dictyostelium*: primary structure comparisons between a cytoplasmic motor enzyme and flagellar dynein. J Cell Biol 1992; 119:1597–1604.
17. Mikami A, Paschal BM, Mazumdar M, Vallee RB. Molecular cloning of the retrograde transporter cytoplasmic dynein (MAPIC). Neuron 1993; 10: 787–796.
18. Zhang Z, Tanaka Y, Nonaka S, Aizawa H, Kawasaki H, Nakata T, Hirokawa N. The primary structure of rat brain (cytoplasmic) dynein heavy chain, a cytoplasmic motor enzyme. Proc Nat Acad Sci USA 1993; 90:7928–7932.
19. Cheney RE, Riley MA, Mooseker MS. Phylogenetic analysis of the myosin superfamily. Cell Motil Cytoskel 1993; 24:215–223.
20. Goodson HV, Spudich JA. Molecular evolution of the myosin family: relationships derived from comparisons of amino acid sequences. Proc Nat Acad Sci USA 1993; 90:659–663.
21. Goodson HV, Kang SA, Endow SA. Molecular phylogeny of the kinesin family of motor proteins. J Cell Sci 1994; 107:1875–1884.
22. Eshel D, Urrestarazu LA, Vissers S, Jauniaux J-C, van Vliet-Reedijk JC, Planta RJ, Gibbons IR. Cytoplasmic dynein is required for normal nuclear segration in yeast. Proc Nat Acad Sci USA 1993; 90:11172–11176.
23. Neely MD, Erickson HP, Boekelheide K. HMW-2, the Sertoli cell cytoplasmic dynein from rat testis, is a dimer composed of nearly identical subunits. J Biol Chem 1990; 265:8691–8698.
24. Loomis WF, Smith DW. Molecular phylogeny of *Dictyostelium discoidium* by protein sequence comparison. Proc Nat Acad Sci USA 1990; 87: 9093–9097.
25. Wilkerson CG, King SM, Witman GB. Molecular analysis of the γ heavy chain of *Chlamydomonas* flagellar outer arm dynein. J Cell Sci 1994; 107: 497–506.
26. Myster SH, Knott JA, O'Toole E, Porter ME. The *Chlamydomonas* DHC1

gene encodes a dynein heavy chain subunit required for assembly of the I1 inner arm complex. Mol Biol Cell 1997; 8:607–620.
27. Piperno G, Ramanis Z, Smith EF, Sale WS. Three distinct inner dynein arms in *Chlamydomonas* flagella: molecular composition and location in the axoneme. J Cell Biol 1990; 110:379–389.
28. Vaisberg EA, Grissom PM, McIntosh JR. Mammalian cells express three distinct dynein heavy chains that are localized to different cytoplasmic organelles. J Cell Biol 1996; 133:831–842.
29. Habermacher G, Sale WS. Regulation of flagellar dynein by phosphorylation of a 138-kD inner arm dynein intermediate chain. J Cell Biol 1997; 136: 167–176.
30. Vale RD, Toyoshima YY. Rotation and translocation of microtubules induced by dyneins from *Tetrahymena* cilia. Cell 1988; 52:459–469.
31. Hyman AA, White JG. Determination of cell division axes in the early embryogenesis of *Caenorhabditis elegans*. J Cell Biol 1987; 105:2132–2135.
32. Horwich A, Brueckner M. Left, right and without a cue. Nature Genetics 1993; 5:321–322.
33. Strome S. Determination of cleavage planes. Cell 1993; 72:3–8.
34. Freeman G, Lundelius JW. The developmental genetics of dextrality and sinistrality in the gastropod *Lymnaea peregra*. Wilhelm Roux's Arch Dev Biol 1982; 191:60.
35. Margulis L. Symbiosis in Cell Evolution. 2nd Ed. New York: Freeman, 1993.
36. Krull FJ, Sablin EP, Lau R, Fletterick RJ, Vale RD. Crystal structure of the kinesin motor domain reveals a structural similarity to myosin. Nature 1996; 380:550–554.

# 2

# Modeling Outer Dynein Arm Activity and Its Relation to the Ciliary Beat Cycle

**Peter Satir and Toshikazu Hamasaki**
*Albert Einstein College of Medicine, Bronx, New York*

**Michael Holwill**
*Kings College, London, United Kingdom, and Albert Einstein College of Medicine, Bronx, New York*

Ciliary motility powers the mucociliary escalator of the respiratory tract; A. Silberberg and colleagues made important contributions to our knowledge of mucus propulsion [1,2]. It has become clear that ciliary activity in the respiratory tract is stimulated by the presence of mucus, so that more and more cilia are recruited to beat, and beat frequency increases. Unlike cilia of Paramecium or mussel gill, or even ependymal cilia in the brain, all of which beat continuously to move fluid, cilia lining the respiratory tract are mainly quiescent until stimulated. Despite the considerable advances in our knowledge of fundamental details of the mechanochemistry of ciliary motility, the overriding questions of motility production and its control remain largely unanswered, and hypotheses that attempt conceptual, unifying answers remain controversial. One difficulty in posing such hypotheses in a useful way is that the complexity of ciliary structure makes it very hard to visualize how local force-producing events, such as dynein arm activity, act on the entire structure to produce coordinated microtubule sliding and bending. How do we go from simplified mechanisms to real-time ciliary beating? We believe that a major impediment lies in visualizing events in the axoneme in three-dimensions. Science has always relied on model building; recently material models have been replaced by computer models.

Computer modeling of a complex cell organelle, such as the ciliary axoneme, represents a new technique by which to examine the changing interrelationships of the structural components of the organelle in different states of activity. The modeling requires accurate information about the structure of the components of interest at high resolution, approaching molecular sizes. Experimental information about conformational changes, force-producing elements, and other physical principles involved in changes in activity state is also required. In essence, this approach is a higher order application of the crystallographic modeling of molecular activity that has proven so successful in increasing our understanding of the activity of myosin [3] or kinesin [4] and related P-loop proteins [5]. Our modeling procedure should permit us to arrive at a new synthesis between structural states, in this case of axonemal components, and biophysical properties of the organelle, i.e., the force-generating mechanisms leading to microtubule sliding and repetitive bend formation and propagation in the cilium. A modeling method is also one of the best ways to develop new hypotheses relating protein structure to organellar physiology.

The ciliary axoneme seems ideally suited to computer modeling. Despite the important information available, many specific details of three-dimensional axonemal structure are either unresolved or just being approached. Examples include the structure of the inner dynein arms and circumferential links [6,7], new information on kinesin localization [8], and structural information on the tektins [9]. Molecular localization of perhaps some 40% of the approximately 250 polypeptides in the axoneme is related to the major structures, largely through analysis of mutant axonemes where these structures do not assemble—but the locations of the remaining polypeptides have not been mapped. Most importantly, the mechanisms of bend formation and propagation remain known only in the most narrow sense. For example, we know that the mechanochemical cycle of the outer dynein arms regulates microtubule sliding rate and beat frequency [10] and that a signal transduction pathway from the radial spokes to the inner dynein arms [11,12,13] regulates the amplitude of a forming bend. The physical mechanisms by which regulation of bend formation is accomplished are obscure. Previous mathematical or computer modeling approaches to ciliary motility [14,15] simplify axonemal structure and complexity drastically. While they give a reasonable fit to the form of motion, they do not have the molecular resolution necessary for the type of structural biology approach suggested here.

## I. COMPUTER MODELING OF THE THREE-DIMENSIONAL AXONEME AND MICROTUBULE SLIDING

At the start of these studies, Sugrue et al. [16] completed the initial version of the 3D structural model required at a macromolecular resolution approaching 4 nm. Major axonemal structures were modeled based on detailed information obtained from negative stain, thin section, and rapid freeze-etch electron microscopy, principally of Tetrahymena axonemes, as well as on optical diffraction of such micrographs. The software package developed for this purpose is called SURREAL (SURface REndering Algorithms) and was written by A.N. Glazzard specifically to model axonemal structures. A feature of the computer model as designed is that the structural features of the various axonemal constituents can be readily altered to simulate what happens dynamically within the axoneme as it is induced to slide, to split, or to beat. An updated version is shown in Figure 1.

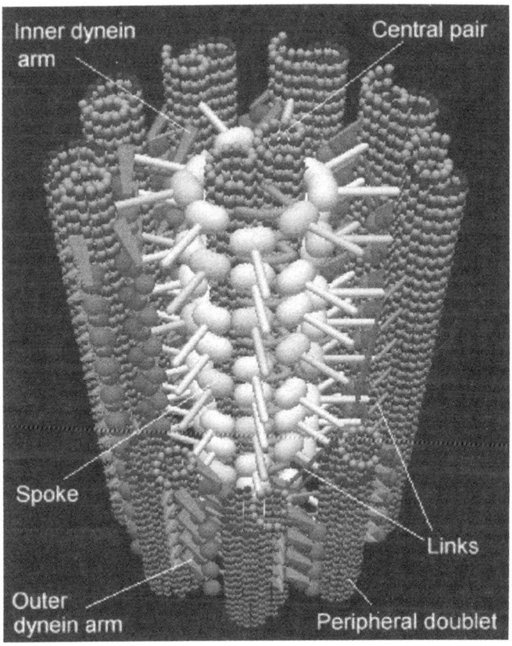


**Figure 1** Computer-generated model of the axoneme, cut away to show the internal structures and incorporating new information that has become available since Sugrue et al. [16]. Image kindly supplied by D. Helen Taylor, Department of Physics, Kings College, London, England.

As a preliminary step to understanding how dynein arm activity could operate within the axoneme to generate regulated, propagated, and repetitive bending, we [17] began a stepwise, systematic analysis of microtubule sliding within the axoneme that resulted in the telescoping extension of doublets. As is well known, sliding in lieu of bending occurs in axonemes after a brief protease treatment that degrades specific structures, particularly the radial spokes [18]. In the simplest sense, sliding may be modeled by a pair of doublets N and N + 1, where the dynein arms on doublet N propel doublet N + 1 in a tipward (+) direction [19]. We have begun our functional analysis with this situation, with the aim of (a) defining the duty cycle and other mechanochemical parameters of the arm cycle that produce sliding and (b) defining the sequence of arm activity along doublet N with time. Because the outer and inner dynein arms act differently, we have originally confined our modeling to the outer dynein arm, although either arm alone can produce sliding.

## II. A STOCHASTIC MODEL OF OUTER DYNEIN ARM ACTIVITY

Earlier models of outer dynein arm activity during a ciliary beat propose that every dynein arm along the microtubule is active sequentially from base to tip during a single beat. We refer to this as the metachronal model of arm activity. This model is a residue from a compelling measurement of energy consumption during sea urchin sperm axoneme swimming [20] in which evidently one ATP molecule was hydrolyzed per dynein arm per beat. The complexity of axonemal dynein ATPases now known suggests that this measurement, if correct, is coincidental. Furthermore, it is very difficult to see how such sequential activation would operate in the continuous in vitro translocation of microtubules over a dynein substratum.

Additional considerations, such as those discussed by Satir [21], also cast doubt on the metachronal model. For example, at a beat frequency of 25 Hz, bending during the axonemal effective stroke takes about 10 msec and produces active sliding of about 0.1 μm between pairs of doublets. Dynein step size ($d$) is estimated to be 16 nm from geometrical considerations [22]; a minimal value for $d$ is 4 nm. Therefore, no more than 25 dynein hydrolysis events (and possibly as few as 6) per doublet are needed to produce the sliding required for bend formation. However, there are over 400 outer arms on a 10 μm long axonemal doublet. The metachronal model requires special adjustments to account for the discrepancy between arm steps required and the numbers of active outer arms available.

These considerations have led us to propose a different model for dynein arm activity during in vitro translocation and sliding and bending in the axoneme, which we refer to as the stochastic model of arm activity [23]. In this model, we assume that individual dynein arms cycle randomly both with respect to time and to position on the microtubule. Further, we assume that the activity of each dynein arm is independent of every other arm. When two or more dynein molecules interact simultaneously with the microtubule, the velocity produced is that associated with a single molecule and when the duty phases overlap, the translocation distance is reduced in proportion to the overlap.

## III. MODELING MICROTUBULE TRANSLOCATION BY OUTER DYNEIN ARMS

Holwill et al. [23] have initially used the stochastic model to create the force generating steps of outer dynein arms in simulations of the behavior of a sliding system comprised of two microtubules. These simulations yield results that are directly comparable to the results derived from experiments examining in vitro microtubule translocation over outer arm dynein, as shown by Hamasaki et al. [10]. However, they also may apply to axonemal sliding experiments, such as those of Takahashi et al. [24], with minor modifications. In such simulations, the duty phase of the outer dynein arm—where the arm activity generates the sliding force—is very short, only a small fraction ($f = 0.01$) of the total mechanochemical cycle time.

The stochastic model of dynein arm activity proposes that when turned on, all outer arms are potentially capable of hydrolyzing ATP to produce force, but only a few random arms actually do this in the short time available for an effective stroke—a different number with each beat. This provides an explanation for the small number of steps taken by each doublet in the beating axoneme. In addition, the hypothesis predicts that beat frequency will vary slightly from beat to beat, a prediction that can be examined experimentally.

## IV. CONCLUSION

In summary, previous work has produced a comprehensive computer model of the three-dimensional structure of the axoneme at a resolution sufficient to examine the interdependent dynamic changes in individual components during the production of a beat. The model permits us to focus systematically on the mechanical and structural properties of these components that are involved in force generation during sliding and in bend production and propagation during beating. Systematic analysis has begun with an examination of the nature of the sliding interaction between two adjacent doublets driven by the activity of the outer dynein arm. The parameters defined by this analysis produce a functional motile system of computer images that can be directly compared to video images of dynein-driven microtubule motility and that can be used to predict variations in beat frequency of a single cilium. In principle, this method can be extended to encompass the modeling of the full range of axonemal activity.

## REFERENCES


1. Sade J, Eliezer N, Silberberg A, Nevo C. The role of mucus in the transport by cilia. Amer Rev Respir Dis 1970; 102:48–52.
2. Silberberg A. Rheology of mucus, mucociliary interaction and ciliary activity. Cell Motility Suppl 1982; 1:25–28.
3. Rayment I, Rypniewsky WR, Schmidt-Bass K, Smith R, et al. Three-dimensional structure of myosin subfragment–1: a molecular motor. Science 1993; 261:50–58.
4. Kull FJ, Sablin EP, Lau R, Fletterick RJ, Vale RD. Crystal structure of the kinesin motor domain reveals a structural similarity to myosin. Nature 1996; 380:550–555.
5. Smith CA, Rayment I. Active site comparisons highlight structural similarities between myosin and other P-loop proteins. Biophys J 1996; 70: 1590–1602.
6. Porter M. Axonemal dyneins: assembly, organization and regulation. Current Opinion Cell Biol 1996; 8:10–17.
7. Woolley DM. Studies on eel sperm flagellum. I. The structure of the inner dynein arm complex. J Cell Sci 1997; 110:85–94.
8. Bernstein M, Beech PL, Katz SG, Rosenbaum JL. A new kinesin like protein (Klp1) localized to a single microtubule of the *Chlamydomonas* flagellum. J Cell Biol 1994; 125:1313–1326.
9. Norrender JM, Perrone CA, Amos LA, Linck RW. Structural comparisons of tektins and evidence for their determination of complex spacings in flagellar microtubule. J Mol Biol 1996; 257:385–397.
10. Hamasaki T, Holwill MEJ, Barkalow K, Satir P. Mechanochemical aspects of axonemal dynein activity studies by in vitro microtubule translocation. Biophys J 1995; 69:2569–2579.
11. Smith EF, Sale WS. Regulation of dynein-driven microtubule sliding by the radial spokes in flagella. Science 1992; 257:1557–1559.

12. Habermacher G, Sale WS. Regulation of dynein-driven microtubule sliding by an axonemal kinase and phosphatase in *Chlamydomonas* flagella. Cell Motil Cytoskel 1995; 32:106–109.
13. Habermacher G, Sale WS. Regulation of flagellar dynein by an axonemal type-1 phosphatase in *Chlamydomonas*. J Cell Sci 1996; 109:1899–1907.
14. Brokaw CJ. Models for oscillation and bend propagation by flagella. SEB Soc Symp 1982; 35:313–338.
15. Brokaw CJ. Computerized analysis of flagellar motility with digitization and fitting of film images with straight segments of equal length. Cell Motil Cytoskel 1990; 17:309–316.
16. Sugrue P, Avolio J, Satir P, Holwill MEJ. Computer modelling of *Tetrahymena* axonemes at macromolecular resolution: Interpretation of electron micrographs. J Cell Sci 1991; 98:5–16.
17. Holwill MEJ, Satir P. A physical model of microtubule sliding in ciliary axonemes. Biophys J 1990; 58:905–918.
18. Summers KE, Gibbons IR. Effects of trypsin digestion on flagellar structures and their relationship to motility. J Cell Biol 1973; 58:618–629.
19. Sale WS, Satir P. Direction of active sliding of microtubules in *Tetrahymena* cilia. Proc Natl Acad Sci USA 1977; 74:2045–2049.
20. Brokaw CJ. Adenosine triphosphate usage by flagella. Science 1967; 156: 76–78.
21. Satir P. Biomechanical events in the production of ciliary movement. In: Akkas N ed. Biomechanics of Active Movement and Division of Cells, NATO ASI series, Vol H84, Heidelberg, Berlin-N.Y: Verlag, 1994; 465–470.
22. Satir P, Wais-Steider J, Lebduska S, Nasr A, Avolio J. The mechanochemical cycle of the dynein arm. Cell Motil 1981; 1:303–327.
23. Holwill MEJ, Foster GF, Hamasaki T, Satir P. Biophysical aspects and modelling of ciliary motility. Cell Motil Cytoskel 1995; 32:114–120.
24. Takahashi K, Shingyoji C, Kamimura S. Microtubule sliding in reactivated flagella. Symp Soc Exp Biol 1982; 35:159–177.

# 3

# Regulation of Outer Arm Dynein Activity via Light Chain Phosphorylation

**Toshikazu Hamasaki, Jason H. Nielson, and Peter Satir**
*Albert Einstein College of Medicine, Bronx, New York*

## I. INTRODUCTION

Activity of cilia in airway tract epithelium has a significant role in mucus transport [1]. Patients with primary ciliary dyskinesia (PCD) show respiratory symptoms including lung congestion, chronic otitis media, and sinusitis. Respiratory cilia are comprised of a microtubule-based cytoskeleton that is arranged characteristically in a "9 + 2" pattern called the axoneme, on which many additional components are attached for providing power for ciliary motility or its regulation [see Chapter 2]. The major components that can be found on the outer microtubules, both structurally and functionally, are outer and inner rows of dynein arms. Dynein is a mechanochemical enzyme, or molecular motor ($M_r$ of 1500 kDa to 2000 kDa) that consists of 15 to 20 polypeptides [2]. The major polypeptide in the dynein complex is a large-$M_r$ polypeptide, the so-called heavy (H) chain, which has a peptide sequence that carries $Mg^{2+}$-ATPase activity [see Chapter 1]. Outer arm dynein possesses two or three of these different H chains. Other polypeptides include intermediate (I) chains and light (L) chains. Through its mechanochemical cycle, the head portion of dynein attaches to an adjacent microtubule and pushes it a short distance at a time, while dynein undergoes its molecular structural changes accompanied by ATP hydrolysis similar to what myosin does in muscle contraction. The structure of the ciliary axoneme as well as the mechanochemical activity of dynein is carried over from protozoa such as *Paramecium* or *Chlamydomonas* to the mammalian airway tract or middle ear.

Ciliary beat frequency (CBF) is modulated according to external stimuli. In airway cilia, for instance, a mechanical stimulus to the cell increases CBF [3, see Chapter 8]. Neurotransmitters such as serotonin (5HT) also increase CBF. The protozoan *Paramecium* has been used to study signal transduction by which external stimuli, such as a mechanical or a chemical stimulus, lead to changes in CBF or beat pattern [4]. For instance, a mechanical stimulus to the posterior end of a *Paramecium* stimulates beat, so the cell swims faster because of elevated CBF. In this case, a stimulus activates the mechanoreceptor channels that evoke cellular hyperpolarization, which in turn leads to an activation of adenylyl cyclase, which produces cAMP via voltage-sensitive $K^+$-channel activation. The cAMP then activates cAMP-dependent protein kinase, which phosphorylates ciliary proteins [4,5,6].

## II. REGULATION OF CILIARY BEAT FREQUENCY

Our search for a specific regulatory molecule that is responsible for an elevation of CBF via cAMP-dependent protein phosphorylation yielded a specific *Paramecium* axonemal protein, $M_r$, of 29 kDa (p29) [7]. Phosphorylation of this protein in vitro takes place in the presence of micromolar concentrations of cAMP, while micromolar-concentrations of $Ca^{2+}$ suppress the cAMP-dependent phosphorylation. This result is consistent with the changes in swimming speed of Triton X-100-permeabilized, $Mg^{2+}$-ATP-reactivated paramecia when cAMP or cAMP and $Ca^{2+}$ are added to the medium [7,8,9,10]: cAMP increases swimming speed, but $Ca^{2+}$ added together with cAMP quenches the cAMP effect. Further studies revealed that this protein is a component of outer arm dynein (22S dynein) [9,10,11]. Phosphorylation of p29 increases in vitro microtubule translocation activity, which is 40% of the activity related to in situ microtubule sliding [10,11]. Hence this protein, p29, is regarded as a regulatory light chain of ciliary outer arm dynein. Regulation of outer arm dynein is particularly important for regulation of ciliary beat frequency because the outer arm provides the chief sliding force [12], and the speed produced directly reflects the beat frequency [5].

We have further undertaken a kinetic analysis approach to microtubule translocation activity of outer arm dynein based on concurrent measurements of the length of a microtubule that is translocating over a dynein-coated surface and its translocation velocity [13]. Only 2–10% of p29 is phosphorylated in vitro in the dynein preparations or isolated p29 preparations [10,11]. The analysis estimates that the frequency will increase by about 40% from 6.7 to 9.4 Hz under these conditions. The results of the kinetic analysis provides an estimate of the maximum ciliary beat frequency if all the p29 becomes phosphorylated. Frequency is then predicted to increase to 34–140 Hz which covers the physiological range of an elevated beat frequency [13]. It is yet to be explained why little or no increase in dynein ATPase activity has been detected with cAMP-phosphorylated dynein [14, Hamasaki, unpublished].

## III. FREQUENCY REGULATION IN MAMMALIAN AIRWAY CILIA

In mammalian airway cilia, Salathe et al. [15] reported a cAMP-dependent phosphoprotein of $M_r$ 26 kDa (p26) in isolated and permeabilized ovine tracheal cilia. Results from our preliminary experiments show that p26 is

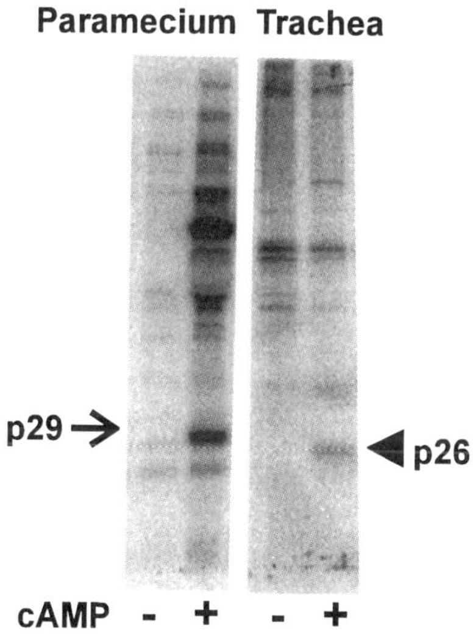


**Figure 1** In vitro phosphorylation of isolated *Paramecium* and ovine tracheal axonemes. Aliquots of isolated axonemes were incubated with $\gamma$-$^{32}$P-ATP either in the presence or absence of 10 $\mu$M cAMP for 5 min on ice, followed by SDS-sample buffer denaturation. Proteins were then separated in a 10% SDS-PAGE and the $^{32}$P-labeled proteins were visualized by autoradiography. Arrow indicates position of *Paramecium* p29 and arrowhead indicates that of tracheal p26.

a structural component of the ciliary axoneme. Tracheal cilia p26 becomes phosphorylated in a similar manner in time and cAMP concentration to *Paramecium* p29 (Figure 1). Although further extensive research will be required to characterize p26, this protein is a likely candidate for a mammalian homologue of *Paramecium* p29 as a mammalian ciliary outer arm dynein regulatory light chain, which regulates CBF.

## IV. CONCLUSION

Although regulation of CBF in protozoan or mammalian cilia depends primarily upon regulation of outer-arm dynein activity, other aspects of ciliary beat such as the bend amplitude may be regulated by changes in other axonemal components including inner dynein arms, radial spokes, and the central microtubule pair complex. This regulation probably affects beat form. For instance, phosphorylation of inner arm component(s) seems to be involved in microtubule sliding velocity in mutant *Chlamydomonas* axonemes [16]. It is important to dissect these regulatory changes in detail in order to understand the changes that may occur during normal or abnormal mucociliary transport.

## ACKNOWLEDGMENT

This work is supported by the American Heart Association, New York City Affiliate to T.H.

## REFERENCES

1. Satir P, Sleigh MA. The physiology of cilia and mucociliary interactions. Ann Rev Physiol 1990; 52:137–155.
2. Holzbaur ELF, Vallee RB. Dyneins: molecular structure and cellular function. Ann Rev Cell Biol 1994; 10:339–372.
3. Sanderson MJ, Dirksen ER. Regulation of ciliary beat frequency in respiratory tract cilia. Chest 1992; 101(suppl):69s–71s.
4. Bonini NM, Evans TC, Miglietta LAP, Nelson DL. The regulation of ciliary motility in *Paramecium* by $Ca^{2+}$ and cyclic nucleotides. Adv. Second Messeng Phospho Res 1991; 23:227–272.
5. Satir P, Barkalow K, Hamasaki T. The control of ciliary beat frequency. Trends in Cell Biol 1993; 3:409–412.
6. Walczak CE, Nelson DL. Regulation of dynein-driven motility in cilia and flagella. Cell Motil Cytoskel 1994; 27:101–107.

7. Hamasaki T, Murtaugh TJ, Satir BH, Satir P. *In vitro* phosphorylation of *Paramecium* axonemes and permeabilized cells. Cell Motil Cytoskel 1989; 12:1–11.
8. Bonini NM, Nelson DL. Differential regulation of *Paramecium* ciliary motility by cAMP and cGMP. Cell Biol 1988; 106:1615–1624.
9. Bonini NM, Nelson DL. Phosphoproteins associated with cyclic nucleotide stimulation of ciliary motility in *Paramecium*. J Cell Sci 1990; 95:219–230.
10. Hamasaki T, Barkalow K, Richmond J, Satir P. A cAMP–stimulated phosphorylation of an axonemal polypeptide that copurifies with the 22S dynein arm regulates microtubule translocation velocity and swimming speed in *Paramecium*. Proc Nat Acad Sci (USA) 1991; 88:7918–7922.
11. Barkalow K, Hamasaki T, and Satir P. Regulation of 22S dynein by a 29 kDa light chain. J Cell Biol 1994; 126:727–735.
12. Brokaw CJ, Kamiya R. Bending patterns of *Chlamydomonas* flagella IV. Mutants with defects in inner and outer dynein arms indicate differences in dynein arm function. Cell Motil Cytoskel 1987; 8:68–75.
13. Hamasaki T, Holwill MEJ, Barkalow K, Satir P. Mechanochemical aspects of axonemal dynein activity studied by *in vitro* microtubule translocation. Biophys J 1995; 69:2569–2579.
14. Walczak CE, Nelson DL. *In vitro* phosphorylation of ciliary dyneins by protein kinases from *Paramecium*. J Cell Sci 1993; 106:1369–1376.
15. Salathe M, Pratt MM, Wanner A. Cyclic AMP-dependent phosphorylation of a 26 kD axonemal protein in ovine cilia isolated from small tissue pieces. Am J Respir Cell Mol Biol 1993; 9:306–314.
16. Habermacher G, Sale WS. Regulation of dynein-driven microtubule sliding by an axonemal kinase and phosphatase in *Chlamydomonas* flagella. Cell Motil Cytoskel 1995; 32:106–109.

# 4

# Inhibitory and Excitatory Neural Regulation of the Mucociliary Transport System

**Donovan B. Yeates**
*University of Illinois at Chicago, and VA Chicago Health Care System, Chicago, Illinois*

## I. INTRODUCTION

Inhibitory and excitatory neural regulation of ciliary activity has been demonstrated in gastropods, molluscs, and frogs. More recently, evidence has been presented for inhibitory and excitatory neural regulation of mucociliary wave frequency in the rabbit maxillary sinus. We contend that in the mammalian tracheobronchial airways, ciliary activity, and mucociliary transport are homeostatically maintained at a low basal level by the dominance of inhibitory neural transmission over excitatory neural transmission. This neural traffic is transmitted via both parasympathetic and sympathetic neural pathways. Inhibitory regulation is affected by two mechanisms: (a) the suppression of the release of excitatory mediators in the mucosa, such that delivery of their antagonists does not decrease ciliary beat frequency nor mucociliary clearance in the unchallenged lungs, and (b) the release of an inhibitory mediator(s) in the mucosa (such as neuropeptide Y), which decreases ciliary activity below its basal level. Stimulation of the mucociliary transport system can be achieved by either activation of an excitatory reflex (with ammonia vapor) or suppression of inhibitory efferent neurotransmission. Conversely, depression of the mucociliary transport can be achieved by activation of the inhibitory reflex (with $SO_2$) or suppression of the excitatory neurotransmission. These autonomic neural systems provide robust regulatory mechanisms to control the tracheobronchial mucociliary transport system during homeostasis as well as provide the means whereby it can respond to irritants and antigens to affect their removal and maintain airway patency.

The mucociliary transport system is responsible for cleansing the conducting airways from the terminal respiratory bronchioles through the bronchioles, bronchi, and trachea to the glottis. As the total circumference of these airways decreases by 4 orders of magnitude, inherent regulatory mechanisms must control its components at each airway level in an energy efficient manner while preventing the accumulation of excess mucus. The major components regulating mucociliary transport are (a) ciliary beat—the sole mechanism propelling the overlying mucus—(b) the secretion of viscoelastic mucus into the airway, and (c) the depth of the periciliary layer. In this paper, I will take the position that the mucociliary transport system is regulated by inhibitory and excitatory neural control. I will provide the background and discuss the preliminary data that supports this hypothesis. In so doing, I will demonstrate a model of how this mucociliary transport system maintains airway patency by operating at a low basal level during homeostasis, yet has intrinsic and redundant mechanisms to afford it protection when faced by an insult. Inherent in such a system, however, is the potential for a mismatch between the inhibitory and excitatory neural balance such that mucus transport is impaired with a consequent occlusion of the airway.

## II. HISTORY

A historical perspective of the neural regulation of ciliary activity has been presented previously [1] as has some data supporting the thesis outlined in this paper [2–5]. In healthy persons, homeostasic regulation of the mucociliary transport system maintains a low basal level as evidenced by the scant airway secretions and the inability of a cough to effect their removal from the airway. These observations imply that in the absence of a challenge, very little viscoelastic mucus is secreted and fluid is absorbed from the airway as mucus is transported proximally. From the results of mathematical modeling of mucus transport in the airway [6], either the rate of mucus transport decreases distally [7] and/or the transport of mucus is intermittent. These constraints, in turn, predict that ciliary beat frequency (CBF) decreases distally or that there are periodic changes in ciliary activity, perhaps analogous to that observed in the trachea [8]. Assuming the presence of excitatory and inhibitory neural regulation of the mucociliary transport system, this suggests that the homeostatic regulation of each of these functions is provided by the predominance of inhibitory neural activ-

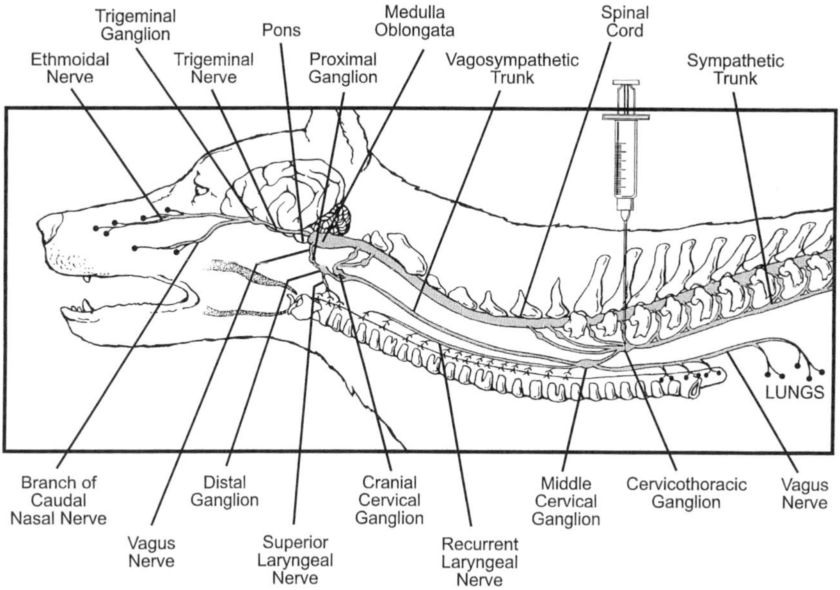


**Figure 1** The neural pathways innervating the respiratory tract are shown. It can be seen that the airways of the lungs are innervated by nerves traversing the vagosympathetic trunk and by nerves from the dorsal root ganglia traversing the cervicothoracic ganglia.

ity as predicted by Grey in 1928 [9]. Evidence for inhibitory neural regulation has been reported in the rabbit maxillary sinus [10,11].

Anatomically the mucosa of the trachea and bronchial airways are innervated by nerves traversing the vagosympathetic trunk, VST, and the sympathetic dorsal root ganglia and cervicothoracic ganglia, CTG (Figure 1). Information from sensory nerves in juxtaposition to the bronchial [12–14] and alveolar [15] epithelial cells can be transmitted via vagal [16] and sympathetic [17,18] afferents with the efferent nerves of these reflexes regulating mucosal function. Access to neural transmission in the vagosympathetic trunk was achieved by bilateral elevation of the vagosympathetic trunks and their enclosure in skin tubes. Neural transmission was abolished by cooling the skin tubes to 0°C with thermoelectric coolers. Neural activity transmitted via the sympathetic trunks was regulated by bilateral placement of needles in juxtaposition to the cervicothoracic ganglia (Figure 1) and injection of a local anesthetic to abolish all neurotransmission or local injection of test agonists and inhibitors of ganglionic transmission of the excitatory and inhibitory neural activity.

## III. REGULATION OF THE MUCOCILIARY TRANSPORT SYSTEM DURING HOMEOSTASIS

In dogs anesthetized with thiamylal sodium, bronchial mucociliary clearance measured using radioaerosol techniques was increased when the vagosysmpathetc trunks, VST, were cooled to 0°C but was not changed following injection of the local anesthetic lidocaine in the vicinity of the cervicothoracic ganglia, CTG [19]. However when both the VST were cooled to 0°C and lidocaine administered to the CTG, there was a marked stimulation of bronchial mucociliary clearance (BMC) that was much larger than that obtained with vagosympathetic block alone [19]. This indicates that the major neural regulatory control of the mucociliary transport system during homeostasis is inhibitory and that the inhibitory efferent neurotransmission via the CTG modulates that transmitted via the VST. At this juncture we do not know whether this inhibitory neural transmission is decreasing ciliary beat, mucus secretion, periciliary fluid depth, or all three of these.

In the airway, mucosa autonomic excitatory reflexes—be they cholinergic or adrenergic—are transmitted in the sympathetic ganglia dorsal root, cervicothoracic ganglia, and parasympathetic airway ganglia via nicotinic ($N_1$) receptors. Thus the contribution of excitatory neural pathways in the maintenance of mucociliary transport during homeostasis was ascertained by IV injection of 2 mg/kg of hexamethonium in one dog. In this case bronchial mucociliary clearance was totally abolished indicating the presence of a low level of excitatory autonomic tone during homeostasis. In the presence of VST blockade, local injection of 1 mg of hexamethonium in the vicinity of the CTG also ablated bronchial mucociliary clearance in one dog. In this case, it is assumed that the parasympathetic ganglia are not blocked. This can be interpreted to indicate that in the absence of excitatory neural transmission through the CTG, the dominance of inhibitory signals suppresses peripheral excitatory mechanisms.

Thus inherent in this model is the assumption that the inhibitory neurotransmission is not mediated via nicotinic ganglionic transmission. Inhibitory neurotransmittors of the autonomic nervous system are potential candidates. These include nitric oxide (NO), γ-aminobutyric acid (GABA), and neuropeptide Y (NPY): NO is the major putative inhibitory neurotransmitter regulating airway smooth muscle function [20]. Nitric oxide synthase (NOS) is likely located on preganglionic fibers in the CTG

[21]. Administration of an inhibitory agonist to the CTG should impair bronchial mucociliary clearance and an agent that inhibits the production of this agonist should stimulate mucociliary transport. In one dog, when 1.2 μg of nitroglycerin (an NO donor) was administered to each CTG not only was the stimulation of BMC induced by cooling the VST together with anesthetization of the CTG inhibited, but the resultant clearance was slower than the baseline. In a separate experiment, following administration of L-NAME (a competitive substrate for nitric oxide synthase, NOS, to each CTG in the presence of total vagal blockade, the increase in mucociliary clearance was identical to that obtained with vagosympathetic blockade and anesthetization of the CTG. These observations are consistent with the premise that NO is an inhibitory neurotransmitter or modulator in the CTG.

These data raise questions as to the location and mechanisms whereby efferent inhibitory neurotransmission maintains the observed low basal level of mucociliary clearance. Two possibilities are, firstly, that efferent inhibitory neural activity suppresses the release of excitatory muscarinic and adrenergic mediators of the excitatory neural pathways and/or, secondly, that the release of inhibitory neurotransmitter(s) acts directly on mucosal cells to decrease their basal rate of ciliary beat, mucus production, and/or transepithelial ion and water transport. In the absence of inhibitory and excitatory neurotransmission via the VST and CTG, blockage of excitatory transmission in the airway ganglia by administration of 2 mg/kg IV nicotine did not inhibit bronchial mucociliary clearance. This indicates that the stimulation of the mucociliary transport system following removal of inhibitory efferent neural activity cranial to the airway ganglia is not mediated via excitatory nicotinic neurotransmission in the peripheral ganglia. This suggests that the inhibition of efferent inhibitory activity allows the release of mediators in the mucosa that stimulate the mucociliary transport system. Co-administration of the muscarinic antagonist, ipratropium bromide, and the adrenergic antagonist, propranolol, by aerosol totally abolished the stimulation of BMC induced by inhibition of neurotransmission via the vagosympathetic trunk and the CTG. This implies that the mediators released by inhibition of the inhibitory transmission are the classical adrenergic and muscarinic agonists, norepinephrine and acetylcholine. It is notable that administration of these antagonists in the presence of efferent inhibitory neural activity also caused a slight decrease in basal mucociliary clearance, consistent with the presence of a small amount of adrenergic and cholinergic tone during homeostasis. The inhibitory neuropeptide NPY has been shown to be released from peripheral sympathetic nerves. Thus, this peptide may act directly on ciliated cells to decrease ciliary activity. Recently, Wong and Yeates

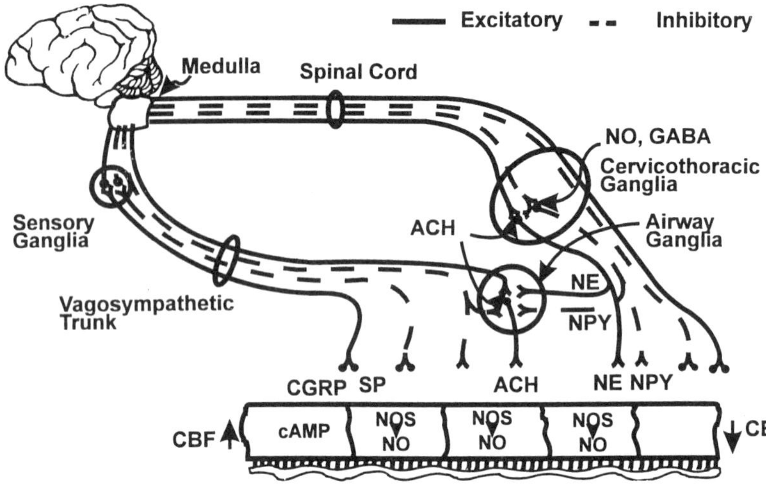


**Figure 2** A model of the proposed excitatory and inhibitory neural pathways regulating mucosal function. Candidate neurotransmitters or modulators in the cervicothoracic ganglia, airway ganglia as well as those of the efferent and sensory nerves in the mucosa are shown. In the epithelium, it is proposed that increases in CBF are mediated via increases in cAMP (or $IP_3$) and nitric oxide and decreases in CBF induced by NPY are caused via a decrease in intracellular calcium.

presented preliminary data showing that 1 and 10 $\mu M$ NPY administered to human epithelial cell cultures suppresses CBF below baseline levels [22]. Thus mucociliary transport appears to be maintained at a low basal level by both an inhibition of excitatory mediators and the release of an inhibitory neuropeptide.

These data, together with that in the literature and some further preliminary studies, were integrated resulting in the following model (Figure 2) for the neural regulation of the mucociliary transport system.

## IV. ACTIVATION OF EXCITATORY AND INHIBITORY PATHWAYS

Thus far I have considered the regulation of the mucociliary transport system during homeostasis as represented in the anesthetized dog. The next step is to determine whether these inhibitory and excitatory neural pathways form the efferent pathways of inhibitory and excitatory neural reflexes that are activated by inhaled irritants and/or allergens to defend the airway.

## A. Excitatory Neural Pathways

Excitatory neural pathways can be activated by the release of endogenous mediators and their interactions with receptors on peripheral sensory nerves [23]. Inhalation of less that 2 μg of ragweed allergen deposited in the alveolar and/or bronchial regions of the lungs of ragweed-sensitized dogs precipitates bradycardia and apnea followed by tachypnea [24], responses characteristic of the pulmonary neurally transmitted reflex [25]. Concurrent with these responses is a marked increase in bronchial mucociliary clearance [24]. The allergen-induced increases in bronchial mucociliary transport were abolished by vagal blockade and mucociliary clearance was impaired when both the transmission via the vagosympathetic trunk and cervicothoracic ganglia were abolished. This preliminary data is consistent with the increase in mucociliary clearance being mediated through excitatory neural transmission via both the vagosympathetic trunk and the cervicothoracic ganglia.

There is a potential that several neurotransmitters associated with the stimulatory (excitatory) responses of the mucociliary transport system activate their respective receptors on the cell plasma membrane to affect an increase in a common intracellular messenger. Nitric oxide is a possible candidate. Nitric oxide synthase in the airway epithelium has been shown to be upregulated by ovalbumin sensitization [26]. Nitric oxide has been implicated in the stimulation of ciliary beat frequency induced by adrenergic agonists [27] but not in the maintenance of basal ciliary beat frequency [28]. On the assumption that the efferent mediators released following ragweed allergen challenge induced an increase of epithelial NOS and NO, L-NMMA (50 mg/5 mL) a competative nitric oxide synthase inhibitor was administered prior to inhalation of the ragweed allergen. This competitive inhibition of NOS abolished the increase in bronchial mucociliary clearance induced by deposition of ragweed into the alveolar regions. This is consistent with the potential of NO as a common intracellular mediator whereby the action of several mediators activating their various receptors affect a stimulation of ciliary beat and consequently mucociliary transport.

## B. Inhibitory Neural Pathways

Inhibitory neural pathways are likely activated by acids. It appears that protons activate receptors on sensory nerves [29]. Such inhibitory sensory nerves may also be activated by dehydration of the airway epithelia. In a series of experiments in beagle dogs where the lumen of the airways was separated from that of the bronchial airways by a cuff inflated in the lower trachea, CBF was measured in the trachea and either dry air or dry

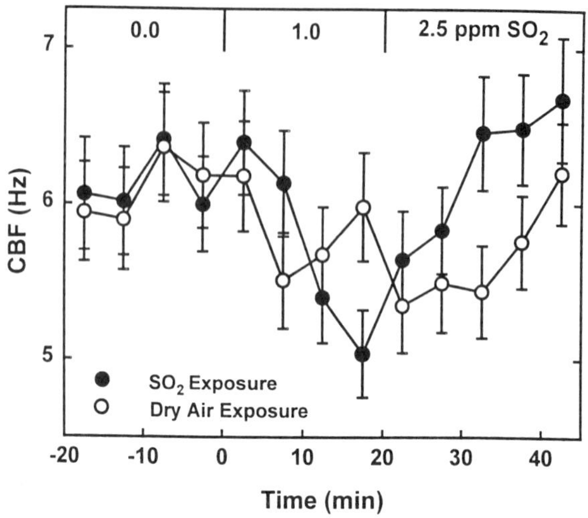


**Figure 3** The temporal changes in tracheal CBF in a cohort of 6 beagles in which the bronchial airways and not the trachea were exposed to either dry air or dry air together with low concentrations of $SO_2$. It can be seen that when $SO_2$ was combined with the dry air, CBF decreased more than when dry air was administered alone and that the recovery of CBF occurred earlier than when only dry air was administered. (Extracted from the thesis of J. Daugird [30].

air with a low concentration ($<2$ ppm) of $SO_2$ was administered exclusively to the bronchial airways. As can be seen in Figure 3, the decrease in CBF in the trachea due to dry air in the bronchial airways was exacerbated by the delivery of $SO_2$. The recovery also appears more rapid when $SO_2$ was delivered concurrently [30]. It is notable that the time course and recovery of CBF is similar to that obtained in a separate series of studies where the trachea as well as the bronchial airway was exposed to dry air [31]. Inhalation of a slightly higher concentration of $SO_2$ caused a marked impairment of bronchial clearance [5]. The impairment of CBF by 5 ppm $SO_2$ was inhibited by delivery of aerosolized cromoyln sodium [5] consistent with the desensitization of sensory nerves by cromoyln sodium [32].

## V. CONCLUSION

The model presented has been based, in large part, on preliminary data. Further validation, modification and extension of the model is required.

It is important to test the premises outlined herein. The extent to which inhibitory and excitatory neurotransmission regulates each of the components of the mucociliary transport system will be the basis for future investigations. Excitatory neural regulation of mucus glands [33] and goblet cells [34] has been demonstrated. The model presented in this paper provides the underlying rationale for investigators to elucidate the complementary inhibitory counterpart(s). It is likely that ion (and water) transport is neurally regulated as it is in the gut [35]. The inhibitory and excitatory pathways regulating ciliary beat frequency can be further delineated in vivo using heterodyne laser light scattering [36]. It is quite possible that impaired and "normal" mucociliary transport can not be fully defined as being just slow or fast transport, respectively. Rather impaired mucociliary clearance can be predicted to result from a relative imbalance of the excitatory and inhibitory neural pathways on each of the components of the mucociliary transport system leading to the accumulation of mucus due to the dehydration or hyperhydration of the airway together with inadequate propelling forces of the ciliary activity to transport mucus along the airway. The elucidation of these pathways and mechanisms will not only provide a basis for underlying pathophysiology of diseases of the airway—such as COPD, asthma, cystic fibrosis, bronchiectasis, and bronchorrhea—but also will provide novel approaches to the pharmacologic prevention and treatment of these diseases.

## ACKNOWLEDGMENTS

I am indebted to Anthony V. Daza who conducted the dog studies to obtain most of the preliminary data and Mark Pelnar who so competently translated my sketches of the neural pathways to provide the artistry contained within Figures 1 and 2. This conceptual model also benefited from constructive discussions with my colleagues Lid B. Wong and Shu-Fang Liu. This work was supported by NIEHS of NIH, grant ES04317, and the Medical Research Service of the Department of Veterans Affairs.

## REFERENCES

1. Wong LB, Yeates DB. Dynamics of the regulation of ciliated epithelial function. Comments in Theoretical Biology 1997; 4(2–3):183–208.
2. Yeates DB, Eljamal M, Wong LB, Miller IF. Cellular–neural–cellular pathways mediating the response of tracheal ciliary beat frequency to inhaled capsaicin. Chest 1992; 101(3):72S–73S.
3. Yeates DB, Eljamal M, Wong LB, Miller IF. Cellular and neural pathways

mediating the response of tracheal ciliary beat to inhaled irritants. In: Eds. Dodgson J, McCallum RI, Inhaled Particles VII. Ann of Occup Hygiene. Oxford: Pergamon, 1994; 939–949.
4. Eljamal M, Wong LB, Yeates DB. Capsaicin activated bronchial- and alveolar-initiated pathways regulating tracheal ciliary beat frequency. J Appl Physiol 1994; 77(3):1239–1245:736–744.
5. Yeates DB, Katwala S, Daurgird J, Daza AV, Wong LB. Excitatory and inhibitory neural regulation of tracheal ciliary beat frequency. Inhaled Particles VIII. Ann Occup Hyg 1997; 41(Suppl 1):736–744.
6. Yeates DB, Aspin N. A mathematical description of the airways of the human lungs. Respir Physiol 1978; 32:91–104.
7. Yeates DB, Gerrity TR, Garrard CS. Characteristics of tracheobronchial deposition and clearance in man. Inhaled Particles V. Ann Occup Hyg 1982; 26 (nos 1–4):245–257.
8. Chandra T, Yeates DB, Miller IF, Wong LB. Stationary and nonstationary correlation-frequency analysis of heterodyne mode laser light scattering: magnitude and periodicity of canine tracheal ciliary beat frequency in vivo. Biophysical Journal 1994; 66:878–890.
9. Grey J. Ciliary Movement, New York: Macmillian Co., 1928.
10. Cervin A, Lindberg S, Merke U. The effect of neuropeptide Y on mucociliary activity in the maxillary sinus. Acta Otolaryngol (Stock) 1991; 111:960–966.
11. Cervin A, Lindberg S, Merke U, Uddman R. Neuropeptide Y in the rabbit maxillary sinus modulates cholinergic acceleration of mucociliary activity. Acta Otolaryngol (Stock) 1992; 112:872–881.
12. Laitinen A. Ultrastructural organization of intraepithelial nerves in the human airway tract. Thorax 1985; 40:488–492.
13. Baluk P, Gabella G. Innervation of the guinea pig trachea: a qualitative morphological study of intrinsic neurons and extrinsic nerves. J Comp Neurol 1989; 285:117–132.
14. Kummer W, Fisher A, Kurkowski R, Heym C. The sensory and sympathetic innervation of the guinea–pig lung and trachea as studied by retrograde neuronal tracing and double–labelling immunohistochemistry. Neuroscience 1992; 49:715–737.
15. Fox B, Bull TB, Guz A. Innervation of alveolar walls in the human lung: an electron microscopic study. J Anatomy 1980; 131:683–692.
16. Coleridge HM, Coleridge JCG. Pulmonary reflexes: neural mechanisms of pulmonary defense. Ann Rev Physiol 1994; 56:69–91.
17. Coon RL, Clifford PS, Hopp FA, Zuperku EJ. Reflex ventilatory effects of KCl stimulation of lung receptors with sympathetic afferents. Respir Physiol 1990; 82:349–358.
18. Kostreva DR, Hopp FA, Zuperku EJ, Igler FO, Coon RL, Kampine JP. Respiratory inhibition with sympathetic afferent stimulation in the canine and the primate. J Appl Physiol 1978; 44:718–724.
19. Yeates DB, Katwala SP, Daza AV, Wong LB. Inhibition of neural transmission via the vagosympathetic trunk and cervicothoracic ganglia stimulates

tracheobronchial mucociliary clearance. Amer Res Respir Dis 1995; 151(4 part):A819.
20. Barnes PJ. Is asthma a nervous disease? Chest 1995; 107(3):119S–124S.
21. Ceccatelli S, Lundberg JM, Zhang X, Aman K, Hokfelt T. Immunohistochemical demonstration of nitric oxide synthase in the peripheral autonomic nervous system. Brain Research 1994; 656:381–395.
22. Wong LB, Yeates DB. Neuropeptide Y inhibits ciliary beat frequency (CBF) in cultured human ciliated cells via protein kinase C (PKC) dependent mechanism. Am J Respir Critical Care Med 1997; 155(Part 2):A432.
23. Greene R, Fowler J, MacGlashan D, Weintreich D. IgE-challenged lung mast cells excite vagal sensory neurons in vitro. J Appl Physiol 1988; 64(5): 2249–2253.
24. Yeates DB, Daza AV, Mussatto DJ. Bronchial and alveolar allergen-induced anaphylaxis and the stimulation of bronchial mucociliary clearance in ragweed-sensitized dogs. Proceedings of the Association of American Physicians 1997; 109(4):1–13.
25. Cotton DJ, Bleeker ER, Fisher SP, Graf PD, Gold WM, Nadel JA. Rapid, shallow breathing after *Ascaris suum* antigen inhalation: role of the vagus nerves. J Appl Physiol: Respirat Environ Exercise Physiol 1977; 42(1): 101–106.
26. Liu SF, Haddad E-M, Adock I, Salmon M, Kato H, Gibey T, Barnes PJ, Chung KF. Inducible nitric oxide synthase after sensitization and allergen challenge in Brown–Norway rat lungs. Br J Pharmacol 1997; 121: 1241–1246.
27. Jain B, Rubinstein I, Robbins RA, Leise KL, Sisson JH. Modulation of airway epithelial cell ciliary beat frequency by nitric oxide. Biochem and Biophysical Res Comm 1993; 191(1):83–88.
28. Sisson JH. Ethanol stimulates apparent nitric oxide-dependent ciliary beat frequency in bovine airway epithelial cells. Am J Physiol 1995; 268 (Lung Cell Mol Physiol 12):L596–L600.
29. Fox AJ, Urban L, Barnes PJ, Dray A. Effects of capsazepine against capsaicin- and proton-evoked excitation of single airway C-fibers and vagus nerve from the guinea-pig. Neuroscience 1995; 67(3):741–752.
30. Daugird J. The effect of sulphur dioxide on the mucociliary transport system. MS Thesis, University of Illinois at Chicago, 1994.
31. Winters SL, Yeates DB. The interaction between ion transporters and the mucociliary transport system in the dog and baboon. J Applied Physiol 1997; 83(4).
32. Dixon M, Jackson DM, Richards IM. The action of sodium cromoglycate on 'C' fibre endings in the dog lung. Br J Pharmac 1980; 70:11–13.
33. Fung DCK, Beacock DJ, Richardson PS. Vagal control of mucus glycoconjugate secretion into the feline trachea. J Physiol (Lond) 1992; 453:435–447.
34. Kuo HP, Rohde JAL, Tokuyama K, Barnes PJ, Rogers DF. Capsaicin and sensory neuropeptide stimulation of goblet cell secretion in guinea-pig trachea. J Physiol 1990; 431:629–641.

35. Cooke HJ. Neuroimmune signaling in regulation of intestinal ion transport. Am J Physiol 1994; 266 (Gastrointest Liver Physiol 29):G167–G178.
36. Chandra T, Yeates DB, Miller IF, Wong LB. Stationary and nonstationary correlation-frequency analysis of heterodyne mode laser light scattering: magnitude and periodicity of canine tracheal ciliary beat frequency *in vivo*. Biophysical J 1994; 66:878–890.

# 5

# The Nucleoside Diphosphate Kinase and Chloride-Sensitive GTP-Dependent Phosphorylation in Apical Membrane from Ciliated Airway Cells

**R. Muimo, S. J. Banner, Lindsay Marshall, and Anil Mehta**
*Ninewells Hospital and Medical School, and*
*University of Dundee, Dundee, United Kingdom*

## I. INTRODUCTION

In airway epithelium, the intracellular chloride concentration $[Cl^-]_i$ plays a key role in regulating a number of cellular functions such as transepithelial ion transport [1], cell volume homeostasis [2], and G protein activity [3]. Although the mechanisms by which the cell senses $[Cl^-]_i$ to regulate these various cellular activities are obscure, it is clear that the chloride ion modulates the conductance of both sodium and chloride ion channels [1].

## II. CHLORIDE AS A SECOND MESSENGER

Recently, we have shown that $[Cl^-]$ (and other anion species) determine the phosphorylation state of apical membrane proteins from human airway epithelium in vitro [4]. We found that $Cl^-$ regulated the intensity of phosphorylation of a 37 kDa protein (p37) with GTP as phosphate donor. Furthermore, when ATP was the phosphate donor, a different pattern of chloride dependent phosphorylation was generated within the apical membrane proteins. We postulated that chloride ions were acting as second messengers to the apical membrane and that their concentration altered membrane protein structure by modulating the activity of membrane-bound kinases/phosphatases endogenous to the apical membrane. The aim of the present study was to define a suitable model system for the human airway epithelium that could be used to generate increased amounts of tissue to characterize this $Cl^-$-sensitive phosphorylation cascade.

## III. SHEEP TRACHEAL EPITHELIUM IS A GOOD MODEL FOR THE HUMAN AIRWAY

Our new data show that apical membrane preparations from sheep tracheal epithelium also contain a 37 kDa protein (p37s) whose phosphorylation state is sensitive to $[Cl^-]$. Analysis of phosphorylation of p37s, using sheep apical membranes prepared by the $Cl^-$-free sucrose gradient technique and incubated with increasing concentrations of the chloride salt of *N*-methyl-*D*-glucamine (NMDGCl), showed that maximal phosphorylation occurred at 5 mM NMDGCl. Interestingly, NMDGCl also affected the phosphorylation of two smaller proteins p19s/p21s. Higher concentrations of NMDGCl resulted in increased phosphate incorporation whereas lower concentrations of NMDGCl prevented their phosphorylation. We have identified and characterised these proteins as isoforms of the nucleoside diphosphate kinase (NDP Kinase, NDPK), a kinase that appears to play a role in regulating the phosphorylation of p37s. Our data show that the autophosphorylation of NDPK in the presence of $Cl^-$ or $Mg^{2+}$ is reciprocally related in phosphate incorporation to the phosphorylation of p37s. When apical membrane proteins prepared by the $Mg^{2+}$ precipitation technique were incubated with $\gamma[^{32}P]$ GTP, we found that both p19s and p21s were transiently phosphorylated before p37s. This data suggested an association between p19/p21s and p37s and that $Mg^{2+}$ played a role in the dephosphorylation of p19s/p21s. The inhibition of dephosphorylation of p19s/p21s by EDTA confirmed dephosphorylation of p19s/p21s is divalent ion dependent. In addition, we found that divalent ions $Mn^{2+}$ and $Co^{2+}$ but not $Ca^{2+}$ or $Zn^{2+}$ also increased dephosphorylation of p19s and p21s. The threshold for the effect of $Mg^{2+}$ on the dephosphorylation of p19/p21s was 200 $\mu M$. It is, at present, not clear whether loss of phosphate from p19/p21s was due to direct transfer to p37s or due to dephosphorylation resulting from protein phosphatase activity.

## IV. IDENTIFICATION OF NDPK

In order to establish whether p19s and p21s were protein kinases or substrate proteins, membrane proteins were separated and renatured within polyacrylamide gels. Autoradiography analysis of the polyacrylamide gels showed that p19s and p21s autophosphorylated with either $\gamma[^{32}P]$GTP or $\gamma[^{32}P]$ATP as substrates, suggesting the two proteins were kinases. Western blotting and immunoprecipitation with a specific polyclonal antibody showed that p21s was a membrane-bound isoform of nucleoside diphosphate kinase (NDPK, EC 2.7.4.6), a protein kinase that catalyses the formation of GTP from ATP and is, among other functions, essential for UTP synthesis and cell secretion.

Our previous data from human nasal epithelium showed that phosphorylation of p37h was more conspicuous in the presence of GTP as phosphate donor [4]. Since NDPK catalyses transfer of $\gamma$ phosphate predominantly from ATP to diphosphate nucleotides [5], we postulated that incubation of an apical membrane preparation in the presence of $\gamma[^{32}P]$ATP and GDP would enhance phosphorylation of p37s by promoting production of $\gamma[^{32}P]$GTP. Maximal phosphorylation of p37s occurred when apical membrane proteins prepared by the sucrose gradient technique were incubated with $\gamma[^{32}P]$ATP at 37°C in the presence of GDP (500 n*M*) and was approximately 10 fold greater than buffer control. Thin layer chromatography analysis of the incubation mixture showed that GTP was produced. Our working model of the relationship between NDPK and p37s is that NDPK synthesizes GTP from GDP and ATP and channels this GTP either to a second kinase targeting p37s or directly to p37s for autophosphorylation. We cannot discriminate between these two mechanisms until we have purified p37s to homogeneity.

In summary, our data shows that we have identified a chloride-dependent protein phosphorylation cascade in sheep airway epithelium. We have shown that sheep airway epithelium is a good model for further characterization of the chloride-dependent phosphorylation cascade [6].

## ACKNOWLEDGMENT

Supported by the Wellcome Trust.

## REFERENCES

1. Dinudom A, Young DI, Cook DI. Sodium and chloride conductances are controlled by cytosolic chloride concentration in the intralobular duct cells of mouse mandibular glands. J Memb Biol, 1993; 135:289–295.

2. Foskett JK. $[Ca^{2+}]_i$ modulation of $Cl^-$ content controls cell volume in single salivary acinar cells during fluid secretion. Am J Physiol., 1990; 259: C998–C1004.
3. Higashijima T, Ferguson KM, Sternweis PS. Regulation of hormone–sensitive GTP-dependent regulatory proteins by chloride. J Biol Chem. 1987; 262: 3597–3602.
4. Treharne KJ, Marshall LJ, Mehta A. A novel chloride-dependent, GTP–utilising protein kinase in plasma membranes from human respiratory epithelium. Am J Physiol. (Lung Cell. Mol. Physiol) 1994; 267:L592–L601.
5. Parks, RE Jr, Agarwal RP. Nucleoside diphosphokinases. In: Boyer PD ed. The Enzymes. New York: Academic Press, 1973; 307–331.
6. Muimo et al. Am J Resp Cell Mol Biol 1997 (in press).

# 6

# Regulation of Ciliogenesis and Ciliated Cell Specific Genes

**Lawrence E. Ostrowski, Kaya L. Andrews, Pravin D. Potdar,* and Paul Nettesheim**
*National Institutes of Health, Research Triangle Park, North Carolina*

## I. INTRODUCTION

Ciliated cells line the surface of mammalian airways and by the coordinated beating of their cilia clear the mucus and entrapped material from the pulmonary system. The percentage of ciliated cells varies both among species and different levels of the respiratory tract. It has long been recognized that ciliated cells are essential for mucociliary clearance; however, accumulating evidence suggests that ciliated cells have other important functions as well. For example, Diamond et al. [1] reported that bovine ciliated cells produce tracheal antimicrobial peptide (TAP), a member of the beta-defensin family. These natural antibiotics are likely to help prevent respiratory infections, and recent studies demonstrating their inactivity in high salt conditions may help explain the recurrent infections common in cystic fibrosis patients [2,3]. Ciliated cells are also known to express the cystic fibrosis transmembrane conductance regulator protein (CFTR), and this implies a role for ciliated cells in the regulation of ion and fluid transport. Further studies are likely to reveal additional functions of ciliated cells in such areas as immunity and inflammation. However, the most widely studied aspect of ciliated cells is their role in mucociliary clearance (for a recent review, see Ref. 4). The importance of this defense mechanism is illustrated by individuals who suffer from primary ciliary dyskinesia (PCD). Primary ciliary dyskinesia is an inherited disorder in

* *Current affiliation:* Cancer Research Institute, Parel, Bombay, India

which a genetic defect, usually observed in the structure of the cilium itself, results in absent or impaired mucociliary clearance. These individuals suffer from repeated respiratory infections. Mucociliary clearance can also be impaired in many other situations. These can be chronic, as in cystic fibrosis, or acute, as in viral or bacterial infections. In addition, ciliated cells are sensitive to many environmental pollutants, such as ozone, sulfur dioxide, and nitrogen dioxide. In many of these conditions, ciliated cells are lost from the airways. In order for mucociliary clearance to be restored, ciliated cells need to be regenerated. Since under most conditions it appears that ciliated cells themselves do not divide, the replacement of ciliated cells most likely occurs through the differentiation of precursor cells. Recent work in our laboratory has focused on understanding the regulation of ciliated cell differentiation.

Using an in vitro model system in which primary cultures of undifferentiated rat tracheal epithelial (RTE) cells can be induced to undergo mucociliary or squamous differentiation [5], several factors have been identified that influence the extent of ciliated cell differentiation. In these experiments, RTE cells are isolated from adult Fisher 344 rats and plated on a collagen gel-coated porous membrane. The cells are cultured in a complete growth media (which is present above and below the membrane) for seven or eight days, at which time the cultures are near confluence. An air/liquid interface is formed by removing the media from the apical surface of the cultures, and over the next seven to eight days, the cultures undergo mucociliary differentiation. If RTE cells are cultured in the absence of exogenous retinoic acid, the cells undergo squamous differentiation. Using a monoclonal antibody that reacts specifically with rat ciliated cells to quantitate ciliated cell differentiation, conditions were identified that inhibited or stimulated ciliated cell differentiation [6]. For example, maintaining the cultures submerged in media inhibited ciliated cell differentiation by approximately 90%. In contrast, removing epidermal growth factor (EGF), bovine pituitary extract (BPE), or cholera toxin (CT) from the media at the time the air/liquid interface was formed increased ciliated cell differentiation by 2–3 fold. The removal of both EGF and CT from the media at the time the air/liquid interface was formed has consistently induced high levels of ciliated cell differentiation. The ability to regulate, both positively and negatively, the extent of ciliated cell differentiation in this system makes it an attractive model to study the regulation of this process.

## II. MECHANISMS REGULATING TRANSCRIPTION

To understand ciliated cell differentiation at the molecular level, we wished to study the mechanisms regulating transcription of ciliated cell specific genes. Because no mammalian ciliated cell specific genes had yet been described, information obtained from lower organisms was used to clone partial cDNAs of rat axonemal dynein heavy chains using RT-PCR (for a discussion of axonemal dyneins, see Gibbons, this volume). Primers were designed based on highly conserved sequences around the catalytic site, or first P-loop [7]. These primers were designed to specifically amplify fragments from either axonemal or cytoplasmic dynein heavy chains. The axonemal-specific primers only produced a product with RNA isolated from rat tissues that contains axonemal structures, such as brain, lung, testis, and trachea. Heart and liver were negative by this assay. As a control, primers specific for the cytoplasmic dynein heavy chain, which is expressed in all cells types, produced a product from all tissues analyzed. Using this procedure, seven different axonemal dynein heavy chains (axo a–axo g) were identified and cloned. Southern analysis indicated six of the seven probes corresponded to a unique, single-copy gene, while one probe (axo a) hybridized to several fragments. Evolutionary analysis demonstrated that each of the axonemal dynein heavy chains is more closely related to axonemal dynein heavy chains of other species than to any cytoplasmic dynein heavy chains. Based on Southern analysis using a pooled probe, it is likely that ciliated RTE cells express additional axonemal dynein heavy chains. Northern analysis demonstrated that each of the seven axonemal dyneins hybridized to a large (approximately 15 kb) message in cultures allowed to undergo mucociliary differentiation, while in cultures grown in the absence of retinoic acid (i.e., squamous differentiation) expression of the axonemal dyneins was not detectable. Cytoplasmic dynein heavy chain was expressed at similar levels under both conditions. Axonemal dynein expression was then analyzed at different times during the differentiation of RTE cells cultured under three different conditions. In cultures grown in complete media at an air/liquid interface, the expression of axonemal dynein heavy chains was absent at day 7 and only appeared weakly at day 13, increasing in intensity at day 15. Rat tracheal epithelial cells cultured while continuously submerged in media did not express axonemal dyneins, with the exception of axo f, which was expressed in a pattern similar to the control condition. This suggests that the regulation of axo f may be unique from the other axonemal dyneins. In cultures of RTE cells grown in the absence of EGF and CT, a condition that stimulates ciliated cell differentiation, the expression of all the axonemal dynein heavy chains was strongly induced by day 13 and remained high at day 15. These studies demonstrate that the expression of axonemal dynein heavy chain mRNA is regulated coordinately by conditions that influence ciliated cell differentiation.

We have also examined the expression in this culture system of a unique dynein heavy chain called 1b [8]. Dynein heavy chain 1b has sequence similarity to cytoplasmic dynein, but was observed to increase during ciliogenesis of sea urchin embryos. Evolutionary analysis places dynein 1b between the other known axonemal and cytoplasmic dynein heavy chains. Using primers specific for the rat dynein 1b, RT-PCR reactions produced a product in all tissues examined, consistent with a cytoplasmic function for 1b. In addition, 1b was detected at both early and late time points in RTE cultures undergoing mucociliary differentiation, whereas axonemal dyneins were only detected late in culture. This again suggests dynein 1b is a cytoplasmic dynein. However, Northern analysis demonstrated that 1b expression was also increased by conditions that stimulated ciliogenesis, in a pattern similar to the axonemal dynein heavy chains. Finally, using antibodies raised against peptide fragments of dynein 1b, it was shown that dynein 1 b is strongly expressed in ciliated cells and is localized just below the apical surface. The cilia themselves were not strongly positive with this antibody. Taken together, this data suggests that dynein heavy chain 1b is a cytoplasmic dynein with a special function in ciliated cells.

## III. DIFFERENTIAL DISPLAY

To identify additional ciliated cell specific genes, the technique of differential display was used. Rat tracheal epithelial cells were cultured under the three conditions described above and RNA was isolated. Differential display was performed using RNA isolated from two independent experiments and radioactive bands corresponding to differentially expressed genes were isolated, cloned, and used as probes on northern blots to confirm their differential expression. Two novel genes have been identified whose expression is increased/decreased in parallel with ciliated cell differentiation [Andrews et al., Ostrowski et al., in preparation]. One of these hybridizes to a mRNA of about 2 kb and is expressed in many tissues, while the other hybridizes to a message of about 7 kb and appears to be only expressed in tissues with axonemal structures. Sequencing analysis of both genes has not revealed any significant homology to any proteins of known function. Cloning of the full-length sequence for each of these novel genes is in progress, as are additional expression studies. Additional experiments will be required to determine the function of the proteins encoded by these genes in ciliated cell differentiation.

## IV. CONCLUSION

In summary, we have developed a system in which ciliated cell differentiation can be regulated both positively and negatively by varying the culture conditions. We have identified seven rat axonemal dynein heavy chain genes and shown that their expression is regulated coordinately with ciliated cell differentiation in this model. It was demonstrated that dynein heavy chain 1b is a cytoplasmic dynein and that the expression of 1b is increased during ciliated cell differentiation. In addition, we have identified several novel genes whose expression also parallels ciliated cell differentiation. To further understand the regulation of ciliated cell specific gene transcription, and also ciliated cell differentiation, the promoters of several of these genes are now being isolated and characterized.

## REFERENCES


1. Diamond G, Zasloff M, Eck H, Brasseur M, Maloy WL, Bevins CL. Tracheal antimicrobial peptide, a cysteine-rich peptide from mammalian tracheal mucosa: peptide isolation and cloning of a cDNA. Proc Natl Acad Sci USA 1991; 88:3952–3956.
2. Smith JJ, Travis SM, Greenberg EP, Welsh MJ. Cystic fibrosis airway epithelia fail to kill bacteria because of abnormal airway surface fluid. Cell 1996; 85:229–236.
3. Goldman MJ, Anderson GM, Stolzenberg ED, Kari UP, Zasloff M, Wilson JM. Human beta-defensin-1 is a salt-sensitive antibiotic in lung that is inactivated in cystic fibrosis. Cell 1997; 88:553–560.
4. Wanner A, Salathe M, O'Riordan TG. Mucociliary clearance in the airways. Am J Respir Crit Care Med 1996; 154:1868–1902.
5. Kaartinen L, Nettesheim P, Adler KB, Randell SH. Rat tracheal epithelial cell differentiation in vitro. In Vitro Cell Dev Biol 1993; 29A:481–492.
6. Clark AB, Randell SH, Nettesheim P, Gray TE, Bagnell B, Ostrowski LE. Regulation of ciliated cell differentiation in cultures of rat tracheal epithelial cells. Am J Respir Cell Mol Biol 1995; 12:329–338.
7. Andrews K, Nettesheim P, Asai D, Ostrowski LE. Identification of seven rat axonemal dynein heavy chain genes: expression during ciliated cell differentiation. Mol Biol Cell 1996; 7:71–79.
8. Criswell PS, Ostrowski LE, Asai DJ. A novel cytoplasmic dynein heavy chain: expression of DHC1b in mammalian ciliated epithelial cells. J Cell Sci 1996; 109:1891–1898.

# 7

# Mucus Transporting Cilia: A Model System for Investigating Calcium Signaling in Electrically Inexcitable Cells

**Alon Korngreen and Zvi Priel**
*Ben-Gurion University of the Negev, Beer-Sheva, Israel*

## I. INTRODUCTION

In a 1993 perspective in *Science* Putney argued that there is much "Excitement about calcium signaling in inexcitable cells" [1]. Indeed, in the last decade growing interest has developed in the signaling processes in many types of cells that were previously less fashionable. From these studies emerged a general signal-transduction pattern. In electrically inexcitable cells, the predominant second messenger responsible for release of calcium from intracellular stores is Inositol 1,4,5-trisphosphate ($IP_3$) [2]. Following this release a prolonged increase in the calcium influx through the plasma membrane is observed [3]. The cellular mechanisms regulating this calcium influx vary between cell types. It can be the result of the activation of a receptor operated calcium channel [4], of a second messenger-regulated calcium channel [5], or of a second messenger generated by the depletion of the calcium stores [6,7]. It was suggested that the calcium concentration in the vicinity of the plasma membrane increased due to the increase in the influx of calcium while the rest of the cytoplasm remained at a lower calcium concentration [3,8]. Thus the physiological importance of this prolonged calcium influx is large. The cell utilizes this mechanism to retain a high level of activation without sustaining a harmfully high cytoplasmic calcium concentration.

To fully investigate this biphasic cellular activation it is necessary to evaluate the signaling mechanism and the physiological response. However in most of the electrically inexcitable cells it is hard to observe the physiological response. The ciliary cell is an exception to this rule. The beating of the cilia can be observed with a regular light microscope and quantified with relatively simple electro-optic instruments. Moreover, since the cilia are juxtaposed to the plasma membrane a signaling cascade taking place in this membrane or in close proximity will have a different effect on the cilia than on the cytoplasm. We have developed a unique system to measure simultaneously the intracellular calcium concentration ($[Ca^{2+}]_i$) and the ciliary beat frequency (CBF) thus providing for the first time a tool for observing the physiological cause and effect in the ciliary cell [9,10].

## II. METHODS

### A. Chemicals and Solutions

All chemicals, unless stated otherwise, were obtained from Merck (Darmstadt, Germany). All tissue culture reagents were from Biological Industries (Beit Haemek, Israel). The pentapotassium salt ($K_5$fura-2) and the acetoxymethyl ester (fura-2/AM) forms of the calcium indicator fura-2 were purchased from Molecular Probes (Eugene, OR). Fresh solutions were made each day in dry dimethyl sulfoxide (DMSO). The sodium salt of ATP and thapsigargin were obtained from Sigma Chemical Co. (St. Louis, MO). Prior to the experiments, ATP was dissolved in physiological solution and the pH of the solution was corrected if necessary by addition of NaOH. The physiological solution contained 140 m*M* NaCl, 1 m*M* $Na_2HPO_4$, 6 m*M* KCl, 1.5 m*M* $CaCl_2$, 1 m*M* $MgCl_2$, 5 m*M* D-glucose, and 3 m*M* HEPES. The pH of the solution was adjusted to 7.4 with NaOH or HCl as required. Physiological solutions with low concentrations of $Ca^{2+}$ were prepared by adding 0.5 m*M* EGTA to a solution without $CaCl_2$, together with a predetermined volume of a 1 *M* $CaCl_2$ solution, utilizing a computer program to calculate the free calcium concentration [11].

### B. Tissue Culture

Adult white rabbits were killed by exposure for several minutes to carbon dioxide in a rising concentration. Tracheas were removed, the ciliary epithelium was pealed off the cartilage rings and cut into small pieces. Two to three pieces were placed on a glass cover slip that was pre-incubated for 24 hours in RPMI-1640 growth medium supplemented with 10% fetal calf serum, 20 units/mL penicillin, 2.5 units/mL nyastatin and 20 μg/mL streptomycin. The glass cover slips were then placed inside 35 mm plastic Petri dishes with a few drops of growth medium and incubated at 37°C under 5% $CO_2$. After 24 hours, growth medium was added to a total volume of 0.5 mL. It was replaced every other day. The cells were used during their second week in culture.

## III. RESULTS AND DISCUSSION

### A. Simultaneous Measurement of $[Ca^{2+}]^i$ and CBF

The simultaneous measuring apparatus was built around a Zeiss Axioskope fluorescence microscope, displayed schematically in Figure 1. Ex-

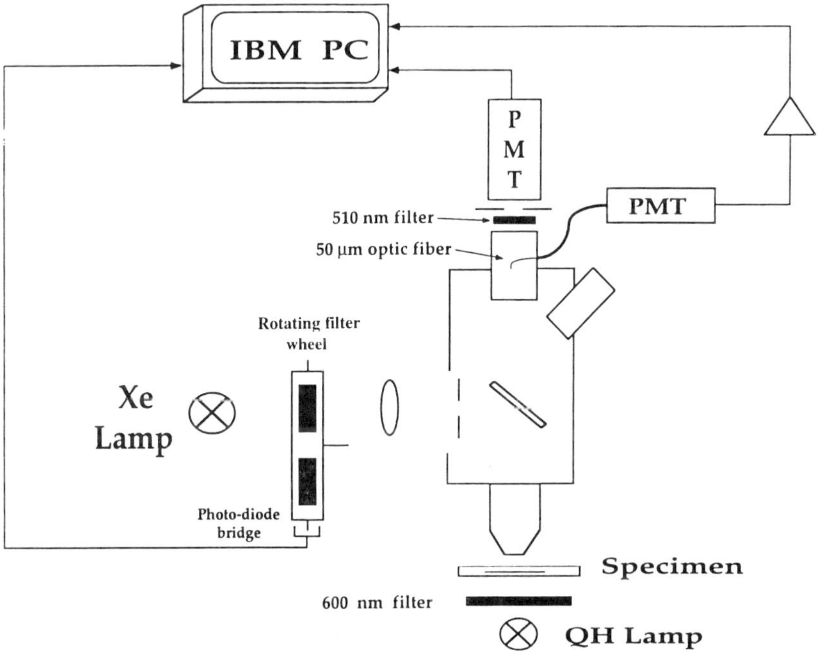


**Figure 1** Schematic drawing of the simultaneous measurement apparatus. The description of parts displayed in this figure is given in the text. (Reproduced from Korngreen and Priel, 1994, with permission.)

citation light for the fura-2 was received from a 150 W Xenon lamp (Oriel Corp., Stanford CT) filtered by interference filters (340 and 380 nm, full-width at half maximum = 10 nm, Oriel Corp., Stanford CT) mounted on a four-position filter wheel. The wheel's rotation speed was monitored by a photodiode bridge mounted on its rim. The light was attenuated by an iris diaphragm in order to reduce the illuminated field to the size of three to four cells, reflected by a dichroic mirror and focused on the culture by a 40× objective (Zeiss Acroplan, NA = 0.6). The emitted light was directed to the photo-eyepiece, filtered by a 510 nm interference filter (Oriel Corp., Stanford CT) and passed into a photon counting head (H3460-53, Hamamatsu, Japan) mounted directly on the photo-eyepiece. The photon counting head produced a standard logical pulse (TTL) which was counted by a computer counter board (PCL-780, Contech) in a 486 PC computer. The 340/380 fluorescent ratio was calculated for every turn of the filter wheel and averaged over a period of one second.

Ciliary beat frequency was measured by transilluminating the cells with light from a Quartz Halogen lamp that was filtered by a 600 nm interference filter (full-width at half maximum = 10 nm, Oriel Corp., Stanford CT). This wavelength of light was chosen because it did not participate in either the excitation or emission processes of fura-2. After passing through the sample and the objective this light was also directed to the photo-eyepiece that contained at the focal plane as illustrated in Figure 1, a 50 μm diameter optic fiber (EG&G Gamma scientific). This fiber, at the magnification of 40×, collected light from a surface of the equivalent diameter of 1.25 μm on the tissue culture [12]. The light was transmitted by an optic fiber bundle to a photomultiplier (PMT) and the resulting electrical signal was amplified and low-pass filtered to reduce noise. One second intervals were digitized at a sampling rate of 1024 Hz and the CBF was calculated using a Fast Fourier Transform (FFT). The details of the fura-2 loading procedure and the calibration of the intracellular calcium concentration are presented elsewhere [9,10].

## B. CBF Activation by Calcium Ions

By simultaneous measurements of CBF and $[Ca^{2+}]_i$ it was established that the resting values of CBF and $[Ca^{2+}[_i$ measured by fura-2 did not correlate [9]. This possibly indicates that the calcium ions are not involved in ciliary regulation at rest. On the other hand, a more attractive assumption is that the calcium concentration is not homogeneous throughout the cell. This hypothesis agrees with the high degree of correlation observed between the maximal rise in CBF and the maximal rise in $[Ca^{2+}]_i$ following stimulation (See Figure 2 in Ref. 9). The maximal rise in CBF was sigmoidally dependent on the maximal rise in $[Ca^{2+}]_i$ with an $EC_{50}$ of 420 ± 30 n$M$. Moreover, this correlation was independent of the method that was used to generate the rise in $[Ca^{2+}]_i$ (stimulation with extracellular ATP, store depletion with thapsigargin, or flooding the cell with calcium using the calcium ionophore ionomycin). Therefore, all the stages involved in the process of ciliary stimulation appear to require the presence of calcium. In other words, the findings suggest that a rise in CBF may serve as a measure of a rise in $[Ca^{2+}]_i$. Thus it is possible to use this curve as a qualitative calibration curve for the calcium concentration near the membrane at various stages of ciliary stimulation.

### C. CBF is Activated by Calcium Release from Internal Stores and by Calcium Influx

Based on the simultaneous measurement of $[Ca^{2+}]_i$ and CBF, we proposed that the biphasic effect of extracellular ATP is due to the release of calcium from two distinct calcium stores [9]: a small thapsigargin insensitive store situated in close proximity to the cilia and a larger thapsigargin sensitive store within the cell cytoplasm. The system is maintained at a high level of activation by an influx of calcium. This combined effect is best seen in Figure 2. Corresponding with the rise in $[Ca^{2+}]_i$ (Figure 2A), induced by the bath application of 100 μM ATP, there was a strong rise in CBF (Figure 2B) consisting of two stages as previously described by us [10]. While the calcium concentration decayed to an elevated plateau above its resting value (Figure 2A) the CBF decayed slightly to a high level of sustained activation (Figure 2B) that appeared at all concentrations of extracellular ATP. Removal of the ATP from the solution resulted in a decay in $[Ca^{2+}]_i$ to initial values and a slower decay in CBF to the basal beating frequency (Figures 2A and 2B). The biphasic rise in CBF seen in Figure 2B was even more evident when the rise phase of CBF was plotted as a function of corresponding $[Ca^{2+}]_i$ values (Figure 2C). During the initial jump in CBF, due to the release of calcium from the small thapsigargin insensitive calcium store, only a small increase in $[Ca^{2+}]_i$ was observed. Following this jump the rate of the CBF ascent slowed down due to the combined actions of the two calcium stores [9]. This resulted in the CBF reaching its maximal value after the $[Ca^{2+}]_i$ had already begun to decay. This was emphasized in Figure 2C by fitting a smooth curve through the data points. The time course of the primary and secondary rises in CBF and the time required for the $[Ca^{2+}]_i$ to reach its maximal value were independent of the concentration of extracellular ATP used.

No effect was observed on the initial biphasic rise pattern of CBF when the $[Ca^{2+}]_0$ was lowered to 0.1 μ$M$ (Figure 2F) indicating that it

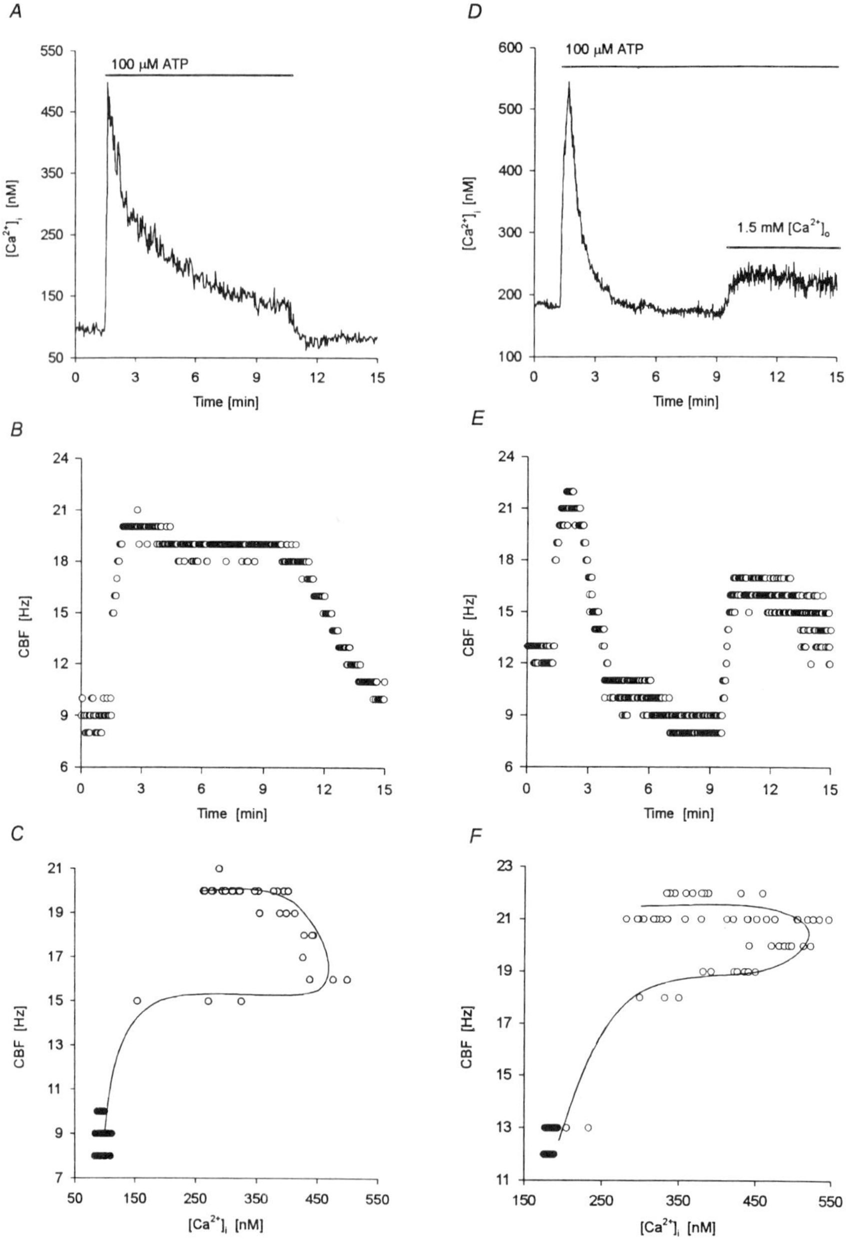


**Figure 2** The effect of ATP on $[Ca^{2+}]_i$ and CBF. The rises in $[Ca^{2+}]_i$ (A) and CBF (B) induced by 100 μ*M* extracellular ATP at 1.5 m*M* $[Ca^{2+}]_0$. Application and removal of the ATP from the solution are indicated by the solid bar. The

was due to intracellular events. The magnitude of the CBF rise was also not affected. The sustained ciliary activation was, however, completely eliminated (Figure 2E). Furthermore, in the displayed experiment the CBF decayed past its initial value (Figure 2E) without a corresponding decay in the $[Ca^{2+}]_i$ (Figure 2D). This observation agreed with the finding that the CBF was not correlated with the $[Ca^{2+}]_i$ in the resting state [9]. Increasing the calcium concentration in the solution to 1.5 m*M*, in the presence of ATP in the solution, caused a rise in $[Ca^{2+}]_i$ to a plateau and a marked rise in CBF (Figures 2D and 2E). Washing the ATP from the solution at this stage resulted in a decay in $[Ca^{2+}]_i$ and a return of CBF to basal values (data not shown). Increasing the $[Ca^{2+}]_0$ from 0.1 $\mu M$ to 1.5 m*M* prior to the addition of ATP to the solution did not cause any change in the CBF or in the $[Ca^{2+}]_i$. The sustained CBF activation displayed a sigmoidal dependence on the concentration of ATP in the solution with an $EC_{50}$ of 16 ± 5 $\mu M$ (Figure 3A). The elevated calcium plateau also displayed a sigmoidal dependence on the concentration of ATP in the solution with an $EC_{50}$ of 15 ± 3 $\mu M$ (Figure 3B). These $EC_{50}$ values are similar to the one found for the maximal rise in CBF [9]. According to the calibration curve we obtained [9], a CBF of 23 Hz corresponds to a $[Ca^{2+}]_i$ of about 500 n*M*. However, for a similar CBF during the plateau phase we obtained a value 240 n*M* of calcium (Figure 3B). Using the rationale that the CBF can serve as a measure for the calcium concentration in the vicinity of the cilia it is possible to estimate that the calcium concentration remains high juxtaposed to the cilia throughout the stimulation.

These experiments lead us to conclude that an influx of calcium is responsible for sustaining the cilia at a high level of activation for the duration of the stimulus. To the best of our knowledge, this is the first time that calcium influx has been noted to participate in prolonged ciliary activation. Also we were able to show that extracellular ATP triggers prolonged calcium influx in ciliary cells from frog palate [13,14] and human

---

biphasic rise is emphasized in the CBF versus $[Ca^{2+}]_i$ plot of the control period and the time directly following the addition of ATP (C) (control: solid points, ATP: empty points). The time between adjacent points, in this experiment, was two seconds. When the same experiment was carried out at 100 n*M* extracellular calcium concentration the rise in $[Ca^{2+}]_i$ (D) and CBF (E) retained its biphasic pattern (F) (control: solid points, ATP: empty points). Horizontal bars represent the times of addition of 100 $\mu M$ ATP and 1.5 mM $[Ca^{2+}]_0$. The time between adjacent points, in this experiment, was one second. The continuous lines in the CBF vs. $[Ca^{2+}]_i$ plots (C and F) were drawn by hand as a visual aid for the time flow in the experiments. (Reproduced from Korngreen and Priel, 1996, with permission.)

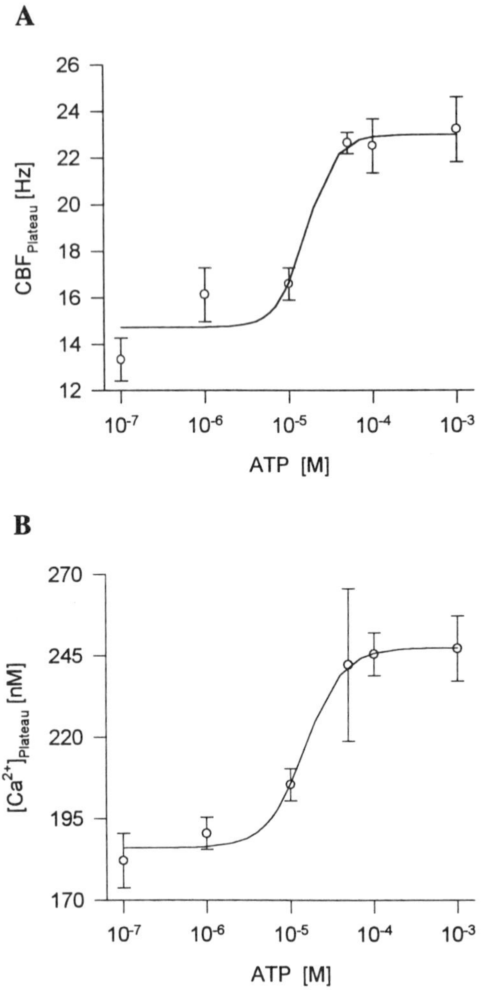


**Figure 3** The dependence of the elevated levels of CBF (A) and $[Ca^{2+}]_i$ (B) on the concentration of extracellular ATP. The values of $[Ca^{2+}]_{Plateau}$ and $CBF_{Plateau}$ were determined as previously described [9]. The curves were fitted by a Hill equation (solid line) to determine the value of the $EC_{50}$ reported in the text. Data is presented as means ± SE.

nasal polyps [15]. Based on the experimental findings we were able to propose a mathematical model describing the dynamics of calcium signal transduction in electrically inexcitable cells [16]. The model predicted a dependence, similar to that obtained experimentally [9], of the maximal rise in the cytoplasmic calcium concentration, the level of the elevated calcium plateau, and the rate of calcium decay on the level of cell excitation. Using the model it was possible to simulate the experimental procedures that were carried out by us in ciliary cells such as the washout of the agonist from the bathing solution and the depletion of the calcium store by thapsigargin. Moreover, the model fit well to the experimental results. From this curve fitting it was learned that extracellular ATP increased the calcium influx into the cell by only 20% above the resting value. This finding fit well with our observation that the calcium concentration remained elevated only in the vicinity of the cilia. A larger increase in the conductance of calcium might have caused a larger increase in $[Ca^{2+}]_i$, which could have impaired the normal function of the cell. This prediction of the model might also assist in future attempts to characterize the calcium influx pathway.

## IV. CONCLUSION

The implications of our findings in ciliary cells to the field of electrically inexcitable cells are evident. The ciliary cells of the rabbit trachea display both the release from internal stores and a prolonged influx of calcium. The immediate benefit to the ciliary system is the physiological response. This enables us to achieve what is still difficult to achieve in other types of electrically inexcitable cells: an investigation of the signaling cascade that occurs between the rise in $[Ca^{2+}]_i$ and the physiological response. Most investigations are of the pathways that exist between the activation of the receptor and the rise in $[Ca^{2+}]_i$, simply because of the hardships of measuring the physiological response of the given cell. Thus we would like to propose that the mucociliary system can serve as a good model system for investigation of signaling processes taking place in electrically inexcitable cells.

## REFERENCES

1. Putney Jr JW. Excitement about calcium signaling in inexcitable cells. Science 1993; 262:676–678.
2. Berridge MJ. Inositol phosphate and calcium signaling. Nature 1993; 361: 315–325.

3. Rasmussen H, Rasmussen JE. Calcium as intracellular messenger: from simplicity to complexity. Current Topics In Cellular Regulation 1990; 31:1–109.
4. Benham CD, Tsien RW. A novel receptor operated $Ca^{2+}$-permeable channel activated by ATP in smooth muscle. Nature 1987; 328:275–278.
5. Lückhoff A, Clapham DE. Inositol 1,3,4,5-tetrakisphosphate activates an endothelial $Ca^{2+}$-permeable channel. Nature 1992; 355:356–358.
6. Putney Jr JW. Capacitative calcium entry revisited. Cell Calcium 1990; 11: 611–624.
7. Randriamampita C, Tsien RY. Emptying of intracellular $Ca^{2+}$ stores releases a novel small messenger that stimulates $Ca^{2+}$ influx. Nature 1993; 364: 809–818.
8. Alkone DL, Rasmussen H. A spatial–temporal model of cell activation. Science 1988; 239:998–1005.
9. Korngreen A, Priel Z. Simultaneous measurement of ciliary beating and intracellular calcium. Biophysical Journal 1994; 67:377–380.
10. Korngreen A, Priel Z. Purinergic stimulation of rabbit ciliated airway epithelia: control by multiple calcium sources. Journal of Physiology (Lond) 1996; 497:53–66.
11. Fabiato A. Computer programs for calculating total from specific free or free from specific total ionic concentrations in aqueous solutions containing multiple materials and ligands. Methods in Enzymology 1989; 157:378–417.
12. Eshel D, Priel Z. Characterization of metachronal wave of beating cilia on frog's palate epithelium in tissue culture. Journal of Physiology (Lond) 1987; 388:1–8.
13. Tarasiuk A, Bar-Shimon M, Gheber L, Korngreen A, Grossman Y, Priel Z. Extracellular ATP induces hyperpolarization and motility stimulation of ciliary cells. Biophysical Journal 1995; 68(3):1163–1169.
14. Levin R, Braiman A, Priel Z. Protein kinase C induced calcium influx and sustained enhancement of ciliary beating by extracellular ATP. Cell Calcium 1997; 21:103–113.
15. Korngreen A, Priel Z. Signal transduction in ciliary systems: the effect of exogenous ATP on intracellular free calcium in human nasal epithelium. Biophysical Journal 1993; 64(2):A76.
16. Korngreen A, Gold'shtein V, Priel Z. Realistic modeling of biphasic calcium transients in electrically nonexcitable cells. Biophys J 1997; 73:659–673.

# 8

# Intercellular Communication in Mammalian Airway-Ciliated Epithelia

**Ellen Roter Dirksen**
*University of California, Los Angeles, Los Angeles, California*

## I. INTRODUCTION

The mechanisms by which cells in the tissues of multicellular organisms coordinate their function is of considerable physiological importance. In mammalian airway epithelia, cell communication is particularly important in mucociliary clearance mechanisms since mucus transport requires the cooperative activity of cilia across many cells. We have demonstrated, using primary monolayer cultures [1] from rabbit tracheal epithelium (Figure 1) and a transilluminating computer-assisted photoelectronic technique [2], that mechanical stimulation of a single cell with a glass microprobe, analogous to the weight that would be produced by the arrival of a mucus load at that point, increases the ciliary beat frequency (CBF) of that cell [3,4]. By using computer-assisted, frame-by-frame analysis of high-speed film, which allowed us to analyze ciliary activity across the cells (in a microscope field) of a monolayer culture, we further demonstrated that the CBF increase is not limited to the stimulated cell, but that after a short delay it is communicated to surrounding cells with an increased delay proportional to the distance of the neighboring cells to the stimulated cell [5]. This localized CBF increase effectively increases the rate at which mucus would be moved towards the oropharynx for elimination.

Further evidence for cell communication was derived from the presence of gap junctions in the airway epithelium, both in culture and in the intact tissue, as demonstrated by ultrastructural freeze-fracture tech-

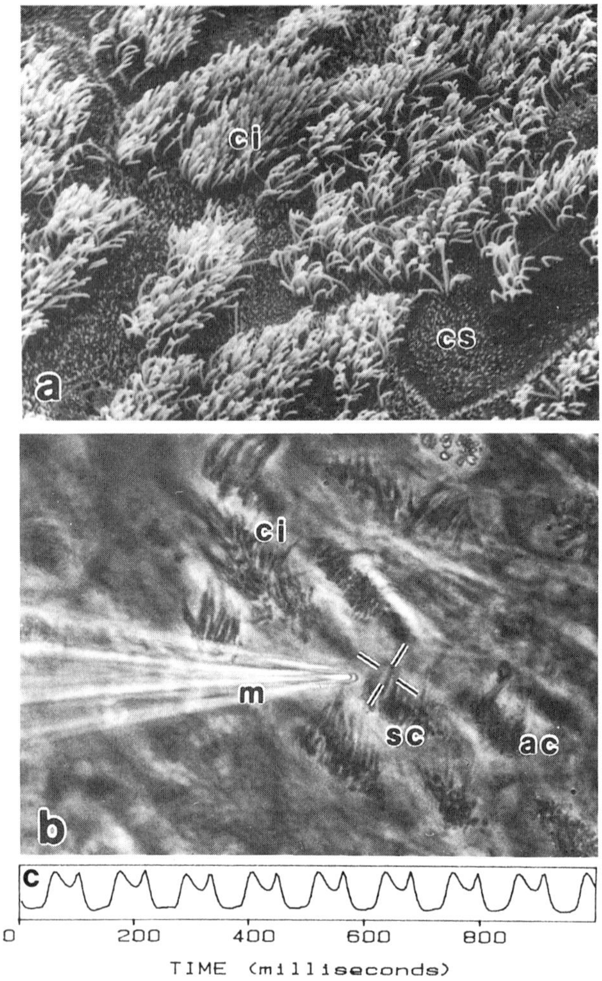


**Figure 1** Tracheal epithelial cells in monolayer culture. (a) Scanning electron micrograph (× 1,250) showing cilia (ci) of a cell and the cell surface (cs) of a nonciliated cell. (b) Phase-contrast micrograph (× 750) of rabbit tracheal epithelial cells in primary monolayer cultures. The cell to be stimulated (sc) or the adjacent cell (ac) is aligned with the photodetector with the aid of the cross hairs. A glass microprobe (m), driven by the displacement of a piezoelectric crystal, stimulates the surface of a single cell or a group of cilia. The recorded photoelectronic signal (c) displays the triphasic nature of the ciliary beat cycle; left to right, the trough, first peak, and second peak represent the rest phase, recovery phase, and effective phase, respectively [2,3].

niques and from electrical coupling associated with both the unciliated and ciliated cells [5].

Computer digital imaging microscopy with the $Ca^{2+}$-sensitive dye fura-2 revealed that the mechanically induced increase in CBF and its propagation in tracheal epithelial cells in culture involved an intracellular $Ca^{2+}$ increase, which first occurred in the stimulated cell and was briefly delayed at the cell border before being propagated to the surrounding adjacent cells as a "$Ca^{2+}$ wave" (Figure 2). These results clearly showed that the propagation of the intracellular $Ca^{2+}$ concentration ($[Ca^{2+}]_i$) increase was communicated intercellularly to the neighboring cells and, parenthetically, that it was the cause of the rise in CBF of the airway ciliated cells [6]. However, an interesting phenomenon presented itself: in $Ca^{2+}$-free tissue culture medium, the cell that was mechanically stimulated did *not* increase its $[Ca^{2+}]_i$, while the surrounding neighboring cells did, although with slightly modified temporospatial characteristics compared to

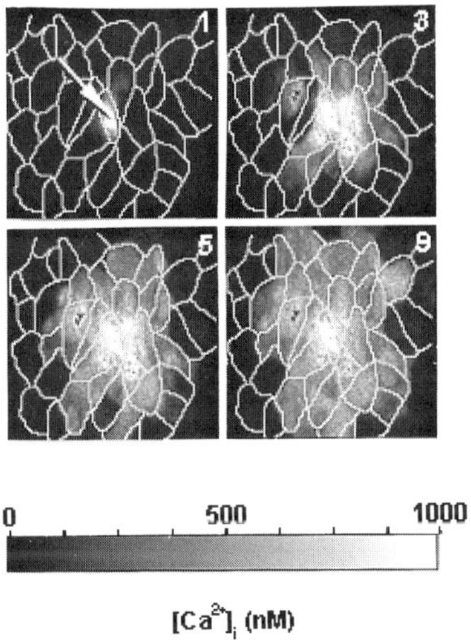


**Figure 2** Fluorescence images demonstrating changes in intracellular $Ca^{2+}$ ($[Ca^{2+}]_i$) of tracheal cells in monolayer cultures, consisting of both ciliated and nonciliated cells. Cell boundaries are highlighted in white. Following mechanical stimulation (arrow), pixel intensity brightens at that point, the brightness first filling the stimulated cell before spreading to adjacent cells. The time elapsed (sec) after the stimulus is indicated in the upper right-hand corner. The brightening of the pixel intensity represents an increase in $[Ca^{2+}]_i$ (n*M*), as shown in the "grayscale" bar.

cells in $Ca^{2+}$-containing medium [6]. There appeared to be two components to the communicated $Ca^{2+}$ wave: (a) the requirement by the stimulated cell for extracellular $Ca^{2+}$ and (b) the dependence of the neighboring cells for a source of $Ca^{2+}$ from intracellular stores, analogous to the $Ca^{2+}$ stores of the sarcoplasmic reticulum of muscle cells.

## II. THE STIMULATED CELL REQUIRES EXTRACELLULAR $Ca^{2+}$

When a cell is mechanically stimulated in $Ca^{2+}$-containing medium, there is a $Ca^{2+}$ influx into the cell due to the positive electrochemical $Ca^{2+}$ gradient; however, under $Ca^{2+}$-free conditions there is a $[Ca^{2+}]_i$ decrease because of cellular $Ca^{2+}$ efflux from the cell due to a negative $Ca^{2+}$ electrochemical gradient. In order to ascertain that any available intracellular $Ca^{2+}$ pools in the cell could not contribute to the $Ca^{2+}$ increase of the stimulated cell, the intracellular $Ca^{2+}$ stores were first depleted by thapsigargin, a sarcoplasmic/endoplasmic reticulum $Ca^{2+}$-channel blocker that prevents the refilling of $[Ca^{2+}]_i$ stores. Mechanical stimulation in $Ca^{2+}$-medium containing thapsigargin immediately increased $[Ca^{2+}]_i$ in the stimulated cell, which quickly returned to resting levels without, however, initiating an intercellular $Ca^{2+}$ wave. When thapsigargin-treated cells were mechanically stimulated under $Ca^{2+}$-free conditions no $[Ca^{2+}]_i$ increase was evident, indicating that extracellular $Ca^{2+}$ influx accounted for the $Ca^{2+}$ increase of the stimulated cell [7].

The stimulated cell's extracellular $Ca^{2+}$ requirement provided evidence for the possible presence of mechanically sensitive, $Ca^{2+}$-conducting ion channels in the cultured cells. In the absence of extracellular $Ca^{2+}$, together with the mechanosensitive $Ca^{2+}$-channel blocker gadolinium, the mechanically stimulated cell increased its $[Ca^{2+}]_i$ [7]. Under the same conditions, and with lanthanum and nickel or with the dihydropyridine (DHP) $Ca^{2+}$-channel antagonists nifedipine and nimodipine, the stimulated cell also exhibited a $[Ca^{2+}]_i$ increase. Under $Ca^{2+}$-free conditions and in the absence of $Ca^{2+}$-channel blockers, not only does the mechanically stimulated cell not increase its $Ca^{2+}$, but it loses whatever basal $Ca^{2+}$ is present to the extracellular space. Therefore, the prevention of a $Ca^{2+}$ loss from the stimulated cell by the $Ca^{2+}$-channel blockers used in these experiments is evidence that cultured airway epithelial cells possess $Ca^{2+}$-conducting channels, allowing for $Ca^{2+}$ fluxes across the cell membrane. In addition, the direction of this flux is determined in part by the extracellular $Ca^{2+}$ concentration [7].

Evidence has been found [8] that mechanical stimulation may activate $Ca^{2+}$ fluxes across the cell membrane by causing a membrane potential change, which would in turn be capable of activating voltage-sensitive $Ca^{2+}$-conducting channels. In the presence of extracellular $Ca^{2+}$ (without mechanical stimulation) addition of BAY-K8644, a DHP agonist, increased $[Ca^{2+}]_i$ in many cells; in the absence of the agonist there was no increase. The DHP-sensitivity of this response and the earlier studies on the effect of DHP antagonists [7] suggested that the mechanically induced $Ca^{2+}$ flux was probably due to the presence, on the cell membrane, of DHP-binding sites associated with $Ca^{2+}$-conducting ion channels [8].

## III. THE SOURCE OF $Ca^{2+}$ FOR THE NEIGHBORING CELLS DERIVES FROM INTRACELLULAR STORES AND IS $IP_3$-ACTIVATED

The experimental evidence outlined above established that the source of $Ca^{2+}$ for the mechanically stimulated cell was derived from the extracellular environment; another second messenger was required to account for the $[Ca^{2+}]_i$ increase propagated as a wave from cell to cell through the monolayer cultures. In the absence of extracellular $Ca^{2+}$, the $[Ca^{2+}]_i$ increase of the neighboring cells had to originate from intracellular stores, mediated by a second messenger, inositol 1,4,5-trisphosphate ($IP_3$). Evidence for this derives from experiments where $IP_3$, injected iontophoretically into a cell, produces a $Ca^{2+}$ increase, which is then communicated to neighboring cells [6]. This response is similar to a $Ca^{2+}$ wave produced by mechanical stimulation of a single cell.

Pulsed high-frequency electroporation was used to selectively load one section of a monolayer cell culture with the large glycosaminoglycan, heparin (known to block the $IP_3$ membrane receptor), while leaving adjacent sections heparin-free. Under these conditions it was shown that mechanical stimulation of a cell in a heparin-loaded area prevented the $[Ca^{2+}]_i$ increase in that cell and the subsequent $[Ca^{2+}]_i$ increase in adjoining cells, effectively blocking the propagation of the $Ca^{2+}$ wave. A mechanically stimulated cell in the heparin-free area of the same coverslip still increased its $[Ca^{2+}]_i$, and communication of the $Ca^{2+}$ wave was not inhibited. The heparin does not prevent the $IP_3$ from diffusing from cell to cell, but by blocking the $IP_3$ receptor, located in the endoplasmic reticulum in which $Ca^{2+}$ is stored, heparin prevents the release of $Ca^{2+}$ from these intracellular stores [9].

Phospholipase C, the membrane-bound enzyme that generates $IP_3$, can be inhibited by the aminosteroid U73122. In the presence of this inhibitor, mechanical stimulation of a single cell in a monolayer culture did not induce a $[Ca^{2+}]_i$ increase in that cell or initiate a $Ca^{2+}$ wave propagation to neighboring cells [10]. Presumably this occurs because the inactivation of the $IP_3$-generating enzyme prevents the formation of $IP_3$, which is no longer available to release $Ca^{2+}$ from intracellular stores or to diffuse from cell to cell.

Although evidence had accumulated for the involvement of $IP_3$ in the mechanically induced $Ca^{2+}$ wave, it was necessary to demonstrate that mechanical stimulation could produce an actual increase in $IP_3$. By growing confluent monolayers of tracheal epithelial cells on flexible substrates and utilizing an $IP_3$ competition assay, we were able to show that stretching of the substrates with their attached cells, equivalent to mechanical stimulation with a microprobe of a single cell, increased the cellular $IP_3$ concentration threefold [11]. The cells maintained this concentration for approximately 18 sec, although it is likely that the $IP_3$ concentration is considerably higher in the cells than could be demonstrated by this assay.

## IV. THE ROUTE OF $IP_3$ IN THE PROPAGATION OF THE $Ca^{2+}$ WAVE

The experimental evidence provided above strongly suggested that it is the intercellular movement of $IP_3$, rather than $Ca^{2+}$, that is responsible for the propagation of the $Ca^{2+}$ wave. The temporospatial characteristics of the $Ca^{2+}$ wave, which often followed a nonlinear pathway throughout the culture, further suggested that gap junctions were the most likely route by which $IP_3$ was being communicated from cell to cell and several lines of evidence confirmed this. When a single cell of the epithelial monolayer was mechanically stimulated in the presence of the anesthetic halothane, known to block gap junctions, a weak $Ca^{2+}$ response was obtained across the cultured cells, implying that the drug blocked $IP_3$ diffusion through gap junctions [6].

Glial cells in primary culture also respond to mechanical stimulation by propagating intercellular $Ca^{2+}$ waves [12,13]. A glioma cell line (C6) is available that has few gap junctions, providing a tool with which to further demonstrate the role of gap junctions in intercellular communication of the $Ca^{2+}$ wave. The C6 glioma cell line expresses low levels of connexin43, a gap-junction protein; when these cells were transfected with the cDNA encoding connexin43, a series of glioma cell lines were obtained with differential expression of this protein. When a single cell in a monolayer culture was mechanically stimulated, it propagated a $Ca^{2+}$ wave, obtained with $Ca^{2+}$-imaging methods, whose degree of communication correlated with the degree of expression of the connexin43 in that particular transfected cell line. Data obtained from dye coupling and Northern blot and immunoblot analysis [14] corroborated the relationship between the degree of connexin43 transfection in a particular glioma cell line and the extent of cell–cell communication.

A third line of evidence that more directly addresses the issue of the role of gap junctions in intercellular communication of the $Ca^{2+}$ wave derives from recent studies on antibodies raised against specific groups of linear amino acids of the connexin32 protein. Pulsed high-frequency electroporation was used to introduce antibodies, generated to peptide sequences within the extracellular loop, the cytosolic loop, the carboxyl tail, or the amino terminus, into the airway cell monolayer cultures. Electroporated antibodies specific to a peptide sequence of the cytosolic loop or to the carboxyl tail of connexin32 inhibited $Ca^{2+}$-wave propagation, whereas antibodies to extracellular loop peptides or amino terminus sequences did not inhibit the intercellular $Ca^{2+}$ wave [15]. As an internal control, a peptide antibody that inhibited intercellular $Ca^{2+}$ propagation was co-loaded with a noninhibitory peptide, eliminating the original inhibition. Thus, blocking of gap junctions by antibodies prepared to specific connexin32 peptides provides further evidence that mechanically induced and intercellularly propagated $Ca^{2+}$ waves involve gap junctions (for review see Ref. 16).

## V. A MODEL FOR $Ca^{2+}$ SIGNALING

Because the number of cells in the airway epithelial monolayer cultures that have been experimentally demonstrated to participate in the propagation of the $Ca^{2+}$ wave is limited to about 5 or 6 cells in each direction from the mechanically stimulated cell, the question had arisen as to whether the amount of $IP_3$ generated by the mechanical stimulus of a single cell is sufficient to diffuse through the gap junctions of all the cells in order to activate their intracellular $Ca^{2+}$ stores. Mathematical modeling [17] attempts to answer this question by postulating that only a very small amount of $IP_3$ is necessary to release large amounts of $Ca^{2+}$ from each cell because of the amplification process: $Ca^{2+}$ release is amplified through a $Ca^{2+}$-induced $Ca^{2+}$-release positive feedback mechanism in which $Ca^{2+}$, once released by $IP_3$ receptor activation, stimulates its own further release. This process continues until the finite amount of $IP_3$, produced in the stimulated cell and diffusing via gap junctions from the mechanically stimulated to the neighboring cells, is no longer sufficient to activate $IP_3$ receptors to release $Ca^{2+}$ from internal stores (Figure 3).

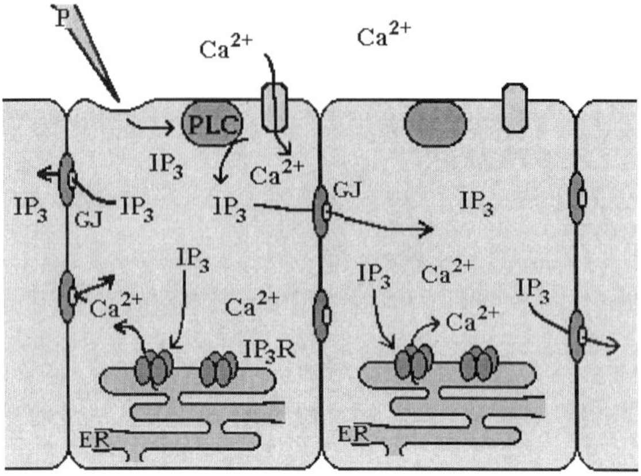


**Figure 3** Perturbation of the cell membrane by a glass pipette (p) initiates two important events: (a) activation of the enzyme phospholipase C (PLC) and (b) activation of mechanosensitive $Ca^{2+}$-conducting channels. PLC generates $IP_3$, a small portion of which diffuses through the cytoplasm of the stimulated cell to act upon the $IP_3$ receptors ($IP_3R$) on the endoplasmic reticulum to release intracellular $Ca^{2+}$ stores; the major portion of the $IP_3$ diffuses through gap junctions (GJ) to neighboring cells to release their intracellular $Ca^{2+}$ stores. $Ca^{2+}$ in the cell is further increased by a positive feedback amplification mechanism in which small amounts of the $Ca^{2+}$ released from stores acts upon the $IP_3$ receptor to release more $Ca^{2+}$. At the same time, the activation of the $Ca^{2+}$-conducting channels of the stimulated cell also allows extracellular $Ca^{2+}$ to enter. In the meantime, the larger amount of $IP_3$ released by the activation of PLC diffuses from cell to cell, generating a $Ca^{2+}$ wave.

It is of further interest to note that $Ca^{2+}$ signaling as a response to mechanical stimulation has, by now, been shown to occur in a vast array of tissues, both in culture and in the in vivo state [16]. Mechanosensitivity as a general phenomenon of cells can be viewed as an experimental tool for analyzing the cellular and molecular mechanisms by which cells integrate their activities, necessary to the proper functioning of multicellular organisms [18].

## VI. SIGNALING BY $Ca^{2+}$ IS NOT INITIATED BY AN EXTRACELLULAR MESSENGER

Ciliated cells of the mammalian respiratory tract increase their ciliary beat frequency and $[Ca^{2+}]_i$ by extracellular adenosine 5′-triphosphate

(ATP), which is often released by cells and diffuses through the medium to act, presumably, through $P_2$-purinergic plasma membrane receptors. To ascertain that the signal that propagates the $Ca^{2+}$ wave in airway cells in culture did not have an extracellular component, we compared intercellular $Ca^{2+}$-wave communication after a local application of ATP, with and without unidirectional biased fluid flow. Upon local application of ATP with a pipette, the cells nearest the pipette were the first to increase their $[Ca^{2+}]_i$, followed by a $[Ca^{2+}]_i$ increase in more distantly located cells, with a time delay that was proportional to the distance from the pipette. The $Ca^{2+}$ wave produced by the addition of ATP was not unlike the propagated wave produced by the stimulation of a single cell. However, when ATP was applied together with a unidirectional fluid flow across the culture, the $Ca^{2+}$ response was biased in the direction of those cells downstream of the flow, which were the cells that primarily increased their $[Ca^{2+}]_i$. Mechanical stimulation of a single cell together with biased fluid flow resulted in a normally propagated $Ca^{2+}$ wave and this is evidence that an extracellular messenger was not responsible for this phenomenon. Although the ATP-stimulated $Ca^{2+}$ wave propagation was attenuated in the presence of the $P_2$-purinergic receptor antagonist, suramin, its presence did not inhibit the $Ca^{2+}$ wave normally produced by mechanical stimulation. Since suramin had no effect on the normal propagation of the mechanically produced $Ca^{2+}$ wave, we must conclude that extracellular ATP (or other compounds that might be released by cells due to mechanical stimulation) have no effect on intercellularly propagated $Ca^{2+}$ waves [19].

## VII. $Ca^{2+}$ SIGNALING IN INTACT TRACHEAL EPITHELIUM

Intercellular $Ca^{2+}$ communication as a response to mechanical stimulation of a single cell has been demonstrated utilizing monolayer cultures, where the cells tend to be flattened and about 2 μm thick (Figure 1). This morphology is considerably different from that found in vivo, where the respiratory epithelium of the upper airways is characterized as pseudostratified, ciliated columnar with goblet cells. With laser scanning confocal microscopy and the $Ca^{2+}$-indicator fluo-3 we investigated the effect of mechanical stimulation on freshly microdissected tissue (or explant), consisting of the tall (ca 15 μm) epithelium and some of the underlying connective tissue. When a single cell of the explant was stimulated by a glass microprobe there was a $[Ca^{2+}]_i$ increase in this cell, which then spread radially as a wave of increased $Ca^{2+}$ through several cells (Figure 4) [20]. Addition of ATP to the bath also had effects similar to those obtained for

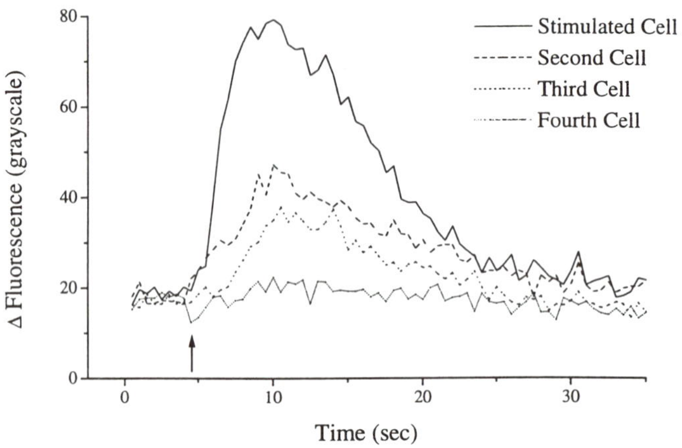


**Figure 4** Mechanical stimulation of a single cell in the intact tissue immediately increases its relative $Ca^{2+}$ fluorescence, which is then propagated to surrounding cells. The fluorescence increase, after first appearing in the stimulated cell (arrow), increases in a second, third, and fourth cell, with a time delay relative to the distance from the stimulated cell.

the monolayer cultures [19] in that both $[Ca^{2+}]_i$ and ciliary beat frequency were increased over the whole explant [21]. While these are preliminary results, there is considerable significance in the evidence that $Ca^{2+}$ signaling in airway epithelium is also a phenomenon that occurs in vivo.

## VIII. CONCLUSIONS

The pseudostratified ciliated columnar epithelium that lines the conducting passages of the upper respiratory tract is the first line of defense that rids the airways of airborne contaminants before they reach the lungs. The demonstration that the epithelium, both in culture and in vivo, displays complex patterns of $Ca^{2+}$ signaling and intercellular communication suggests that airway contaminants that may interfere with $Ca^{2+}$ signaling, and hence cell communication, could be responsible for a diversity of cell dysfunctions, including tumor promotion. This latter event can be related to the loss of cell communication and cooperative behavior of the tissue, resulting in the loss of contact inhibition, a prerequisite to tumor formation. It is of utmost importance, therefore, that we identify the components of intercellular $Ca^{2+}$ signaling that are involved in cell–cell communication.

## ACKNOWLEDGMENTS

I wish to thank my many collaborators whose dedication made these investigations possible. I owe much especially to Dr. Sanderson and to Drs. Boitano, Woodruff, Charles, Hansen, and Chaban. Particular mention must be made of Ms. Felix, whose recent efforts in developing the explant and simulated microgravity studies have added considerably to the work of the laboratory. At different times, this work has been supported by the Cystic Fibrosis Foundation, the Smokeless Tobacco Research Council, the National Institutes of Health and the University of California Tobacco Health-Related Disease Research Program. More recently, support has been available from NASA's Microgravity Research Division.

## REFERENCES

1. Dirksen ER, Felix JA, Sanderson MJ. The preparation of explant cultures, organ cultures and single cells from airway epithelium. In: Dentler W, Witman G, eds. Cilia and Eukaryotic Flagella, Methods in Cell Biology. Academic Press, San Diego, CA, 1995:65–74. Vol. 47.
2. Sanderson MJ, Dirksen ER. A versatile and quantitative computer-assisted photoelectronic technique used for analysis of ciliary beat cycles. Cell Motility 1985; 5:293–310.
3. Sanderson MJ, Dirksen ER. Mechanosensitivity of cultured ciliated cells from the mammalian respiratory tract. Implications for the regulation of mucociliary transport. Proc Natl Acad Sci USA 1986; 83:7302–7306.
4. Dirksen ER, Sanderson MJ. Regulation of ciliary activity in the mammalian respiratory tract. Biorheology 1990; 27:533–545.
5. Sanderson MJ, Chow I, Dirksen ER. Intercellular communication between ciliated cells in culture. Am J Physiol 1988; 254 (Cell Physiol 23) C63–C74.
6. Sanderson MJ, Charles AC, Dirksen ER. Mechanical stimulation and intercellular communication increases intracellular $Ca^{2+}$ in epithelial cells. Cell Regulation 1990; 1:585–596.
7. Boitano S, Sanderson MJ, Dirksen ER. A role for $Ca^{2+}$ conducting ion channels in mechanically induced signal transduction of airway epithelial cells. J Cell Sci 1994; 107:3037–3044.
8. Boitano S, Woodruff ML, Dirksen ER. Evidence for voltage-sensitive calcium-conducting channels in airway epithelial cells. Am J Physiol 1995; 269 (Cell Physiol 38):C1547–C1556.
9. Boitano S, Dirksen ER, Sanderson MJ. Intercellular propagation of calcium waves mediated by inositol trisphosphate. Science 1992; 258:292–295.

10. Hansen M, Boitano S, Dirksen ER, Sanderson MJ. A role for phospholipase C activity but not ryanodine receptors in the initiation and propagation of intercellular calcium waves. J Cell Sci 1995; 108:2583–2490.
11. Felix JA, Woodruff ML, Dirksen ER. Stretch increases inositol 1,4,5-trisphosphate concentration in airway epithelial cells. Am J Respir Cell Mol Biol. 1996; 14:296–301.
12. Charles AC, Merrill JE, Dirksen ER, Sanderson MJ. Intercellular signaling in glial cells: communicated calcium waves and asynchronous calcium oscillations in response to mechanical stimulation and glutamate. Neuron 1991; 6:983–992.
13. Charles AC, Dirksen ER, Merrill JE, Sanderson MJ. Mechanisms of intercellular calcium signaling in glial cells studied with dantrolene and thapsigargin. Glia 1993; 7:134–145.
14. Charles AC, Naus CCG, Zhu D, Kidder GM, Dirksen ER, Sanderson MJ. Intercellular calcium signaling via gap junctions in glioma cells. J Cell Biol 1992; 118:195–201.
15. Boitano S, Evans WW, Dirksen ER. Use of sequence specific antibodies to study gap junction-mediated intercellular $Ca^{2+}$ signalling in airway epithelial cells. Mol Biol Cell 1996; 7(Suppl):92a.
16. Sanderson MJ. Intercellular waves of communication. News Physiol Sci 1996; 11:262–269.
17. Sneyd JM, Wetton BTR, Charles AW, Sanderson MJ. Intercellular calcium waves mediated by diffusion of inositol trisphosphate: a two-dimensional model. Am J Physiol 1995; 268 (Cell Physiol 37):C1537–C1545.
18. Sanderson MJ, Charles AC, Boitano S, Dirksen ER. Mechanisms and function of intercellular calcium signaling. Molecular and Cellular Endocrinology 1994; 98:173–187.
19. Hansen M, Boitano S, Dirksen ER, Sanderson MJ. Intercellular calcium signaling induced by extracellular adenosine 5-triphosphate and mechanical stimulation in airway epithelial cells. J Cell Sci 1993; 106:995–1004.
20. Felix JA, Chaban V, Dirksen ER. Mechanical stimulation initiates intercellular $Ca^{2+}$ signaling in intact tracheal epithelium maintained under normal gravity and simulated microgravity. FASEB J 1997; 11:A309.
21. Felix JA, Chaban V, Dirksen ER. Unpublished results.

# 9

# $\beta_2$ Agonist Terbutaline Makes the Ciliary Beat Frequency of Freshly Isolated Human Airway Cells Refractory to Calcium Signaling

**Gordon Honeyman, P. J. Kemp, C. Cobain, and Anil Mehta**
*Ninewells Hospital and Medical School, and University of Dundee, Dundee, United Kingdom*

**D. P. Dhillon and John H. Winter**
*Kings Cross Hospital, Dundee, United Kingdom*

## I. INTRODUCTION

Despite a vigorous debate about their safety $\beta_2$ agonists are widely prescribed medications in lung disease and remain first-line therapy for the relief of intermittent bronchoconstriction, e.g., in mild asthma. Recent studies have cast further doubt on their efficacy as sole therapy because regular use of inhaled $\beta_2$-agonists is associated with a deterioration of asthma control [3], an increase in airway responsiveness [4,5], and promotion of drug dependence [6]. Here we add a further cautionary note about their indiscriminate use.

There is a widespread belief based largely on in vitro data that $\beta_2$ agonists augment mucus clearance because, at concentrations in the $\mu$M range, they increase ciliary beat frequency (CBF) in human airway epithelial cells [7]. These findings have been extrapolated to suggest that these drugs will improve mucus clearance in vivo. There is scant supportive evidence for this notion other than in studies of the upper airway. In volunteers subjected to a nasal turbinate biopsy, there is a correlation between the in vivo rate of transport of nasal particulate matter [8] and the co-temperaneous (agonist-free) in vitro CBF (here defined as the intrinsic CBF, i.e., CBF in the absence of $\beta_2$ agonists maintained by intracellular calcium release). Respiratory epithelial cells exposed to mucus or particulate matter also increase intrinsic CBF in response to the mechanical load via this calcium-dependent pathway [9] with the calcium acting via calmodulin–dependent phosphorylation [10].

In this study, we challenge the notion that $\beta_2$ agonists increase CBF because, when applied at the p*M* and n*M* concentrations that are likely to be found in the peripheral lung following inhalation therapy, we show that a widely used $\beta_2$ agonist (terbutaline) inhibits the calcium-dependent mechanism that controls intrinsic CBF.

## II. METHODS

### A. Collection of Human Nasal Respiratory Epithelial Cells

Cells were obtained from patients undergoing routine operations (Ethical Committee approved) by brushing the inferior turbinate. The cells were dislodged into culture medium (medium 199), transferred on ice to the laboratory and aliquots placed in a perfusion chamber containing tissue culture medium.

### B. CBF Measurement

A video-based contrast enhancement system (Brian Reece Scientific, Newbury Berks) was used to measure CBF as previously described [11]. Briefly, the chamber was placed on a platform of an inverted microscope connected to a video camera and VDU. The contrast signal was digitized and the CBF displayed on a computer screen as a wave form and as a numerical value. The system samples contrast changes at 50 Hz, which is well above the range at which cilia beat in this model (~4.5–10 Hz at 23°C). The protocol for taking CBF readings involved three stages: (a) A 30-minute period in which the cells were allowed to settle in the chamber (no CBF readings were taken during this period), (b) Perfusion with control solution and CBF readings taken every 5 minutes for 30 minutes, and (c) Blind 5-min CBF readings in the presence of agonists/inhibitors.

### C. Measurement of Cell Calcium

Cell calcium was measured with Fura-2 in separate experiments as previously described [12].

### D. Statistical Analysis

The CBF values were expressed as the mean ± SEM percentage change from baseline CBF. Differences between CBF values within groups at the times stated were determined by a paired Student's *t* test, statistical significance was accepted when $p < 0.05$.

## III. RESULTS

### A. Terbutaline Inhibits $Ca^{++}$ Stimulation of CBF

Figure 1 shows the effect of terbutaline on the CBF response to 1 μ*M* 4Br-A23187 in the presence of increasing concentrations of terbutaline. Cells from 10 different individuals were first exposed to either no terbutaline or $10^{-12}$–$10^{-6}$ M terbutaline for 30 minutes (separate experiments were conducted for each of the terbutaline concentrations tested). A subsequent perfusion with 1 μ*M* 4Br-A23187 was applied in the presence of the terbutaline. The percentage change in CBF represents the CBF response at 15 minutes post A23187 (expressed relative to the CBF value at 30 min post terbutaline alone = 100%).

In the absence of terbutaline, 1 μM 4Br-A23187 induced a 14.7% ± 6.69% increase in CBF similar to that found in our earlier studies [11,12]. However, in the presence of 1 μ*M*, 100 n*M*, 10 n*M*, 100 p*M*, and 10 pM terbutaline, the subsequent addition of 1 μM 4Br-A23187 resulted in a significant ($p < 0.05$) decrease in CBF of 9.86% ± 7.12%, 3.69% ± 5.75%, 6.08 ± 6.09%, 3.41% ± 6.59%, 3.05% ± 8.54%, and 0.42% ± 5.92% respectively compared to the CBF response to A23187 alone. In separate experiments, terbutaline did not induce any rise in cell calcium above baseline. Thus we could not explain the loss of the response to the calcium ionophore on the basis of terbutaline acting via a release of cell calcium.

### B. Inhibition of Protein Kinase A Reverses Terbutaline Inhibition

When cells were preincubated with H7 (an inhibitor previously shown by us to act via inhibition of PKA), then exposed to 100 n*M* terbutaline + 1 μ*M* 4Br-A23187, the original response to 4Br-A23187 was restored (Figure 1).

By reversing the order of perfusion, we also tested the notion that the cells failed to increase their CBF in response to calcium stimulus because they were already beating at a maximal CBF. When perfused in the reverse order (cells first exposed to 1 μ*M* 4Br-A23187 and then 100

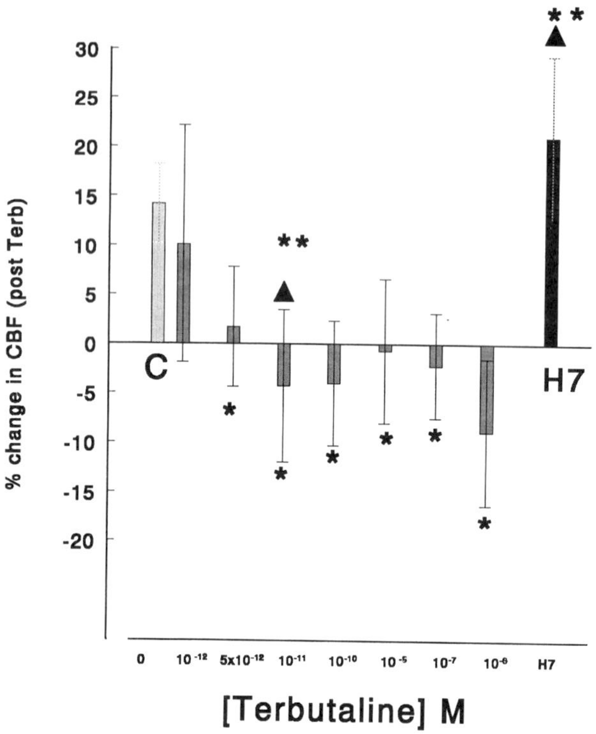


**Figure 1** Percentage CBF in response to 1 μ*M* 4BrA23187 ± terbutaline in first infusion. In the no terbutaline control (C), the expected CBF response of circa 14% is present. In contrast, at concentrations of terbutaline above 1 p*M*, a significant (*) inhibition of the A23187 response occurs that is reversed (closed triangles above standard error bars, $n = 10$) when H7 is present prior to the addition of 10 n*M* terbutaline (** refers to a significant difference compared to the 10 n*M* control—closed triangle).

n*M* terbutaline in the presence of 4Br-A23187), the addition of 100 n*M* terbutaline caused a further significant rise in CBF from 107.1 ± 2.3% (5.75 ± 0.39 Hz) to 111.14 ± 2.98% (5.96 ± 0.41 Hz) within 15 minutes ($p < 0.05$). We therefore discarded this notion that CBF was already maximal and tested the idea that terbutaline might also inhibit an alternative calcium-dependent ciliary stimulus induced by the mechanical effect of flow.

## C. Flow Rate Causes a Transient Decrease in CBF in Cells Previously Exposed to Terbutaline

When cells were perfused with medium 199, the CBF increased with flow rate as previously shown [13]. However in the above experiments, when the cells were first exposed to terbutaline and then perfused with either 4Br-A23187 + terbutaline or terbutaline alone, CBF declined for up to 5 min (Figure 2). In the presence of 1 μ*M*, 100 n*M*, 10 n*M*, 100 p*M*, 10 p*M*, and 50 p*M* terbutaline, CBF decreased by 13.01 ± 6.99%, 4.42 ± 5.75%, 17.73 ± 12.6, 1.7 ± 6.25%, 2.44 ± 7.4%, and 6.63 ± 5.64% respectively (significant at 1 μ*M* and 100 n*M* but not at 50 p*M*). This decline was not an experimental artifact because no decrease was seen with the second perfusion when no $\beta_2$ agonist was present.

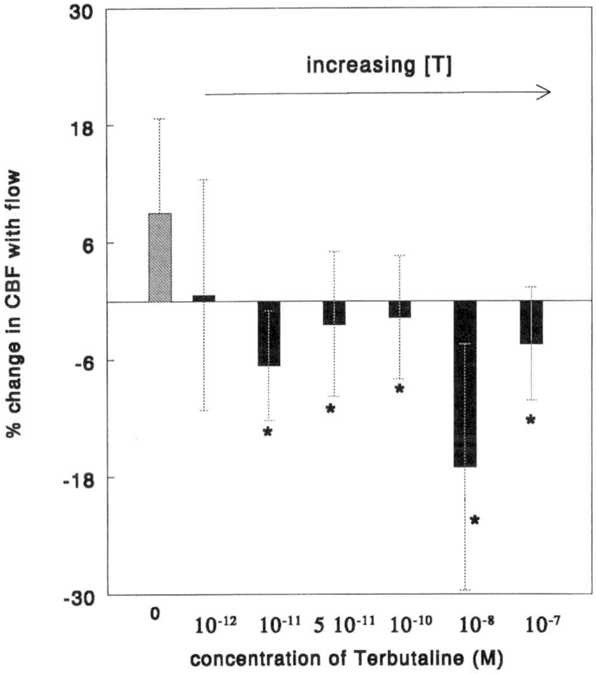


**Figure 2** Terbutaline inhibits the ciliary response to flow rate. As in Figure 1, except that the response measured is to flow rate ± terbutaline. As little as 50 pM terbutaline is sufficient to induce a significant (*) transient decline in CBF instead of the expected transient rise (see 0 lane) in response to the mechanical stimulus.

### D. Removal of the Terbutaline Does Not Relieve Antagonism of the $Ca^{++}$ Pathway

When cells were exposed to 100 n*M* terbutaline, CBF increased to 108.34 ± 2.4% within 15 minutes. When 1 μ*M* 4Br-A23187 was added to these cells, it caused a significant decrease in CBF to 105.66 ± 2.64% by 5 minutes, (107.26 ± 4.27% by 15 minutes). Removal of terbutaline, by washing the cells with 1 μM 4Br-A23187 alone caused an immediate and significant decline in CBF to 103.2 ± 4.05% towards the original preterbutaline baseline. However, although the CBF then oscillated between this and a peak value of 109.31 ± 4.36% (significantly different than baseline) throughout the remainder of the experiment, $n = 12$ (data not shown), the expected A23187 response (circa + 15%) did not occur.

## IV. DISCUSSION

It has been claimed that the increase in asthma related deaths when fenoterol was introduced was more likely due to increased use of $\beta_2$ agonists by patients with more severe asthma, rather than by a physiological mechanism. However, an investigation into this confounding by severity theory found that for patients whose asthma was severe enough to require hospital admission, there was little evidence for this notion [2,14]. Others have invoked either downregulation or decoupling of $\beta_2$ receptors [15,16] as an explanation for the excess morbidity associated with the use of these drugs. In this study we suggest a novel way in which $\beta_2$ agonists could promote mucus retention.

Effective mucociliary clearance is dependent upon the ciliary response to a varying mucus load induced, for example, by inhaled particulate matter. We have recently shown that cyclic AMP (acting via PKA) and $Ca^{++}$ (acting via calmodulin-dependent phosphorylation) can no longer be considered to be independent regulatory pathways for the control of CBF in human nasal airway in vitro [11]. We found that a cell permeant analogue of cAMP activated PKA, which in turn inhibited the calcium-dependent CBF response but not vice versa. In this study, we show that the same phenomenon happens at the receptor level. We find that the $\beta_2$ agonist terbutaline inhibits the intrinsic, $Ca^{++}$-dependent regulation of CBF. Furthermore, this inhibition both persists after removal of the $\beta_2$ agonist and is also seen following a mechanical stimulus applied as an increase in flow rate. Support for this notion comes from our recent data with Substance P. The stimulatory effects of SP on CBF [17] are also inhibited when PKA is active (unpublished data). The reversibility of the inhibitory effect to terbutaline in the presence of the PKA antagonist H7 adds weight to the idea that these drugs inhibit the mechanism that controls intrinsic CBF. The observation that this inhibition is maintained in the p*M* range of terbutaline concentration (which is not stimulatory towards intrinsic CBF) is of interest because such concentrations are likely to be present in the lungs of patients treated with $\beta_2$ agonists. The mechanism underlying our findings is unknown but the work of Salathe et al. [18] points the way forward towards a molecular understanding.

We suggest the following model: $\beta_2$ agonists render the mucus escalator refractory to an increased mucus load. Should subsequent mucus accumulation or further inflammation occur, a clinically unstable state is predicted where clearance is dependent on the further use of a $\beta_2$ agonist. Teshima T. Isawa (personal communication) supports our model because he is unable to demonstrate enhanced mucus clearance with these drugs in vivo.

## ACKNOWLEDGMENTS

Supported by the Anonymous Trust, the Chest Research and Education Fund, and the European Social Fund.

## REFERENCES

1. Isawa TT, Hirano T, Anazawa Y, Miki M, Konno K, Motomiya M. Does a $\beta_2$ stimulator really facilitate mucociliary transport in the human lungs in vivo? Am Rev Respir Dis 1990; 141:715–720.
2. Tattersfield AE. Use of $\beta_2$ agonists in asthma: much ado about nothing? Still a cause for concern. British Med J 1994; 309:794–796.
3. Sears MR, Taylor DR, Print CG, Lake DC, Qingqing L, Flannery EM, Yates DM, Lucas MK, Herbison GP. Regular inhaled beta-agonist treatment in bronchial asthma. Lancet 1990; 336:1391–1395.
4. Cockcroft DW, McParland CP, Britton SA, Swystun VA, Rutherford BC. Regular inhaled salbutamol and airway responsiveness to allergen. Lancet 1993; 342:833–837.
5. Cockcroft DW, O'Byrne PM, Swystun VA, Bhagat R. Regular use of inhaled albuterol and the allergen-induced late asthmatic response. Journal of Allergy and Clinical Immunology 1995; 96:44–49.
6. Inman MD, Byrne PM. The effect of regular inhaled albuterol on exercise-induced bronchoconstriction. American Journal of Respiratory Critical Care Medicine 1996; 153:65–69.
7. Devalia JL. Sapsford RJ, Rusnak C, Toumbis MJ, Davies RJ. The effects

of salmeterol and salbutamol in ciliary beat frequency of cultured human bronchial epithelial cells, in vitro. Pulmonary Pharmacology 1992; 5:257–263.
8. Duchateau G, S, M, J, E, Graamans K, Zuidema J, Merkus F, W, H, M. Correlation between nasal ciliary beat frequency and mucus transport rate in volunteers. Laryngoscope 1985; 95:854–859.
9. Sanderson MJ, Dirksen ER. Mechanico–sensitivity of cultured ciliated cells from the mammalian respiratory tract: implications for the regulation mucociliary transport. Cell Biology 1986; 83:7302–7306.
10. Di Benedetto G, Manara-Shediac FS, Mehta A. Effect of cyclic AMP on ciliary activity of human respiratory epithelium. European Respiratory Journal 1991(a); 4:789–795.
11. Smith RP, Shellard R, Dhillon DP, Winter J, Mehta A. Asymmetric interactions between phosphorylation pathway regulating ciliary beat frequency in human nasal respiratory epithelium in vitro. Journal of Physiology 1996a; 496.3:883–88.
12. Di Benedetto G, Magnus CJ, Gray PTA. Mehta A. Calcium regulation of ciliary beat frequency in human respiratory epithelium in vitro. Journal of Physiology 1991(b); 439:103–113.
13. Di Benedetto G, Lopez-Vidriero MT, Clarke SW. The effect of perfusion flow-rate on ciliary activity in vitro. Bulletin European Physiopathology Respiration 1987; 23 (suppl 12):364S.
14. Beasley, Burgess C, Pearce N, Woodman K, Crane J. Confounding by severity does not explain the association between fenoterol and asthma death. Clinical and Experimental Allergy 1994; 24 (7):660–668.
15. Lipworth BJ, Struthers AD, McDevitt DG. Tachyphylaxis to systemic but not airway responses during prolonged therapy with high dose inhaled salbutamol in asthmatics. American Review of Respiratory Disease 1989; 140: 586–592.
16. Nijkamp FP, Engels F, Henricks AJ, Van Oosterhout AJM. Mechanisms of β-adrenergic receptor regulator lungs and its implications for physiological responses. Physiological Reviews 1992; 72:323–367.
17. Smith RP, Shellard R, Di Benedetto G, Magnus CJ, Mehta A. Interaction between calcium, neutral endopeptidase and substance P mediated ciliary response in human respiratory epithelium. *European Respiratory Journal* 1996b; 9:86–92.
18. Salathe M, Pratt MM, Wanner A. Cyclic AMP-dependent phosphorylation of a 26kDa axonemal protein in ocilia isolated from small tissue pieces. American Journal Respiratory Cell Molecular Biology 1993; 9:306–314.

# 10

# Ionic Mechanisms in Olfactory Reception: Perspective Through Ion Channels in the Vertebrate Olfactory Cilia

**Tadashi Nakamura, Tomo-Oki Satoh,* Hyung-Hwan Lee,† Hirokatsu Kobayashi,‡ Satoru Noguchi§**
*University of Electro-Communications, Tokyo, Japan*

## I. INTRODUCTION

When vertebrate olfactory receptor cells receive at their cilia the odorant molecules dissolved in the olfactory mucus, they generate a depolarizing receptor potential [1]. The last step of this transduction is the opening of the ion channels. Our laboratory has made an effort to identify those channels to elucidate this transduction system.

Active development of olfactory research started when guanosine triphosphate (GTP)-dependent odor-induced production of cyclic adenosine 3′5′ cyclic monophosphate (cAMP) was discovered in the laboratory of Lancet [2]. The suggestion by Lancet and coworkers that cAMP might work as the second messenger to generate the receptor potential was soon strongly supported by the discovery of cyclic nucleotide gated (CNG) channels in the cilia [3].

At first, the ciliary channels were studied on the supermicro inside-

*Current affiliations:*
* Yamanouchi Pharmaceutical Company, Tsukuba, Japan
† Daewoo Electronics Co., Ltd., Seoul, Korea
‡ Tokyo Electron, Tokyo, Japan
§ ASM Japan, Tokyo, Japan

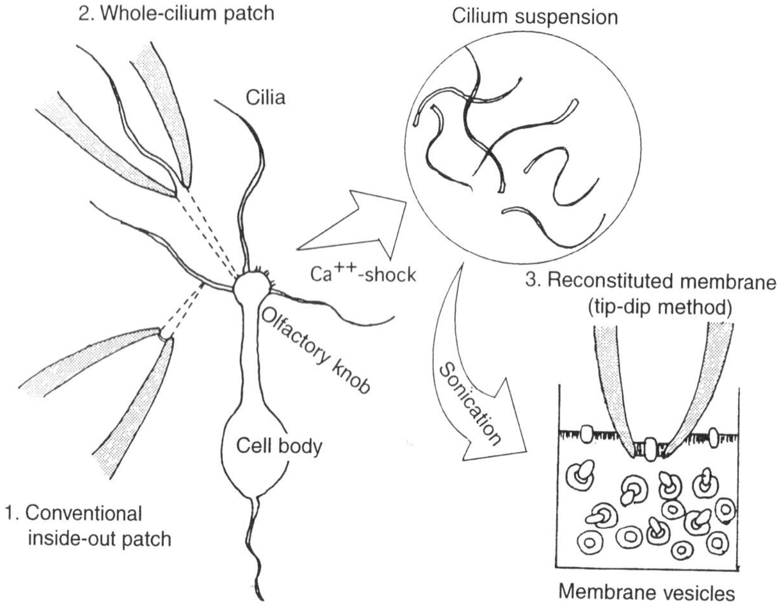


**Figure 1** Three types of membrane prepared from olfactory cilia.

out membrane patches excised from the cilia (Figure 1). Then a method called "whole cilium patch" was introduced by Kleene and Gesteland [4] (Figure 1). On the other hand, lipid bilayers reconstituted from membrane suspensions of the isolated olfactory cilia were also used to detect the channels [5]. The CNG channels were detected in these 3 types of membrane and were supposed to be the main channels for the transduction.

It was subsequently reported that the olfactory cilia contain the $Ca^{2+}$-activated chloride channels [6] and $IP_3$-gated $Ca^{2+}$ channels [5]. Subsequently there has been a question as to what those channels' specific roles are. Biochemical observations showed that some odorants do not activate the adenylate cyclase in the cilium suspension [7], and that some odorants activate the phospholipase C to increase the intracellular $IP_3$ [8,9]. Thus $IP_3$ was proposed to be an additional second messenger working in parallel with cAMP.

Lowe, Nakamura, and Gold [10] recorded the electroolfactograms (EOG) induced by many odorants as the macro responses from many cilia to compare to the biochemical data of activation of the cyclase by the same odorants using bull frogs. They found that the EOG amplitudes are linearly correlated with the cyclase activity for the same concentration of each odorant. They proposed that when the population of receptor cells specific for some odorants in the olfactory epithelium is small, both the capacity to activate the cyclase and to induce the EOG appeared to be low even if the capacity is high in each receptor cell. There has been some doubt, however, if EOG really represents a physiological response.

On the other hand, because the $IP_3$-gated channel is the $Ca^{2+}$ channel, measurement of the intracellular concentration of $Ca^{2+}$ was of interest. In fact, an odor-induced increase was detected and this was proposed as evidence for the transductory role of $IP_3$ [5]. However, dialysis of $IP_3$ through patch pipettes did not cause a clear increase in $Ca^{2+}$ near the cilia in newt olfactory cells, while dialysis of cAMP brought both increases of intracellular $Ca^{2+}$ and membrane conductance [11]. These data suggested that $Ca^{2+}$ entered the cell through the CNG channels.

Although such studies did not support the $IP_3$-pathway, it has been argued that those results were simply due to experimental failures. Thus it is worth testing the technique used for detecting the $IP_3$-gated channels and to examine their existence and their localization. The evidence for the $IP_3$-gated channels mostly depends on the reconstituted membrane from isolated ciliary membranes. Therefore we tried to compare the conductances detected from the native ciliary membranes and the reconstituted membrane from an isolated membrane suspension of bull frog cilia.

## II. RESULTS AND DISCUSSION

Most of the ciliary patch preparations showed the conductance increase induced by the cAMP introduced by means of the perfusing solution from the intracellular side of the membrane (Figure 2). In addition, conductance increased when $Ca^{2+}$ was applied in the same way. Both conductances were detected together in most cilia tested. Chloride ion ($Cl^-$) conductance was confirmed by detecting the shift of reversal potential when applying a $Cl^-$ concentration gradient (Figure 2b, trace 4): $Ba^{2+}$ showed a similar effect to $Ca^{2+}$ [12].

However, regardless of the presence or absence of divalent cations, the $IP_3$-gated conductance was not clearly detected. Though trace 3 in Figure 2a seems to show a little conductance increase, it was within the noise level, and other membranes showed the change in the opposite direction at the same level.

Reconstituted membranes also showed the conductance increase induced by cAMP. Figure 3a shows typical IV-curves measured from the reconstituted membrane. The response to cAMP indicates that this membrane contained the ciliary plasma membrane. In addition, many of those membranes responded to $IP_3$ by increasing the conductance. This result is consistent with the previous reports, though inconsistent with the results in native membrane. The question as to whether the $IP_3$-gated chan-

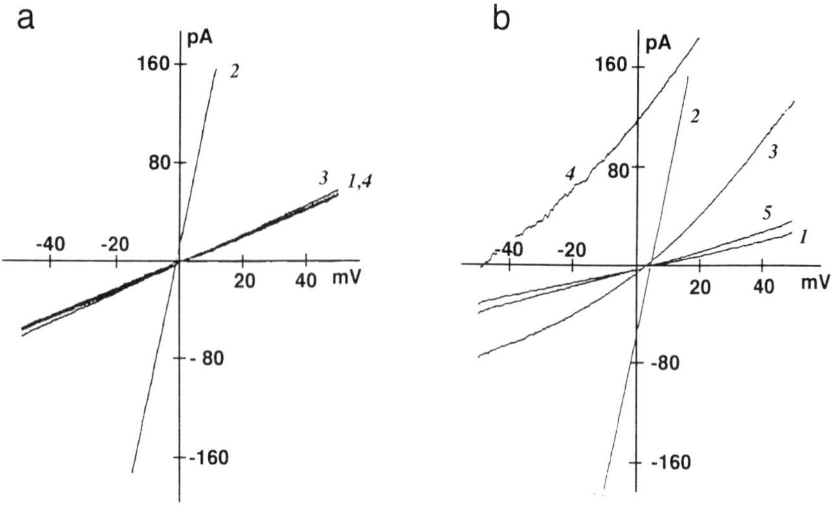


**Figure 2** Conductances of whole cilium patches prepared from bull frog olfactory cells. IV-curves were measured in perfusing solutions containing (a) 1, 4: control; 2: 200 $\mu M$ cAMP; 3: 50 $\mu M$ $IP_3$ and (b) 1, 5: control; 2: 200 $\mu M$ cAMP; 3: 1 m$M$ $Ca^{2+}$; 4: 1 m$M$ $Ca^{2+}$. Perfusing solutions contained the same concentrations of $Cl^-$ as the electrode solution. But most of the $Cl^-$ in the perfusing solution for Figure b trace 4 was replaced with gluconate. (Reproduced with modification from Ref. [13], with permission from Elsevier Science.)

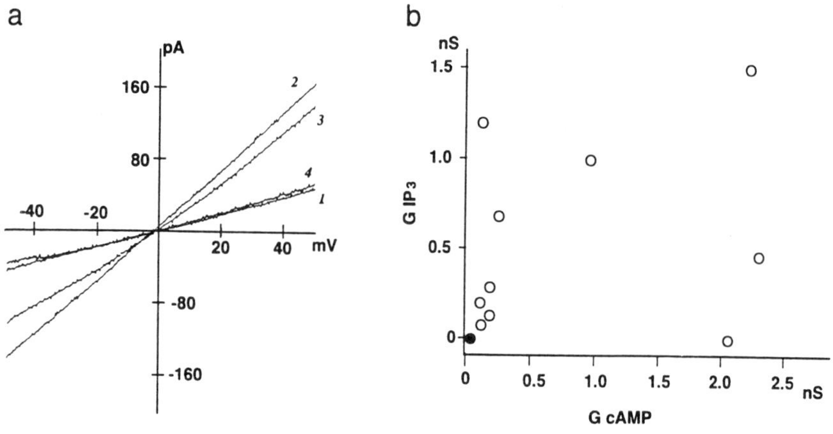


**Figure 3** Conductances of membrane reconstituted from the cilia suspension of bull frogs. a) IV-curves were measured in solutions containing 1, 4: control; 2: 1 m$M$ cAMP; 3: 50 $\mu M$ $IP_3$. b) Relationship between saturated conductances due to cAMP (G cAMP) and $IP_3$ (G $IP_3$). A closed circle represents two points within a very short distance. (Reproduced from Ref. [12], permission from Biophysical Society.)

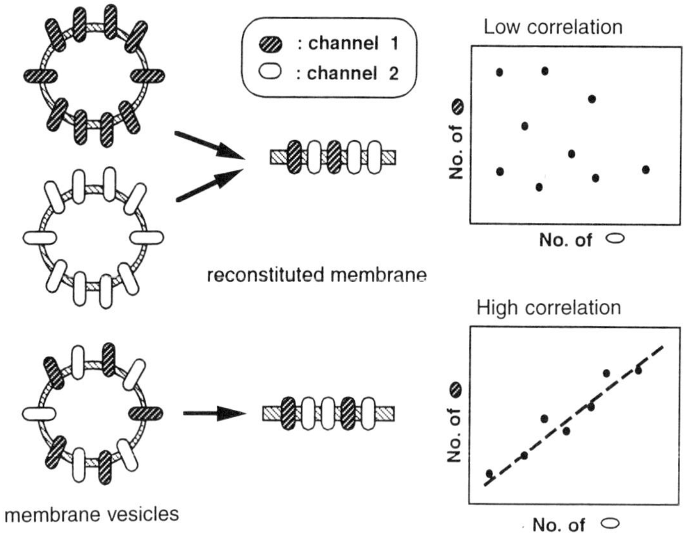


**Figure 4** Correlation of two channels detected in the reconstituted membrane. (See text for details.)

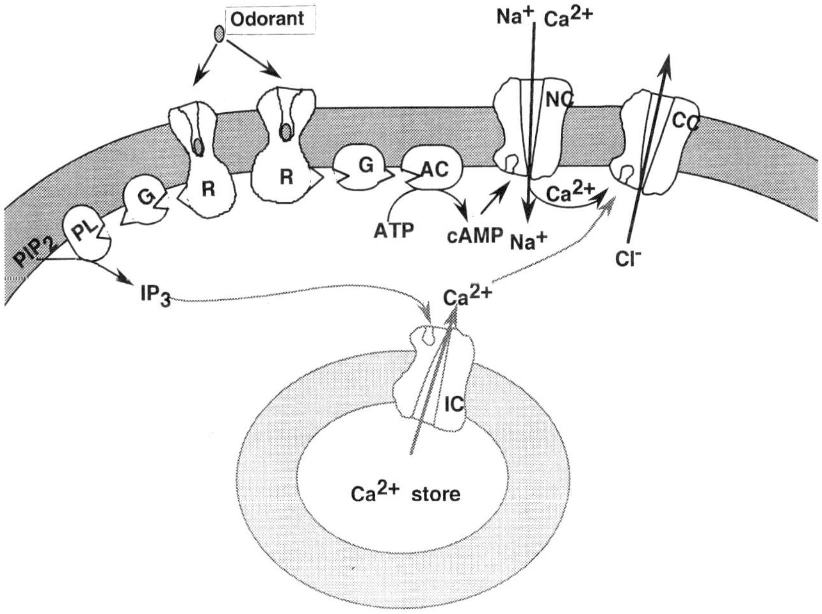


**Figure 5** Current model for olfactory transduction in vertebrate olfactory cilia.

nels reside in the same ciliary membrane was approached by making a scatter plot of the two conductances of 12 membranes (Figure 3b). In this figure, points were rather randomly scattered and the correlation factor between the two conductances was about 0.3. If two channels originate from the same membrane, those points should be on a straight line and their correlation factor should be closer to 1, as illustrated in Figure 4. Therefore Figure 3b suggests that the two channels originate in different membranes.

Thus, although $IP_3$-gated conductance was detected by the previous method, it seemed reasonable to attribute its origin to membranes other than the ciliary membrane. This suggests that the CNG channels and the $Cl^-$ channels seem to be responsible for the response to all the odorants. The schematic transduction should be like Figure 5, which also describes the role of $Cl^-$ channels in amplifying the initial response by the CNG channels (review Ref. 14). Participation of $IP_3$ described in this figure is still to be examined.

Recently, it was reported that mice whose olfactory CNG channels were disrupted did not respond to any kind of odorants by generating a negative EOG. This conclusion is consistent with the present results [15].

## ACKNOWLEDGMENTS

We thank Mr. H. Fujii for his early trial of the lipid bilayer experiment. This work was partly reported elsewhere [12,13].

## REFERENCES

1. Lancet D. Molecular components of olfactory reception and transduction. In: Margolis FL, Getchell TV, eds. Molecular Neurobiology of the Olfactory System, New York: Plenum Press, 1988:25–50.
2. Pace U, Hanski E, Salomon Y, Lancet D. Odorant-sensitive adenylate cyclase may mediate olfactory reception. Nature 1985; 316:255–258.
3. Nakamura T, Gold GH. A cyclic nucleotide-gated conductance in olfactory cilia. Nature 1987; 325:442–444.
4. Kleene SJ, Gesteland RC. Transmembrane currents in frog olfactory cilia. J Membrane Biol 1991; 120:75–81.
5. Restrepo D, Miyamoto T, Bryant BP, Teeter JH. Odor stimuli trigger influx of calcium into olfactory receptor neurons. Science 1990; 249:1166–1168.
6. Kleene SJ, Gesteland RC. Calcium-activated chloride conductance in frog olfactory cilia. J Neurosci 1991; 1:3624–3629.
7. Sklar PB, Anholt RRH, Snyder SH. The odorant-sensitive adenylate cyclase of olfactory receptor cells. Differential stimulation by distinct classes of odorants. J Biol Chem 1986; 261:15538–15543.

8. Huque TH, Bruch RC. Odorant- and guanine nucleotide-stimulated phospho-inositide turnover in olfactory cilia. Biochem Biophys Res Commun 1986; 137:36–42.
9. Boekhoff I, Tareilus E, Strotmann J, Breer H. Rapid activation of alternative second messenger pathways in olfactory cilia from rats by different odorants. Embo J 1990; 9:2453–2458.
10. Lowe G, Nakamura T, Gold GH. Adsenylate cyclase mediates transduction for a wide variety of odorants. Proc Natl Acad Sci USA 1989; 86:5641–5645.
11. Nakamura T, Tsuru K, Miyamoto S. Regulation of $Ca^{2+}$ concentration by second messengers in newt olfactory receptor cell. Neurosci Lett 1994; 171: 197–200.
12. Nakamura T, Lee HH, Kobayashi H, Satoh TO. Gated conductances in native and reconstituted membranes from frog olfactory cilia. Biophys J 1996; 70:813–817.
13. Nakamura T, Satoh TO, Miyamoto S. Ligand gated ionic conductances in the olfactory cilia. Biophysics 1995; 40:185–191.
14. Kurahashi T, Yau KW. Tale of an unusual chloride current. Current Biol 1994; 4:256–258.
15. Brunet LJ, Gold GH, Ngai J. General anosmia caused by a targeted disruption of the mouse olfactory cyclic nucleotide-gated cation channel. Neuron 1996; 17:681–693.

# 11

# Coupling the Internal Axoneme Mechanics of Cilia and Flagella with External Fluid Dynamics

**Robert H. Dillon*** **and Lisa J. Fauci**
*Tulane University, New Orleans, Louisiana*

## I. INTRODUCTION

Cilia and flagella are the prominent organelles associated with the motility of microorganisms. In invertebrates, the cilia and flagella are also associated with the generation of water currents and mucus flow. In mammals, cilia are responsible for the movement of mucus and foreign particles out of the airways and help defend the respiratory system from pathogenic organisms. Although the patterns of flagellar movement are distinct from those of ciliary movement, and the flagella are typically much longer than cilia, their basic ultrastructure is identical. In fact, although the two flagella of the algal cell Chlamydomonas exhibit a cilia-like beat during normal swimming, they undergo sinusoidal flagellar beating in a fast-swimming avoidance response. Considerable interest has been focused on the development of mathematical models for the hydrodynamics of individual as well as rows of cilia and on individual flagellated organisms. In recent decades, the ultrastructure of the cilia and flagellar axoneme has been well described, and with the development of relatively inexpensive high speed computers, it is now possible to construct a mathematical and computational model based on a detailed description of the axonemal physiology. Such a model would include an explicit representation of the force

---

* *Current affiliation*: Washington State University, Pullman, Washington

generation and activation dynamics of the dynein molecular motors, the passive structure of the microtubules, nexin links, and radial arms, as well as the hydrodynamics of the fluid media.

The coupled system of a viscous, incompressible fluid and a single, force-generating organism is difficult to analyze, even when the kinetics of an organism's waveform is taken as given. In the past decades, the efforts to describe quantitatively the fluid dynamics of spermatozoa and ciliary propulsion have been very successful. Since the Reynolds number is quite small and inertial effects may be neglected, the linear Stokes flow assumption has been used to investigate the hydrodynamic consequences of flagellar undulations [1]. These investigations have been both analytical and computational. Resistive-force theory, initially developed by Gray and Hancock [5] makes use of the linear Stokes flow assumption and constructs the flow field by means of distributions of fundamental singularities. Central to the success of the original Gray and Hancock theory was the determination of tangential and normal resistive force coefficients. However, the certainty to which these could be computed was a questionable point, even in the case of a single organism swimming in an infinite fluid domain. Lighthill later improved this theory by incorporating slender body approximations [12], since the diameter of a flagellum is much smaller than its length. More detailed hydrodynamic analysis, such as refined slender body theory and boundary element methods, have produced excellent simulations of both two- and three-dimensional flagellar propulsion in an infinite fluid domain or in a domain with a fixed wall [9,10,11,16]. In addition, ciliary motion, where the effect of a single plane surface at the base of the cilia is included, was studied by Gueron and Liron [7,8], and they also presented a thorough survey of flagellar hydrodynamics.

More difficult is the study of the fluid dynamic interactions of organisms with each other and surrounding geometry. Examples of these interactions include sperm motility in the reproductive tract and metachronal waves exhibited by rows of beating cilia. Recent advances in the mathematical modeling of these complex systems have been described [3,6].

In both cilia and flagella, the typical 9 + 2 axoneme consists of a central pair of singlet microtubules surrounded by nine outer doublet microtubules and encased by the cell membrane [13,18]. The microtubules are composed of $\alpha$ and $\beta$ heterodimers that form protofilaments running the length of the microtubules. The A tubule of the outer doublet is made up of 13 protofilaments, the B tubule of 10 or 11 protofilaments. The nine outer doublets are connected by radial spokes to a sheath surrounding the central pair. In addition, the outer doublets are connected by nexin links between adjacent pairs of doublets. Two rows of dynein arms extend from the A tubule of an outer doublet toward the B tubule of an adjacent doublet at regularly spaced intervals. The bending of the axoneme is caused by sliding between pairs of outer doublets. Active sliding is due to the unidirectional ATP-induced force generation of the dynein power stroke. Backward, passive sliding is due to the active sliding of other pairs of doublets within the axoneme. The precise nature of the spatial and temporal control mechanisms regulating the various waveforms is still unknown.

The object of our work is to develop a model for the individual cilium or flagellum that incorporates discrete representations of the dynein arms, the passive elastic structure of the axoneme including the microtubules and nexin links, as well as the surrounding fluid. This model will couple the internal force generation of the molecular motors through the passive elastic structure with the external fluid mechanics. The ciliary beat or flagellar waveform is not preset, but is an emergent property of the interacting components of the coupled fluid-axoneme system. The forces due to activation of dynein motors (that cause the microtubules to slide), the passive forces due to the stiffness of the microtubules, and the passive forces due to intermicrotubule linking proteins (radial spokes and nexin) must balance with the viscous forces due to the surrounding incompressible fluid.

In this article, we will describe a preliminary model in two space dimensions of the coupled fluid–axoneme system, which is based upon the immersed boundary method. The immersed boundary method was originally developed by Peskin [15] to study blood flow in the heart, but has been used to study many other problems in biological fluid dynamics. We present results of the two-dimensional model in which the spatial-temporal activation of the individual dynein arms is partially imposed. The goals of the model development are to replace the imposed dynein activation dynamics with a locally controlled mechanism based on the local axonemal configuration and to develop a fully three-dimensional model of the axonemal structure.

## II. FORMULATION

The structure of the two-dimensional model axoneme consists of three model microtubules (see Figure 1). Each model microtubule consists of two parallel filaments. (In a full three-dimensional model, these filaments would be the helical protofilaments that make up the microtubule.) Each filament is described by a discrete set of nodes (immersed boundary points) connected by linear elastic spring elements. These elastic springs

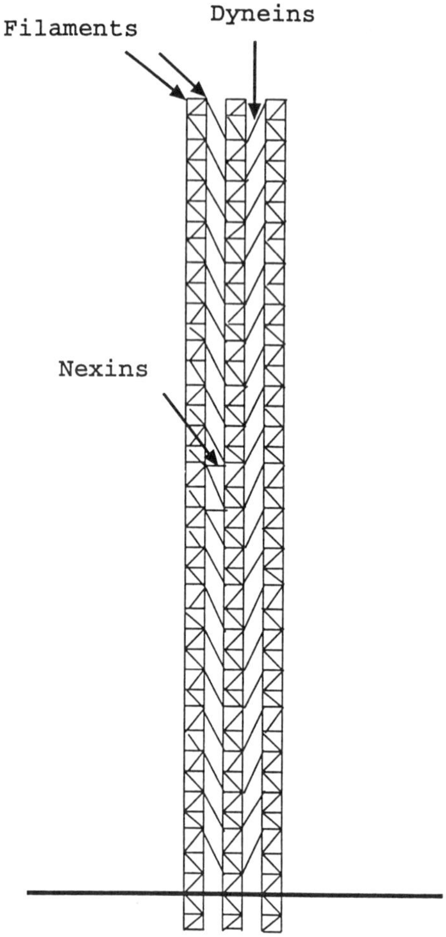


**Figure 1** Schematic of two-dimensional model axoneme.

form vertical connections between neighboring nodes of a filament, as well as horizontal and diagonal connections across the two filaments of the microtubule. Adjacent pairs of microtubules are interconnected by linear elastic springs representing the nexin and/or radial links of the axoneme. In addition, active dynein motor forces are represented as diagonal forces between adjacent pairs of microtubules. The axoneme is tethered to fixed points in space via strong elastic springs at the base. If the dynein arms between the right-hand pair of microtubules are simultaneously activated, the outer microtubule can slide upward relative to the central microtubule causing the axonemal structure to bend toward the right. Simultaneous activation of the left-hand pair causes bending toward the left. The entire structure is embedded in a viscous incompressible fluid.

## A. Representation of Fluid and Axoneme

The fluid is regarded as viscous and incompressible, and the microtubules that comprise the axoneme as elastic boundaries immersed in this fluid. The other protein structures within the axoneme will be represented as forces that act upon the microtubules, as described above. We assume that the flow is governed by the Navier–Stokes equations:

$$\rho\left(\frac{\partial u}{\partial t} + u \cdot \nabla u\right) = -\nabla p + \mu \Delta u + F(x, t) \tag{1}$$

$$F = \sum_k F^k \tag{2}$$

$$\nabla \cdot u = 0$$

Here $\rho$ is the fluid density and $\mu$ is the viscosity, $u = (u, v)$ is the fluid velocity, p denotes the pressure, and F is the force per unit area that is exerted on the fluid by the axoneme. This force is split up into the contributions from each of the filaments comprising the microtubules. These forces $F^k$ due to the *k*th filament include the elastic forces due to the individual microtubule structures, the passive elastic forces due to intermicrotubule links, and the active forces due to dynein arms. The force F is a $\delta$-function layer of force supported only by the region of fluid that coincides with material points of the microtubules; away from these points the force is zero.

The state of the fluid at time t is given by its velocity field; the state of each microtubule is given by the configuration of its one-dimensional filaments. Let $X^k(s,t)$ denote the *k*th filament in our set of microtubules and let $f^k(s,t)$ denote the boundary force per unit length along the *k*th filament. (The form of this boundary force will be discussed below.) The mass of the neutrally buoyant elastic boundary is attributed to the mass of the fluid in which it sits, and thus the forces are transmitted directly to the fluid. The force field $F^k$ from the filament $X^k(s,t)$ is therefore

$$F^k(x, t) = \int f^k(s, t)\delta(x - X^k(s, t))\, ds \tag{3}$$

Here the integration is over the *k*th one-dimensional filament comprising an immersed microtubule, and $\delta$ is the two-dimensional Dirac delta-function. The total force F(x,t) is calculated by adding the forces from each of the filaments.

## B. Derivation of Force Density

Each filament of the microtubule is approximated by a discrete collection of points $X_1, X_2, \ldots, X_n$. This boundary exerts elastic forces on the fluid near each of these points as described above. We imagine that between each pair of successive points $X_j$, $X_{i+1}$ on a microtubule filament, there is an elastic spring, or "link", which generates forces to push the link's length toward a specified resting length. The force density vector arising from this spring at the point $X_j$ has the form

$$-S(\|X_{j+1} - X_j\| - \Delta s)\frac{X_j - X_{j+1}}{\|X_j - X_{j+1}\|} \tag{4}$$

This force is, therefore, a stiffness constant (S) times the deviation from rest length ($\Delta s$) that is applied tangent to the spring. In addition to the forces due to springs along the parallel filaments, forces due to horizontal and vertical crosslinks also contribute to the force density. These springs maintain the integrity of an individual microtubule.

Our model also accounts for the passive intermicrotubule connections, such as nexin or radial spokes, by applying spring forces due to connections between nodes of adjacent microtubules. The spring constants of these connections can be varied to study their effect upon bending behavior.

Most importantly, we need to achieve the relative sliding between adjacent microtubules so that a bend can be initiated. This is achieved by representing the dynein motors as diagonal springs that connect adjacent microtubules. Unlike the microtubule springs or nexin springs, these springs are not passive, but dynamically increase their resting lengths as they are activated. In addition, their connectivity may be reassessed during the course of a simulation. For instance, as the microtubules slide, the endpoint of a dynein spring may "ratchet" from one node of the microtubule to another.

The forces due to all of the springs contribute to the force density function at each of the nodes. A discrete version of Eq. (3) is used to spread this force density defined along microtubules to a force field on the fluid domain.

Our coupled fluid–axoneme system is closed by requiring that the velocity of a material point of a microtubule must be equal to the fluid velocity evaluated at that point:

$$\frac{\partial X_k(s, t)}{\partial t} = u(X_k(s, t), t) = \int u(x, t)\delta(x - X_k(s, t))\, dx \qquad (5)$$

Here the integration is over the entire fluid domain.

### C. Algorithm

We can summarize the immersed boundary algorithm as follows: Suppose that at the end of timestep n we have the fluid velocity field $u^n$ on a grid and the configuration of the immersed boundary points comprising the filaments of the microtubules $(X_k)^n$. Then to advance the system by one timestep we

1. Calculate the force densities $f^k$ from the boundary configuration,
2. Spread the force densities to the grid to determine the forces $F^k$ on the fluid,
3. Solve the Navier-Stokes equations for $u^{n+1}$, and
4. Interpolate the fluid velocity field to each immersed boundary point $(X_k)^n$ and move the point at this local fluid velocity.

The crucial feature of this model is that the immersed boundary is a dynamic force field that influences the fluid motion through the force term in the fluid equations. Since the computational domain is a fixed rectangle, we can use an efficient fluid-solver designed for a regular grid with simple boundary conditions. For step (3) we use the projection method of Chorin [2] with periodic boundary conditions. Steps (2) and (4) involve the use of a discrete delta function that communicates information between the grid and the immersed boundary points (which do not necessarily coincide with grid points). We refer the reader to [4,15] for details.

## III. NUMERICAL RESULTS

Figure 2 shows a sequence from a numerical simulation representing one ciliary beat. During the effective stroke, the dyneins on the right hand side of the axoneme are all active while those on the left are inactive. A recovery stroke is initiated when one of the nexin links on the right-hand side becomes stretched beyond a prescribed length. Thus, the duration of the effective stroke is not imposed, but is dependent upon the local geometry of the cilium. At the beginning of the recovery stroke, the dyneins on the right are inactivated and a travelling wave of dynein activation is imposed on the left from base to tip. A new effective stroke is initiated when one of the nexin links on the left is stretched beyond a prescribed

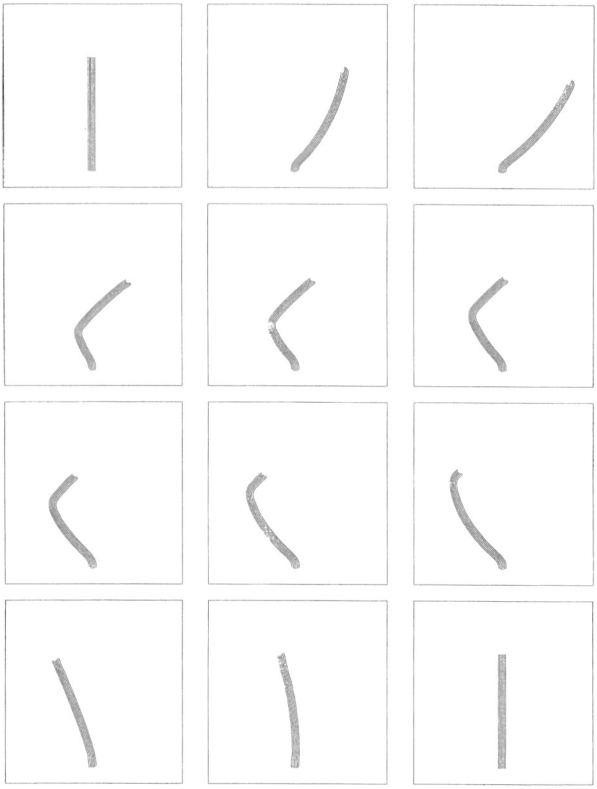

**Figure 2** Sequence of snapshots from a computer simulation of one ciliary beat. The cilium is approximately 10 μm in length and the fluid viscosity is approximately that of water. The computational domain is a 64 × 64 grid.

length. Since the frames in Figure 2 represent equal time intervals, we see that the effective stroke in this simulation has a shorter duration than the recovery stroke. The duration of the recovery stroke depends upon the speed of the travelling wave of dynein activation. Figure 3 is a snapshot depicting a model cilium during the effective stroke along with the fluid velocity field. In the results shown here, we do not include the effect of a boundary or wall to which the cilium is attached, but this may be included easily within our framework by including a wall of immersed boundary points [3].

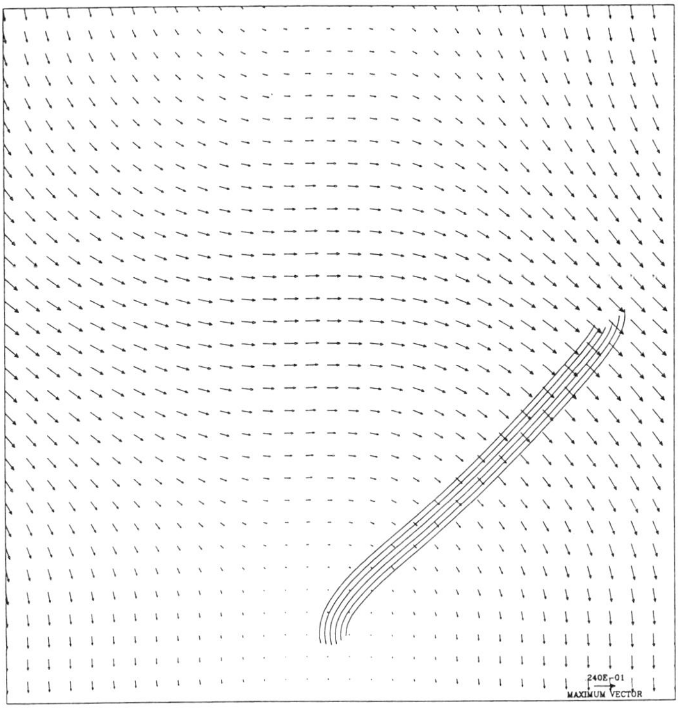


**Figure 3** Snapshot from a computer simulation during the effective stroke. The fluid velocity field is represented by vectors.

## IV. DISCUSSION

The numerical results demonstrate the feasibility of the model to simulate a ciliary beat. In future models, the activation of individual dyneins will be dependent upon the configuration of the axoneme at the dynein site. Information concerning the local curvature and spacing between the microtubules is available at each dynein site and can be included in models based on local dynein activation kinetics. In addition, since this model represents a discrete collection of dynein motors, stochastic models of dynein activation can be implemented. A similar model will also be used to study flagellar wave forms. A parallel effort is the construction of a three dimensional axoneme based upon the familiar 9 + 2 ultrastructure.

## ACKNOWLEDGMENTS

The work of R. Dillon was supported in part by a NSF Postdoctoral Research Fellowship DMS-9508815. The work of L. Fauci was supported in part by NSF grant DMS-9501048.

## REFERENCES

1. Brennen C, Winet H. Fluid mechanics of propulsion by cilia and flagella. Ann Rev Fluid Mech 1977; 9:339–398.
2. Chorin AJ. Numerical solution of the Navier–Strokes equations. Math Comp 1968; 22:745.
3. Fauci LJ, McDonald A. Sperm motility in the presence of boundaries. Bull Math Biol 1995; 57:679–699.
4. Fauci LJ, Fogelson AL. Truncated Newton methods and the modeling of complex immersed elastic structures. Comm Pure and Appl Math 1993; 46: 787–818; J Comput Phys 1984; 56:111–134.
5. Gray J, Hancock G. The propulsion of sea-urchin spermatozoa. J Exp Biol 1995; 32:802–814.
6. Gueron S, Levit-Gurevich K, Liron N, Blum J. Cilia internal mechanism and metachronal coordination as the result of hydrodynamical coupling. To Appear.
7. Gueron S, Liron N. Ciliary motion modeling, and dynamic multicilia interaction. Biophys J 1992; 63:1045–1058.
8. Gueron S, Liron N. Simulations of three-dimensional ciliary beats and cilia interactions. Biophys J 1993; 65:499–507.
9. Higdon JJL. A hydrodynamic analysis of flagellar propulsion. J Fluid Mech 1979; 90:685–711.
10. Higdon JJL. The hydrodynamics analysis of flagellar propulsion: helical waves. J Fluid Mech 1979; 94:331–351.
11. Higdon JJL. The generation of feeding currents by flagellar motion. J Fluid Mech 1979; 94:305–330.
12. Lighthill JL. Flagellar hydrodynamics. SIAM Review 1976; 18:161–230.
13. Murase M. The Dynamics of Cellular Motility. Chichester: John Wiley, 1992.
14. Pedley TJ, Kessler JO. Hydrodynamic phenomena in suspensions of swimming microorganisms. Annu Rev Fluid Mech 1992; 24:313–358.
15. Peskin CS. Numerical analysis of blood flow in the heart. J Comp Phys 1977; 25:220–252.
16. Phan-Thien N, Tran-Cong T, Ramia M. A boundary-element analysis of flagellar propulsion. J Fluid Mech 1987, 184:533–549.
17. Pozrikidis C. Boundary integral and singularity methods for linearized viscous flow. Cambridge Texts in Applied Mathematics, 1992.
18. Witman GB. Introduction to Cilia and Flagella in Ciliary and Flagellar Membranes Bloodgood RA, ed. New York: Plenum, 1990; 1–30.

# 12

# Metachronal Coordination as the Result of Hydrodynamical Coupling

**Shay Gueron, Konstantin Levit-Gurevich, and Nadav Liron**
*Technion–Israel Institute of Technology, Haifa, Israel*

**Jacob J. Blum**
*Duke University Medical Center, Durham, North Carolina*

## I. INTRODUCTION

Ciliary motion and particularly the metachronism phenomenon have attracted a great deal of research effort both experimentally and theoretically. The reason arrays of cilia beat in a metachronal pattern is not fully understood. Some investigators [1] speculate that membrane voltage and calcium levels affect the direction of the metachronal wave, while others speculate that metachronism is possibly the result of hydrodynamic coupling [2,3,4].

We describe here a theoretical model that supports the latter conjecture. Gueron and Liron (GL) introduced an improved technique for describing the hydrodynamics of moving cilia based on a refined slender body theory [5]. Their technique offers an alternative for the simplistic Resistive Force Theory (known also as the Gray and Hancock approximation). This improves the consistency and accuracy of the model for cilia beating. More important, the GL equations are the first published method for dynamic simulations of multicilia configurations that account for the effects of neighboring cilia and the effect of the surface from which the cilia emerge. The GL model was originally applied to a two dimensional setup and later extended to describe three dimensional beating [6].

The work of GL was mainly oriented towards developing the framework for dynamic modeling of multicilia configurations. In their model the internal mechanism of the cilia (engine) is a separate building block that can be replaced by any other engine. So far, no reliable model for the internal mechanism has been proposed [5,7]. A simple plausible functional form for the engine that is related to the dyenin arms and the radial spokes system has been recently proposed by us [7,8]. It has a small number of parameters, whose values are obtained from data on real cilia beats of Paramecium.

The new engine was tested by incorporating it into the GL dynamic equations for two-dimensional beats. The combined model produced realistic beats and reproduced experimental results such as an exponential decrease in beat frequency with increased viscosity. We report here briefly on the results concerning multicilia configurations and metachronal coordination that evolves autonomously due to the hydrodynamic interaction between the cilia. This supports the conjecture that metachronism can be explained as the result of hydrodynamic coupling.

## II. MODEL AND RESULTS

We describe briefly our approach for modeling the active shear force (engine) generated by the internal mechanism of the cilia, and how it was applied with specific data. The beat cycle is separated into four distinct phases, occurring in the following order.

Phase 1: The effective stroke. The cilium beats approximately as a straight rod.
Phase 2: A geometrically controlled switch from the effective to the recovery stroke, initiating a change in the direction of the motion and a bend at the basal end.
Phase 3: The recovery stroke. The bend propagates along the cilium until it becomes straight again.
Phase 4: A second, geometrically controlled switch from the recovery to the effective stroke, starting a new effective stroke.

We represent the forward (effective) and backward (recovery) motions by two different simple models because of observations that show completely different behavior during these strokes and evidence that these phases are related to different connections between the nine filaments. In our model, the transitions from one phase to another occur due to switches activated by the momentary geometric configuration of the cilium.

The details of the motion of a single cilium are difficult to separate since experiments describe the motion of an ensemble of cilia. Thus, we cannot determine whether the principles we propose here apply in general. However, the models application to a specific case, based on available data, proves to be successful.

In this study we considered cilia that beat in a two-dimensional pattern. For data we used the cilium of Paramecium, whose beat pattern is

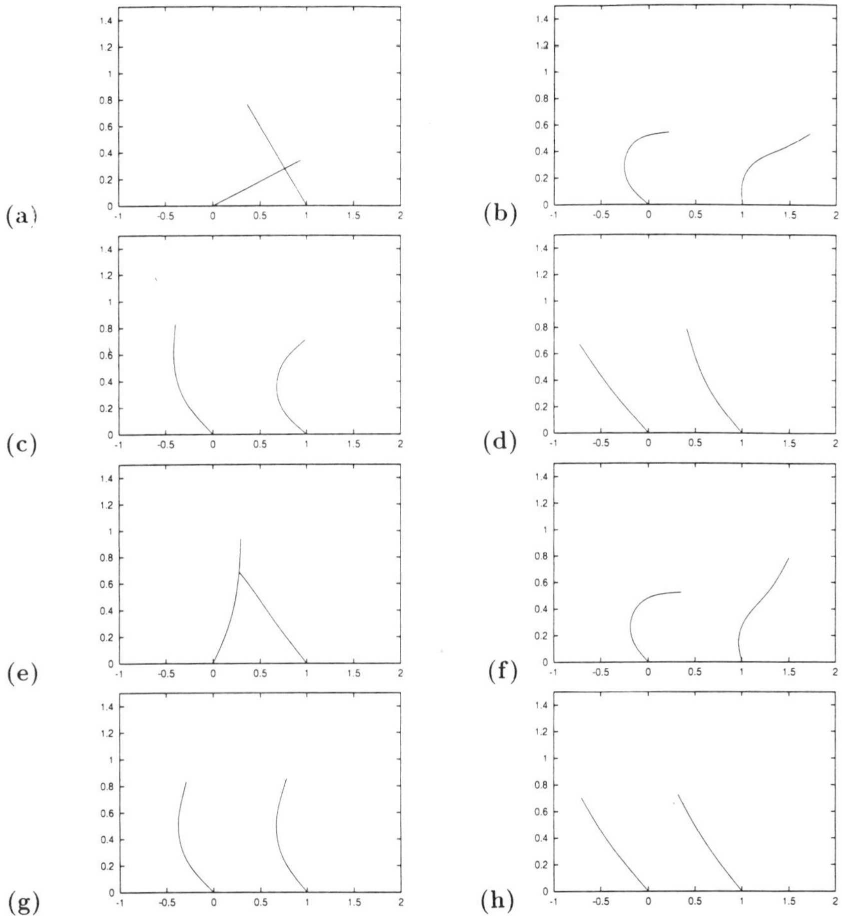


**Figure 1** Autonomous synchronization between two cilia starting at opposite phases (the left cilium starts at the recovery stroke and the right cilium starts at the effective stroke). Synchronization is achieved after two cycles. Cilia spacing is 1 cilium length. The illustration labeled successively from a to h, are separated in time by 9ms. The resulting beat frequency is $\approx$ 31 Hz. The units of the axes are nondimensional length.

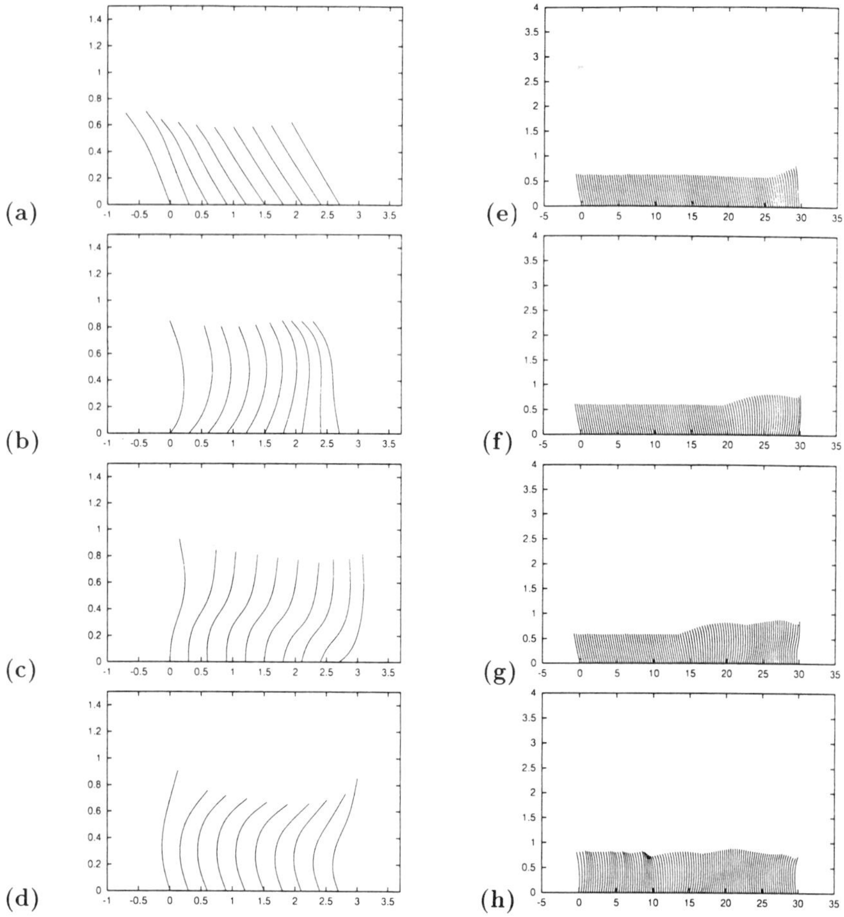


**Figure 2** (a) - (d): Self organized metachronism in a row of ten cilia. (e) - (f): A hundred cilia configuration. Metachronal wave pattern is generated by hydrodynamic coupling. In both configurations the cilia spacing is 0.3 cilium length, the illustrations are separated in time by 2ms, the resulting beat frequency is $\approx$ 42Hz, and the units of the axes are non-dimensional length.

very close to two dimensional [9]. Observed beat patterns are used for computing locations and velocities at various times during the beat cycle. The geometric and hydrodynamic equations of the model were used to recalculate the forces, which were fitted into a simple and plausible functional form. Based on this, all parameters were obtained from actual data.

The resulting model was first tested with a single isolated cilium. Realistic beats were reproduced, and the response of the model beat frequency to a change in the viscosity of the surrounding fluid matched experimental data. At this point we were ready to start investigating multicilia configurations.

When neighboring cilia beat close enough to each other, each one is subjected to the flow generated by the others. This influences their beat pattern and beat frequency. The simplest multicilia configuration consists of two cilia. Gray [10] reported on experiments with two flagella, beating initially at different frequencies and having different wave lengths. When these are put close enough to each other, they tend to synchronize quickly. An analogous phenomenon is self-synchronization and phase locking between rings and lines of coupled oscillators, and this has been investigated experimentally and theoretically [11,12].

Figure 1 displays illustrations of the beat cycles of two cilia separated by one cilium length, beginning their beat in opposite phases. They synchronize completely within two cycles. Their mutual beat frequency increases slightly from 29.5 Hz (for a single cilium) to 31 Hz. This is some indication of the advantage of multicilia configurations. The presence of neighboring cilia (one in this case) helps them beat faster.

Metachronal coordination is characteristic of ciliary arrays. The cilia on the same line (perpendicular to the direction of the effective stroke) are synchronized, and adjacent rows of cilia (parallel to the direction of the effective stroke) beat with a phase difference. We used multicilia configurations where the cilia start simultaneously with the same initial conditions. Figure 2 displays illustrations of ten cilia and one hundred cilia configurations. Antiplectic metachronal waves that are typical for the cilia of Paramecium appear autonomously due to the hydrodynamic interactions.

## ACKNOWLEDGMENTS

This chapter is a short account of the model and the results that are reported in detail in our papers [7,8]. Figures 1 and 2 appear in Reference 8.

This research was supported by the U.S. - Israel Binational Science Foundation Grant number 94-242, by the Technion V.P.R. fund, and by the Fund for the Promotion of Research at the Technion. S. Gueron acknowledges the support of the Henri Gutwirth Fund for the Promotion of Research. K. Levit-Gurevich acknowledges the support of the Technion Graduate School.

## REFERENCES


1. Machemer H. Ciliary activity and the origin of metachronism in paramecium: effects of increased viscosity. J Exp Biol 1972; 57:239–259.
2. Gheber L, Priel Z. Synchronization between beating cilia. Biophys J 1989; 55:183–191.
3. Sleigh MA. Coordination of the rhythm of beat in some ciliary systems. Int Rev Cytol 1969; 25:31–54.
4. Sleigh MA. Cilia and Flagella. London: Academic Press, 1974.
5. Gueron S, Liron N. Ciliary motion modeling and dynamic multicilia interactions. Biophys J 1992; 63:1045–1058.
6. Gueron S, Liron N. Simulations of three-dimensional ciliary beats and cilia interactions. Biophys J 1993; 65:499–507.
7. Gueron S, Levit-Gurevich K. Computation of the internal forces in cilia: Application to the ciliary motion, the effects of viscosity, and cilia interactions. Biophys J 1997; Submitted.
8. Gueron S, Levit-Gurevich K, Liron N, Blum JJ. Cilia internal mechanism and metachronal coordination as the result of hydrodynamical coupling. Proc Natl Acad Sci USA 1997; 94:6001–6006.
9. Sleigh MA. The Biology of Cilia and Flagella. Oxford: Pergamon Press, 1962.
10. Gray J. Ciliary Movement. Cambridge: Cambridge University Press, 1928.
11. Ermentrout GB. The behavior of rings of coupled oscillators. J Math Biol 1985; 23:55–74.
12. Ermentrout GB, Kopell N. Frequency plateaus in a chain of weakly coupled oscillators i. SIAM J Math Anal 1984; 15:215–237.

# 13

# Metachronal Wave, Ciliary Stroke, and Mucus Propulsion

**Larisa Gheber**
*The Johns Hopkins University, Baltimore, Maryland*

**Zvi Priel**
*Ben-Gurion University of the Negev, Beer-Sheva, Israel*

## I. INTRODUCTION

Mucus-propelling cilia interact with the overlying blanket of secreted mucus to achieve transport of loads over epithelial surfaces [1]. In order to transport mucus, cilia beat in an asymmetric, periodic, and coordinated manner. The form of the ciliary beat and the properties of coordination were studied mainly in water-propelling cilia, since mucus-propelling cilia are shorter and more dense. In addition, their activity is highly fluctuative [2,3]. These are the main reasons for difficulties in the study of mucus-propelling cilia, and there is also a relatively small amount of data describing the characteristics of their metachronal coordination (see below) and the form of the ciliary beat. The results obtained in studies carried out by us in the past few years are summarized in this chapter.

## II. RESULTS AND DISCUSSION

### A. Form of the Ciliary Beat

The ciliary beat consists of a fast effective stroke performed in a plane perpendicular to the cell surface and of a slow recovery stroke that is performed in a clockwise direction (top view) during which the cilium returns to its initial position moving in a plane inclined to the cell surface.

Occasionally a pause exists after the effective stroke. The propulsion of mucus takes place during the fast effective stroke when ciliary tips penetrate the mucus layer and push it in the direction of the effective stroke [4,5]. Due to the direct interaction between cilia and mucus, changes in the ciliary beat form are most likely to influence the properties of mucociliary transport.

Over the years attempts have been made to measure the parameters of the ciliary beat both in water- [6–10] and in mucus-propelling cilia [4,5,11–14]. Most of these studies were performed with a light microscope supplemented by fast cinematography that photographs the ciliary motion at a rate of a few hundred pictures per second. This technique requires a reconstruction of three dimensional motion from a two dimensional picture and requires large magnifications. Using this technique, a general description of the ciliary beat cycle was performed for mucus-propelling cilia from frog oro-pharyngeal [4], cultured rabbit tracheal [5], and human respiratory cells [11,13,14]. In newt lungs, well-defined parameters of the ciliary beat were quantitatively assessed [12]. However, the time consuming frame-by-frame procedure resulted in a small number of cycles for analysis [12].

The photoelectric techniques that are used currently in the studies of ciliary activity are based on the collection and digitization of the light that is reflected by, or transmitted through, an active ciliary epithelium. The photoelectric signals obtained are created by the beating of the cilia and are therefore a summation of the ciliary activity in the observed area. In fact, characteristics of the beating cilia, namely frequency, phase shifts, wavelength, and direction of the metachronal wave (see below), the form of the beat of a single cilium, their length and arrangement are all reflected in the shape of the photoelectric signals. This implies that by using an appropriate approach, the ciliary beat parameters can be extracted from the signals. Because the photoelectric measurements are relatively easy to carry out, the success of this approach will lead to a relatively easy assessment of the ciliary beat parameters. This idea of the reconstruction of the ciliary beat cycle from the photoelectric signal was previously addressed both theoretically [15] and experimentally [16,17]. However, due to the large number of free parameters in this problem, important features of ciliary activity such as the existence of metachronal wave (see below) and the three dimensional ciliary beat were neglected.

## Simulation Approach

We have approached the assessment of the ciliary beat from photoelectric measurement by a simulation and correlation approach. Simulated photoelectric signals are created taking into account a *full set* of parameters defining ciliary activity and geometry. First we have studied how the different parameters of ciliary beat and geometry influence the shape of the simulated photoelectric signal. An array of the simulated signals is then created using a series of beating parameters and compared to the measured photoelectric signals.

The simulations of the signals are based on the generation of a series of time-frozen top-view frames of an active ciliary area and the determination of the amount of light passing through an observation area in each one of these frames. The signals are then generated by taking into account all the factors that may contribute to their shape, namely partial ciliary transmitance of light; three-dimensional ciliary beat, which is composed of recovery, effective, and pause phases; phase distribution on the ciliary surface, and the large number of cilia that contribute to the photoelectric signal. The values of ciliary beat frequency and phase shift between adjacent cilia in two directions are obtained from the three-point simultaneous photoelectric measurement of ciliary activity developed in our laboratory [18]. Using these data and assigning the values of ciliary length and spacing as constant for a given ciliary area, the number of free parameters in the simulations was only four [19].

### Influence of the Ciliary Beat Parameters on the Shape of the Simulated Signal

The influence of the ciliary beat parameters on the photoelectric signal shape was studied by generating a series of signals in which one parameter was varied, while all the other parameters remained constant (Figure 1). The free parameters are Rt, the ratio of the duration of the recovery stroke to the duration of the effective stroke, representing the degree of temporal asymmetry of the beat; $\tau_p$ the duration of the pause phase; $\theta_e$, the angle of the arc swept by the cilium during the effective stroke; $\theta_r$, the angle of the arc projected by the cilium on the cell surface during the effective stroke; and Ra, the degree of spatial asymmetry of the beat, defined as $Ra = \theta_e/\theta_r$. (See more detailed description in reference 19 and 20.) The values of free parameters were chosen to cover the physically reasonable range in ciliary systems [4,5,11,16]. In Figure 1 the variable parameters are indicated on the top of each signal, while the values of the constant parameters are indicated in the figure legend.

As can be seen in Figure 1, all of the ciliary beat parameters influence the shape of the simulated signal. There are some general conclusions that can be made regarding the influence of the ciliary beat parameters on the shape of these signals. For example, an increase in the portion of the cycle occupied by the recovery stroke causes an increase in the duration of

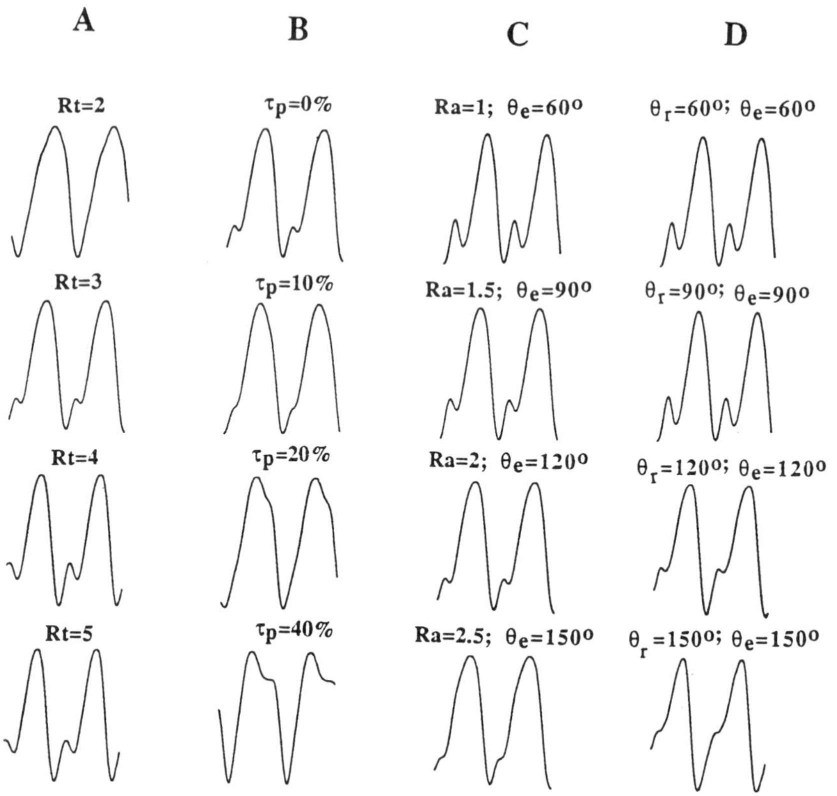


**Figure 1** Series of simulative signals generated with various ciliary beat parameters. In each column (A-D) one ciliary beat parameter was changed, while the other parameters remained constant. For more detailed description of generating the simulated signals see Reference 19. The values of the changing parameters are indicated on the top of each signal. Since the angle of the effective stroke ($\theta_e$) changes as a function of Ra and/or $\theta_r$, its values are also indicated when Ra or $\theta_r$ were changed. The values of the constant parameters are beat frequency = 10 Hz, ciliary length = 6 μm, spacing between cilia = 0.5 μm, Rt = 2, $\tau_p$ = 0%, $\theta_r$ = 60°, Ra = 2, metachronal wavelength = 5.7 μm, wave direction = 110° relative to the effective stroke direction, observation area diameter = 2.5 μm. (Reproduced from Gheber and Priel, 1997, with permission)

the low-intensity part of the signal (the lower part of the signal). During the recovery stroke the cilia are closer to one another and therefore prevent more light from coming through the area of observation. Such a trend can be observed when the temporal asymmetry of the beat is increased and/or the duration of the cycle occupied by the pause is decreased (Figure

1A and B). Both effects result in increasing the portion of the cycle occupied by the recovery stroke. However except for these general observations it is clear that the influence of the ciliary parameters on the signal shape is very complicated and cannot be predicted a priori. This implies that the extraction of the ciliary beat parameters from the measured photoelectric signals cannot be done analytically. However, we have evaluated the parameters of the ciliary beat based on correlation analysis between measured and simulated photoelectric signals.

### Correlation Between Simulated and Measured Photoelectric Signals

Forty-five simulated signals were generated with changing beat parameters in order to calculate their correlation with the measured signals. The fits to signals of 30 cycles, each measured under normal conditions, were studied. The sets of beat parameters of the best-fitting simulated signal to each cycle were averaged. The values of these averages are (mean ± SE) Rt = 3.4 ± 0.3; $\tau_p$ = 15.3% ± 1.5%; Ra = 2.1 ± 0.4; $\theta_e$ = 120° ± 10°. It is accepted that in mucociliary systems, under normal conditions, the effective stroke is performed 2–5 times faster than the recovery stroke, the duration of the pause is not zero, and the angle of the effective stroke is larger than 100° [4,5,11]. The ciliary beat parameters obtained from the best-fitting signals agree well with these values. These findings and the fact that variations in the beat parameters induce a change in the shape of the simulated photoelectric signals (Figure 1), strongly support the assumption that correlation analysis between simulated and measured photoelectric signals can be used in order to measure the parameters of the ciliary beat.

## B. Metachronal Coordination of Mucus-Propelling Cilia

On an active ciliary area there is usually a phase shift between cilia. This phenomenon is responsible for the existence of a metachronal coordination or metachronal wave [1,7,21]. Light microscopy observations have revealed that in mucociliary systems the metachronal wave propagates backwards and to the right relative to the direction of the effective stroke. In other systems (including water-propelling cilia) the direction of wave propagation varies and often differs from that of the effective stroke [22]. The factors that determine the direction of the wave propagation and wavelength in different systems are unknown.

As a result of metachronal coordination cilia join their propulsive efforts to create a continuous movement of mucus. Consequently, characteristics of the metachronal wave determine the number of cilia that interact with mucus instantaneously per given area. The number of cilia actually involved in the transport and the energy transferred by the cilium to the mucus dictate the speed of mucus propulsion. Therefore, metachronal characteristics may play an important role in transport of mucus. Unfortunately, due to technical difficulties the metachronism of mucociliary systems under stimulated or under loaded conditions has not been examined. Moreover, the potential role of metachronism in mucociliary transport has been ignored. Obviously, changes in metachronal wave properties (similar to what was observed for water-propelling cilia) may change considerably the number of cilia that are simultaneously involved in the transport process.

## Theoretical Model

Recently we have proposed an analytical model to explain the dependence of metachronal wave properties on parameters that characterize the ciliary beat, the dimensions of the cilia, and the geometry of their arrangement on the ciliated surface [20]. This model, which connected for the first time the properties of the metachronal wave with the properties of ciliary beat by means of mathematical equations, provides an explanation for the variations in the wave directions in different systems [20]. According to the model, the direction of the recovery stroke (clockwise or counterclockwise) and the type of wave component in the plane of the effective stroke determines within 90° the direction of the wave propagation. Within this range the wave direction is determined by the degree of asymmetry of the beat. The higher the degree of asymmetry, the more perpendicular is the wave direction relative to the effective stroke direction. As the degree of asymmetry of the beat decreases, the wave direction becomes more parallel to the plane of the effective stroke.

The model also explains the behavior of metachronal wavelength as a function of ciliary parameters. The wavelength was suggested to be strongly dependent upon the ciliary length and was expected to be in the range that corresponds to ciliary length of a particular system. For example, in mucus-propelling cilia, the metachronal wavelength is expected to be in the range of 5–8 μm [20]. Other ciliary beat parameters that are expected to influence the metachronal wavelength are the degree of asymmetry in time and space of the ciliary beat, duration of the rest phase, and the angle of the arc circled by the ciliary tip during the effective stroke. On the other hand, the metachronal wave is expected to be only weakly dependent on the ciliary beat frequency. The limited amount of data about ciliary beat and metachronal coordination characteristics in several systems was found to be consistent with the predictions of the model [20]

One of the main predictions of the model is that in all ciliary systems, including mucociliary epithelia, the characteristics of the metachronal wave may be altered by changes in experimental conditions. Although in water-propelling cilia changes in metachronal wave parameters were observed [6–8] there is no data available to indicate that changes in metachronal coordination of mucus-propelling cilia can also occur as a function of time or under various experimental conditions. To address this issue, we have developed a photoelectric method that enabled the measurement of the metachronal wave properties in real time.

## C. Photoelectric Measurement of the Metachronal Properties in Cultured Mucociliary Epithelium

### Normal Conditions

The photoelectric approach to performing a real-time assessment of the metachronal wave in tissue culture is based upon the measurement of phase shifts between beating cilia at a known separation. Since, in mucociliary epithelia, the wave direction is difficult to evaluate, and since it can change in the course of an experiment, measurement of phase shifts in only one direction may result in measurement of the length of the metachronal wave component, rather than the true metachronal wavelength. To avoid this problem a method was developed that is based on simultaneous recording of scattered light from three small active ciliary areas at known separation (Figure 2A and Reference 18). This technique allows the simultaneous measurement of the ciliary beat frequency, degree of correlation, and the phase shift between cilia in two directions: parallel and perpendicular to the effective stroke. As a result, the apparent lengths of two wave components and consequently the metachronal wavelength and direction are assessed (Figure 2B). Since the direction of the effective stroke is easily seen under the microscope, the evaluation of the two wave components and the metachronal wave characteristics becomes straight forward [18,20].

Average values of ciliary beat frequency (A), phase shifts (B), and the Pearson correlation factor (C), obtained by three points simultaneous measurement as a function of the distance between each pair of measured signals, are shown in Figure 3. As can be seen, the phase shift between cilia increases with an increase in the distance between areas of observation in the two directions. However, we have found that the phase shift between cilia in the direction parallel to the effective stroke is considerably smaller than in the direction perpendicular to the effective stroke (Figure 3B). This indicates that the length of the wave component in the direction parallel to the effective stroke is longer. In all the ciliary areas that were

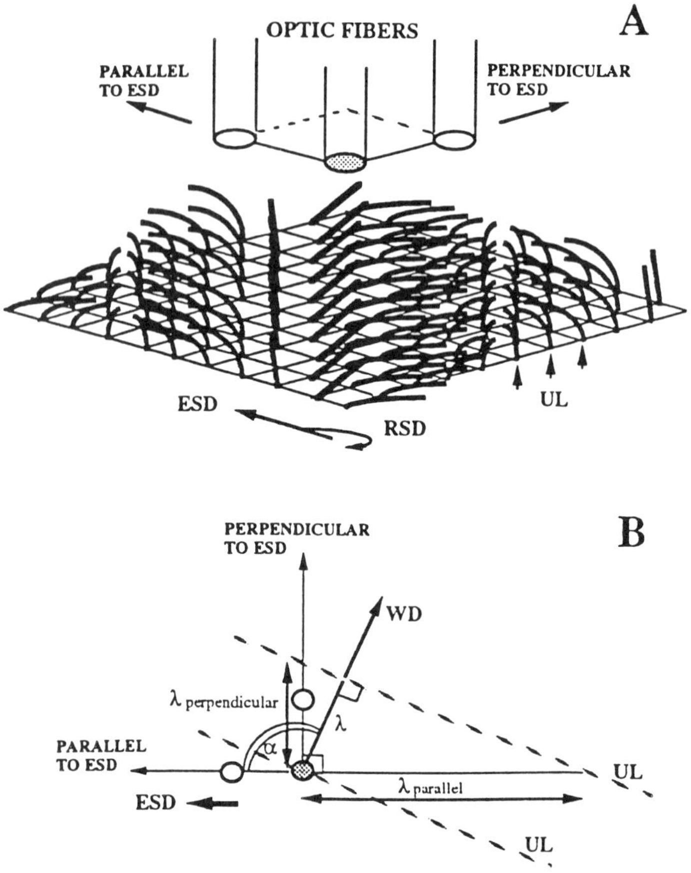


**Figure 2** A demonstration of the three-points measurement and the assessment of the metachronal wave parameters. (A) Three-dimensional representation of the orientation of the fibers above an active ciliary area involved in metachronal coordination. The fibers are represented by the cylinders while the cilia are shown as bold, bent lines. The different shapes of these lines represent the different stages of the cilia during the beat cycle. The lines which connect cilia in the same stage during the beat cycle are the uniphase lines (UL). The directions of the effective stroke direction (ESD) and the clockwise recovery stroke direction (RSD) are shown at the bottom of the panel. The directions of the movement of the two portable fibers, parallel and perpendicular to the effective stroke direction, are indicated by arrows at the upper part of the panel. The phase difference between the beating cilia bring about the metachronal wave that is clearly seen in this schematic representation. (B) A schematic representation (top view) of the metachronal wave parameters of an active ciliary area observed by three optical fibers. The parameters are wavelength ($\lambda$), the angle ($\alpha$) between the wave direction (WD) and the effective

measured under normal conditions the length of the component parallel to the effective stroke was more than twice as long as the wave component perpendicular to the effective stroke. Based on the lengths of the two components, metachronal wavelength and wave direction were calculated using simple trigonometric equations [18,20] Figure 2B. We have found that in cultured frog palate and esophagus the metachronal wavelength is in the range of 5–9 μm, while the wave direction is found to be 125°–90° relative to the effective stroke [18]. These values for metachronal wave characteristics were predicted by the theoretical model [20]. In addition, similar wavelengths were measured using digitized video photography supplemented by a parametric estimation technique [23].

By calculating the Pearson correlation factor between each pair of photoelectric signals, the degree of correlation between beating cilia can be evaluated [24]. This correlation is expressed by a single number in the range between 0–1. The closer the Pearson correlation factor to 1, the better is the correlation between cilia in the two measured areas [24]. Results obtained with the three points simultaneous measurement revealed that at the same separation, the Pearson correlation factor is higher when measured parallel to the direction of the effective stroke (Figure 3C). This indicates that cilia are more coordinated in the direction parallel to the effective stroke, which is also the direction of mucus propulsion. The higher coordination between cilia in the direction of mucus propulsion and the longer wave component in this direction are metachronal coordination features that are likely to contribute to the efficiency of mucus propulsion.

## Viscous Load

The metachronal wave properties have been suggested to be controlled by hydrodynamic interactions between beating cilia [25], implying that

---

stroke direction (ESD), and the lengths of the wave components, λ parallel and λ perpendicular (represented by the double arrows). The optic fibers are shown as three filled circles. Two uniphase lines (UL) in which the cilia are in the same stage during the cell cycle are represented by the dashed lines. The distance between these two lines is, by definition, the true metachronal wavelength (λ), while the wave direction (WD) is perpendicular to these lines. Knowing the lengths of the components parallel and perpendicular to the effective stroke direction (ESD) and according to the geometry of the ciliary area, the metachronal wavelength and wave direction are assessed by

$$\lambda = \lambda_{parallel} \cdot \sin [\text{acrtan} (\lambda_{perpendicular}/\lambda_{parallel})$$

$$\alpha = 90° + \arctan(\lambda_{perpendicular}/\lambda_{parallel})$$

(Reproduced from Gheber and Priel, 1994, with permission)

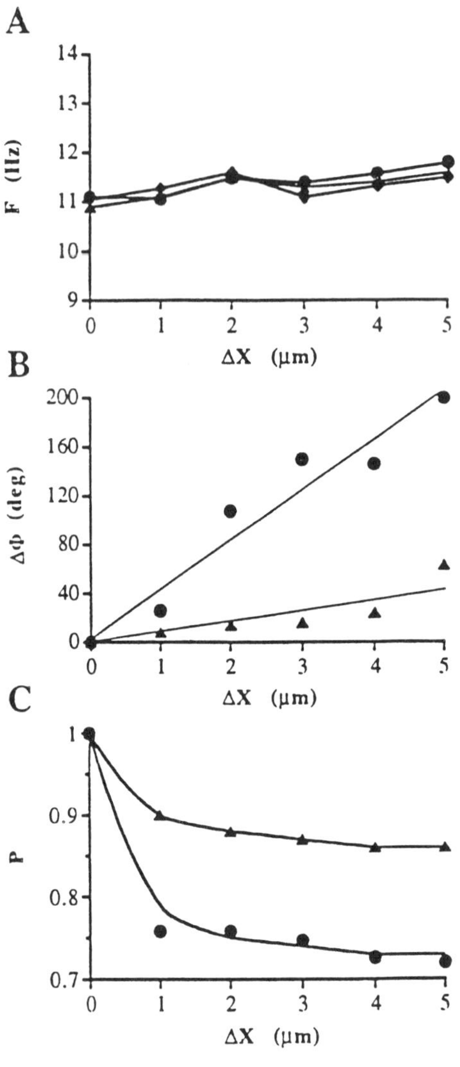


**Figure 3** Average frequency, phase shift, and Pearson correlation factor as a function of the distance between the observed ciliary areas, parallel and perpendicular to the effective stroke direction. In all the panels each point represent an average over 40 consequent seconds. (A) The ciliary beat frequency (F) of the three measured areas on an active ciliary domain. The frequency was assessed from the FFT of the photoelectric signals obtained from the fixed fiber (●) and from the fibers moved parallel (▲) and perpendicular (◆) to the effective stroke direction. The maximal error of these measurements is up to 10%. (B) The phase shift (ΔΦ) between two small ciliary areas parallel (▲) and perpendicular (•) to

the form and timing of the ciliary beat determine the properties of ciliary coordination. This hydrodynamic coupling hypothesis relies on experiments examining the effect of increasing viscosity on metachronal wave properties in water-propelling cilia [7,26]. Since Mucus-propelling cilia are considerably shorter and believed to be less coordinated than water-propelling cilia [1,3,4], it was interesting to determine whether the hydrodynamic hypothesis also applies to mucus-propelling cilia.

We have found that with increased viscosity (20–1500 cp), the wave component perpendicular to the effective stroke direction increased, reflecting both an increase in metachronal wavelength [3] and a change in wave direction towards the plane of the effective stroke [3], which results in more orthoplectic metachronism. In water-propelling cilia metachronal wavelength increased [7,26] and metachronal coordination also became orthoplectic [7] as viscosity increased. Although these changes were obtained at lower viscosities than those employed in our present study on mucus-propelling cilia, they are in the same direction as the present results.

Changes in metachronal wave properties may be due to changes in coupling conditions of the medium and/or to changes in the shape of the ciliary beat. In most ciliary systems the effective stroke is considerably faster than the recovery stroke, which results in a relatively high temporal asymmetry of the beat [7,8]. According to a theoretical model that connects the properties of the metachronal wave with the characteristics of ciliary beat, the only beat parameter that can induce both an increase in metachronal wavelength and more orthoplectic metachronism is a decrease in the degree of temporal asymmetry of the beat [20], meaning that the effective stroke occupies a larger portion of the cycle. This implies that increasing the viscosity of the medium induces change in the beat parameters, decreasing the degree of temporal asymmetry. Indeed, in water-propelling cilia of *Paramecium* a decrease in temporal asymmetry of the beat was observed under a viscous load of up to 40 cp [7], an observation that strongly supports this conclusion. The intraciliary control

---

the effective stroke direction. The phase shifts were assessed by maximization of the correlation between the photoelectric signals. The maximal error of the average phase shift measurement was up to 20% mainly due to the phase fluctuations on the time scale of seconds. (C) Normalized Pearson correlation factor (P) parallel (▲) and perpendicular (•) to the effective stroke direction. At each separation the values of the Pearson factor were normalized to the values obtained at the distance of 0 μm. The absolute values of the Pearson factor at this separation were 0.97 and 0.95 for the parallel and perpendicular directions respectively. The maximal error was up to 5%. (Reproduced from Gheber and Priel, 1994, with permission.)

mechanism of the three-dimensional ciliary beat form is unknown. Therefore, at this stage, the mechanism by which increasing viscosity of the medium induces reduction of the temporal asymmetry of the beating is still unclear. However, it is reasonable to assume that increased viscosity would slow the fast effective stroke, where the cilium is projecting nearly straight out into the medium, more than the slow recovery stroke, where the cilium is bent and therefore closer to the cell surface.

We have found that mucus-propelling cilia in cultured frog palate and esophagus are active under a quite large viscous load. This result indicates that these cilia are well adapted to their physiological role of activity under load. However, the fact that similar changes in metachronal wave coordination are induced by applying viscous load on both water- and mucus-propelling cilia, indicates that the basic forces governing the properties of the metachronal wave in both systems are similar.

### Stimulation of Ciliary Activity

Mucus-propelling cilia can respond to a large spectrum of stimuli by a rapid and a dramatic increase in the ciliary beat frequency. We have studied the effect of stimulation on metachronal coordination by two extracellular agents, acetylcholine and ATP. Within a few seconds both ATP and acetylcholine caused a dramatic enhancement of the ciliary beat frequency. The frequency enhancement caused by ATP was more pronounced (~ threefold) and longer lasting than the effect of acetylcholine (~ twofold) [18].

The degree of synchronization as reflected in the value of the Pearson correlation factor varied slightly between ciliary areas [18]. However, regardless of the degree of synchronization of a particular ciliary area, the cilia parallel to the effective stroke direction were more synchronized than the cilia at the same separation perpendicular to the effective stroke direction (Figure 3C). In most cases enhancement of the ciliary beat frequency did not change this general trend.

In 80% of the cases, in which the ciliary beat frequency was increased by extracellular agents, there was an increase in the length of the component parallel to the effective stroke direction. The degree of this enhancement varied from two- to tenfold and was independent of the nature of the stimulant. On the other hand, the apparent wavelength of the perpendicular component remained almost unchanged [18]. The velocity of the wave propagation, being the product of the wavelength and the ciliary beat frequency, was also dramatically elevated in the direction parallel to the effective stroke direction. This enhancement was quite large, 6–30 fold. The response of the wave components to stimulation was immediately translated to a change in the direction of wave propagation. As the component parallel to the effective stroke direction increased while the perpendicular component remained unchanged, the obtuse angle between the metachronal wave and effective stroke direction decreased, approaching 90°. According to accepted terminology, the type of metachronal coordination became more diaplectic. Typically, the change in the wave direction was in the range of 10°–20°. The same phenomenon was also observed when the ciliary beat frequency was increased spontaneously or by other stimulants. It therefore seems that an immediate rise in ciliary beat frequency is often accompanied by an increment of the component parallel to effective stroke direction resulting in a more diaplectic type of metachronism.

Unlike the direction of wave propagation, the metachronal wavelength was found to be independent of ciliary stimulation [18]. The values of the wavelength were similar to those of the ciliary length in these systems [1,7]. This also agrees with the predictions of our model of a strong dependence of the metachronal wavelength on the ciliary length [20]. The variations that occur within this range may be the result of variations in the beat parameters such as the portion of the cycle occupied by the pause, the degree of asymmetry at the time of the beat, and the angle of the arc projected on the cell surface during the recovery stroke. This implies that in ciliary cells from cultured frog esophagus epithelium the parameters of the ciliary beat vary in time and that this causes time-dependent fluctuations of the metachronal wavelength within the range of ± 30%.

The initial lengths of the two metachronal wave components in our system provide an explanation for the response of the metachronal wave characteristic to ciliary stimulation by extracellular ATP or acetylcholine. While the dramatic rise in the frequency was accompanied by a considerable increase in the length of the component parallel to the effective stroke direction, the perpendicular component was not affected. Since prior to stimulation, the parallel component was already quite long as compared to the perpendicular component, no further influence on the metachronal wavelength is expected by an additional increase of the component parallel to the effective stroke direction. On the other hand, the direction of the wave propagation is more sensitive to this change and it approaches 90° as the ratio between the lengths of the parallel and perpendicular components increases.

The basic assumption that the metachronal wave characteristics are influenced by the ciliary beat parameters [7,20] supports the suggestion that the change in the wave direction as a response to ciliary stimulation indicates that some of the features of the ciliary beat were altered. Our model predicts that a rise in the ciliary beat frequency alone cannot explain the change in metachronal coordination. On the other hand, an increase in the asymmetry of the ciliary beat, in time and in space, will lead to a more diaplectic type of metachronism [20]. It is therefore possible that the degree of beat asymmetry increases as a result of ciliary stimulation by extracellular ATP or acetylcholine.

## III. CONCLUDING REMARKS

In mucociliary systems ciliary beat frequency is a relatively easily measured parameter. This is probably why it is almost the only parameter that has been studied in these systems. However, it is certainly not the only parameter that is related to mucus propulsion. The form of the ciliary beat and the characteristics of metachronal coordination are features that also determine the extent of mucociliary interaction and coupling and therefore are expected to contribute to mucociliary clearance.

We have demonstrated that photoelectric methods can be used to assess in real time metachronal coordination and the form of the ciliary beat. Although photoelectric assessment of ciliary beat form requires further development, our preliminary results indicate that when cilia are stimulated by extracellular ATP, the angle swept by the cilium during the effective stroke increases, which increases the velocity of the ciliary tip during the effective stroke (Gheber and Priel, unpublished results). Since mucus is transported by cilia performing effective strokes, this effect is very likely to cause more rapid propulsion of mucus.

We have shown that when the cilia are stimulated, the increase in the ciliary beat frequency is accompanied by a dramatic increase in the wave component parallel to the effective stroke direction, while metachronal wavelength remains unchanged. We have also shown that the metachronal wavelength is increased by loading with a highly viscous medium. An increase in the number of cilia involved simultaneously in cooperative pushing of mucus can be achieved by two mechanisms. One mechanism is to increase the number of cilia involved in propulsion per given row (rows are defined parallel to the effective stroke direction) by causing a decrease in the delay time between cilia parallel to the effective stroke direction. The second mechanism is to increase the number of rows involved in the synchronous effort which is achieved by an increase in metachronal wavelength. Since the first effect is observed when cilia are stimulated and the second is observed when the load is increased, we conclude that the net result of these changes in metachronal properties may result in a considerable increase in the number of cilia that are simultaneously involved in the transport process per given area. This change is expected to increase the efficiency of mucus propulsion and may therefore reflect the physiological importance of the metachronal wave to the function of mucociliary systems.

Based on these findings and our previous publications, it is clear that the properties of the metachronal wave and probably those of the ciliary beat form change under stimulating and loading conditions in ways that would improve the efficiency of mucus propulsion. It is our belief that more effort should be invested in investigating the behavior of metachronal coordination and form of the ciliary beat in real time in other mucociliary systems. This information would establish whether the described phenomena represent general features of mucus-propelling cilia.

## ACKNOWLEDGMENT

This work was supported by a grant from the Israel Science Foundation.

## REFERENCES


1. Sleigh MA, Blake JR, Liron N. The propulsion of mucus by cilia. Am Rev Respir Dis 1988; 137:726–741.
2. Eshel D, Grossman Y, Priel Z. Spectral characterization of ciliary beating: variations of frequency with time. Am J Physiol 1985; 249:C160–C165.
3. Gheber L, Priel Z. Ciliary activity under normal conditions and under viscous load. Biorheology 1990; 27:547–557.
4. Aiello E, Sleigh M. Ciliary function of the frog oro-pharyngeal epithelium. Cell Tissue Res 1977; 178:267–278.
5. Sanderson MJ, Sleigh MA. Ciliary activity of cultured rabbit tracheal epithelium: beat pattern and metachrony. J Cell Sci 1981; 47:331–347.
6. Aiello E, Sleigh MA. The metachronal wave of lateral cilia of *Mytilus edulis*. J Cell Bio 1972; 54:493–506.
7. Machemer H. Ciliary activity and the origin of metachrony in *Paramecium*: effects of increased viscosity. J Exp Biol 1972; 57:239–259.
8. Machemer H, Eckert R. Electrophysiological control of reversed ciliary beating in *Paramecium*. J Gen Physiol 1973; 61:572–587.
9. Mogami Y, Pernberg J, Machemer H. Ciliary beating in three dimensions: steps of a quantitative description. J Math Biol 1992; 30:215–249.
10. Teunis PF, Machemer H. Analysis of three-dimensional ciliary beating by means of high-speed stereomicroscopy. Biophys J 1994; 67:381–394.
11. Marino MR, Aiello E. Cinemicrographic analysis of beat dynamics of human respiratory cilia. Prog Clin Biol Res 1982; 80:35–39.
12. Weaver A, Hard R. Newt lung ciliated cell models: effect of MgATP on beat frequency and waveforms. Cell Motil 1985; 5:377–392.

13. Rautiainen M, Matsune S, Shima S, Sakamoto K, Hanamure Y, Ohyama M. Ciliary beat of cultured human respiratory cells studied with differential interference microscope and high speed video system. Acta Otolaryngol (Stockh) 1992; 112:845–851.
14. Rautiainen M, Matsune S, Yoshitsugu M, Ohyama M. Degeneration of human respiratory cell ciliary beat in monolayer cell cultures. Eur Arch Otorhinolaryngol 1993; 250:97–100.
15. Rabinovitch A, Rabinovitch R. A simple model of light transmission through metachronally moving cilia. J Appl Phys 1990; 67:1108–1112.
16. Sanderson MJ, Dirksen ER. A versatile and quantitative computer-assisted photoelectronic technique used for the analysis of ciliary beat cycles. Cell Motil 1985; 5:267–292.
17. Lansley AB, Sanderson MJ, Dirksen ER. Control of the beat cycle of respiratory tract cilia by Ca2+ and cAMP. Am J Physiol 1992; 263:L232–L242.
18. Gheber L, Priel Z. Metachronal activity of cultured mucociliary epithelium under normal and stimulated conditions. Cell Motil Cytoskeleton 1994; 28: 333–345.
19. Gheber L, Priel Z. Extraction of cilium beat parameters by the combined application of photoelectric measurements and computer simulation. Biophys J 1997; 72:449–462.
20. Gheber L, Priel Z. On metachronism in ciliary systems: a model describing the dependence of the metachronal wave properties on the intrinsic ciliary parameters. Cell Motil Cytoskeleton 1990; 16:167–181.
21. Sleigh MA. Ciliary function in transport of mucus. Eur J Respir Dis Suppl 1983; 128:287–292.
22. Knight-Jones EW. Relations between metachronism and the direction of the ciliary beat. Q J Microsc Sci 1954; 95:503–521.
23. Ben-Shimol Y, Dinstein I, Meisels A, Priel Z. Ciliary motion features from digitized video photography. J Comp-Assist Microsc 1991; 3:103–116.
24. Gheber L, Priel Z. Synchronization between beating cilia. Biophys J 1989; 55:183–191.
25. Gray J. The mechanism of ciliary movement. VI Photographic and stroboscopic analysis of ciliary movement. Proc R Soc Lond (Biol) 1930; 107: 313–332.
26. Aiello E. Factors affecting ciliary activity of gill of the mussel *Mytilus edulis*. Physiol Zool 1960; 33:120–135.

# 14

# Roles of Cilia in Filter Feeding and Mucous Transport

**John R. Blake**
*The University of Birmingham, Birmingham, United Kingdom*

**Michael A. Sleigh**
*University of Southampton, Southampton, United Kingdom*

**Nadav Liron**
*Technion–Israel Institute of Technology, Haifa, Israel*

## I. INTRODUCTION

It seems certain that cilia were first used by unicellular organisms to enable them to swim. Subsequently cilia have been adapted in a great number of ways, notably for locomotion by swimming in water or gliding over surfaces and for the collection and transport of particles of food or unwanted debris. Three classes of roles for cilia can be recognized when these functions are being performed:

1. the propulsion of water,
2. the manipulation of particles, and
3. relative motion between a ciliated surface and a solid or visco-elastic (mucus) surface.

In some cases the function involves only one of these roles, as in the swimming of ciliated protozoan cells or comb jellies (only water propulsion): In the gliding of flatworms, the cleansing of respiratory surfaces in mammals, or the collection of particles from surfaces (the ciliated surface interacting with surface and/or mucus). Where organisms are using cilia to collect food particles that are suspended in water by a form of filter feeding, cilia perform two or three different roles. Cilia propel water to bring particles within reach of cilia that can intercept them and then further manipulate the particles to pass directly to a mouth opening or to bring particles into contact with mucus that other cilia convey along with the entrapped particles to the mouth for ingestion. Water propulsion and particle interception and manipulation may be performed by the same or different cilia in such filter feeding organisms; the transport of mucus is always performed by specialized cilia.

## II. BASIC FLUID MECHANICS

The three roles of water propulsion, filtration of particles, and propulsion of mucus (or a solid surface) require quite different types of interaction between the cilia and the surrounding fluids and exploit different features of low-Reynolds-number fluid dynamics. We shall consider first the requirements of cilia performing these functions in relation to fluid dynamic principles, and then demonstrate the application of these by considering a small number of examples.

### A. Beat Pattern Characteristics

Cilia propel fluids because the cyclical beating patterns are asymmetric. There are two components of the beat cycle—a fast, almost pendular rigid effective stroke extends from the cell surface whereas, in the recovery stroke, a wave is propagated along the length of the cilium from base to tip, keeping the cilium close to the epithelium until it reaches the start of the next effective stroke. The fluid mechanics of the ciliary beat cycle are now well understood. The force on two identical needle-like bodies at the same velocity of translation is almost twice as much for one moving perpendicular to its axis versus parallel to its axis. It would therefore be highly desirable to have a perpendicular mode of motion during the effective stroke and a tangential mode during the recovery stroke. The force is linearly dependent on the velocity; thus, the faster motion in the effective strokes leads to a greater force being exerted on the fluid. The epithelial surface may be exploited to differentiate fluid motion between the effective and recovery strokes; the more outstretched the cilium, the greater the volume of fluid under its influence.

### B. Filtration of Particles from Water and the Role of Eddies

The filtration of particles at very low Reynolds number is not well understood because there is no capture through inertial impact. More sophisticated behaviour is required that is in part based on the thickness, spacing, motion, and design of the ciliary organelle (e.g., rakes, paddles, cirri; see [1]). One approach may be through the action of the cilia to compress a volume, thus squeezing water through the filter. Others may exploit the prevalence of eddies to bring food particles close to a surface so that electrostatic or other short-range forces may take over. The coordination of the different cilia may play an important part in particle transfer.

### C. Mucociliary Transport and Mucus Rheology

The fluid mechanical principles of mucociliary transport revolve around the characteristics of several length scales namely (a) molecular length scales associated with the biochemical structure of mucus, (b) length scales associated with the cilium tip (~ 0.1 μm), (c) length scales associated with the cilium length, cell size, metachronal wave length, and mucus plaque (5–50 μm), and (d) length scales associated with the length of an airway (5–10 mm). The key length scale with regard to cilium-mucus interaction is the coupling between the cilium and the entangled network of molecules [2]. Does the cilium "see" a continuous viscous liquid or a loosely coiled network of molecules? For the shorter cilia in a mucus-propelling environment, there is not such a distinction between effective and recovery strokes so another mechanism for transport is required. Mucus behaves like an elastic structure on the macro scale but has a more liquid like behaviour on the microscale of the cilium tip [3]. The viscoelastic properties of mucus occur because of the coiled macromolecular network that provides the consistency and integrity for effective transport. Mucus rheology also plays a key role in muco-ciliary transport with the properties of the mucus determining the effective functioning of the clearance mechanism. Theoretical models can provide useful information on the mechanical principles of the beat pattern as well as providing reliable estimates of the transport rate.

## III. KINEMATICS

While the above discusses our current understanding of the basic fluid mechanics—the kinematics of beat shape—coordination and location also play an important role in determining the efficiency of a particular function. At present we only have a qualitative understanding of the links between beat frequency, cilia length, beat amplitude, changes in beat shape, position of cilium on surface, and relationship with other cilia. The metachronal wave is one easily observable outcome allowing a continuous fluid movement over the epithelium.

## IV. FILTER FEEDING

### A. General Observations

Many animals collect food particles from suspension in water. Some animals merely hold out some form of trap or net and wait for prey to be caught, but we are interested here in the active suspension feeders. There are four stages in the active process:

1. creation of a current of water bearing particles,
2. separation of particles from the water,
3. transfer of the particles to the mouth, and
4. ingestion.

As the suspension feeding involves cilia, stages 1, 2, and 3 may all depend upon cilia, with the participation of mucus in stage 3 in many cases. There are two main ways to separate particles from the water. The more obvious way is to mechanically filter particles from the water by some sort of mesh (of cilia or occasionally mucus) that intercepts particles that do not pass through the holes in the mesh. A second method referred to as "scan and trap" [4,5] involves the predator detecting the presence of a prey item in the water flow and altering its behaviour so as to isolate a small parcel of fluid containing the particle and transfer it to the stream of food being taken to the mouth.

Filter feeding typically employs bands of cilia that beat perpendicular to the long axis of the band to propel water across the surface. Particles may be separated from the water on the upstream side of the band, with or without the participation of other cilia, or on the downstream side of the band, in a system that involves a second set of cilia. Then the cilia used in collection and capture of food may themselves move the food particles to the mouth, or the particles may be passed to a mucociliary epithelium for transfer to the mouth. Examples of these patterns are described below in more detail to emphasise features of fluid dynamic interest. Qualitative descriptions of the scan and trap type have been given [6,7], but quantitative details are not available, and these types are not considered further.

## Vorticella

This stalked ciliate protozoon filters particles using an array of cilia held at the downstream side of a set of water-pumping cilia [8]. Metachronal waves are propagated along the upper, close-set double row of water-propelling (polykinety) cilia towards the cell mouth. The cilia move through about 90° in their effective stroke, which is directed obliquely downwards towards the lower row of filtering (haplokinety) cilia. The filtering cilia are spaced less than 1 μm apart and appear to be held out stiffly from the body, perpendicular to the flow of water driven by the polykinety cilia. The haplokinety cilia perform only a very low-amplitude beat cycle, approximately in phase with the water-propelling ones. When the polykinety cilia beat as an undulating sheet at frequencies of 30–50 Hz, a water current is created. According to Fenchel [9], the main part of the flow that is created travels along the ciliary tracts towards the mouth between the polykinety and the haplokinety, and progressively the water escapes through the haplokinety filter. Bacteria-sized particles in this water flow are retained by the filtering cilia, although the filter has an open margin. It is thought that the action of the polykinety cilia is to press water through the filter rather than permitting it or the particles to escape at the free margin. Particles are retained mechanically by the filter, without any participation by mucus, and are bounced along between the bands of cilia to the mouth by a succession of metachronal waves. Vorticella might also exploit the facility to alter the length of the stalk, allowing a more efficient filtering process of the surrounding fluid. Recently, Blake and Otto [10] have termed this behaviour chaotic filtration.

## Stentor and Euplotes

Spirotrich ciliates such as Euplotes and Stentor feed with the aid of a single row of compound cilia (membranelles) that collect particles on the upstream face of the water-propelling cilia. Metachronal waves pass along the row of membranelles, starting from the mouth at one end. Each membranelle is formed of 50 or more cilia, arranged in 2–3 rows, all of which move together as a single plate. The plane of this plate is perpendicular to the row of membranelles. In Stentor each membranelle twists slightly as it performs an effective stroke that is nearly perpendicular to the row of membranelles and then moves in an anticlockwise loop during its recovery stroke. Although the membranelle appears to move as a single paddle, it can be seen in flash photographs that the individual cilia spread apart in the distal part of the plate [11]. The beating of membranelles, at rates of 30 Hz or more, draws a current of water across the edge of the cell, where the row stands, at speeds of the order of 1 mm $s^{-1}$ [12], and particles are caught against the row of membranelles on the upstream side. Because the current is somewhat oblique to the row of membranelles, particles mechanically filtered from the water flow are swept along the row to the mouth.

Investigations of the movements of membranelles of Euplotes by Fenchel [9] suggest that their main motion is in the plane of the row of membranelles, and that by twisting during their to-and-fro movement the membranelles propel water through the row by a turbine-like action. Particles larger than the spacing between the membranelles (2 μm at the base) are caught at the upstream side of the row and because of the mouthward component of the overall water flow are carried along to the mouth. It is not clear that this story is fully complete, because the membranelles of Euplotes appear to have a more complex pattern of beat than Fenchel describes and Wilks and Sleigh (submitted) have found that particles as small as 0.6 μm are retained quite efficiently by Euplotes mutabilis although higher retention is shown for particles 2–3 μm in diameter. A high proportion of the particles must be lost at the free edge of the membranelle band in types like Stentor and Euplotes, not only because water spills over the margin of the band but also because the actual filter elements propel the water current.

## Mytilus

This bivalve mollusc uses an array of parallel gill filaments to collect food particles on a filter placed on the upstream side of the ciliary pump. Each filament bears three types of cilia that perform different functions. Cilia of the lateral bands are about 15 μm long and beat in prominent metachronal waves at a frequency of 15–20 Hz [13]. These cilia act as a pump that propels water through the interfilament canals at 1–2 mm $s,^{-1}$ see [12,14,15] and references therein.

Laterofrontal cirri extend from either side of each filament across the entrance to the interfilament canal. They are spaced at intervals of about 3 μm, and each of them is composed of 40–50 cilia closely packed in two rows that lie perpendicular to the filament. The component cilia are of different lengths, with the shortest ones at the frontal side and those nearest the lateral cilia being 20–30 μm in length. Successive component cilia of each row bend laterally in a feather-like array, with a perpendicular distance between the component "branches" of about 1 μm [16]. The lateral branches of alternate cirri are of such a length that they can form an almost complete sieve if these cirri lie across the entrance to the interfilament canal. The beat of these laterofrontal cirri is coordinated in two sets: alternate cirri beat approximately in phase, and the beat of adjacent cirri is therefore about 180° out of phase. These cirri perform a slow effective stroke in which they bend across the frontal face of their filament and then move more quickly back to a rest position across the interfilament canal, remaining more or less still while the cirri of the other set perform their beat cycle. When laterofrontal cirri behave in this way, they form a mechanical filter with a mesh of about 1 μm across the water flow. This filtration function is consistent with the finding that Mytilus does not retain particles less than 1 μm in diameter, but retains 50% of 1.5 μm particles. It is assumed therefore that the laterofrontal cirri intercept larger particles from the flow. Most of these particles are then carried to the frontal face of the gill as the cirri perform their effective strokes—with the particles possibly being concentrated as water is squeezed through the filter as the cirri press down towards the frontal face of the filament during the later stages of the effective stroke.

A broad tract of short cilia occupies the frontal face of each filament between the rows of laterofrontal cirri. These cilia beat along the long axis of the filament. They show characteristics typical of mucus-propelling cilia, such as weak coordination and short length (5–7 μm). Mucus-secreting goblet cells occur among the frontal ciliated cells, and mucus released from these goblet cells is propelled along the frontal face of the filament by these cilia, taking with it any particles that become entangled with the mucus.

There is a school of thought, whose ideas have been described in detail by Jörgensen [17], which suggests that particles are moved from the main flow between the filaments into a current along the frontal ciliary tract by hydrodynamic means that do not involve either direct filtration by the laterofrontal cirri or the participation of mucus in the trapping and transport of particles along the frontal face of the filament. The evidence supporting these alternative ideas is not easy to evaluate, and opinions are still sharply divided, although the fluid dynamical theory of Blake and Fulford [18] does provide an explanation for some of Jörgenson's observations.

The beat of both lateral and laterofrontal cilia is modified by changes in concentration of the hormone serotonin (5-hydroxytryptamine or 5-HT). With a higher concentration of $10^{-5}$. *M* serotonin the laterofrontal cirri remain stationary at the end of their effective stroke, bent over the frontal cilia. Under these conditions the lateral cilia beat at a frequency of 19 Hz and propel water around the ciliary tips at up to 1.9 mm $s^{-1}$, generating a mean flow through the interfilament canal of 1.7 mm $s^{-1}$ [14]. The same authors reported that with serotonin at $10^{-6}$ *M* the laterofrontal cirri beat actively in the manner described above and the lateral cilia beat at only 16 Hz. The water flow under these conditions was not given, but both the resistance of the laterofrontal filter and the lower frequency of beat of the lateral pumping cilia must mean that it is reduced.

## REFERENCES


1. Cheer AYL, Koehl MAR. Paddles and rakes: Fluid flow through bristled appendages of small organisms. J Theor Biol 1987; 129:17–39.
2. Sleigh MA, Blake JR, Liron N. The propulsion of mucus by cilia. Am Rev Resp Dis 1988; 137:726–741.
3. Silberberg A. Biorheological matching: muco-ciliary interaction and epithelial clearance. Biorheology 1983; 20:215–222.
4. Koehl MAR, Strickler JR. Copepod feeding currents: food capture at low Reynolds number. Limnol Oceanogr 1981; 26:1062–1093.
5. LaBarbera M. Feeding currents and particle capture mechanisms in suspension feeding animals. Amer Zool 1984; 24:71–84.
6. Strathmann RR, Jahn TL, Fonseca JRC. Suspension feeding by marine invertebrate larvae: clearance of particles by ciliated bands of a rotifer, pluteus and trochophore. Biol Bull Mar Biol Sta Woods Hole 1972; 142:505–519.
7. Nielsen C. Strict and function of metazoan ciliary bands and their phylogenetic significance. Acta Zoologica (Stockh) 1987; 68:205–262.
8. Sleigh MA, Barlow D. Collection of food by *Vorticella*. Trans Amer Microsc Soc 1976; 95:482–486.
9. Fenchel T. Protozoan filter feeding. Prog Protistol 1986; 1:65–113.
10. Blake JR, Otto SR. Ciliary propulsion, chaotic filtration and a 'blinking' stokeslet. J Eng Math 1996; 30:151–168.
11. Blake JR, Sleigh MA. Hydro–mechanical aspects of ciliary propulsion. In: Wu TY, Brokaw CJ, Brennen C. Plenum, eds. Swimming and Flying in Nature. New York: 1975:185–210.
12. Sleigh MA, Aiello E. The movement of water by cilia. Acta Protozool 1972; 1:265–277.
13. Aiello E, Sleigh MA. The metachronal wave of lateral cilia of *Mytilus edulis*. J Cell Biol 1972; 54:493–506.
14. Nielsen NF, Larsen PS, Riisgard HU, Jörgensen CB. Fluid motion and particle retention in the gill of *Mytilus edulis*: video recordings and numerical modelling. Marine Biology 1993; 116:61–71.
15. Silvester NR. Hydrodynamics of flow in *Mytilus* gills. J Exp Mar Biol Ecol 1988; 120:171–182.
16. Silvester NR, Sleigh MA. Hydrodynamic aspects of particle capture by *Mytilus*. J Mar Biol Ass UK 1984; 64:859–879.
17. Jörgensen CB. Bivalve filter feeding: hydrodynamics, bioenergetics, physiology and ecology. Fredensborg, Denmark: Olsen and Olsen, 1990.
18. Blake JR, Fulford GR. Hydrodynamics of filter feeding. In: Ellington CP, Pedley TJ, eds. Biological Fluid Dynamics, 1995:183–197.

# 15
# Digital Microrheometry

**John Hamilton**
*University of Liverpool, Liverpool, United Kingdom*

## I. INTRODUCTION

The volume of mucus secreted by many organs is so small that it is often extremely difficult to obtain samples adequate for study. A rheometer that can analyse minute volumes of mucus is of obvious value. Such an instrument was developed at the University of Pennsylvania over a quarter of a century ago [1]. The rheometer was initially used in creep mode. Fourier transformation of the resulting material provided rheological data as a function of stress frequency. The method was laborious and provided little useful information beyond very low frequencies. The same group modified the instrument to deliver a variable simple harmonic magnetic driving force [2]. The resulting oscillating sphere microrheometer, although modified, remains the prime instrument for the rheological analysis of microlitre quantities of mucus [3,4].

## II. PRINCIPLE

The scheme of the microrheometer is shown in Figure 1. The instrument may be considered to consist of four main parts: the magnetic field generator, the well for the mucus and magnetic probe, the optical tracking system, and the computer. Each of these contributes to the accuracy of the rheometer.

A combination of alternating and direct currents generates a magnetic field that varies in simple harmonic fashion. A magnetic sphere is placed in the field and moves under its influence. The sphere is immersed

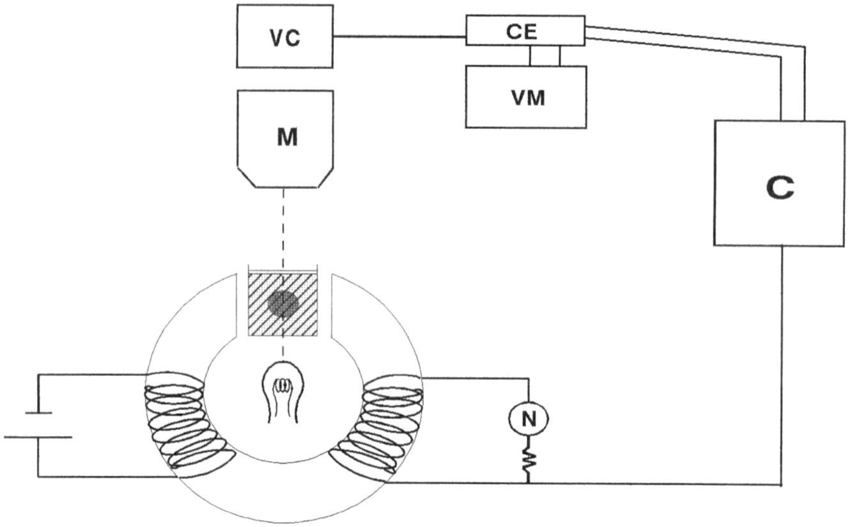


**Figure 1** Scheme of the digital microrheometer: M, microscope; VC, video camera; CE, contrast enhancer; VM, monitor; C, computer.

in the test volume of mucus. The mucus retards the motion of the sphere and this effect is revealed by comparison of the driving force and the resultant movement of the sphere. Assuming that the behaviour of the mucus in this system is linear, this impedance can be expressed in terms of the dynamic viscoelastic properties of the mucus.

## III. THEORY

The oscillating sphere microrheometer uses a magnetic driving force that varies in simple harmonic fashion. For a particle that is assumed to be sufficiently small as to not distort a magnetic field, the force per unit volume on the particle is determined by the field strength alone. This in turn is proportional to the current passing through the solenoids. If the direct current and sphere size are held constant for all experiments, they can be incorporated into the proportionality constant:

$$F_{\rm m} = cI_{\rm a}$$

where $F_{\rm m}$ is the magnetic force, $I_{\rm a}$ is the alternating current, and $c$ is the constant of proportionality.

The motion of the sphere is retarded by drag and inertial forces. For steady flow the drag force is given by Stokes' Law. For a force applied in a simple harmonic fashion to a linear system, the equation of motion is described by a Fourier transform of Stokes' equation (assuming that the sphere is sufficiently small and accelerated sufficiently slowly that inertial effects may be ignored) [1]:

$$F_0 = 6\pi r \eta^* \omega X_0 j e^{jd}$$

where $\eta^*$ is the complex dynamic viscosity, $F_0$ is the amplitude of the magnetic force, $X_0$ is the amplitude of probe displacement, $d$ is the phase angle in radians by which the simple harmonic movement of the sphere lags the magnetic force, $r$ is the radius of the sphere, $\omega$ is the angular frequency in rad $\sec^{-1}$, and $j = \sqrt{-1}$.

Calculation of the dynamic viscoelastic properties of the mucus is thus dependent on $F_0$, $X_0$, and $d$. Although $F_0$ cannot be measured directly, it is a function of the alternating current. The unknown constant relating the current to the magnetic force can be eliminated by comparison of the behaviour of the mucus sample with that of a nonelastic calibrating fluid of known viscosity. The phase lag of such a fluid is $\frac{\pi}{2}$ radians. Thus

$$F_N = 6\pi r \eta_N \omega_N X_N = c I_N$$

where the subscript N refers to values pertinent to the Newtonian fluid

Comparing the equations for the nonelastic calibrating fluid and the mucus:

$$\eta^* = (I_m/I_N)(\omega_N/\omega_m)(X_N/X_m)\eta_N j e^{jd}$$

where the subscript m refers to values pertinent to the mucus sample.

The complex dynamic viscosity may be expressed in terms of real and imaginary components:

$$\eta^* = \eta' - j\eta''$$

The component of the complex dynamic viscosity in phase with the velocity of the sphere, $\eta'$, is a measure of the viscous behaviour of the mucus and is termed the dynamic viscosity. The component of the complex dynamic viscosity in phase with the displacement of the sphere, $\eta''$, is a measure of the elastic property of the mucus. It is usually expressed in terms of a corresponding parameter given by $G' = \omega\eta''$. This parameter, $G'$, is termed the dynamic elasticity. Comparing real and imaginary components of the equation for $\eta^*$:

$$\eta' = (I_m/I_N)(\omega_N/\omega_m)(X_N/X_m)\eta_N \sin d$$

$$G' = (I_m/I_N)(X_N/X_m)\omega_N \eta_N \cos d$$

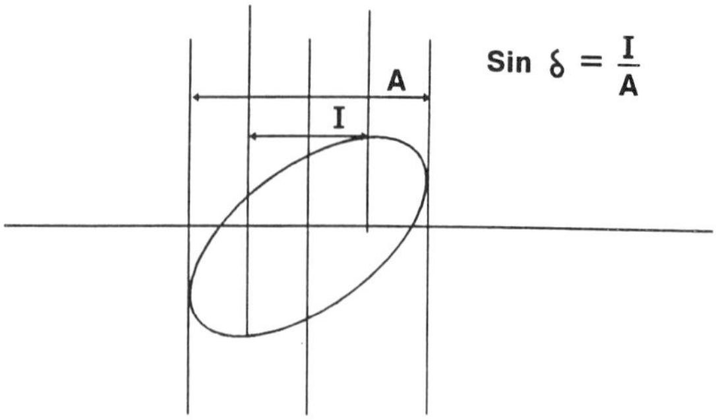


**Figure 2** Graphical determination of the phase angle between the applied force and the resultant displacement.

This means that it is not necessary to determine the absolute values of current or displacement but simply their unitless values on a ratio scale. This saves considerable time with calibration. For instance, $X_0$ is measured in terms of pixels on a monitor, and $d$ is calculated from a Lissajous figure constructed from an $x$–$y$ plot of $I_m$ against $X_m$ (see Figure 2).

The small size of the samples and well used with the oscillating sphere microrheometer mean that correction factors for deviation from Stokes' law are required. The correction is on the order of 20% [3]. At higher frequencies of sphere oscillation, the effect of the inertia of the system can no longer be neglected. Nonetheless, the error at 10 rad $sec^{-1}$ remains on the order of 1% [1].

## IV. ADVANTAGES OF THE MICRORHEOMETER

The paramount advantage of this instrument is, of course, the ability to analyse very tiny samples of mucus. The size of the probe also means that, except for very high frequencies, inertial effects may be neglected. In addition, results are presented as dynamic viscosity and elasticity. The value of these parameters is suggested by their correlation with mucociliary transport rates [4].

## V. DISADVANTAGES AND ASSOCIATED ERRORS

A chief disadvantage of the rheometer is that in samples of low dynamic elasticity, the probe tends to drift. This can be caused both by gravity and by lack of uniformity in the applied magnetic field. This can lead to difficulty calibrating the instrument with the propagation of systematic errors in test samples. It can also give rise to random errors in weakly cross-linked mucus samples. Sakakura and colleagues approached this problem by aligning the applied magnetic field with that of gravity and using a compensatory force to diminish drift [5]. Since the position of the probe requires optical tracking, the optical properties of the immersing mucus can influence the precision of this measurement. Unfortunately, many mucus samples are far from clear and some may be almost opaque. Under such circumstances, the measurement of probe displacement may approximate, if not be impossible.

Analysis of the resulting data using the geometrical technique is prone to large errors. Saldiva and colleagues approached this problem using a computerized curve-fitting technique [6]. The combination of these difficulties means that microrheometry is a time-consuming process. A minute sample is prone to deteriorate through dehydration during the period of analysis. In addition, the analysis of samples may be deferred until time permits. This introduces problems with the storage of samples.

## VI. THE DIGITAL MICRORHEOMETER

The system we have used to generate and measure the driving force is essentially similar to that of King and Macklem [3]. The magnetic field is the result of alternating and direct currents fed through solenoids. The alternating current is measured as the potential difference across an in-series 1 ohm resistor. A steel toroid is used as a field guide and the well is placed in a small gap between the planar ends of the electromagnet. The base and walls of the well are made of 0.5-mm thick quartz glass. The internal dimensions of the well are 1.5 mm × 1.5 mm × 3 mm. The steel sphere has a radius of 150 microns. To minimise evaporation of the test sample, it is covered with a layer of light paraffin and the well is closed with a glass coverslip. Mucus samples are obtained by allowing mucociliary transport to cause mucus to accumulate on a cytology brush left in contact with the test mucosa for a few minutes. The mucus is then collected on a needle and passed to the well.

By contrast, the arrangement we have developed to track the sphere is novel and has several advantages over earlier systems. It is comprised of the following elements: microscope optics and lighting, monochrome video camera, video contrast enhancer and monitor, analog to digital converter, computer with video tracking card and operating and analysis software. The video signal is inverted by the contrast enhancer to provide a white target against a black background. The gain of the enhancer is adjusted to precisely define the edge of the image of the sphere and in this manner tracking through optically dense mucus is improved.

The tracking card is capable of reporting the displacement of the edge of a white target at fifty readings a second. The camera is operated in shuttered mode and the shutter is opened at the end of each field. During this interval the operating software simultaneously reads the ball displacement in terms of monitor screen pixels and the value of the forcing-function current. These are written to a data array as unitless ratio-scale data.

The measurement software calculates the amplitude of both the current and the displacement from the array. The software also plots the data to the computer graphics screen and produces a Lissajous figure as the experiment proceeds. At the end of the experiment the measurement software constructs *x* and *y* axes based on the extreme data points. Intercepts on the *y* axis are calculated by interpolation from the nearest data points and are also displayed. The symmetry of the resulting display provides an immediate guide to the validity of the results. The measurement software calculates the phase angle of the two signals producing the figure from the ratio of the magnitudes of the *y*-axis intercept and the y-axis extent

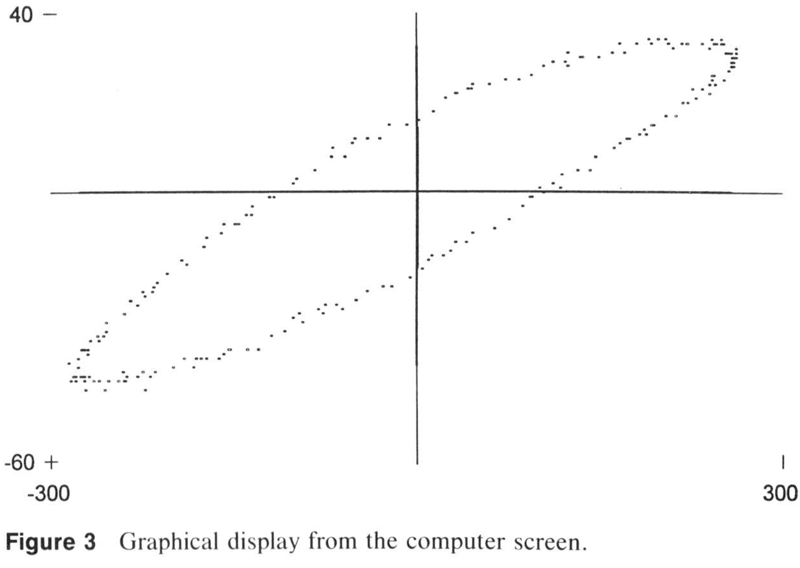


**Figure 3** Graphical display from the computer screen.

(see Figure 3). The computer requires the specification of whether a test is a calibration with a nonelastic medium or an experiment with mucus. Only when calibrated will the system allow experimental runs to proceed. Once both calibration and experiment are completed, the measurement software immediately calculates and displays the dynamic viscosity and the dynamic elasticity.

Automation of measurement and analysis means the system is both rapid and accurate. In turn, these improvements reduce the problems due to sample deterioration and eliminate the requirement for sample storage.

## VII. RESULTS

Typical results from normal nasal mucus obtained using the digital microrheometer are shown in Figure 4. The full set of results was obtained in a few minutes. The reproducibility of the results is good and suggests that the random errors due to sample deterioration are small. Pooled results from ten patients are shown in Figure 5. This graph is remarkably similar to one produced by King and Macklem two decades ago [3] suggesting that the systematic errors of testing are also low.

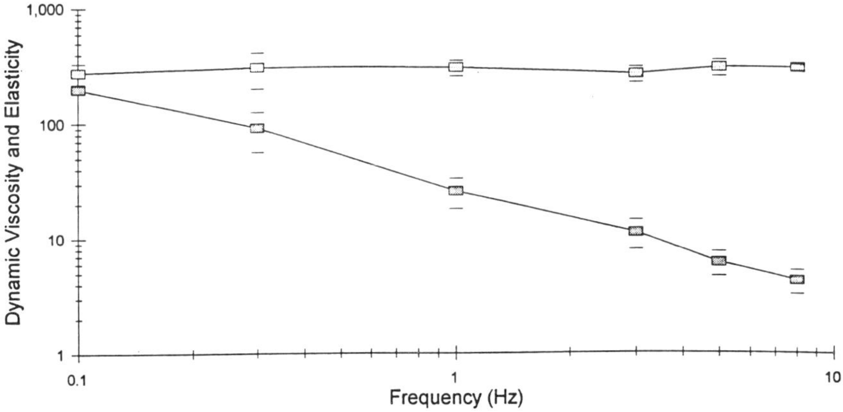


**Figure 4** Dynamic viscoelastic behaviour of a single sample of normal nasal mucus.

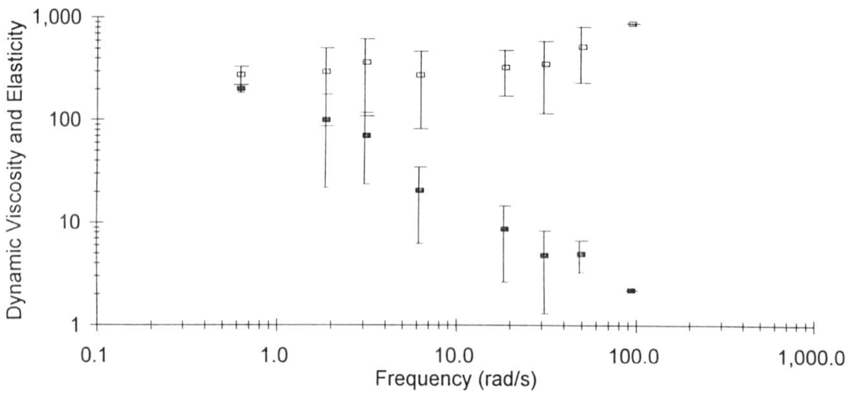


**Figure 5** Pooled data of the dynamic viscoelastic behavior of normal nasal mucus from ten patients.

## VIII. CONCLUSIONS

The oscillating sphere microrheometer is a valuable tool for the analysis of the dynamic mechanical properties of microlitre volumes of mucus. It has been used for the study of mucus simulants, animal model mucus, and tracheobronchial secretions. It remains underused for the analysis of fluids such as middle ear effusions, nasal mucus, synovial fluid, and other tissue fluids that are only available in minute quantities. This may be due to the hitherto time-consuming nature of the measurements and analysis. The ease with which these measurements may be performed with the digital microrheometer may encourage more widespread use of this instrument.

## REFERENCES

1. Hwang SH, Litt M, Forsman WC. Rheological properties of mucus. Rheologica Acta 1969; 8:438–448.
2. Lutz RJ, Litt M, Chakrin LW. Physical–chemical factors in mucus rheology. In: Gabelnick HL, Litt M, eds. Rheology of Biological Systems, Illinois: Charles C Thomas, 119–157.
3. King M, Macklem PT. Rheological properties of microlitre quantities of normal mucus. J Appl Physiol 1977; 42:797–802.
4. Shih CK, Litt M, Khan MA, Wolf DP. Effect of non-dialysable solids concen-

tration and viscoelasticity on ciliary transport of tracheal mucus. Am Rev Respir Dis 1977; 115:989–995.
5. Hirata K. Dynamic viscoelasticity of nasal mucus from children with chronic sinusitis. Mie Medical J 1985; 34:205–219.
6. Silveira PSP, Boehm GM, Yang HM et al. Computer assisted rheological evaluation of microsamples of mucus. Computer Methods and Programs in Biomedicine 1992; 39:51–60.

# 16
# Regulation of Differentiation and Mucin Secretion in Tracheobronchial Epithelial Cell Cultures

**Paul Nettesheim, K. Guzman, Ja Seok Koo, J.-Y. Yoon, and T. Gray**
*National Institutes of Health, Research Triangle Park, North Carolina*

## I. INTRODUCTION

The epithelium of the conducting airways is frequently severely damaged by microbial agents, chemical insults, and disease states such as asthma. Epithelial necrosis is followed by dedifferentiation, regenerative proliferation, hyperplasia, and redifferentiation. Inflammation and hypersecretion are invariably part of the structural and functional pathology. While we have learned a great deal over the last 10 years about the various paracrine and autocrine mediators involved in airway diseases, many questions regarding the control of differentiation of the epithelium, the regulation of its ion transport and its various secretions, remain unresolved. Mucins are the major secretory product of the surface epithelium and the submucosal glands; however, we should not forget that the conducting airways secrete other important proteins such as lysozyme and lactoferrin, which are part of a nonimmunological defense system, and antiproteases (e.g., SLPI), which are important as antiinflammatory agents. The overall purpose of our studies is two-fold: (a) to develop an understanding of basic principles governing secretory and ciliated cell differentiation and (b) to elucidate mechanisms involved in the regulation of mucin and nonmucin secretions.

This chapter is focussed on the role of retinoic acid (RA) in regulating differentiation of airway epithelium and on the effects of retinoic acid on mucin secretion and mucin gene expression. Mucin secretion is our main marker for mucous cell differentiation, thus regulation of mucous cell differentiation and of mucin secretion are at this point in our studies closely entangled subjects.

Why did we choose to study retinoic acid (RA)? Vitamin A (retinol) and it's key metabolic derivatives, all-*trans*-, 13-*cis*-, and 9-*cis*-retinoic acid, are important morphogens in embryonic development and regulators of growth and differentiation of numerous tissues including bone marrow, the immune system, cartilage, epidermis, and many mucous membranes (for a recent review see Ref. [1]). Most important in the context of this discussion is that vitamin A is essential for normal lung development and for differentiation and maintenance of the mucociliary epithelium of the conducting airways. This was first shown by Wolbach and Howe in 1925 [2] who described massive squamous metaplastic changes in the respiratory tract of vitamin A deprived rodents. Since then innumerable studies have been conducted on the subject including several studies by E.M. McDowell and her colleagues [3] who showed that in advanced stages of vitamin A deficiency the hyperplastic/metaplastic lesions are heavily infiltrated by inflammatory cells. She attributed the increased proliferation in these lesions to growth factors, cytokines, and free radicals released by the inflammatory cells. The metaplastic changes have been shown to be reversible by reintroducing vitamin A into the diet [4].

The mechanisms of action of retinoids are the subject of intensive investigation. Retinoic acid binds to nuclear receptors, RARs and RXRs, which belong to the family of steroid hormone receptors. These receptors, in contrast to membrane receptors, act as transcription factors that bind to response elements in the promoter regions of target genes (for a review see Ref. [1]).

## II. THE MODEL

To study the role of RA in the regulation of mucous differentiation and mucin secretion we used so-called air–liquid interface cultures, a cell culture system that promotes normal mucociliary differentiation of early passage normal human tracheobronchial epithelial (NHTBE) cells [5], as well as airway cells from other species [6,7]. The effect of growth factors, cytokines, and hormones on the development and function of the epithelium can be readily analyzed in this in vitro system.

The cultures are seeded with second passage NHTBE cells, which remain undifferentiated during the early logarithmic growth phase. Mucous differentiation sets in during the later part of the growth phase, as indicated by mucin secretion. The plateau phase cultures (day 10 to 14) secrete large amounts of mucin measured with monoclonal antibodies directed against human mucin. Ciliated cells appear between day 18–21. Immunocytochemically, about 35% of the cells can be identified as mucous cells and 8–10% as ciliated cells [5].

## III. THE CONTROL OF DIFFERENTIATION BY RETINOIC ACID

The first questions we address are what is the role of RA in differentiation of NHTBE cultures and can we reproduce the squamous metaplastic changes occuring in vivo during chronic vitamin A deficiency? Mucin secretion was used as the marker for mucous differentiation and cornifin gene expression as the marker for squamous differentiation. Cornifin, together with other crosslinked envelope proteins such as involucrin, is expressed during terminal squamous differentiation [8,9]. Cultures were grown for 2 weeks in the presence or absence of 50 n*M* RA. We found that RA-sufficient cultures produced large amounts of mucin (usually $> 100\ \mu g/10^6$ cells); in contrast RA-deficient cultures produced no mucin or only trace amounts of mucin and expressed high levels of cornifin mRNA and protein. The RA-sufficient NHTBE cultures also secreted small amounts of lysozyme (LZ), lactoferrin (LF), and the antiprotease SLPI. To our surprise these nonmucin proteins were greatly increased in the RA-deficient cultures, and LZ in particular was greatly increased, as much as 10-fold, compared to RA-sufficient controls [5]. Morphologically the RA-sufficient, polarized, mucous cultures became completely stratified and squamous when RA was deleted from the media. Immunohistochemically the epithelium reacted strongly with anticornifin antibody. Subsequent studies showed that intracellular LZ greatly increased in the RA-deficient cultures apparently reaching cytotoxic levels resulting in exaggerated cell death and release of LZ extracellularly. It appeared that lysozyme degradation and/or secretion was impaired in the retinoid deficient cultures since we found no evidence of increased LZ synthesis. Whether similar abnormalities in LZ metabolism also occur in vivo in vitamin A deficiency induced squamous metaplastic epithelium is currently being investigated.

To date we still know very little about whether and under what circumstances airway secretions are transcriptionally or translationally regulated. A number of different mucin genes have been identified; some of them are preferentially expressed in the respiratory tract, others are expressed in the intestinal tract or other mucous membranes (for a review see Ref. [10]). Recent studies have shown that several agents that induce mucin hypersecretion including bacterial lipopolysaccharides (LPS, [11]) and tumor necrosis factor alpha (TNF-$\alpha_1$, [12,13]) can increase mRNA levels of the mucin gene MUC2. The LPS were reported to induce MUC2 transcription [11]. We wanted to determine whether the effect of RA on mucin secretion involved changes in mRNA levels. This seemed to be particularly important because of a recent report describing that MUC2 message levels were markedly downregulated by retinol treatment of tracheobronchial cell cultures [14]. Since the cultures nevertheless produced mucin, these investigators concluded that the major secretion produced is probably not MUC2 mucin.

Measuring human mucin transcript levels by Northern blot analysis is not very satisfactory because of the large smears produced by most mucin mRNAs. Therefore we adapted the Clontech PCR MIMIC technique for measuring message levels of the mucin genes MUC2 and MUC5AC [7]. Based on published MUC2/MUC5AC gene sequences, primers were designed to detect a 482 bp 5′ upstream MUC2 and a 680 bp 3′ downstream MUC5AC sequence. The identity of the products was verified by partial sequencing of both ends of the PCR products. Using these primers we found that MUC2 was expressed in trachea, lung, uterus, stomach, and NHTBE cultures. Other tissues such as skeletal muscle, mammary, and salivary glands were negative. To estimate message levels, so-called MIMICs were added to the PCR mixture as competitive internal standards [15]. Each MIMIC consists of a nonhomologous (stuffer) DNA fragment with primer templates that mimic and are thus recognized by a pair of gene-specific primers.

The PCR MIMIC and target templates compete for the same primers in the amplification reaction. Since the amount of MIMIC is known, the amount of mRNA can be determined. When the ratio of the mRNA signal and the MIMIC signal are 1.0, then the amount of mRNA equals the amount of MIMIC added to the reaction. Using this technique we found that retinoid-deficient cultures did not express MUC2 or MUC5AC message; these cultures also did not secret mucin. However cultures grown in media containing either RA or retinol expressed both mucin genes and secreted mucin [7]. We concluded that MUC2 and MUC5AC as well as mucin secretion are upregulated by retinoic acid and retinol in NHTBE cultures. It is currently not known which mucin is the predominant secretory product in these cultures.

In vivo vitamin A deficiency induced squamous metaplasia is reversible. Upon addition of retinoids to the diet the squamous epithelium is rapidly replaced by mucociliary epithelium [4]. What is not known is whether mucin gene expression and mucin secretion are restored early during this conversion process or whether they are late manifestations of a new cellular phenotype emerging under the influence of vitamin A. We cultured NHTBE cells in RA-free media and on day 7 added RA to the cultures. Secretions as well as RNA were collected every 24 hours. We found that mucin secretion increased substantially within 48 to 72 hours after the start of RA treatment, MUC2 message levels were increased at 24 hours, and MUC5AC mRNA was upregulated 24 to 48 hours later. The squamous marker cornifin was downregulated at 24 hours. These studies indicated rapid induction of the MUC2 gene and inhibition of cornifin expression in response to RA treatment suggesting that activation and repression, respectively, of these genes may be early events in the reprogramming of the differentiation pattern.

We have started to explore the biochemical and molecular mechanisms involved in the RA effect on mucin gene expression and determined which retinoid receptors are critical for mediating the RA response. Using receptor-selective retinoids [16] we found that RARs but not RXRs are likely to be the most important receptors involved in transduction of the signals regulating mucin gene expression and mucin secretion.

## IV. SUMMARY

In air–liquid interface cultures of NHTBE cells the development of mucociliary epithelium occurs in three phases: rapid growth of undifferentiated cells, mucous differentiation during the mid and late logarithmic-growth phase, and ciliated cell differentiation during the plateau phase of growth. The NHTBE cells cultured in complete serum-free medium require retinoids for mucous differentiation. In the absence of retinoids the cultures undergo squamous differentiation. Retinoid-sufficient cultures secrete mucins and nonmucin proteins including lysozyme, lactoferrin, and the antiprotease SLPI. In retinoid-deficient cultures mucin secretion ceases, but production of lysozyme, lactoferrin, and SLPI increases. Retinoid-deprived cultures can be rescued by RA from undergoing squamous differentiation, provided RA treatment is initiated early during the differentiation process. Addition of RA to retinoid-deficient cultures coordinately upregulates MUC2 and MUC5AC gene expression and mucin secretion, and RA-induced mucin gene expression is mediated by RARs.

## REFERENCES


1. The FASEB Journal 1996; Reviews 9, 10:940–1049.
2. Wolbach SB, Howe PR. Tissue changes following deprivation of fat-soluble A vitamin. J Exp Med 1925; 42:753–777.

3. Zhang X-M, McDowell EM. Vitamin A deficiency and inflammation: the pivotol role of secretory cells in the development of atrophic, hyperplastic, metaplastic changes in tracheal epithelium in vivo. Virchows Archiv B Cell Pathol 1992; 61:375–387.
4. McDowell EM, Keenan KP, Huang M. Restoration of mucociliary tracheal epithelium following deprivation of vitamin A; a quantitative morphologic study. Virchows Archiv 1984; 45:221–240.
5. Gray TE, Guzman K, Davis CW, Abdullah LH, Nettesheim P. Mucociliary differentiation of serially passaged normal human tracheobronchial epithelial cells. Am J Respir Cell Mol Biol 1996; 14:104–112.
6. Guzman K, Bader TH, Nettesheim P. Regulation of MUC5 and MUC1 gene expression: correlation with airway mucous differentiation. Am J Physiol 1996; 270 (Lung Cell Mol Physiol, 14):L846–L853.
7. Guzman K, Gray TE, Yoon JH, Nettesheim P. Quantitation of mucin RNA by PCR reveals induction of both MUC2 and MUC5 mRNA levels by retinoids. Am J Physiol 1996; (Lung Cell Mol Physiol) rapid communication, 271 (LMC, 15):L1023–L1028.
8. Jetten AM, Nervi C, Volberg TM. Control of squamous differentiation in tracheal and epidermal cells: role of retinoids. J Natl Cancer Inst Monogr 1992; 13:93–100.
9. Marvin KW, George MD, Fujimoto W, Saunders NA, Bernacki S, Jetten AM. Cornifin, a cross-linked envelope precursor in keratinocytes that is down-regulated by retinoids. Proc Natl Acad Sci USA 1992; 89:11026–11030.
10. Rose MC. Mucins: structure, function and role in pulmonary diseases. Am J Physiol 1992; 263:L413–L429.
11. Jian-Dong Li, Dohrman AF, Gallup M, et al. Transcriptional activation of mucin by *Pseudomonas aeroginosa* lipo-polysaccharide in the pathogenesis of cystic fibrosis lung disease. Proc Natl Acad Sci USA 1997; 94:967–972.
12. Levine SJ, Larivée P, Logun C, Angus CW, Ognibene FP, Shelhamer JH. Tumor necrosis factor-α induces mucin hypersecretion and MUC-2 gene expression by human airway epithelial cells. Am J Respir Cell Mol Biol 1995; 12:196–204.
13. Bader T, Nettesheim P. Tumor necrosis factor-α modulates the expression of its receptors and several cytokines in rat tracheal epithelial cells. J Immunol 1996; 13:41–58.
14. An G, Luo G, Wu R. Expression of MUC2 gene is down-regulated by vitamin A at the transcriptional level in vitro in tracheobronchial epithelial cells. Am J Respir Cell Mol Biol 1994; 10:546–551.
15. Bouaboula M, Legoux P, Pessegue B, Delpech B, Dumont X, Piechaczyk M, Casselas P, Shire D. Standardization of mRNA titration using a polymerase chain reaction method involving coamplication with a multispecific internal control. J Biol Chem 1992; 267:21830–21838.
16. Fanjul A, Dawson MI, Hobbs PD, Jong L, Cameron JF, Harlev E, Graupner G, Lu X-P, Pfahl M. A new class of retinoids with selective inhibition of AP-1 inhibits proliferation. Nature 1994; 372:107–111.

# 17

# Intracellular Regulation of Mucin Secretion in Airway Epithelial Cells In Vitro

**Bernard M. Fischer,* Chengming Li, Hongfei Li, Nonghoon Choe,† David T. Wright, Lori G. Rochelle,‡ and Kenneth B. Adler**
*North Carolina State University, Raleigh, North Carolina*

## I. INTRODUCTION

The airway epithelium is one of the first tissue layers to encounter environmental stimuli such as inhaled allergens, particulates, or microbes. In addition, there are a variety of both resident and infiltrating cells in the airways—including macrophages, mast cells, and neutrophils—that interact with the airway epithelium during disease. Consequently, the airway epithelium plays a pivotal role in the pathophysiology of pulmonary/airway diseases, acting as both a target and effector tissue. The epithelial responses associated with pulmonary diseases such as asthma, cystic fibrosis, and chronic bronchitis include mucus hypersecretion, changes in mucociliary escalator activity, and release of a variety of inflammatory mediators including cytokines and lipid-derived mediators. The mechanisms associated with the mucin hypersecretory response are characterized by a complex network of intracellular signaling pathways. In this report, we begin to delineate some of these signal transduction pathways. Specifically, our results suggest a central signaling pathway mediated by nitric oxide (NO) in response to a variety of inflammation-associated stimuli to enhance mucin secretion.

*Current affiliations*:
*Duke University Medical Center, Durham, North Carolina
†Department of Defense, Bethesda, Maryland
‡The University of North Carolina at Chapel Hill, Chapel Hill, North Carolina

The production of NO is catalyzed by isoforms of nitric oxide synthase (NOS), an enzyme found in a variety of tissues and cells including the epithelium. Nitric oxide, also known as endothelial derived relaxing factor (EDRF), is a potent vasodilator and smooth-muscle relaxant. Macrophages produce large quantities of NO and utilize it as part of their bacteriocidal activity [1–3]. At lower concentrations, NO is thought to act as a signaling molecule, primarily through activation of soluble guanylate cyclase, in many cell types [4]. In the airways, NO has been reported to modulate a variety of physiological processes, including ciliary beat frequency [5] and submucosal gland mucus secretion [6]. In addition, NO in expired air and elevated NOS activation have been associated with airway inflammation and asthma [7–10].

In this report, we examine the effects, on mucin secretion from guinea pig airway epithelium, of four diverse inflammatory mediators known to be important in the pathophysiology of allergic and inflammatory airway diseases: the mast cell derived peptide histamine (HIS), the lipid mediator platelet-activating factor (PAF), the proinflammatory primary cytokine tumor necrosis factor-alpha (TNF-α), and reactive oxygen species (ROS). In addition, the signal transduction mechanisms associated with mucus hypersecretion resulting from these stimuli are delineated. A common signal transduction pathway involving NO and cyclic GMP (cGMP) is suggested by these studies to mediate mucus hypersecretion in response to all these stimuli.

## II. METHODS AND RESULTS

For these studies, guinea pig tracheal epithelial (GPTE) cells were cultured in an organotypic air/liquid interface system that maintains the functional and structural characteristics of the epithelial cells [11]. A monoclonal antibody-based sandwich ELISA specific for guinea pig tracheal mucins was used for the detection of mucin secretion [12,13]. Noncytotoxic concentrations of each agonist and inhibitor were used for these studies. The GPTE cells were exposed to each agonist and/or inhibitor as previously reported [14].

### A. Mechanisms of TNF-α–Induced Mucin Secretion

The GPTE cells were continuously exposed to human recombinant TNF-α for different amounts of time. Optimum concentrations were determined to be 10–15 ng/mL (specific activity $2.87 \times 10^7$ U/mg protein). Peak mucin secretory activity was seen at eight hours of stimulation with a plateau through 24 hours of stimulation. The TNF-α–induced mucin hypersecretion was inhibited by the competitive NOS inhibitor $N^G$-monomethyl-L-arginine (LNMA), but not by its inactive stereoisomer, DNMA. In addition, LY83583, an inhibitor of soluble guanylate cyclase (GC-S), inhibited TNF-α–induced mucin secretion. To verify the importance of NO in this pathway, intracellular cGMP production, in response to TNFα, was measured as an indicator of the NO–cGMP pathway using a commercially available enzyme-linked immunoassay (EIA). Similar to mucin secretion, LNMA and the GC-S LY83583 inhibited TNF-α–induced cGMP production. Moreover, TNF-α–induced an increase in intracellular nitrate/nitrite ($NO_x$). In addition, cGMP is known to activate an enzyme system called cGMP-dependent protein kinase(s) (PKG) [4]. The TNF-α–induced mucin secretion was inhibited by the PKG inhibitor, KT5823. The cell permeable cGMP analogue, dibutyryl-cGMP, also stimulated GPTE mucin hypersecretion. This stimulation was inhibitable by KT5823. These results indicate the importance of NO and cGMP as second messengers mediating TNF-α–induced mucin hypersecretion and suggest the following pathway:

$$\text{TNF}\alpha \Rightarrow \text{NOS} \rightarrow \text{NO} \rightarrow \text{GC-S} \rightarrow \text{cGMP} \rightarrow \text{PKG} \Rightarrow \text{mucin hypersecretion}$$

The TNF-α–induced phospholipase C (PLC) activity was investigated as a possible mechanism leading to NOS activation. In other cellular systems, both phosphatidylcholine- and phosphatidylinositol-specific phospholipase C (PC-PLC, PI-PLC, respectively) activity appear to mediate certain aspects of TNF-α activity [15,16]. The PC-PLC inhibitor, D609, effectively inhibited TNF-α–induced mucin secretion. Stimulation by TNF-α did not induce inositol polyphosphate turnover, suggesting PC-PLC rather than PI-PLC involvement. In our studies with GPTE cells, PC-PLC activity has also been shown to produce diacylglycerol, which in turn may activate other enzyme systems such as protein kinase C (PKC). To investigate the importance of PKC in our hypothetical signal transduction pathway, two different PKC inhibitors with different mechanisms of action, calphostin C and Ro31-8220, were utilized. Both agents inhibited TNF-α–induced mucin hypersecretion. Similar to mucin secretion, TNF-α–induced cGMP production was inhibited by both the PC-PLC inhibitor D609 and the PKC inhibitor calphostin C. Collectively, these results suggest the following hypothetical pathway:

$$\text{TNF-}\alpha \Rightarrow \text{PC-PLC} \rightarrow \text{DAG} \rightarrow \text{PKC} \rightarrow \text{NOS} \rightarrow \text{NO} \rightarrow \text{GC-S} \rightarrow \text{cGMP} \rightarrow \text{PKG} \Rightarrow \Rightarrow \text{mucin hypersecretion}$$

### B. Mechanisms of HISTAMINE (HIS)-Induced Mucin Secretion

Samples of GPTE cells were exposed to HIS at varying time points and concentrations. For the studies described in this report, 100 $\mu M$ HIS for a period of 30 minutes was used. Similar to TNF-α, HIS-stimulated mucin hypersecretion was inhibited by the NOS inhibitor, LNMA, but not by its stereoisomer, DNMA. In addition, HIS-induced mucin secretion was inhibited by the GC-S inhibitor, LY83583, and the PKG inhibitor, KT5823. The HIS addition also stimulated intracellular cGMP production in the GPTE cells that was inhibited by LNMA. These results suggest a common pathway mediating mucin secretion for HIS and TNF-α involving NO–cGMP–PKG dependent pathway. Unlike TNF-α, HIS induces PI-PLC activity resulting in inositol phosphate turnover. As previously described, HIS provokes ionositol phosphate turnover by an $H_1$-receptor and pertussis toxin-insensitive, G protein-dependent mechanism [17]. The relationship between PI-PLC and mucin secretion will be discussed later in this report.

### C. Mechanisms of PAF-Induced Mucin Secretion

A lipid mediator that interacts with the airway epithelium, PAF is produced by the airway epithelium. Previously, PAF-induced mucin secretion has been demonstrated to be mediated by lipoxygenase products (i.e., HETEs) [18] or PKC [19]. Paralleling the results seen with HIS and TNF-α, both PAF-induced mucin hypersecretion and cGMP production were inhibited by the NOS inhibitor, LNMA. In addition, PAF induced inositol phospholipid turnover. Both mucin secretion and phospholipid metabolism were mediated by a pertussis toxin-sensitive G protein. However, inhibition of PI-PLC had no effect on PAF-induced mucin secretion. These results implicate NO and cGMP as important mediators of PAF-induced mucin secretion, but this process is independent of PI-PLC activation.

### D. Mechanisms of ROS-Induced Mucin Secretion

The ROS include a large number of highly reactive species such as ozone, superoxide, hydrogen peroxide, and hydroxyl radical. For the studies described in this report, superoxide was enzymatically generated by the combination of 500 $\mu M$ purine and 20 mU/mL of xanthine oxidase (PXO). As previously reported, PXO stimulates GPTE mucin secretion by a mechanism dependent on arachidonic acid metabolism [20]. In addition, this PXO-induced hypersecretory response was inhibited by the hydroxyl radical scavanger, dimethylthiourea (DMTU) [13] and superoxide dismutase (SOD) [20]. Moreover, PXO stimulation of GPTE cells generated some intracellular ROS and reactive nitrogen species (e.g., NO and peroxynitrite [RNS]) as measured by the oxidation of dihydrorhodamine to rhodamine. Dihydrorhodamine is oxidized by hydroxyl radical formation (i.e., ROS) and peroxynitirite formation (i.e., RNS) and can be used as an indicator for intracellular production of either or both of these species (Rochelle and Adler, submitted). These results suggest intracellular ROS and RNS may be important mediators of PXO-induced mucin secretion. Similar to PAF, HIS, and TNFα, PXO-induced mucin secretion and cGMP production were inhibited by LNMA. Interestingly, PXO-induced inositol phospolipid turnover was also inhibited by DMTU or LNMA. However, PI-PLC inhibition had no effect on PXO-induced mucin secretion. These results suggest a NO-dependent process for PXO-induced mucin secretion and inositol phospholipid metabolism, but this mucin secretory response is independent of phospolipid turnover [13].

The effects of exogenous RNS (i.e., NO and peroxynitrite) on the airway epithelial cells was also evaluated. Exogenous RNS were chemically generated by the combination of *S*-nitroso-*N*-acetylpenicillamine (100 μ*M*; SNAP) + L-cysteine (100 μ*M*) or 1 m *M* SIN-1. Both RNS generators enhanced GPTE mucin secretion and inositol phospholipid me-

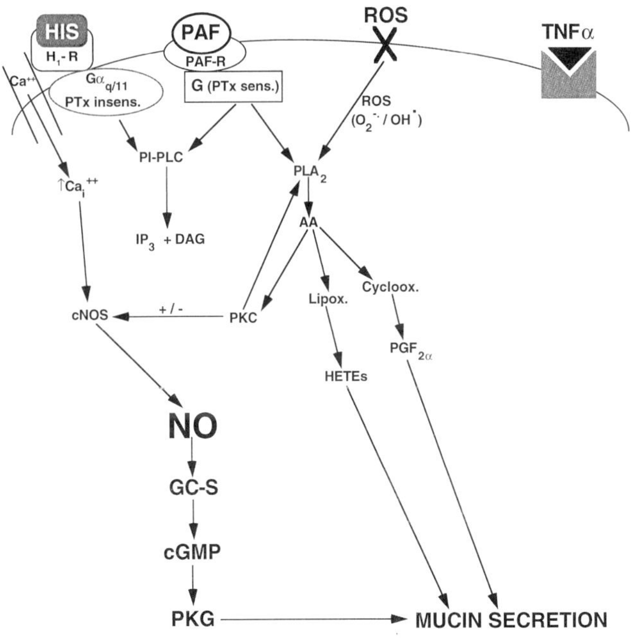


**Figure 1** Histamine (HIS)-, platelet-activating factor (PAF)-, and reactive oxygen species (ROS)-induced signal transduction mechanisms in airway epithelial cells.

tabolism. In addition, SNAP + L-cys caused a dramatic increase in both intracellular cGMP production and cGMP secretion. Mucin secretion induced by SIN-1 and phospholipid turnover were inhibited by the hydroxyl radical scavenger DMTU. Similar to PXO, SIN-1-induced responses could be mediated by ROS.

## III. CONCLUSIONS AND DISCUSSION

These results demonstrate the importance of intracellular NO as a key signaling molecule in airway epithelial cells. For all four inflammatory mediators, TNF-α, HIS, PAF, and ROS, a common signaling pathway consistent with these results would be

$$NO \rightarrow GC\text{-}S \rightarrow cGMP \rightarrow PKG \Rightarrow \text{mucin hypersecretion}$$

It is also possible that NO serves as a pivotal mediator of other signaling pathways in the airway epithelium, such as inositol phospholipid

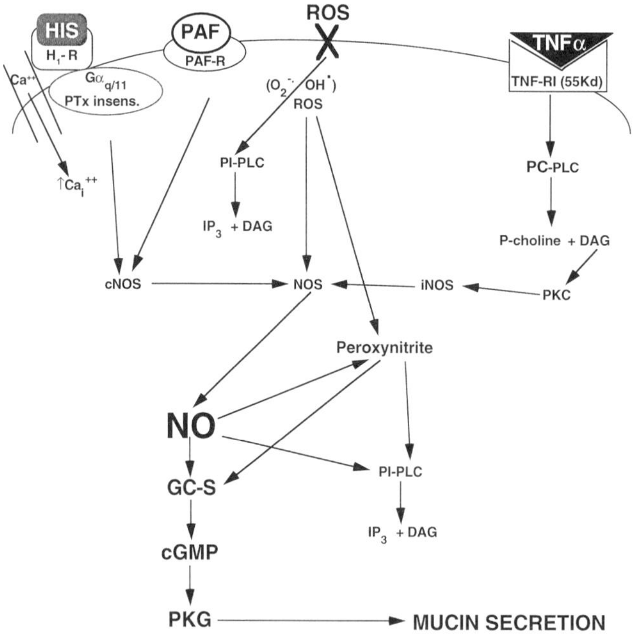


**Figure 2** Tumor necrosis factor-alpha (TNF-α) and reactive oxygen species (ROS)-induced signal transduction mechanisms in airway epithelial cells.

metabolism. However, the mechanism resulting in NOS activation appears to differ for different stimuli (see Figures 1 and 2). This finding is not surprising considering that there are two groups of NOS isoforms: calcium dependent (cNOS = bNOS or eNOS; Type I or Type III) and calcium independent (iNOS = macNOS; Type II) and are activated by different mechanisms. While changes in intracellular calcium and the mechanisms involved were not addressed in these investigations, such studies might delineate the differential pathways mediating NOS activation. In other systems, PKC has been shown to have differing effects on NOS isoforms. In endothelial cells, PKC downregulates eNOS activity [21,22], while in macrophages PKC activity is important for upregulation of macNOS activity [23]. In this laboratory and others, the pathways leading to NOS activation are still under investigation.

This unique phenomenon of a common pathway mediating a biological response such as mucin secretion, can have important implications for our understanding, treatment, and/or prevention of diseases such as asthma, COPD, and ARDS. The exact modulatory role of NO and cGMP in airway epithelium and the lung are still being elucidated. Perhaps understanding of the regulatory functions of these signaling molecules will provide more insight into both normal respiratory physiology and the pathophysiology of many life-threatening diseases.

## REFERENCES

1. Fortier AH, Polsinelli T, Green SJ, et al. Activation of macrophages for destruction of Francisella tularensis: identification of cytokines, effector cells, and effector molecules. Infection and Immunity 1992; 60(3):817–825.
2. Green SJ, Scheller LF, Marletta MA, et al. Nitric oxide: cytokine-regulation of nitric oxide in host resistance to intracellular pathogens. Immunology Letters 1994; 43(1–2):87–94.
3. Vouldoukis I, Riveros-Moreno V, Dugas B, et al. The killing of *Leishmania* major by human macrophages is mediated by nitric oxide induced after ligation of the Fc epsilon RII/CD23 surface antigen. Proc Natl Acad Sci 1995; 92(17):7804–7808.
4. Schmidt HH, Lohmann SM, Walter U. The nitric oxide and cGMP signal transduction system: regulation and mechanism of action. Biochim Biophys Acta 1993; 1178(2):153–175.
5. Jain B, Rubinstein I, Robbins RA, et al. Modulation of airway epithelial cell ciliary beat frequency by nitric oxide. Biochem Biophys Res Comm 1993; 191(1):83–88.
6. Nagaki M, Shimura M, Irokawa T, et al. Nitric oxide regulation of glycoconjugate secretion from feline and human airways in vitro. Respir Physiol 1995; 102(1):89–95.

7. Hamid Q, Springall DR, Riveros-Moreno V, et al. Induction of nitric oxide synthase in asthma. Lancet 1993; 342(8886–8887):1510–1513.
8. Kharitonov S, Yates D, Robbins R, et al. Increased nitric oxide in exhaled air of asthmatic patients. Lancet 1994; 343(8890):133–135.
9. Kharitonov SA, O'Connor BJ, Evans DJ, et al. Allergen-induced late asthmatic reactions are associated with elevation of exhaled nitric oxide. Am J Respir Crit Care Med 1995; 151:1894–1899.
10. Kharitonov SA, Wells AU, O'Connor BJ, et al. Elevated levels of exhaled nitric oxide in bronchiectasis. Am J Respir Crit Care Med 1995; 151: 1889–1893.
11. Adler KB, Cheng P-W, Kim KC. Characterization of guinea pig tracheal epithelial cells maintained in biphasic organotypic culture: cellular composition and biochemical analysis of released glycoconjugates. Am J Respir Cell Mol Biol 1990; 2:145–154.
12. Li C, Cheng P-W, Adler KB. Production and characterization of monoclonal antibodies against guinea pig tracheal mucins. Hybridoma 1994; 13(4): 281–287.
13. Wright D, Fischer B, Li C, et al. Oxidant stress stimulates mucin secretion and PLC in airway epithelium via a nitric oxide-dependent mechanism. Am J Physiol 1996; 271(5 Pt 1):L854–61.
14. Adler KB, Fischer BM, Li H, et al. Hypersecretion of mucin in response to inflammatory mediators by guinea pig tracheal epithelial cells in vitro is blocked by inhibition of nitric oxide. Am J Respir Cell Mol Biol 1995; 13: 526–530.
15. Schutze S, Potthoff K, Machleidt T, et al. TNF Activates NF-kβ by phosphatidylcholine–specific phospholipase C-induced "acidic sphingomyelin breakdown. Cell 1992; 71:765–776.
16. Rapuano B, Bockman R. Tumor necrosis factor-alpha stimulates phosphatidylinositol breakdown by phospholipase C to coordinately increase the levels of diacylglycerol, free arachidonic acid and prostaglandins in an osteoblast (MC3T3-E1) cell line. Biochim Biophys Acta 1991; 1091(3):374–384.
17. Li H, Choe N, Wright D, et al. Histamine provokes turnover of inositol phospholipids in guinea pig and human airway epithelial cells via an H1-receptor/G protein-dependent mechanism. Am J Respir Cell Mol Biol 1995; 12(4):416–424.
18. Adler KB, Akley NJ, Glasgow WC. Platelet-activating factor provokes release of mucin-like glycoproteins from guinea pig respiratory epithelial cells via a lipoxygenase-dependent mechanism. Am J Respir Cell Mol Biol 1992; 6:550–556.
19. Larivee P, Levine SJ, Martinez A, et al. Platelet-activating factor induces airway mucin release via activation of protein kinase C: evidence for translocation of protein kinase C to membranes. Am J Respir Cell Mol Biol 1994; 11:199–205.
20. Adler KB, Holden-Stauffer WJ, Repine JE. Oxygen metabolites stimulate release of high-molecular-weight glycoconjugates by cell and organ cultures

of rodent respiratory epithelium via an arachidonic acid-dependent mechanism. J Clin Invest 1990; 85(1):75–85.
21. Hirata K, Kuroda R, Sakoda T, et al. Inhibition of endothelial nitric oxide synthase activity by protein kinase C. Hypertension 1995; 25(2):180–185.
22. Johnson A, Phelps DT, Ferro TJ. Tumor necrosis factor-alpha decreases pulmonary artery endothelial nitrovasodilator via protein kinase C. Am J of Physiol 1994; 267(3 Pt 1):L318–25.
23. Severn A, Wakelam MJ, Liew FY. The role of protein kinase C in the induction of nitric oxide synthesis by murine macrophages. Biochem Biophys Res Comm 1992; 188(3):997–1002.

# 18

# Cellular Basis for the Purinergic Regulation of Mucin Secretion in the Airways

**C. William Davis, Lubna H. Abdullah, and Richard C. Boucher**
*The University of North Carolina at Chapel Hill, Chapel Hill, North Carolina*

## I. INTRODUCTION

Mucin, the high molecular weight glycoconjugate forming the scaffolding of mucus, plays a prominent role in the normal physiology of the airways and in the various obstructive airway diseases (asthma, cystic fibrosis, bronchiectasis, and chronic bronchitis). Consequently, it is important to understand the mucin secretory mechanism and its regulation. In the airways, mucins are stored in and—upon an appropriate signal—released from goblet cells residing in the superficial epithelium and mucous cells in submucosal glands. We have focussed our attention specifically on mucin secretion by goblet cells for three reasons. First, the small airways generally represent the site of mucus plugging in the obstructive diseases because of low laminar airflow. At this locus, submucosal glands are absent, leaving goblet cells as the most likely source of mucin. Second, recent data suggest that even in the large, cartilaginous airways, goblet cells may be an important source of mucin. Using robust morphometric techniques in the large airways of two species of monkey, Plopper and associates found that goblet cells possess a greater aggregate quantity of mucin than submucosal glands [1,2, see also 3,4]. Although these findings have yet to be repeated in human tissues, they indicate nonetheless that mucins derived from the large airways may emanate to a significant degree from goblet cells. Third, mucin secretion from these two sources may be regulated very differently. Table 1 summarizes the data available for the

**Table 1** Regulation of Mucin Secretion in the Airways

| Submucosal glands[a] | Superficial epithelium[b] |
|---|---|
| Cholinergic stimulation muscarinic ($M_1$ and $M_3$) | no effect (SPOC1 cells) |
| Adrenergic stimulation | |
| $\alpha_1$ - secretagogue (serous $\gg$ mucous ?) - stimulates gland contraction | no effect (explants of superficial epithelium) |
| $\alpha_2$ - inhibits β-adrenergic stimulation | ND[c] |
| $\beta_?$ - secretagogue | no effect (SPOC1 cells) |
| Peptidergic stimulation | |
| VIP - secretagogue - augments muscarinic responses | ND |
| SP - secretagogue - stimulates gland contraction | ND |
| Histaminergic stimulation | |
| $H_2$ - secretagogue | no effect (SPOC1 cells) |
| Leukotriene stimulation | |
| $LTC_4$, $LTD_4$, 15-HETE-stimulatory in intact tissues, but no effect on isolated glands | ND |
| Prostaglandin stimulation | |
| $PGE_2$ -inhibitory | ND |
| $PGA_2$, $PGD_2$, $PGF_{2\alpha}$ - secretagogues | ND |
| PAF stimulation | |
| stimulatory only in presence of platelets | no effect in SPOC1 cells stimulatory in primary cultures, but mediated by prostaglandins and/or leukotrienes |
| Endothelin stimulation | |
| secretagogue | ND |
| Purinergic stimulation | |
| $P_{2U}$ - secretagogue | secretagogue (many preparations) |

[a] After Shimura and Takishima, Ref. 59
[b] After Davis, Ref. 4
[c] Not determined

regulation of mucin release from submucosal glands and the superficial epithelium. It shows that submucosal glands respond to a very large number of agonists, whereas goblet cells, where tested, respond, perhaps uniquely, to purinergic agonists. In this review, we summarize (a) the evidence supporting purinergic activation as playing a major regulatory role in goblet cell physiology and (b) the intracellular messenger pathways underlying this purinergic response.

## II. MATERIALS AND METHODS

### A. Epithelial Explants

Briefly, following an incubation with collagenase and dispase injected into the submucosal tissue, sheets of superficial epithelium of canine trachea or human nasal turbinates were isolated by mechanically scrapping the tissues with a coverslip. The isolated epithelial sheets were subsequently explanted onto transparent nitrocellulose substrata, to which the cells attached during an overnight incubation [5].

### B. Cell Culture

Primary cultures of rat trachea epithelial cells were established on Costar Transwell-Col (TCol) permeable supports and maintained according to the procedures outlined by Kaartinen, et al. [6]. Primary cultures of human bronchial cells were similarly cultured on TCol supports according to the procedures described previously [7]. According to previously published procedures [9–11], SPOC1 cells were cultured on TCols, or plastic, depending upon the experiment.

### C. Mucin Secretion Assays

Mucin secretion by individual goblet cells in epithelial explants was monitored by direct observation of exocytotic events with video microscopy [5,12]. Alternately, epithelial explants or cell cultures were mounted in simple perfusion chambers [11–13], or were incubated in static bathes [34], and the mucins secreted into the luminal perfusate or solution were detected by mucin binding assays [ELLA, 11, or ELISA 7,12].

## III. RESULTS AND DISCUSSION

### A. Purinergic Activation of Mucin Secretion

The extracellular actions of ATP have been appreciated for over 50 years, beginning with efforts to understand the causes of traumatic shock, [60] and ATP and other nucleotides have long been appreciated for effects in the vasculature and other organ systems [14]. It was not until 1986, however, that ATP was discovered to affect the respiratory side of the lung in stimulating surfactant production by alveolar type II cells [15]. Subsequently, ATP and UTP have been shown to elicit an increase in $Cl^-$ secretion by airway epithelia derived from normal individuals and from cystic fibrosis patients [16,17]. Since ATP and UTP were equipotent in their effects on both ion transport [17] and inositol phosphate production in primary cultures of airway cells [18], a novel purinoceptor was suggested. Similar results were reported for a purinoceptor in other cells, and the P2Y2 purinoceptor receptor was finally cloned by Julius from mouse neuroblastoma cells [19–22]. The human homolog of the receptor was cloned and sequenced from an airway cell library [23].

Initially ATP was shown to elicit goblet cell mucin secretion in primary cultures of hamster tracheal epithelial cells [24] and by native goblet cells in epithelial explants from canine trachea [5]. Subsequently, ATP and UTP were shown to stimulate exocytosis in goblet cells from human nasal turbinates [12] (Figure 1). Recent studies have also demonstrated purinergic stimulation of mucin secretion in passaged human bronchial primary cell cultures [7,13], the SPOC1 cell line [11] (Figure 2), and rat tracheal primary cell cultures [11]. Other mucin-secreting cells are also responsive to purinergic agonists, namely, airway submucosal gland mucous cells [25], gastric mucous cells [26], and the colonic mucin-secreting cell line, HT29-Cl.16E [27].

Available evidence supports the notion that ATP and UTP elicit secretion by interacting with the P2Y2 purinoceptor. First, ATP, UTP, and ATPγS appear to be approximately equipotent in stimulating goblet cell mucin secretion in a variety of model systems. In epithelial explants from human nasal turbinates, goblet cell exocytosis and mucin release were induced approximately equally by 100$\mu M$ ATP and UTP [12]. Primary cultures of hamster trachea epithelial cells exhibited similar dose-response relationships between nucleotide dose and mucin secretion when ATP was compared with ATPγS [24] and when ATP was compared with UTP [28]. Similarly, in SPOC1 cells ATP, UTP, ATPγS had virtually superimposable dose-response relationships (Figure 2), as did ATP and UTP in rat tracheal primary cell cultures [11]. Second, human nasal goblet cells [12] and SPOC1 cells [11] failed to respond to the P2Y1 purinoceptor hallmark agonist, 2-methlythio-ATP, and canine tracheal goblet cells [5] and SPOC1 cells [11] did not respond to the $P_1$ agonist, adenosine. Lastly, the P2Y2 purinoceptor mRNA was identified by RT-PCR in SPOC1 cells [11] and by Northern blotting in primary cultures of hamster trachea epithelial cells [28], consistent with pharmacologic data. Since luminal exposures to P2Y2 agonists elicit mucin secretion from a variety of preparations, whereas serosal exposures leads to mixed responses [5], or no response [11], the data are most consistent with this purinoceptor residing

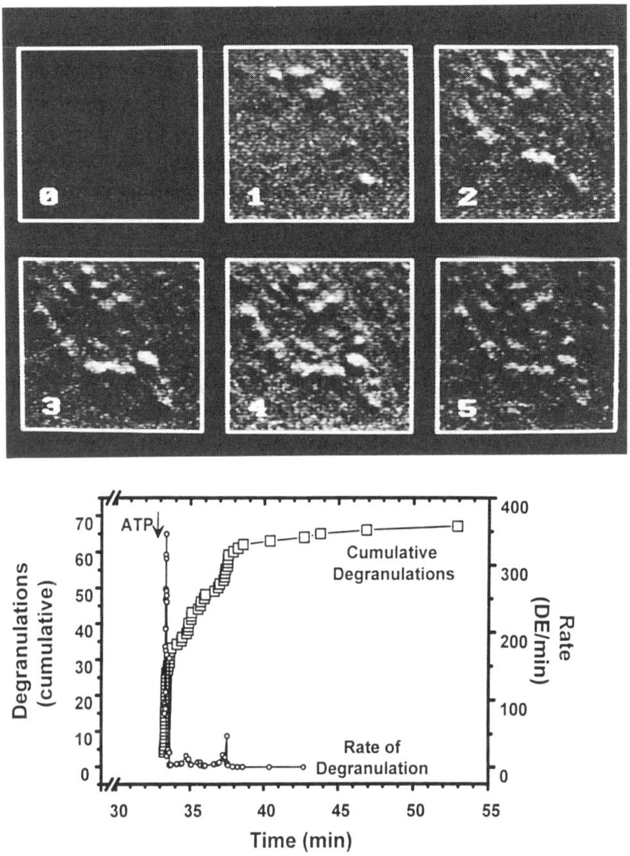


**Figure 1** Mucin granule exocytosis from an individual goblet cell in an epithelial explant from a human nasal turbinate. Top: Visualization of the first 5 seconds of response to ATP (100 $\mu M$) using digitally subtracted images. The images were digitized from video tape at 1 second intervals (2 frame averages), and from each image was subtracted an image obtained 1 second prior to frame 0. The displayed results depict individual degranulations, cumulatively, as large bright spots. Bottom: Kinetics of exocytosis from the same cell. The squares depict the cumulative degranulations and the circles represent the rate calculated from sequential 4-least squares fits to the raw data, (Both figures were reprinted from Ref. 5 with permission from the publisher.)

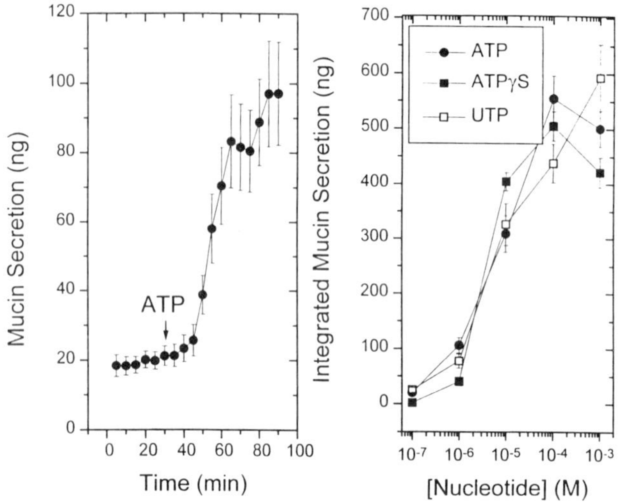


**Figure 2** Effects of purinergic agonists on mucin secretion by cultures of SPOC1 cells. Left: Time course of the response to luminal ATP (100 μ*M*). Right: Dose-response relationships for ATP, UTP, and ATPγS. The data are presented as the suprabasal, temporally integrated responses to nucleotide. (Reprinted from Ref. 11 with permission from the publisher.)

in the goblet cell apical membrane. The receptors populating the goblet cell basolateral membrane are presently unknown, as are other potential receptors in the apical membrane.

## B. Intracellular Messenger Pathways

Very little definitive information is available regarding the cellular messenger pathways underlying mucin secretion in goblet cells. In recent work with the SPOC1 mucin-secreting cell line [34], we investigated whether cyclic nucleotides played such a role and found that cAMP production was indeed stimulated by forskolin. Mucin secretion, however, was stimulated by neither forskolin or the permeant cpt-cAMP. Similarly, the permeant 8-Br-cGMP was without effect. Consequently, and in contrast to the effects of β-adrenergic agents in promoting mucin secretion from submucosal glands (Table 1), goblet cells in the superficial epithelium may not respond to agents that act on adenylate or guanylate cyclase-coupled receptors.

The known P2Y purinoceptors couple to phospholipase C via G-proteins to activate the intracellular production of diacylglycerol (DAG) and inositol phosphates (IP), with the subsequent release of $Ca^{2+}$ from the endoplasmic reticulum [21,22,29–31]. Brown et al. first showed that ATP stimulated CF/T43 epithelial cell cultures to produce IPs [18], and Kim et al. subsequently found that IP production in primary cultures of hamster trachea epithelial cells was similarly stimulated by ATP [32]. There are no published accounts of the effects of nucleotides on $Ca_i^{2+}$ in goblet cells. It is interesting to note, however, that $IP_3$ receptors have been localized to secretory granules in endocrine and neuroendocrine cells [33], leaving open the possibility that secretory granules may generally act as local $Ca^{2+}$ stores and that they may have the potential of participating directly in the regulation of exocytosis.

Our laboratory has recently found that SPOC1 cell mucin secretion is stimulated by the $Ca^{2+}$ ionophore ionomycin with a saturating dose-dependency [34] (Figure 3). Thapsigargin, which mobilizes $Ca^{2+}$ through inhibition of the $Ca^{2+}$ ATPase associated with the ER $Ca^{2+}$ store, also elicited mucin secretion from SPOC1 cells. The EC50 for this response was 300 n*M*, a value that is consistent with other accounts [35]. Hence,

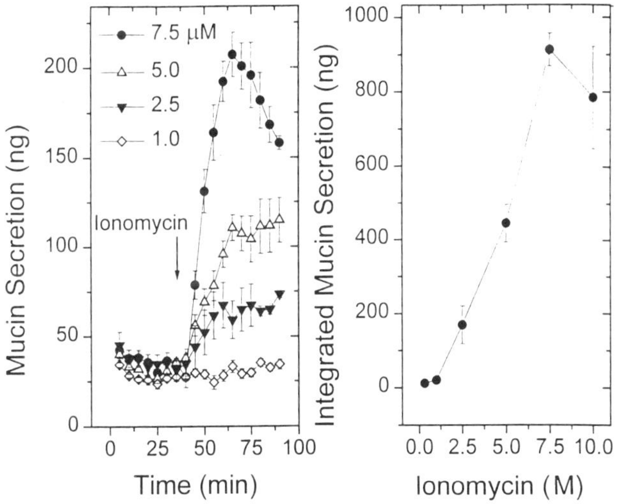


**Figure 3** Effects of ionomycin on SPOC1 cell mucin secretion. Left: Time courses of the responses to several doses of ionomycin. Right: Dose-response relationship between ionomycin and mucin secretion. (Reprinted from Ref. 34 with permission from the publisher.)

these results are consistent with $Ca_i^{2+}$ playing an important role in mediating the effects of purinergic agents on goblet cell mucin secretion.

In recent studies with primary cultures of hamster tracheal epithelial cells, however, Kim and associates concluded that $Ca_i^{2+}$ played no role in the regulation of mucin secretion. This conclusion was based on the following observations: (a) ionomycin, while it stimulated mucin release at 3 and 10 $\mu M$, also caused LDH release, (b) thapsigargin, at doses as high as 10 $\mu M$, was without effect, and (c) treatment of the cells with BAPTA/AM (50 $\mu M$ for 10 min) to buffer $Ca_i^{2+}$ had no effect on the ability of the cells to respond subsequently to ATP. There are at least two explanations for these results. First, the goblet cells in primary cultures of hamster trachea epithelial cells may indeed not respond to elevations in $Ca_i^{2+}$ with mucin secretion. If so, these cells would be different from literally all other secretory cells studied [36–45]. Second, in these experiments, for unknown reasons, effective $Ca_i^{2+}$ mobilization by thapsigargin and effective buffering by BAPTA did not occur (the effect of these maneuvers on $Ca_i^{2+}$ was not measured). It is notable that ionomycin *did* stimulate mucin release in these experiments, however, and the notion that a simultaneous release of LDH indicates an ionomycin-induced, nonspecific release of mucin from secretory granules was found to have no experimental basis at present. Consequently, in the absence of direct $Ca_i^{2+}$ measurements, the conclusion that $Ca_i^{2+}$ plays no role in regulating hamster goblet cell mucin secretion may be premature.

The regulation of goblet cell mucin secretion by protein kinase C [46] has also been the subject of recent investigation, with several laboratories demonstrating a stimulation of mucin secretion with the use of the phorbol ester, PMA. Positive effects of PMA have been achieved using primary cultures of hamster [47,48] and feline [49] trachea epithelial cells, epithelial explants of human nasal turbinates [34], SPOC1 cells [34] (see Figure 4) airway submucosal glands [50], submandibular glands [51], and the colonic cell lines LS174T [52], T84 [53,54], and HT29-18N2 [55]. In most of these studies, the possible nonspecific effects of PMA were investigated using an inactive phorbol isomer (e.g., 4$\alpha$-phorbol), inhibitors of PKC, and/or downregulation of PKC with long-term PMA exposure to show a loss of stimulation.

The family of PKC isoforms, however, is both numerous and diverse, with 12 presently known PKC isoforms assigned to 3 subfamilies [46,56,57]. Both the *conventional* $Ca^{2+}$-dependent isoforms (cPKC $\alpha$, $\beta_1$, $\beta_2$, and $\gamma$) and the *novel* $Ca^{2+}$-independent isoforms (nPKC $\delta$, $\epsilon$, $\eta$, $\theta$ and $\mu$) are activated by DAG and PMA. Consequently, treatment of cells with PMA may force a physiologically-inappropriate activation of all cPKC and nPKC isoforms in a cell. In cardiac myocytes, for instance, PMA has

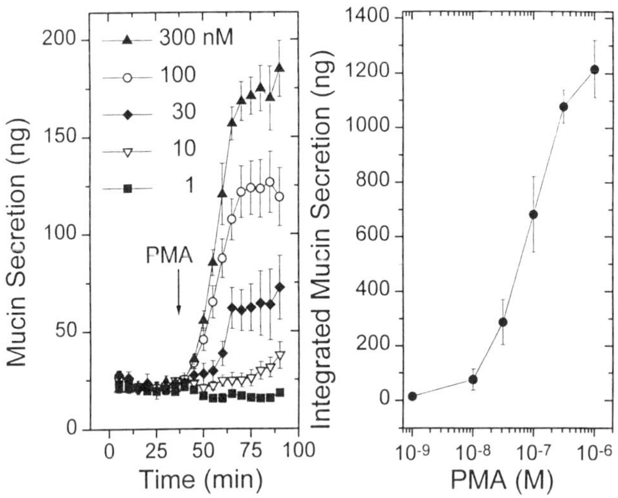


**Figure 4** Effects of PMA on SPOC1 cell mucin secretion. Left: Time courses of the responses to several doses of PMA. Right: Dose-response relationship between PMA and mucin secretion. (Reprinted from Ref. 34 with permission from the publisher.)

been shown clearly to activate cPKCα, nPKCδ, and nPKCε, whereas agonist stimulation of secretion of these cells involves the selective participation of the nPKCs [57,58]. As one approach to determining whether PKC plays a physiologic role in the purinergic regulation of mucin secretion, we tested the effects of UTP, PMA, and ionomycin on measured PKC activity in SPOC1 cells [34]. Figure 5 shows that UTP indeed caused a significant increase in PKC activity (approximately 150% of control) in the membrane fraction of these cells. As expected, PMA had an even larger effect, whereas ionomycin did not affect membrane-associated PKC activity. Thus, purinergic stimulation of SPOC1 cells most likely activates one or more PKC isoforms.

Of the several potential interactions between $Ca^{2+}$ and PKC-dependent pathways, mucin-secreting cells may exhibit a rare form, namely full independence. This has been demonstrated for both SPOC1 [34] and T84 [53] cells with the findings that maximal doses of ionomycin and PMA were additive in their effects to elicit mucin secretion. Additionally, in both SPOC1 [34] and T84 cells [54], calcium ionophores were effective in promoting mucin secretion from PKC-downregulated cells. Thus, these

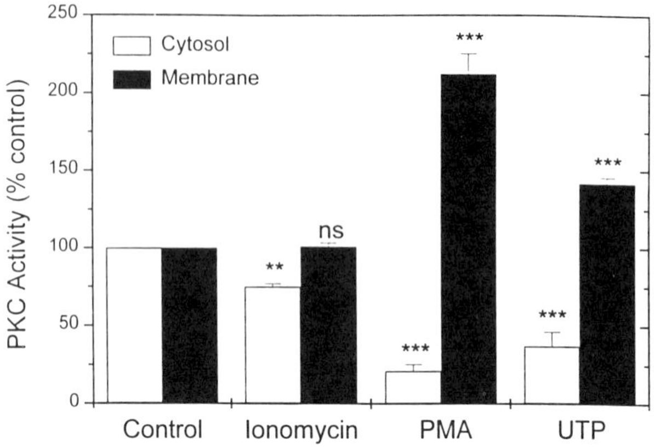


**Figure 5** Effects of UTP, ionomycin, and PMA on PKC activity in SPOC1 cells. After exposure to maximal doses of secretagogue (40 min), SPOC1 cells were harvested, cytosol and membrane prepared, and PKC activities determined. The data were first normalized to the amount of protein in each fraction, and then expressed relative to the respective control activity.

results may indicate that nPKC isoforms act to elicit exocytosis of mucin secretory granules, independently of regulation by $Ca_i^{2+}$.

## C. Conclusions and Future Directions

Relative to the knowledge that has accumulated concerning agonist regulation in other cells, the knowledge for goblet cells has been woefully lacking. Recent work in many laboratories to establish good experimental models, however, has finally paid off and we are beginning to realize the benefits in terms of data relevant to the regulation of goblet cell secretory activity. As reviewed above, it appears that goblet cells are under powerful regulation by purinergic agonists acting through their apical membranes to elicit mucin release, and we have made a good start on understanding the cellular mechanisms underlying this response. At the organ level, the source(s) of luminal ATP, UTP, and/or other P2Y2-interacting nucleotides is an open question. At the cellular level, several classes of agonists need to be tested for effects on goblet cell mucin secretion (all of the NDs on the right-hand side of Table 1). At the molecular level, much work remains in identifying the regulatory protein pathways and their effectors that act to control the process of exocytosis. Thus, we are obviously only at the beginning of a period in which goblet cell function is intensely studied at all levels of organization.

## ACKNOWLEDGMENTS

Support for studies in the authors laboratories was received from the Cystic Fibrosis Foundation, the American Lung Association of North Carolina, the National Institutes of Health, and Glaxo Wellcome Corporation. Figure 1 is reprinted with the permission of the American Journal of Respiratory Cell and Molecular Biology, Figure 2 with the permission of the Biochemical Journal, and Figures 3–5 with the permission of the American Journal of Physiology.

## REFERENCES

1. Heidsiek JG, DM Hyde, Ploppen CG, St. George JA. Quantitative histochemistry of mucosubstance in tracheal epithelium of the macaque monkey. J Histochem Cytochem 1987; 35:435–442.
2. Plopper CG, Heidsiek JG, Weir AJ, George JA, Hyde DM. Tracheobronchial epithelium in the adult rhesus monkey: a quantitative histochemical and ultrastructural study. Am J Anat 1989; 184:31–40.
3. Rogers DF. Airway goblet cells: responsive and adaptable front–line defenders. Eur Respir J 1994; 7:1690–1706.
4. Davis CW. Goblet cells: physiology and pharmacology. In: Rogers DF, Lethem MI, eds. Airway Mucus: Basic Mechanisms and Clinical Perspectives. Basel: Birkhauser, 1997.
5. Davis CW, Dowell ML, Lethem M, Van Scott M. Goblet cell degranulation in isolated canine tracheal epithelium: response to exogenous ATP, ADP, and adenosine. Am J Physiol 1992; 262:C1313–1323.
6. Kaartinen L, Nettesheim P, Adler KB, Randell SH. Rat tracheal epithelial cell differentiation in vitro. In Vitro Cell Dev Biol Animal 1993; 29A:481–492.
7. Gray TE, Guzman K, Davis CW, Abdullah LH, Nettesheim P. Mucociliary differentiation of serially passaged normal human tracheobronchial epithelial cells. Am J Respir Cell Mol Biol 1996; 14:104–112.
8. Matsui H, Johnson LG, Randell SH, Boucher RC. Loss of binding and entry of liposome–DNA complexes decreases transfection efficiency in differentiated airway epithelial cells. J Biol Chem 1997; 272:1117–1126.
9. Doherty MM, Liu J, Randell SH, Carter CA, Davis CW, Nettesheim P, Ferriola PC. Phenotype and differentiation potential of a novel rat tracheal epithelial cell line. Am J Respair Cell Mol Biol 1995; 12:385–395.
10. Randell SH, Liu JY, Ferriola PC, Kaartinen L, Doherty MM, Davis CW, Nettesheim P. Mucin production by SPOC1 cells—an immortalized rat tracheal epithelial cell line. Am J Respir Cell Mol Biol 1996; 14:146–154.

11. Abdullah LH, Davis SW, Burch L, Yamauchi M, Randell SHX, Nettesheim P, Davis CW. P2u purinoceptor regulation of mucin secretion in SPOC1 cells, a goblet cell line from the airways. Biochem J 1996; 316:943–951.
12. Lethem MI, Dowell ML, Van Scott M, Yankaskas JR, Egan T, Boucher RC, Davis CW. Nucleotide regulation of goblet cells in human airway epithelial explants: normal exocytosis in cystic fibrosis. Am J Respir Cell Mol Biol 1993; 9:315–322.
13. Davis CW, Abdullah LH. In vitro models for airways mucin secretion. Pulm Pharmacol. In press.
14. Dubyak GR, Fedan JS, eds. Biological Actions of Extracellular ATP. Vol. 603. New York: The NY Acad Sci 1990.
15. Rice WR, Singleton FM. P2-purinoceptors regulate surfactant secretion from isolated rat alveolar type II cells. Br J Pharmacol 1986; 89:425–491.
16. Mason SJ, Paradiso AM, Boucher RC. Regulation of transepithelial ion transport and intracellular calcium by extracellular ATP in human normal and cystic fibrosis airway epithelium. Br J Pharmacol 1991; 103:1649–1656.
17. Knowles MR, Clarke LL, Boucher RC. Activation by extracellular nucleotides of chloride secretion in the airway epithelia of patients with cystic fibrosis. N Engl J Med 1991; 325:533–538.
18. Brown HA, Lazarowski ER, Boucher RC, Harden TK. Evidence that UTP and ATP regulate phospholipase C through a common extracellular 5′-nucleotide receptor in human airway epithelial cells. Mol Pharmacol 1991; 40: 648–655.
19. Lustig KD, Shiau AK, Brake AJ, Julius D. Expression cloning of an ATP receptor from mouse neuroblastoma cells. Proc Natl Acad Sci USA 1993; 90:5113–5117.
20. Fredholm BB, Abbracchio MP, Burnstock G, Daly JW, Harden TK, Jacobson KA, Leff P, Williams M. Nomenclature and classification of purinoceptors. Pharmacol Rev 1994; 46:143–156.
21. Harden TK, Boyer JL, Nicholas RA. P2-purinergic receptors: subtype-associated signaling responses and structure. Ann Rev Pharmacol Toxicol 1995; 35:541–579.
22. Dubyak GR, el-Moatassim C. Signal transduction via P2-purinergic receptors for extracellular ATP and other nucleotides. Am J Physiol 1993; 265(Pt 1): C577–606.
23. Parr CE, Sullivan DM, Paradiso AM, Lazarowski ER, Burch LH, Olsen JC, Erb L, Weisman GA, Boucher RC, Turner JT. Cloning and expression of a human P2U nucleotide receptor, a target for cystic fibrosis pharmacotherapy. Proc Natl Acad Sci USA 1994; 91:3275–3279.
24. Kim KC, Lee BC. P2 purinoceptor regulation of mucin release by airway goblet cells in primary culture. Br J Pharmaco 1991; 103:1053–1056.
25. Shimura S, Sasaki T, Nagaki M, Takishima T, Shirato K. Extracellular ATP regulation of feline tracheal submucosal gland secretion. Am J Physiol 1994; 267:L159–164.
26. Ota S, Yoshiura K, Takahashi M, Hatal Y, Kohmoto O, Kawabe T, Shimada T, Hiraishi H, Mutoh H, Terano A, et al. P2 purinergic receptor regulation

of mucus glycoprotein secretion by rabbit gastric mucous cells in a primary culture. Gastroenterology 1994; 106:1485–1492.
27. Merlin D, Augeron C, Tien XY, Guo X, Laboisse CL, Hopfer U. ATP-stimulated electrolyte and mucin secretion in the human intestinal goblet cell line HT29-C1.16E. J Membr Biol 1994; 137:137–149.
28. Kim KC, Park HR, Shin CY, Akiyama T, Ko KH. Nucleotide-induced mucin release from primary hamster tracheal surface epithelial cells involves the P2U purinoceptor. Eur J Respir Dis 1996; 9:542–548.
29. Berridge MJ. Inositol trisphosphate and calcium signalling. Nature 1993; 361: 315–325.
30. Clapham DE. Calcium signaling. Cell 1995; 80:259–268.
31. Divecha N, Irvine RF. Phospholipid signaling. Cell 1995; 80:269–278.
32. Kim KC, Zheng QX, Van-Seuningen I. Involvement of a signal transduction mechanism in ATP-induced mucin release from cultured airway goblet cells. Am J Respir Cell Mol Biol 1993; 8:121–125.
33. Blondel O, Bell GI, Seino S. Inositol 1,4,5-triphosphate receptors, secretory granules and secretion in endocrine and neuroendocrine cells. Trends in Neurosciences 1995; 18:157–161.
34. Abdullah LH, Conway JD, Cohn JA, Davis CW. Protein Kinase C and Ca+2 activation of mucin secretion in airway goblet cells. Am J Physiol. 1997; 273: L201–L210.
35. Thastrup O. Role of Ca2(+)-ATPases in regulation of cellular Ca2+ signalling, as studied with the selective microsomal Ca2(+)-ATPase inhibitor, thapsigargin. Agents Actions 1990; 29:8–15.
36. Holz RW. Control of exocytosis from adrenal chromaffin cells. Cellular and Molecular Neurobiology 1988; 8:259–268.
37. Howell TW, Cockcroft S, Gomperts BD. Essential synergy between Ca2+ and guanine nucleotides in exocytotic secretion from permeabilized rat mast cells. J Cell Biol 1987; 105:191–197.
38. Cheek TR, Jackson TR, O'Sullivan AJ, Moreton RB, Berridge MJ, Burgoyne RD. Simultaneous measurements of cytosolic calcium and secretion in single bovine adrenal chromaffin cells by fluorescent imaging of fura-2 in cocultured cells. J Cell Biochem 1989; 109:1219–1227.
39. Hersey SJ, Perez A. Permeable cell models in stimulus–secretion coupling. Ann Rev Physiol 1990; 52:345–361.
40. Gomperts BD, Tatham PE. Regulated exocytotic secretion from permeabilized cells. Methods Enzymol 1992; 219:178–189.
41. Howell SL, Jones PM, Persaud SJ. Regulation of insulin secretion: the role of second messengers. Diabetologia 1994; 37(Suppl 2):S30–35.
42. Augustine GJ, Neher E. Calcium requirements for secretion in bovine chromaffin cells. Journal of Physiology 1992; 450:247–271.
43. Neher E. Spatial and temporal aspects of Ca changes in secretion control. Biochemical Society Transactions 1993; 21:420–423.
44. von Ruden L, Neher E. A Ca-dependent early step in the release of catecholamines from adrenal chromaffin cells. Science 1993; 262:1061–1065.
45. Chow RH, Klingauf J, Neher E. Time course of Ca2+ concentration trigger-

ing exocytosis in neuroendocrine cells. Proc Natl Acad Sci USA 1994; 91: 12765–12769.
46. Nishizuka Y. Studies and perspectives of protein kinase C. Science 1986; 233:305–312.
47. Kai H, Yoshitake K, Isohama Y, Hamamura I, Takahama K, Miyata T. Involvement of protein kinase C in mucus secretion by hamster tracheal epithelial cells in culture. Am J Physiol 1994; 267(Pt 1):L526–530.
48. Ko KH, Jo M, McCracken K, Kim KC. ATP-induced mucin release from cultured airway goblet cells involves, in part, activation of protein kinase C. Am J Respir Cell Mol Biol 1997; 16:194–198.
49. Larivee P, Levine SJ, Martinez A, Wu T, Logun C, Shelhamer JH. Platelet-activating factor induces airway mucin release via activation of protein kinase C: evidence for translocation of protein kinase C to membranes. Am J Respir Cell Mol Biol 1994; 11:199–205.
50. Shimura S, Ishihara H, Nagaki M, Sasaki H, Takishima T. A stimulatory role of protein kinase C in feline tracheal submucosal gland secretion. Respir Physiol 1993; 93:239–247.
51. Fleming N, Bilan PT, Sliwinski-Lis E. Effects of a phorbol ester and diacylglycerols on secretion of mucin and arginine esterase by rat submandibular gland cells. Pflugers Arch-Eur J Physiol 1986; 406:6–11.
52. Keller K, Olivier M, Chadee K. The fast release of mucin secretion from human colonic cells induced by *Entamoeba histolytica* is dependent on contact and protein kinase C activation. Arch Med Res 1992; 23:217–221.
53. McCool DJ, Marcon MA, Forstner JF, Forstner GG. The T84 human colonic adenocarcinoma cell line produces mucin in culture and releases it in response to various secretagogues. Biochem J 1990; 267:491–500.
54. Forstner G, Zhang Y, McCool D, Forstner J. Mucin secretion by T84 cells: stimulation by PKC, Ca2+, and a protein kinase activated by Ca2+ ionophore. Am J Physiol 1993; 264:G1096–102.
55. Phillips TE, Wilson J. Signal transduction pathways mediating mucin secretion from intestinal goblet cells. Dig Dis Sci 1993; 38:1046–1054.
56. Dekker LV, Parker PJ. Protein kinase C—a question of specificity. Trends Biochem Sci 1994; 19:73–77.
57. Steinberg SF, Goldberg M, Rybin VO. Protein kinase C isoform diversity in the heart. J Mol Cell Cardiol 1995; 27:141–153.
58. Puceat M, Hilal-Dandan R, Strulovici B, Brunton LL, Brown JH. Differential regulation of protein kinase C isoforms in isolated neonatal and adult rat cardiomyocytes. J Biol Chem 1994; 269:16938–16944.
59. Shimura S, Takishima T. Airway submucosal gland secretion. In: Takishima T, Shimura S, eds. Airway Secretion: Physiological Bases for the Control of Mucus Hypersecretion. New York: Dekker, 1994:325–398.
60. Green HN, Stoner HB. Biological actions of the adenine nucleotides. London: Lewis HK. 1950.

# 19
# Polymer Biophysics of Mucus in Cystic Fibrosis

**Pedro Verdugo**
*University of Washington, Seattle, Washington*

*To my dear friend Professor Alex Silberberg*

## I. INTRODUCTION

One of the most critical problems in the pathophysiology of cystic fibrosis (CF) is that mucus does not reach a normal level of hydration during the postexocytic swelling equilibrium. This results in a mucus gel of drastically abnormal rheology, defective mucociliary transport contributing to the chronic airway infection usually found in these patients. Mucins, the principal polymeric species in the mucus matrix, are densely packed inside secretory granules and undergo massive swelling upon exocytosis. As mucins are highly polyionic, their swelling is not driven by simple osmosis but rather by a Donnan equilibrium process. It had been thought that decreased mucus hydration might result mainly from altered transepithelial movement of $Cl^-$, water, and other electrolytes, which is defective in CF. However, insufficient mucus hydration could also result from defective properties of the mucin polymer network. The experiments presented here were designed to compare the swelling kinetics of exocytic mucin granules in cultured goblet cells from four CF patients and three non-CF individuals. Swelling kinetics of 694 newly released granules were measured by image processing of video recordings of goblet cells equilibrated in Hanks' solution containing either 1 or 2.5 mM $caCl_2$, pH 7.2, at 37°C. Results show that the diffusivity (D) of mucus from granules of CF goblet cells equilibrated in Hanks' solution containing 1 mM $CaCl_2$ is

77% lower than D from mucus of normal goblet cells. Also when challenged by increased $Ca^{2+}$ concentration in the swelling medium—from 1 to 2.5 mM—D in normal mucus decreased about 19%, while in CF mucus it decreased about 88%. These results show that defective transepithelial movement of water and electrolytes is not solely responsible for the decreased hydration of mucins, suggesting that the mucin polymer network might also be abnormal in CF.

The physicochemical principles that govern mucin hydration and rheology were the focus of Alex Silberberg's work in this subject and are the conceptual foundation for the studies reported here. In what follows, I present a brief review of the polymer physics of mucus that led to the present studies. We start with the tangled conformation of the mucin polymer matrix. Next we focus on the Donnan mechanism that controls mucus hydration and rheology. One of the most important implications of a Donnan hydration mechanism is that the mucin polymer matrix should be condensed while stored inside the granule. Hence, we focus next on the mechanisms of condensation and particularly decondensation of mucin secretory granules, which are at the root of the experiments presented. These experiments were designed to investigate the influence of factors other than the hydroelectrolyte composition of the swelling media on the hydration kinetics of mucus released from goblet cells from CF patients, as compared to goblet cells from normal individuals.

## II. CONFORMATION OF THE MUCUS POLYMER NETWORK

Mucus is a heterogeneous mixture of different secretions that form a hydrogel. The rheological properties of polymer gels like mucus are primarily determined by the conformation, the polyionic charge, and the concentration of their constituent polymers and also by the nature of the interconnections among the polymer chains in the gel's matrix [1,2]. The characteristic non-Newtonian viscoelastic properties of mucus are directly or indirectly determined by mucins. These highly polyanionic polymers of large molecular dimensions form the bulk of the mucus matrix, particularly in chronic inflammation of the airway when mucin secretory cells are increased [3–6].

### A. The Mucin Matrix of the Mucus Gel is a Tangled Polymer Network

Since the early work of Gibbons and Martner [7], the conformation of the mucus gel was thought to be a covanlently crosslinked network, where

S:S bonds served as the crosslinking agents among the mucin polymer chains [6]. Mucus rheology was thought to be controlled by the degree of interchain disulfide bonding. However, the notion of an S:S bonded network was based on results of analytical methods that severely degrade the quaternary structure of the mucus polymer matrix, which is precisely the level of molecular organization at which interchain crosslinks should be verified. Although the dogma of an S:S bonded mucus polymer matrix turned out to be flawed, it has had an enormous influence in mucus research for the past 25 years and can still be found in some textbooks.

The application of nondestructive methods using dynamic laser scattering spectroscopy provided the first indications that the mucin polymers in the mucus are held together by tangles and low-energy bonds rather than by interchain covalent crosslinks [8,9]. Independent verification of the entangled network model for the mucus polymer matrix was provided by the demonstration that mucins have a linear conformation. Electron microscopic images of single linear mucin polymers from cervical, gastrointestinal, and respiratory mucus obtained by rotary shadowing techniques have been published by several groups [10,11,12].

The physics that explain the dynamics of polymer networks has undergone rapid progress, particularly with the introduction of Edwards' theory of tangled polymer networks [1]. A direct implication of this theory is that the rheological properties of gels containing a tangled polymer matrix are primarily determined by the concentrations of tangles among polymer chains [2]. Since tangle concentration decreases with the square of the gel's volume, *hydration* turns out to be a highly efficient vehicle for the physiologic control of mucus rheology [13,14]. Another implication of this argument is that *the mucin matrix must be secreted as a condensed polymer network and become a hydrogel (the mucus gel) only after swelling*. Hydration determines the rheological properties of the gel, hence, by modulating the degree of hydration of the mucus, the mucosa can readily control mucus rheology [15,16]. Another interesting implication is that tangled network gels, unlike covalently crosslinked gels, don't have a swelling limit, i.e., these networks have the potential to virtually swell to complete dispersion. Thus mucus must undergo a progressive change of rheological properties from the moment that it is exocytosed to the time it reaches hydration equilibrium on the surface of the mucosa.

## III. DONNAN EQUILIBRIUM: THE CONTROL MECHANISMS OF MUCUS SWELLING AND RHEOLOGY

All polymer hydrogels swell when immersed in water. Swelling is governed by osmotic forces derived from the concentration gradient between the network and the swelling fluid outside the gel. Because of the large molecular size of the polymers, the osmotic drive is usually small, and swelling is very slow [17]. However, in gels containing a polyionic network, swelling is not driven only by a simple network/solvent concentration gradient, but primarily by the fixed polyionic charges of the polymer chains. Since tangles prevent the polyionic chains from migrating out of the gel, their ionized residues act as fix charges—charges of constrained mobility—i.e., the polymer matrix virtually functions as its own semipermeable membrane. Swelling, in this case, is not driven by the osmotic activity of the polymers, but by a Donnan equilibrium process [18,19]. Owing to the charged residues of mucins and other polyions, mucus is a highly polyionic gel. Thus, the demonstration that mucus swelling has the characteristic features of a Donnan equilibrium was not unexpected [20]. However, it revealed a paradigm of remarkably broad physiological significance. A Donnan process can explain the physicochemical mechanisms that control mucus rheology [20–23] and furnish a powerful framework to understand mucus pathology [15,16].

Since mucus hydration results from a balance between osmotic forces and diffusional hindrance, the polyionic charges of the mucins—and, to a minor extent, those of other fix polyions entrapped inside the gel's matrix—furnish the Donnan driving force for swelling. This driving force is balanced by the internal diffusional hindrance of the mucin chains inside the network.

## A. The Donnan Swelling Drive

The Donnan swelling drive is not constant, but modulated by the concentration of free cations and polycations and the pH of the hydrating fluid. These cations diffuse into the polymer network and screen the fix polyanionic charges of mucin or other polyanions. Consequently, the concentration of small metallic cations or polycations, particularly $Na^+$ and $Ca^{2+}$, and that of polyionic proteins (albumin, lysozyme, proline-rich protein, etc.), as well as the pH and the ionic strength of the fluid on the airway surface can readily modulate the swelling rate of the mucus gel. The characteristic Donnan dependence of mucus hydration on pH, $Na^+$, $Ca^{2+}$, and albumin concentration in the swelling medium has been experimentally validated. When the Donnan drive is neutralized by increased concentration of cations (particularly polycations) or low pH, approaching the isoelectric point of mucins (<3), the diffusivity (D) of the mucus gel—which is the parameter that characterizes the swellability of gels—can decrease up to 95% [20–23].

The respiratory mucosa is equipped with a battery of ion channels and pumps to effectively control the hydroelectrolytic environment of the luminal surface of the epithelium [24]. However, the significance of such a powerful ion and water exchanger, and its responsiveness to neurotransmitters or hormonal mediators, remained speculative for a long time. Emphasis has been placed on the importance of the regulation of fluid volume, which would control the width of the periciliary fluid layer [25]. The idea of a Donnan mechanism for regulating mucus swelling clearly explains the need and the significance of a well-tuned hydroelectrolytic exchanger in the airways and provides verifiable predictions to further investigate this system. For instance, work from the laboratory of John Widdicombe [26] indicated that the electrolytic composition of the airway liquid is not constant. In particular, $Na^+$, $Cl^-$, and $Ca^{2+}$ concentrations can vary under the influence of pharmacological mediators. His observations also suggested that a homeostatic mechanism to control transepithelial pH must be present in the mucosa. Also, his findings that albumin is not only actively transported by the epithelium, but that the transport of albumin is affected by autonomic mediators, suggests a potentially significant physiologic role for this polyion in the control of mucus hydration.

Since the findings of Quinton and Knowles et al. that the transepithelial $Cl^-$ permeability is decreased in the sweat duct and the airway respectively [27,28], increased interest, has resulted in extraordinary advances in the understanding of how the airway epithelium regulates ion transport and in the identification of the defective cystic fibrosis transmembrane regulator (CFTR) gene in CF [29,30]. Although the picture is still incomplete, it shows a remarkable complexity that seems to only increase as investigations reveal an ever more intricate nature and function of the CFTR. At least four classes of mutations [31] have been described: (a) defective protein production; (b) defective protein processing, as in ΔF508, in which the CFTR protein is misfolded and does not target to the lumenal membrane of the cell, but is retained and degraded in the endoplasmic reticulum; (c) defective CFTR that does not respond to normal regulation of the Cl channel, including ATP refractory response; and (d) defective ionic conductance of the Cl channel.

The discovery that CFTR, like other ABC transporters, can permeate ATP [32,33] and the presence of a purinergic receptor in the same protein complex, has brought the idea of a complex transporter:channel: receptor array present in the apical membrane of the airway epithelial cells [34]. Moreover, the interesting finding that the abnormally high absorption of $Na^+$ due to increased open time of the apical membrane Na channel is inversely coupled in a still unknown way with the activation of CFTR by cAMP [35] has added impetus to the notion that the $Cl^-$ conductance is only a small component of a variety of transmembrane proteins with different but highly coordinated functions [36]. Considering the Na/Ca ion exchange properties of mucins [37,38] and the need for Na/Ca ion exchange for mucus swelling [16,23], a decreased $Na^+$ concentration—or increased $Ca^{2+}$ concentration—in the airway surface liquid can be an additional limiting factor in mucus hydration in CF.

In summary, the amount and composition of the airway surface fluid play a critical role in mucus hydration. The defective transepithelial conductances of $Cl^-$ and $Na^+$ ions must play a critical role in the abnormal mucus hydration found in CF patients. However, the precise link between composition of airway surface liquid and mucus hydration still remains unknown.

## B. Diffusional Hindrance and Mucus Swelling

The potential that defective mucin polymers might also result in abnormal mucus hydration in CF has received much less attention. One of the earlier indications of defective processing in CF was the finding of increased mucin sulfation [39–42]. The finding that the acidification of intracellular organelles is abnormal in CF [43] opened the alternative that CFTR might play a role in intracellular organelles [44]. In addition to the electrogenic transport of protons, there are several reports of cation selective ion channels in the membrane of secretory vesicles [35,45]. Chloride channels, on the other hand, provide a switch for counterion movement that serves to maintain electroneutrality and to control pH in intracellular vesicles [46]. Thus, defective acidification of subcellular compartments—where the posttranscriptional processing normally takes place—could indeed result from a faulty or absent CFTR in these compartments. Alkalinization of these organelles may lead to a decrease sialilation and increase fucosylation and sulfation of glycoproteins, including mucins [43,47]. These intriguing observations strongly suggest that the postranscriptional processing of mucins—and probably other glycoproteins—also might be affected in CF. Postranscriptional processing—like sialilation and sulfation of mucins—are important determinants of the ion-exchange properties and of the quaternary conformation and supramolecular conformation (liquid crystalline orders) of polyionic mucins [48]. These changes in the mucins can have drastic effects on the mucus gels. Thus, processing plays a critical role in determining the conformation and physicochemical properties of the mucus gel.

In gels containing a tangled network like mucus, the counter forces that oppose swelling derive from conformational features that increase the reptational hindrance—these are frictional dissipations to the axial translational diffusion of the polymer chains—as the polymers slide past each other while the network swells [2]. Axial hindrance depends critically on the size, the charge, and the conformation (extended/folded) of the mucins, as well as on the interaction of the mucin chains with each other—polymer–polymer affinity—and their interaction with specific counterions, such as Na/Ca ion exchange [1].

Axial–reptative–diffusional times increase with the second power of the polymer length [49]. Thus small changes in mucin length can have striking effects on the postexocytic swelling rate and swelling equilibrium of mucin granules. The length of the mucin polymer had been thought to vary depending on the particular mucin gene that transcribes the apomucin backbone [50–52]. However, studies by John Sheehan and Ingmar Carlstedt (personal communication) indicate that mucins although characteristically polydisperse, are larger in more mature granules—and stored for a longer time in goblet cells—than in granules that expend a shorter time before release. This interesting observation suggests that annealing—end-to-end linking—of mucin chains might be among the many posttranslational processing steps that mucins undergo before they are discharged to the surface of the mucosa. It is not clear how a defective CFTR could result in increased transcription of longer mucin chains or increased annealing of mucin polymers. However, the finding that in CF decreased acidification of subcellular organelles—due to decreased movement of Cl- across intracellular membranes—might be associated with increased sulfation and decreased sialilation of mucins in CF [43] is a strong indication that postranscriptional processing might also be defective in CF.

The increased sulfation and decreased sialilation found in CF mucins [39,41,43] may be particularly important for mucus swelling, since the high affinity of sulfate residues for $Ca^{2+}$ can drastically shift the normal Na/Ca ion-exchange properties of mucins and the swelling of the mucin granule that follows the exocytic release. While stored inside the secretory granule, mucins are in condensed phase, their charge screened by high concentrations of $Ca^{2+}$ that can reach up to 150 mM [53,54]. During exocytosis, once the secretory pore is established, extracellular $Na^+$ diffuses into the granule, driven by its concentration gradient and by the higher ionic affinity of mucins for $Na^+$ over $Ca^{2+}$. However, as explained in detail later, the critical concentration of shielding counterions that is required to induce recondensation—or required to keep the network condensed—is an inverse function of the second power of the counterion valence [55]. Thus, during exocytosis $Na^+$ displaces $Ca^{2+}$ from the mucin matrix by ion exchange, yet the peak concentration it reaches in the mucin network is too low to keep the mucin matrix condensed, triggering the phase transition of the granule matrix from condensed to hydrated phase. If the ion exchange property of the mucin network change due to changes in the ratio of carboxylic/sulfated residues, as found in CF, then it could also stabilize the condensed phase of the gel, inhibiting the decondensation, and/or hydration of mucus upon exocytosis.

Collapsed polymer segments forming supramolecular ordered domains—like the nematic liquid crystalline domains we have described in mucus [56]—must unfold upon mucus hydration, otherwise they can virtually lock tangled networks resulting in gels that mimic a crosslinked gel, but with a high density anisotropy, i.e., with regions that swell and other that remain virtually unswollen [57].

Finally, the hydrophobic/hydrophylic properties of mucins, which also result from the processing of this polymer by the secretory cell, can change the phase transition properties of mucins and drastically inhibit the swellability of mucin gel.

In summary, the low permeability to $Cl^-$, combined with the increased absorption of $Na^+$ could result in abnormal ion exchange and decreased water on the surface of the airway and account for the characteristically deficient hydration and rheology of the mucus in CF. However, decreased hydration of mucus could also be aggravated by a defective processing of the mucins in CF. The later argument is consistent with the results presented here that indicate that when equilibrated in swelling medium of similar composition, the swelling kinetics of exocytosed mucin granules of CF goblet cells is drastically retarded. This result agrees with the evidence—and emphasizes the idea—that abnormal mucin polymers might also be among the multiple phenotypic expressions of defective CFTR in CF.

## IV. MECHANISM OF CONDENSATION AND DECONDENSATION OF SECRETORY GRANULES

The observation that mucin granules of goblet cells—as well as granules of mast cells and other secretory cells—swell upon exocytosis [22,58–60] provided direct support to the idea that secretory material stored inside the secretory granule must be condensed. However, the mechanism of condensation had not been defined. The premise we examine here is that if the mucin polymer network is expanded by swelling upon exocytosis (to form the mucus gel), then it must start from a condensed state within the secretory granule. Two questions follow: What is the mechanism of *condensation* during mucin storage in the granule? Conversely, what is the mechanism of *decondensation of mucin upon release in exocytosis?*

## A. Mucin Condensation in the Secretory Granule

The counter argument of a Donnan-driven swelling is that if electrostatic interactions drive the explosive expansion of the granule during exocytosis, what stops these forces from preventing the condensation of the polyanionic mucin network within the granule? One obvious implication is that the polyionic charges of the mucin network must be shielded by counterions. The search for a cationic shielding agent *verified*, using electron probe microanalysis, that $Ca^{2+}$ is present in large amounts in mucin secretory granules [53,54]. Thus, intragranular $Ca^{2+}$ probably functions as a shielding cation to screen the polyanionic residues of mucins, allowing their condensation. Conversely, if $Ca^{2+}$ ions are exchanged by $Na^+$ and diffuse out of the mucin matrix, the effective cationic shield decreases, and the mutual repulsion of the polyionic charges will quickly drive the expansion of the mucin network. This idea is consistent with observations that variations from 1 to 4 mM $Ca^{2+}$ in the extracellular medium are sufficient to decrease by up to 85% the diffusivity of rabbit goblet cell mucins during exocytosis [23].

However, a critical constraint is that although charge shielding is necessary it is not sufficient to explain hydrogel condensation. For instance, the synthetic polymer networks found in contact lenses are neutral, yet these gels remain fully hydrated. To explain the condensation of mucus the application of a different principle of polymer physics is required.

## B. Polymer Gel Phase Transition, a Powerful Physical Paradigm to Explain Product Storage and Release in Secretion

Polymer gel phase transition (PGPT) is a universal phenomena whereby all polymer gels, regardless of their origin (synthetic or natural, neutral or polyelectrolytic, organic-solvent-based gels or water-based-hydrogels), can change from a condensed, solid-like phase to a decondensed, gel-like phase. Polymer gel phase transition, like other physical phase transitions, is a critical phenomena that exhibits a characteristic discontinuity (infinite cooperativity) upon transition from condensed to decondensed phases and visa versa. Following phase transition, gels undergo drastic changes of volume, dielectric properties, density, etc. The triggering stimuli of these transitions include temperature, solvent composition, pH, ion exchange, electric field, light or interaction with specific molecules [55,63]. This phenomenon is also present in natural gels including the mucin polymer matrix of mucus and the polymer matrix of other secretory granules as well and can readily explain condensation and decondensation of secretory products in secretion [15,59–62].

PGPT results from a balance of forces associated with osmotic drive, the affinity among the polymers that make the matrix of a gel (polymer–polymer affinity), and the affinity between these polymers and the solvent, which in the case of hydrogels is water [63]. The condensation of fully hydrophylic polyelectrolytic polymer gels not only requires the screening of polyionic charges, but also the presence of a nonsolvent to increase the polymer–polymer affinity in the network. However, in polymers containing hydrophobic domains, condensation requires only the shielding of polyionic charges by counterions [64]. As a result, the polymer–polymer affinity among hydrophobic domains dominates, bringing about the transition and the collapse of the network to a condensed phase. The critical concentration of counterion required to induce condensation in these gels is inversely proportional to the second power of the valence of the counterion [55]. Exocytosed mucin gels from goblet cells of the rabbit airway can be readily recondensed by exposing them to a Na-free medium, containing 150–180 mM $Ca^{2+}$ concentrations at pH 3 [62]. These $Ca^{2+}$ concentrations are about equal to those found inside the granule [53,54]. Similarly, the heparin matrix of goblet cells can also be recondensed by exposure to concentrations of cationic histamine that mimic those found in the mast cell secretory granule [59,60]. What is the advantage of product condensation in secretion? All secretory granules, including the mucin granules, the mast cell granule, chromafin granules, and synaptic vesicles, contain a polyanionic polymer matrix and grossly hyperosmotic concentrations of the secretory products, ATP and $Ca^{2+}$. Among those that have been best studied, the mast cell granule contains about 400 mM histamine, 20 mM serotonin, ATP, $Ca^{2+}$, and large amounts of heparin proteoglycan that forms the polymer matrix. The small synaptic vesicle of Torpedo contains up to 500 mM ACh plus high concentrations of $Ca^{2+}$ and a variety of nucleotides among a growing number of other components [65]. It has long been believed that these products are in solution and that the entire cocktail of soluble constituents is released upon exocytosis [65]. However, storing a hyperosmotic mixture inside membrane bound secretory granules in equilibrium with 280 mOsm cytosol poses questions that have long puzzled investigators. An energy consuming uptake mechanism to store secretory products in a complex precipitate has been proposed [66,67]. However, dissolution of a precipitate during release would be very slow. Also, it is unlikely that the secretory products are kept in solution generating large osmotic gradients with the associated continuing energy consumption. A PGPT mechanism [68] accounts for the presence—in all secretory granules—of a condensed polyionic matrix that entraps the secretory products in a polymer cage and is osmotically inert, since it behaves as a single particle. In fact, isolated condensed heparin matrixes of mast cell granules remain condensed even when suspended in distilled water! Trace amounts of $Na^+$ added to the water trigger a transition from the condensed to the hydrated phase and drastic decondensation [59].

### C. Mucin Decondensation and Swelling Kinetics of Mucin Granules upon Exocytosis

The initial decondensing step in PGPT is a change in the dielectric property of the gel prompted by a broad variety of stimuli [63]. Those have been well investigated in synthetic "smart hydrogels," and include temperature, light, electrical fields, or ion exchange, as in exocytic decondensation of secretory granules [60,69]. The change in volume that follows PGPT can be driven by a simple osmotic mass gradient, as in neutral gels, or by a Donnan potential, as in polyelectrolytic gels [19]. Secretory granules from cultured goblet cells undergo dramatic swelling during exocytosis [20–23,62]. This outcome was envisioned by previous findings of Donnan swelling of mucus [20]. However, we did not expect to be able to quantitatively measure, by video enhanced microscopy, the swelling of newly released granules from goblet cells. The finding that, during exocytotic swelling, the expansion of the radius of the secreted material follows first-order kinetics gave the first clue that the swelling of the exocytosed mucin is governed by Tanaka's linear theory of swelling of polymer gels [70]. Although this finding should not have come as a surprise, it was fascinating to verify that a polymer physics theory—formulated and validated in artificial polymer gels—could also apply to the swelling of a natural polymer gel during exocytosis in living cells. Tanaka's theory further predicts that a linear relationship should exist between the square of the final radius of the swollen exocytosed granule and the characteristic time of swelling (the reciprocal of the kinetic constant). The slope of this line has the dimensions of a diffusion ($D = r^2/\tau$ [$cm^2/sec$]) and represents the diffusivity of the polymer network as it expands in the solvent [70]. Thus, this observation confirmed in microscopic scale, previous observations in bulk mucus, i.e., that mucin swelling is driven by a Donnan potential generated by the presence of fix polyions in the granular matrix and it is governed by Tanaka's theory of polymer gel swelling [21–23,62]. As noted earlier, the implication of this notion is that the extracellular concentration of cations and polycations should modulate the swelling rate of mucin granules upon exocytosis.

### D. Effect of $Ca^{2+}$ on the Swelling of Secretory Mucins

According to the Donnan principle, relatively small concentrations of $Ca^{2+}$ should inhibit mucin swelling. Also, from the studies described in the above paragraph we learned that $Ca^{2+}$ is found in large amounts inside mucin secretory granules [71,53,54]. Thus, during exocytosis $Ca^{2+}$ is continuously discharged to the surface of the mucosa. These considerations, along with observations that $Ca^{2+}$ seems to be increased in CF mucus [5], suggested assaying the effect of increased extracellular $Ca^{2+}$ concentration on the swelling kinetics of goblet cell granules in vitro. Results of these studies indicated that variations of $Ca^{2+}$ concentration in the extracellular medium, similar to those found in CF mucus (2 and 4 mM), were sufficient to decrease by 70 to 85% the diffusivity of rabbit goblet cell mucins during exocytosis [23]. These results suggest that $Ca^{2+}$ should normally be chelated and/or removed from the epithelial surface and that increased $Ca^{2+}$ may be one of the many factors that could prevent normal mucus swelling in CF.

In summary we have outlined a conceptual scaffolding to evaluate available evidence and to infer hypothetical alternatives to explain abnormal hydration of mucus in CF based on polymer physics theory. Although there are other puzzling and very important facets in the pathology of CF, the understanding of the pathology of mucus remains an unresolved critical issue in CF research. It is recognized now that dehydrated CF mucus can result not only from defective composition of the swelling medium due to a defective apical CFTR, but also from defective mucin processing due to a defective intracellular membrane CFTR.

## V. MUCUS HYDRATION IN CF

The goal of these experiments was to test for differences in the swelling kinetics of exocytosed granules of cultured goblet cells derived from CF patients as compared to granules from goblet cells derived from non-CF individuals.

### A. Characterization of Swelling Kinetics of Mucin Granules

As discussed earlier, secretory granules undergo massive swelling upon release from the goblet cell [15,22,23,62,72]. For the purpose of establishing differences in the swelling properties of CF and non-CF secretory granules, an elegant and revealing approach is the one derived by Tanaka in his theory of swelling of polymer gels [70]. The application of this theory to exocytic hydration predicts that the swelling kinetics of the newly released granules must have the form of typical first-order kinetics, where the time course of the radial expansion of the exocytosed granule r(t) is related to the initial radius ($r_i$), and final radius ($r_f$) and expressed by the equation

$$r(t) = r_f - (r_f - r_i) \cdot e^{-t/\tau} \tag{1}$$

Tanaka's theory proposes that the diffusivity D [$cm^2$/sec] of the mucin polymer network in the extracellular solution be expressed by the following:

$$D = (r_f)^2/\tau \ [cm^2/sec] \tag{2}$$

In polyionic gels such as mucus, the diffusivity reflects the ability of the polymer network to hydrate, which in this case is driven primarily by the net Donnan potential of the mucin polymer matrix [22}. Thus, D is an extremely sensitive parameter to evaluate the effect of environmental conditions on the swelling of polyionic gels. Those conditions that have been experimentally verified in secretions include the effect of changes of pH, $Ca^{2+}$, and polyionic composition of the extracellular solution in the case of goblet cells [21,23] and pH, histamine, and $Ca^{2+}$ in the case of mast cells [59].

Conversely, if the ionic composition of the extracellular medium is kept constant, changes in swelling kinetics should reflect directly the balance between the effective polyionic drive of the granule and the internal forces that resist the expansion of the network. The forces that hinder expansion in tangle polymer networks result mainly from drag dissipations due to the mutual sliding of polymer chains during swelling [2,49].

Although the swelling kinetics are fairly consistent in each goblet cell, it can vary slightly among different cells. Thus, a more effective way to detect differences in swelling properties of secretory granules is to take advantage of the large number of granules secreted by individual cells and use a challenge test—that is, to compare swelling kinetics of CF and non-CF granules with reference to two or more defined swelling mediums. This protocol allows making observations in single goblet cells using the same cell as a control. We have demonstrated that $Ca^{2+}$ in concentrations of 2 to 4 mM can drastically decrease the diffusivity of the mucin polymer network upon exocytosis in rabbit goblet cells [23]. A similar protocol yielded surprisingly large differences in the swellability of mucin granules of goblet cells from CF patients compared to that of normal individuals.

## B. Primary Tissue Cultures of Human Goblet Cells

Primary tissue cultures of CF goblet cells were grown from three surgery specimens of nasal polyps and one necropsy specimen of trachea. Controls consisted of five nasal polyps from three non-CF individuals. These tissue specimens were kindly provided by the core services of the Cystic Fibrosis Center of the University of Washington. Primary tissue cultures were grown in Rose chambers following procedures previously reported [22]. Briefly, fresh biopsy or necropsy material was quickly transported to the tissue culture laboratory in Hanks' solution at 4°C, containing penicillin 200 IU/mL, streptomycin 150 μg/mL, fungizone 0.5 μg/mL, nystatin 12 IU/mL, vancomycin 12 μg/mL, and gentamicin 3 μg/mL. The tissue was washed in Hanks' solution, cut in small 25 $mm^2$ squares, and passed through three more rinses in sterile Petri dishes filled with antibiotic-containing Hanks' solution. The mucosa was then dissected in pieces of 8–10 $mm^2$, rinsed three more times, and further sectioned in 1 $mm^2$ explants. The explants were set in groups of six on the bottom coverslip of a Rose chamber and covered with an 8-μm polycarbonate membrane. Confluent monolayers enriched in human goblet cells have been grown in collagen with hormone supplemented serum-free medium, using a modified technique based on the methods initially described by Yankaskas et al. [73].

The Rose chambers were filled with 2 mL of hormone-supplemented serum-free Eagle's medium (Steinberger modification) at pH 7.35, following, with some modifications, the method developed by Wu [74,75] and containing the same mix of antibiotic-antimycotics listed above. The cultures were incubated for 14–25 days at 37°C, changing the antibiotic-enriched medium every 24 hrs.

## C. Validation of a Goblet Cell Model to Study Mucus Secretion

One of the paramount problems to the investigation of the polymer physics of mucus, is that even in normal airways mucus is always contaminated by dust and bacteria. The development of tissue culturing methods where mucus microgels can be directly observed as they are released from the cell, allows working with a native, uncontaminated mucus gel. Although these microgels are too small to conduct biochemical studies, they are perfectly suited for the physical study of swelling kinetics, PGPT, or for studies using dynamic laser light scattering for which polymers need to be in exceedingly dilute solutions.

The validation of our experimental model of cultured goblet cells includes (a) the application of quinacrine to characterize the presence of an acidic granular compartment and (b) immunocytochemical methods to establish the presence of membrane epitopes characteristic of human goblet cells.

## A Quinacrine Probe for Imaging Secretory Granules

Quinacrine has been shown to specifically accumulate in the lumen of secretory granules in several types of secretory cells [76]. Quinacrine is a fluorescent weak base with two protonatable groups. As shown in Figure 1, quinacrine that accumulates inside the granules is easily visualized in goblet cells. The application of this probe in tissue cultures of goblet cells derived from human individuals renders fluorescent the characteristic clusters of dark secretory granules observed in these cells by video enhanced phase contrast microscopy.

## Monoclonal Antibodies Tagging of Goblet Cells

Monoclonal antibodies against various types of human epithelial cells have been developed by Drs. Carol Basbaum and Walter Finkbeiner at the Cardiovascular Research Institute, San Francisco [77,78]. These probes

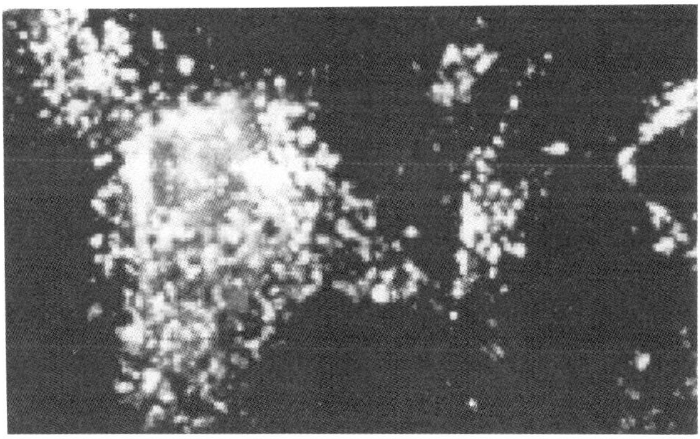

**Figure 1** Fluorescent videomicrograph of human goblet cells in culture. Notice the discrete accumulation of quinachrine in secretary granules.

provide unequivocal identification of cell type and overcome the uncertainties associated with morphological or histochemical identification of cells in culture. Monoclonal antibodies against surface epitopes and secretory granules of human goblet cells and against mucins [78] were kindly provided to us by Dr. Carol Basbaum. Shown in Figure 2 are a phase contrast video enhanced image and an image by fluorescence microscopy of a human goblet cell in culture fixed during degranulation and tagged with these monoclonal probes. Similarly, antibodies against mucins allowed us to confirm that the secretory material release by these cells contains mucins (data not shown).

At the time of the experiments, the cells were equilibrated in Hanks' solution and the cultures were scanned to find actively secreting goblet cells. Although, goblet cells can be stimulated with micromolar concentrations of ATP [79] (we confirmed this in human goblet cells) it was not necessary to do so in our preparation as the cells also seem to be sensitive to mechanical stimulation. The retrieval of the polycarbonate membrane at the beginning of the experiment was usually sufficient to turn on secretion. Both CF and normal goblet cells degranulated for about 6 to 10 min, producing 12 to 35 granules each. Exocytosis of 6 to 10 granules was first monitored and recorded in Hanks' medium (1 mM $Ca^{2+}$). The medium was then changed to Hanks' solution containing 2.5 mM $Ca^{2+}$ while exocytosis of another set of 6 to 10 granules was continually monitored and recorded. The data was obtained from about 12–30 granules/cell, with

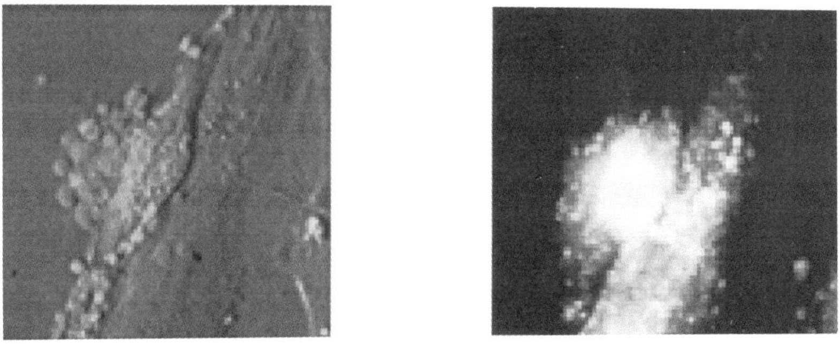

a b

**Figure 2** (a) Phase contrast video enhanced image of human goblet cells. (b) Fluorescent microscopy of a human goblet cell in culture fixed during degranulation and tagged with monoclonal probes

6–10 granule swellings at each $Ca^{2+}$ concentration. Although the swelling kinetics are fairly consistent in each goblet cell, it can vary slightly among cells. Thus, this protocol was designed to make observations in single goblet cells, using the same cell as a control. We collected data from 40 tissue cultures and about 200 goblet cells. Swelling kinetics was recorded by video microscopy using phase contrast optics at a magnification of 40× with a total of about 1500 exocytic events and a total of about 33 hrs of video recording. We present here a randomly selected subset of our data, which was measured in double blind form. Exocytosed mucus microgels swell isotropically forming microspheres of up to 15 μm in diameter. The video images were sampled at 10 fields/sec and transferred to an IBM Pentium PC by a frame grabber board. Diameter of the swelling granules was measured every 200 msec using an image analysis software program developed in our laboratory. Figure 3 below shows typical exocytic swelling kinetics of a mucin granule from a goblet cell.

As shown in Figure 4, the linear relationship between $\tau$ and the second power of the final radius of swollen mucus microgels allows direct calculation of the diffusivity of the mucus granules.

These results show that the diffusivity ($D = r^2/\tau$) of mucus from granules of CF goblet cells equilibrated in Hanks' solution—containing 1 mM $CaCl_2$—is 76% lower than D from mucus of normal goblet cells. Also

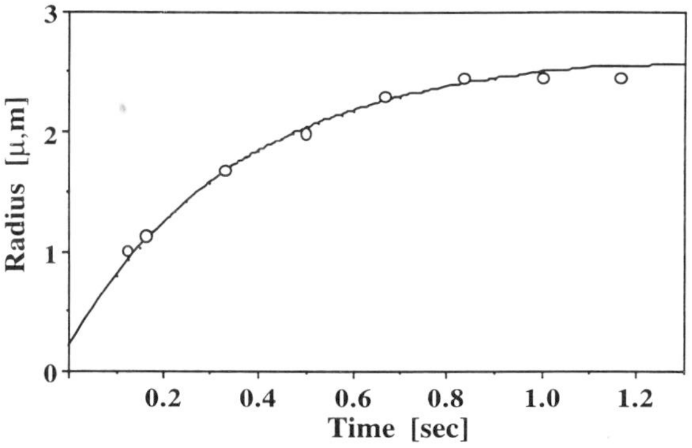


**Figure 3** Typical time course of the swelling of a mucin granule from a cultured goblet cell from airway mucosa. The time course of the radial increase closely follows first-order kinetics of the form $r(t) = r_f - (r_f - r_i)\exp(-t/\tau)$, where $r_f$ and $r_i$ are the initial and final radius and $\tau$ is the characteristic time of swelling. The continuous line is a nonlinear least-squares fitting to the experimental points

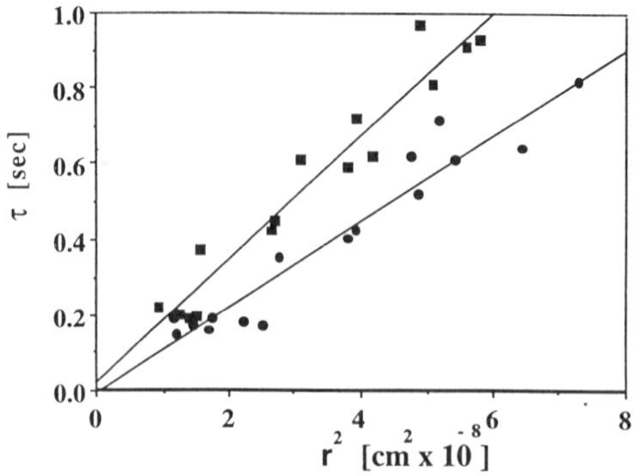


**Figure 4** The characteristic time of swelling $\tau$ of 32 granules equilibrated in 1 mM $Ca^{2+}$ Hanks' solution as a function of the final radius $r^2$ demonstrates a linear relationship. The straight line is a least-squares fitting to the data points. Notice that the slope of this line has the physical dimensions of a diffusion [$cm^2$/sec]. Squares are data from CF granules while circles correspond to normal controls

when challenged by increased $Ca^{2+}$ concentration in the swelling medium (from 1 to 2.5 mM) D in normal mucus decreased about 19%, while in CF mucus it decreased about 88%. These data show that in similar swelling medium, hydration of CF mucins takes place at a rate that is about 4 times lower than normal.

**Table 1** Diffusivity of Mucus

| | 1 mM $CaCl_2$ | Δ % decrease between controls and CF | 2.5 mM $CaCl_2$ | Δ % decrease between 1 and 2.5 mM $CaCl_2$ |
|---|---|---|---|---|
| Controls | D = 3.17 ± 1.23 $\times 10^{-7}$ | 76 | 2.56 ± 0.92 $\times 10^{-7}$ | 33 |
| CF | D = 7.41 ± 2.03 $\times 10^{-8}$ | | 9.11 ± 1.82 $\times 10^{-9}$ | 88 |

The results of 694 swelling events of mucin granules from 35 normal goblet cells and 37 CF goblet cells obtained from nasal polyps of 3 non-CF individuals and 4 CF patients. Numbers represent the average ± SEM.

## VI. CONCLUSION

In summary, we have identified the underlying physical principles that explain mucus hydration and rheology. Mucin secretory granules are condensed, and they decondense by hydration upon exocytosis. We have also shown that swelling of granules during exocytosis is driven by a Donnan potential and governed by Tanaka's theory of swelling of polymer gels. The diffusivity, D, as defined by Tanaka's theory, allows accurate characterization in a simple single parameter of the swelling properties of the mucin network [70]. Both Tanaka's theory and our feasibility studies indicate that the diffusivity of the mucin polymer network is a very sensitive parameter to evaluate the effective balance of ionic and polyionic charges that drive hydration of the mucin network and the internal hindrance of the granule that retards the swelling of the mucin polymer network. Therefore, the dramatic differences of swelling kinetics of newly released granules of CF and normal goblet cells—in an extracellular medium of controlled composition—give an objective indication of defective swelling properties of the mucin granule that could contribute to abnormal mucus hydration in CF.

## ACKNOWLEDGMENTS

Supported by grants from the University of Washington RRF and the USA NSF.

## REFERENCES

1. Edwards SF. The theory of macromolecular networks. Biorheology 1986; 23:589–603.
2. Edwards SF, Grant JWV. The effect of entanglements on viscosity of a polymer melt. J Phys A 1973; 6:1171–1180.
3. Havez R, Roussel P. Bronchial mucus: physical and biochemical features. In: Weiss EB, Sega MS. ed. Bronchial Asthma: Mechanisms and Therapeutics. Boston: Little, Brown and Company, 1976:409–422.
4. Kaliner M, Marom Z, Patow C, Shelhamer J. Human respiratory mucus. J Allergy Clin Immunol 1984; 73:318–23.
5. Potter JL, Matthews LW, Spector S, Lemm J. Studies on pulmonary secretions. II. Osmolality and the ionic environment of pulmonary secretions from patients with cystic fibrosis, bronchiectasis, and laryngectomy. Am Rev Respir Dis 1967; 96:83–97.
6. Roberts GP. Chemical aspects of respiratory mucus. Br Med Bull 1978; 34: 39–41.

7. Gibbons RA, Mattner PE. Some aspects of the chemistry of cervical mucus. Int J Fertil 1966; 11:366–379.
8. Lee WI, Verdugo P, Blandau RJ. Molecular arrangement of cervical mucus: a re-evaluation based on laser scattering spectroscopy. Gynec Invest 1977; 8:254–266.
9. Verdugo P, Tam PY, Butler J. Conformational structure of respiratory mucus studied by laser correlation spectroscopy. Biorheology 1983; 20:223–230.
10. Harding SE, Rowe AJ, Creeth JM. Further evidence for a flexible and highly expanded spheroidal model for mucus glycoproteins in solution. Biochem J 1983; 209:893–896.
11. Sheehan JK, Oates K, Carlstedt I. Electron microscopy of cervical, gastric, and bronchial mucus glycoproteins. Biochem J. 1986; 239:147–153.
12. Slayter HS, Lamblin G, Le Treut A, Galabert C, Houdret N et al. Complex structure of human bronchial mucus glycoprotein. Eur J Biochem 1984; 142: 209–218.
13. Kelemen RJ, Litt M, Holsclaw Jr, DS. Rheological and physicochemical properties of cervical mucus in cystic fibrosis patients. Biorheology 1989; 26:3:503–515.
14. Litt M, Wolf DP, Khan MA. Functional aspects of mucus rheology. Adv Exp Med Biol 1977; 89:91–102.
15. Verdugo P. Goblet cells secretion and mucogenesis. Annu Rev Physiol 1990; 52:157–176.
16. Verdugo P. Molecular biophysics of mucin secretion. In: Takishima T, Shimura S, eds. Airway Secretion: Physiological Basis for Mucus Hypersecretion. Marcel Dekker Inc, New York: 1993:101–121.
17. Flory PJ. In: Flory PJ, ed. Principles of Polymer Chemistry. Ithaca, London: Cornell University Press, 1969:432–493.
18. Donnan FG. The theory of membrane equilibria. Chem Rev 1924; 1:73–90.
19. Katchalsky A, Lifson S, Eisenberg HJ. Equation of swelling for polyelectrolyte gels. J Polymer Sci 1951; 7:571–574.
20. Tam PY, Verdugo P. Control of mucus hydration as a Donnan equilibrium process. Nature 1981; 292:340–342.
21. Aitken ML, Verdugo P. Donnan mechanism of mucin release and conditioning in goblet cells: the role of polyions. In: Chantler EN, ed. Mucus and Related Topics. New York: Plenum Press, 1989:73–79.
22. Verdugo P. Hydration kinetics of exocytosed mucins in cultured secretory cells of the rabbit trachea: a new model. In: Nuget J, O'Connor M, eds. Mucus and Mucosa (Ciba Foundation Symposium). London: Pitman, 1984; 109:212–234.
23. Verdugo P, Aitken M, Langley L, Villalon MJ. Molecular mechanism of product storage and release in mucin secretion. II. The role of extracellular $Ca^{++}$. Biorheology 1987; 24:625–633.
24. Frizzell RA. Role of absorptive and secretory processes in hydration of the airway surface. Am Rev Respir Dis 1988; 138:S3–S6.
25. Kilburn KH. A hypothesis for pulmonary clearance and its implications. Am Rev Respir Dis 1968; 98:449–463.
26. Widdicombe JG. Airway mucus. Eur Respir 1989; 2:107–115.

27. Knowles M, Gatzy J, Boucher R. Relative ion permeability of normal and cystic fibrosis nasal epithelium J Clin Invest 1983; 71:1411–1417.
28. Quinton PM. Cloride impermeability in cystic fibrosis. Nature 1983; 301: 421–422.
29. Kartner N, Riordan J. In vivo mislocation of CFTR in cystic fibrosis sweat duct. J Biol Chem 1992; 12(Suppl 16F). U106.
30. Riordan J, Rommens J, Kerem B-S, Alon N, Rozmahel R, Grzelczak Z, Zielenski J, Lok S, Plavsic N, Chou J-L, Drumm M, Iannuzzi M, Collins F, Tsui L-C. Identification of the cystic fibrosis gene: cloning and characterization of complementary DNA. Science 1989; 245:1066–1073.
31. Welsh M, Smith A. Molecular mechanisms of CFTR chloride channel dysfunction in cystic fibrosis. Cell 1993; 93:1251–1254.
32. Reisin EL, Prat AG, Abraham EH, Amara JF, Gregory RJ, Ausiello DA, Cantiello HF. The CFTR is a dual ATP and chloride channel. J Biol Chem 1994; 269:20584–20591.
33. Schweiber EM, Egan ME, Hwang TH, Fulmer SB, Allen SS, Cutting GR, Guggino WB. CFTR regulates outwardly rectifying chloride channels through an autochrine mechanism involving ATP. Cell 1995; 81:1063–1073.
34. Arispe N, Pollard HB, Rojas E. Calcium-independent K-selective channel from chromaffin granule membranes. J Membrane Biol 1992; 130:191–202.
35. Stutts JM, Canessa CM, Olsen JC, Boucher RC. CFT as a cAMP Regulator of Na channels. Science 1995; 269:447–850.
36. Al-Awqati Q, Barasch J, Landry D. Chloride channels of intracellular organelles and their potential role in cystic fibrosis. J Exp Biol 1992; 172:245–266.
37. Crowther RS, Marriott C. Counter–ion binding to mucus glycoproteins. J Pharm Pharmacol 1983; 36:21–26.
38. Forstner JF, Forstner GG. Calcium binding to intestinal goblet cell mucin. Biochim Biphys Acta 1975; 386:283–292.
39. Boat TF, Cheng PW. Mucous gycoproteins. In: Mangos JA, Talamo RC, eds. Cystic Fibrosis: Projections into the Future. New York: Stratton International, 1976:165–177.
40. Chase KV, Leahy DS, Martin R, Carubelli M, Flux M, Sachdev GP. Respiratory mucus secretions in patients with cystic fibrosis: relationship between levels of highly sulfated mucin components and severity of disease. Clin Chim Acta 1983; 132:143–155.
41. Cheng PW, Boat TF, Cranfill K, Yankaskas JR, Boucher RC. Increased sulfation of glycoconjugates by cultured nasal epithelial cells from patients with cystic fibrosis. J Clin Invest 1989; 84:68–72.
42. Lisle RC. Increased expression of sulfated gp300 and acinar tissue pathology in pancreas of CFTR (−/−) mice. Am J Physiol 1995; 268:G717–G723.
43. Barash J, Kiss B, Prince A, Saiman L, Gruenert D, Al-Awqati Q. Defective acidification of intracellular organelles in cystic fibrosis. Nature 1991; 352: 70–73.
44. Aitken ML, Pier MV, Martin TR, Moriarty A, Leturcq D, Skerrett SJ. Serum–dependent induction of interleukin-8 mRNA and protein in CF epithelial cells. Ped Pulm Suppl 1994; 10:A371.
45. Chin WC, Nguyen T, Verdugo P. ATP induces [$Ca^{2+}$] oscillations and re-

lease in/from secretory granules in respiratory goblet cells. Biophys J 1997; 72:A294.
46. Welsh MJ, Acidification indication. Nature 1991; 352:23–24.
47. Dosanjh A, Lencer W, Brown D, Ausiello DA, Stow J. Heterologous expression of DF508 results in decreased sialylation of membrane glyconjugates Am J Physiol 1994; 266:C360–C66.
48. Verdugo P. Phase Transition in Biopolymric Networks. Macromol Chem Phys Submitted.
49. de Genes PG. Reptation of polymer chain in the presence of fixed obstacles. J Chem Phys 1971; 55:572–579.
50. Gendler SJ, Burchell JM, Duhig T, Lamport D, White R, Parker M, Taylor-Papadimitriou J. Cloning of partial cDNA encoding differentiation and tumor-associated mucin glycoproteins expressed by human mammary epithelium. Proc Natl Acad Sci USA 1988; 84:6060–6064.
51. Gum JR, Hicks JW, Swallow DM, Lagace RL, Byrd JC, Lamport D, Siddiki B, Kim YS. Molecular cloning of the cDNA derived from a novel human intestinal mucin gene. Biochem Biophys Res Commun 1990; 171:407–415.
52. Porchet N, Van Cong N, Dufosse J, Audie JP, Guyonnet-Duperat V, Gross MS, Denis C, Degand P, Bernheim A, Aupert JP. Molecular cloning and chromosomal localization of a novel human tracheo-brochial mucin cDNA containing tandemly repeated sequences of 40 base pairs. Biochem Biophys Res Commun 1991; 175:414–422.
53. Villalon M, Basbaum CB, Johnson DE, Verdugo P. X-ray microanalysis of secretory granules from respiratory goblet cells. FASEB J 1988; 2(5): A958.
54. Villalon M, Johnson DE, Verdugo P. X-ray microanalysis of secretory granules from respiratory goblet cells. Am J Physiol In Press.
55. Tanaka T, Fillmore D, Sun S, Nishio I, Wislow GS, Shah, A. Phase transitions in ionic gels. Physical Review Letters 1980; 45:1636.
56. Viney C, Huber AE, Verdugo P. Liquid cristaline order in mucus. Macromolecules 1993; 26:852–257.
57. Viney C, Huber A, Verdugo P. Processing biological polymers in the liquid crystaline state. In: Ching C, Kaplan D, Thomas E, eds. Biodegradable Polymers and Packaging. Lancaster, Basel Technomics Publishing Co. Inc., 1993: 209–224.
58. Curran MJ, Brodwick MS. Ionic control of the size of the vesicle matrix of beige mouse mast cells. J Gen Physiol 1991; 90:771–790.
59. Fernandez JM, Villalon M, Verdugo P. Reversible recondensation of mast cell secretory products Biophys J 1991; 59:1022–1027.
60. Nanavati C, Fernandez JM. The secretory matrix, a fast acting smart polymer. Science 1993; 259:963–965.
61. Verdugo P, Deyrup-Olsen I, Martin AW, Luchtel DL. Polymer gel phase transition: the molecular mechanism of product release in mucin secretion. In: Karalis E, ed. Swelling of Polymer Networks. NATO ASI Series H. Vol. 64. Heidelberg: Springer-Verlag, 1992:671–681.
62. Verdugo P. Mucin exocytosis. Am Rev Resp Dis 1991; 144:S33–37.

63. Tanaka T. Gels. Sci Am 1981; 244:124–138.
64. Llavsky M. Effect of electrostatic interactions on phase transition in the swollen polymeric network. Polymer 1981; 22:1687.
65. Zimmermann H, Volknandt W, Wittich B, Hausinger A. Synaptic vesicle life cycle and turnover. J Physiol (Paris) 1993; 87:159–170.
66. Palade G. Intracellular aspects of the process of protein synthesis. Science 1975; 189:347–358.
67. Gerdes HH, Rosa P, Phillips E, Baeuerle PA, Frank R, Argos P, Huttner WB. The primary structure of human secretogranin II, a widespread tyrosine-sulfated secretory granule protein that exhibit low pH and calcium-induced aggregation. J Biol Chem 1989; 264:12009–12015.
68. Verdugo P. Polymer gel phase transition in condensatiodecondensation of secretory products Adv Poly Sci 1994; 110:145–156.
69. Osada Y, Okuzaki H, Hori H. A polymer gel with electrically driven motility. Nature 1992; 355:242–244.
70. Tanaka T, Fillmore DJ. Kinetic of swelling of gel J. Chem Phys 1979; 70: 1214–1218.
71. Izutsu K, Johnson D, Schubert M, Wang E, Ramsey B et al. Electron microprobe analysis of human labial gland secretory granules in cystic fibrosis. J Clin Invest 1985; 75:1951–1956.
72. Verdugo P, Deyrup-Olsen L, Aitken M, Villalon MJ, Johnson D. Molecular mechanism of mucin secretion: the role of intragranular charge shielding. J Dent Res 1987; 66:506–508.
73. Yankaskas J, Cotton C, Knowles M, Gatzy J. Culture of human nasal epithelial cells on collagen matrix supports. Am Rev Respir Dis 1985; 132: 1281–1287.
74. Wu R, Smith D. Continuous multiplication of rabbit tracheal epithelial cells in a defined, hormone-supplemented medium. In vitro 1982; 18:800–812.
75. Wu R, Wu M. Effects of retinoids on human bronchial epithelial cells: differential regulation of hyaluronate synthesis and keratin protein synthesis. J Cell Physiol 1986; 127:73–82.
76. Breckenridge LJ, Almers W. Currents through the fusion pore that forms during exocytosis of a secretory vesicle. Nature 1987; 328:814–817.
77. Basbaum CB, Mann JK, Chow AW, Finkbeiner WE. Monoclonal antibodies as probes for unique antigens in secretory cells of mixed exocrine organs. Proc Natl Acad Sci USA 1984; 81:4419–4423.
78. Finkbeiner WE, Basbaum CB. Monoclonal antibodies directed against human airway secretions. Localization and characterization of antigens. Am J Pathol 1988; 131:290–297.

# 20
# Evolution of the Frog Palate Model for Mucociliary Clearance—1970–1997

**Malcolm King and Eliane Festa**
*University of Alberta, Edmonton, Alberta, Canada*

## I. INTRODUCTION

The frog palate has a pseudostratified epithelium composed of both ciliated and mucus-secreting cells and covered with a two-layered periciliary film; since this is quite similar to the situation in human conductive airways, the frog palate has long been considered a useful model to study mucociliary clearance related to human lung disease. The first description of the weight-carrying capacity of frog palate cilia was published by Stewart in 1948 [1]. In 1970, Sadé et al. [2] described the importance of the link between the mucus and the cilia in facilitating particle clearance. During the 1970s and 1980s, a number of authors described relationships between mucociliary clearance on the isolated frog palate and various aspects of the physical and chemical properties of mucus. The mucus-depleted model is used to assess the inherent ciliary transportability of collected mucus from a variety of sources. One waits for depletion of the autogenous mucus in order to test exogenous mucus samples; the relative transport rate of test mucus compared with frog palate mucus is used as an index of mucus transportability [3,4].

More recently, the freshly excised frog palate preparation (*Rana catesbiana*) has been used as a model to study epithelial ion transport. It is used in a fresh preparation to maintain the integrity of the anatomy and the exchange mechanisms of the frog palate. The initial studies involved the analysis of mucociliary velocity (MCV) and transepithelial potential difference (PD) under the effects of amiloride on $Na^+$ transport [5]. This has been followed by studies of MCV, PD, cilia beat frequency, spinnability, rheology (log G*), depth of mucus layer, dry and wet weight, and ion content under the effects of UTP [6], surfactants [7], and benzamil [8]. These studies support the presence of both sodium and chloride channels on the epithelial surface. In order to further explore this model, a control study was performed to assess the effects of vehicle alone (Frog Ringer = ⅔ Ringer + ⅓ distilled water) on mucociliary clearance and related parameters [9]. Seven frog palates were exposed to nebulization with frog Ringer (six exposures of 12 min each using an ultrasonic nebulizer) simulating a "dose-response" curve. Frog Ringer did not significantly alter MCV, PD, mucus rigidity, secretion rate, or solids content over 6 successive vehicle administrations. The $Na^+$ concentration remained between 95–115 mmol/L and the $K^+$ concentration between 7.4–9.4 mmol/L. These results support the usefulness of this model for studying the actions of different drugs on ion transport and mucociliary clearance.

## II. MUCOCILIARY CLEARANCE

Mucociliary clearance, a vital mechanism of pulmonary defense, requires the coordination of many parameters for its effectiveness, such as cilia beat frequency (CBF), mucus secretion rate, the average depth of mucus layer, mucus rheology, and inherent transportability [10]. The properties of the mucus depend mainly on the content of macromolecules and water, and the degree of hydration presumably depends in turn on transepithelial chloride and sodium ion transport. Because of the difficulties in studying mucociliary clearance in intact mammalian airways, investigators long ago turned to more easily accessible systems in which to study the principles that regulate mucociliary transport. The frog palate was perhaps the most obvious choice because of its ready accessibility and its many similarities with higher systems.

The frog palate has a pseudostratified epithelium composed of both ciliated and mucus-secreting cells and covered with a two-layered periciliary film. Since this is quite similar to the situation in human conductive airways, the frog palate has long been considered a useful model to study mucociliary clearance related to human lung disease. The first description of the weight-carrying capacity of frog palate cilia was published by Stewart in 1948 [1]. Stewart demonstrated that the frog palate epithelium has tremendous capacity to carry loads varying in weight from microscopic spores to mercury droplets. In 1970, Sadé et al. [2] described the critical importance of the link between the mucus and the cilia in facilitating particle clearance. They showed that the action of the cilia is translated into effective clearance only in the presence of mucus. In the absence of sufficient mucus, the system fails, but the normal clearance function can be restored by replenishing the supply of mucus. In 1974, King et al. [3] showed that mucous gel simulants with similar rheological properties to mucus were also capable of coupling with the cilia to restore particle transport. These studies [2,3,11] provided the basis for the use of the excised, mucus-depleted frog palate as a practical model for testing the inherent clearability of mucus collected from various sources.

## III. INHERENT CILIARY CLEARABILITY OF MUCUS USING THE MUCUS-DEPLETED FROG PALATE MODEL

The mucus-depleted model is used to assess the inherent ciliary transportability of collected mucus from a variety of sources. One waits for depletion of the autogenous mucus in order to test exogenous mucus samples; the relative transport rate of test mucus compared with frog palate mucus is used as an index of mucus transportability [3,4]. During the 1970s and 1980s, a number of authors described relationships between mucociliary clearance on the isolated frog palate and various aspects of the physical and chemical properties of mucus. These included studies of sputum from chronic bronchitis and cystic fibrosis [12–14], canine tracheal mucus [15–17], bovine cervical mucus [18], and mucus simulants [19]. The existence of a rheological optimum for mucociliary clearance was demonstrated most elegantly by Shih et al. [15], using reconstituted canine mucin gels. Dulfano and Adler [12] and Meyer and Gelman [18] indicated the relative importance of the elasticity of mucus, as opposed to the viscosity, as a determinant of mucociliary clearance. Puchelle and collaborators [13,20] demonstrated that both elasticity and viscosity are important, and that optimal mucociliary clearability was achieved when intermediate values of both parameters were attained. King [17,21] demonstrated that mucus clearability could be best predicted from a combination of the elasticity and the mechanical loss tangent (viscosity/elasticity ratio), both measured at low frequency.

Frog palate clearability of mucus continues to be used as an outcome variable, in both applied and basic research. Normalized frog palate clearance of mucus has been used as a descriptive parameter in comparing disease states [22,23], the effects of drugs [20,23,24], or other interventions [25]. Human, dog, and rat mucus appear to all show qualitatively similar dependencies of clearability on mucus viscoelasticity [26].

## IV. THE FROG PALATE AS AN INTEGRATED MODEL FOR STUDYING MUCOCILIARY CLEARANCE

As stated above, effective mucociliary clearance requires the coordination of many parameters, including the cilia beat frequency, the mucus secretion rate, the average depth of mucus layer, as well as the mucus rheology and inherent transportability. Because of the complexities of the interaction between the mucus and the cilia, the effect on the integrated product, i.e., mucociliary velocity, cannot be readily predicted from measurements of just one parameter. The preparation of nondepleted frog palate allows us to study all of the relevant parameters contributing to mucociliary clearance in a unified system. Drugs to be tested can be nebulized, which permits the whole experiment to be performed with an intact mucociliary transport system. The frog palate is lined with a pseudostratified epithelium, composed of both ciliated and mucus-secreting cells, and covered with a two-layer periciliary fluid, which is similar to the situation found in human conductive airways. Previous studies with amiloride indicated the presence of sodium channels on the apical surface of the frog palate [5,27], and more recently the presence of $Ca^{2+}$-dependent chloride channels has also been demonstrated [6].

### A. Frog Palate Preparation

Frogs (*Rana catesbiana*) were sacrificed using the double pithing method [28]. Immediately after sacrificing the frog, the palate was excised by disarticulation of the jaw and removal of the upper part of the head by cutting with scissors through from the junction of the posterior pharynx and esophagus out to the skin of the back. The excised palate was then placed in the middle of a 20-liter acrylic chamber (20-cm height, 40-cm width, 25-cm depth) at 100% humidity, measured with an aneroid hygrometer (Bacharach 15238, Pittsburgh, PA), and at room temperature (20° to 22°C).

### B. Aerosol Administration

The acrylic chamber was humidified by an ultrasonic nebulizer (ICEL US500, Sto Paulo, Brazil) which gives an output of 1.15 mL/min of the solution nebulized and delivers particles of 2.4 μm (MMD) with a geometric standard deviation of 2.5. The vehicle, which consisted of dilute, non-lactated Ringer injection (⅔ Ringer + ⅓ distilled water), has an osmolarity of 206.5 mOsm/L and contains 98.3 m*M* of NaCl, 2.7 m*M* of KCl, and 1.5 m*M* of $CaCl_2$. The osmolarity of this solution has been shown to be appropriate for the frog palate [4]. The vehicle was nebulized over a period of 12 min, with a delivery rate of solution to the epithelium of 5.8 μL/min, as determined by measuring the increase in weight of the frog palate during several test nebulizations [5].

## C. Mucociliary Velocity

Determination of MCV was made by measuring the rate of displacement of 0.5-mm aluminum disks, with a mean weight of less than $10^{-4}$ mg, placed on the epithelial surface, using a stereomicroscope provided with a reticulated eyepiece [25]. The MCV was calculated by dividing the distance traveled (6 mm) by the time elapsed (min). At least five measurements were made for each nebulization to minimize measurement variability.

## D. Transepithelial Potential Difference

The PD was measured by using two flexible microelectrodes (agar bridges) saturated with KCl and connected to calomel half-cells, one for reference and the other for test, which were connected to the high impedance input of a grounded electrometer [5]. Because of the saturated agar bridges, a correction for diffusion potential was not necessary while measuring at different ion concentrations. The reference electrode was placed in the submucosa, and the test electrode was placed at predetermined locations on the epithelial surface: the intersection of cephalad, intermediate, and caudal regions with the vertical median line of the palate. The PD was computed as the mean of the first two measurements, as the PD measurements from the third area were excluded in order to minimize possible artifacts due to electrical leaks.

## E. Cilia Beat Frequency

The CBF was measured by a modification of the videoscopic technique described by Braga [28]. It consists of visualizing a group of cilia through an optical microscope (25 × objective, 10 × ocular) connected to a video camera. The incident light illuminating the ciliated epithelium is reflected from the cilia packed together and from the thin layer of mucus covering the cilia. This reflection is cyclic, as its direction changes according to the movements of the underlying cilia. These rhythmic changes of flaring and fading intensity are assumed to represent ciliary activity, and each cycle of light fluctuation corresponds to a beat cycle. The CBF was recorded at three locations per test area and then counted directly from the screen at slow motion, from which the real frequency is calculated.

### F. Mucus Collection Rate and Depth

Frog palatal mucus was collected at the cut edge of the palate, as previously described. The weight of mucus collected per unit time was used as an index of the mucus flux [16]. The linear velocity of mucus on the frog palate was determined as described above on MCV determination. From these data, an estimate of the depth of mucus was computed from the following relationship [29]:

$$\text{Mucus depth (mm)} = \frac{\text{Mucus collection rate (mm}^3\text{/min)}}{\text{Linear velocity (mm/min)} \times \text{palate width (mm)}}$$

### G. Mucus Transportability

This was evaluated using the classical in vitro frog palate preparation [3]. In these experiments, the transport velocity of mucus samples placed upon a mucus-depleted frog palate was determined with a reticulated eyepiece, in the same way as described above for aluminum disks. The mucus samples, collected from the frog palate after 12 min of nebulization with vehicle, were stored under light paraffin oil at $-80°C$. Each sample was allowed to reach room temperature before being tested. An aliquot of 1 to 5 μL was cleansed of the surface oil by a brief immersion in petroleum ether, which was permitted to evaporate from the sample before measurements were begun. It was previously demonstrated that this manipulation does not alter mucus transportability and rheology [30].

### H. Magnetic Microrheometry

This technique consists of analyzing the bulk viscosity and elasticity of microliter quantities of mucus [31]. A microscopic steel ball is carefully positioned in a 1–10 μL sample of mucus and the motion of this sphere under the influence of an electromagnet is used to calculate the rheologic properties of the mucus. The moving steel ball inserted in the mucus sample acts as a rheological probe, since its motion is opposed by viscous and elastic forces. By plotting the strain against the sinusoidal stress, it is possible to decompose the overall impedance into its elastic and viscous elements by determining the phase lag between stress and strain. As mucus does not behave as a Newtonian fluid, the measurements were done at three different frequencies (1, 10, 100 rad/s), and the results were expressed in terms of the rigidity modulus ($G^*$, the vectorial sum of elasticity and viscosity), and the loss tangent ($\tan \delta$, the ratio of viscous to elastic deformation). These rheological parameters can be used to predict the effectiveness of mucus in clearance by airflow interaction [21].

## I. Spinnability

The spinnability or filance, which is the ability of mucus to be drawn into long threads, was measured using a filancemeter (SEFAM, Nancy, France). This device measures, in millimeters, the large amplitude elastic deformation (spinnability) as thread formation and also evaluates the internal cohesion forces of the mucus [32]. The measurements were performed with a 15–20 μL oil-free mucus sample, at a retraction velocity of 10 mm/s, and repeated three times for each sample. The recorded spinnability was the average of these three measurements.

## J. Ion Content

Each aliquot of mucus was accurately weighed, suspended in 300 μL of 10% urea, vortexed for 1 min, then incubated for 24 h at 4°C. After incubation, the aliquots were vortexed again for 1 min, and centrifuged for 15 min at 14,000 rpm. The supernatant was analyzed for $[Na^+]$ and $[K^+]$ content by ion-selective electrodes (Sodium/Potassium Analyzer Nova 1), and for $[Cl^-]$ content by means of a direct-reading, digital chloridemeter (Corning Model 920 M), from which data for the ion contents of the original mucus samples were computed.

## K. Dry Weight and Wet Weight

The mucus samples, free of oil, were weighed on previously weighed glass slides. Samples were then dried in a microwave oven (750 watts for 30 min) and allowed to cool. These dried samples were then reweighed, and the percent solids composition (%SC) was calculated from the ratio of dry to wet weight of the mucus [31].

Seven frog palates were exposed to nebulization with frog Ringer (six exposures of 12 min each using an ultrasonic nebulizer, over the course of about 3 hours) simulating a dose-response curve [9]. Frog Ringer did not significantly alter MCV or PD, although both parameters tended to rise in parallel over the course of the experiment (Figure 1). Mucus rigidity, secretion rate, and solids content remained stable over 6 successive vehicle administrations (Figure 2): [$Na^+$] remained between 95–115 mM and [$K^+$] between 7.4–9.4 mM over the course of the experiment.

These results indicate that the freshly excised Rana catesbiana provides a preparation with excellent stability of mucociliary parameters over the course of at least three hours. The results thus support the usefulness of this model for studying the actions of different drugs on ion transport and mucociliary clearance.

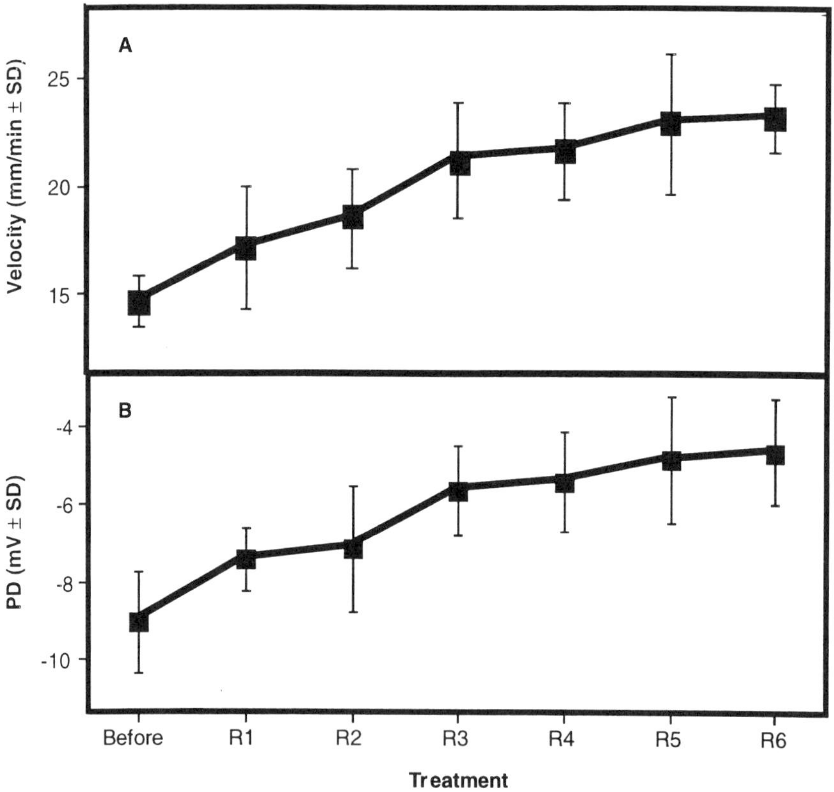


**Figure 1** Mucociliary velocity (A) and transepithelial potential difference (B) for freshly excised *Rana catesbiana* prior to and after six successive nebulizations of frog Ringer.

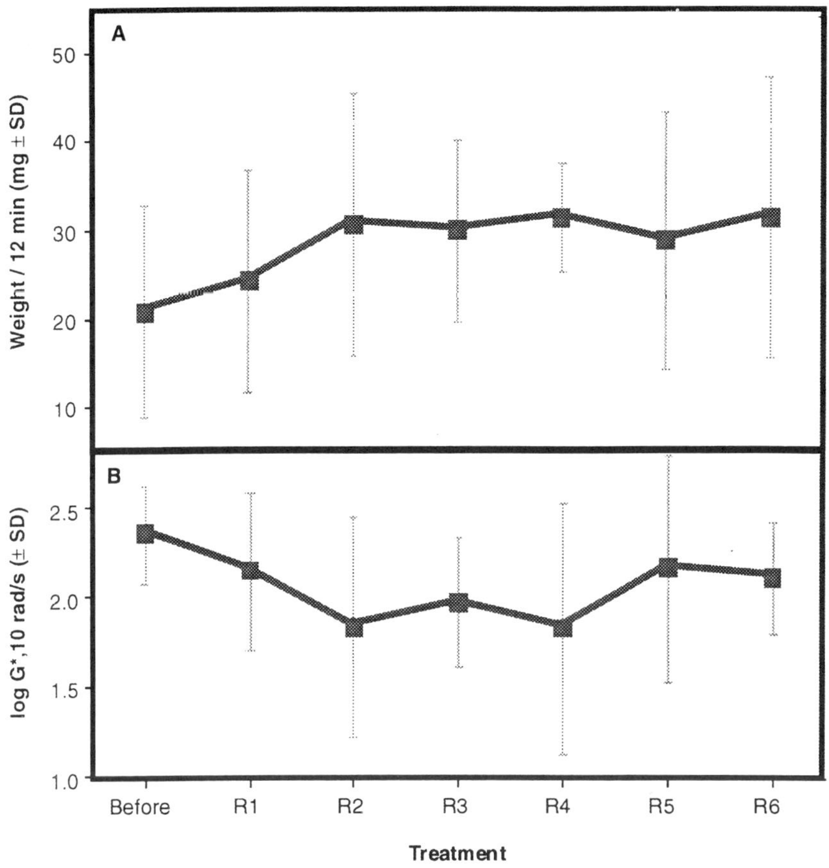


**Figure 2** Mucus collection rate (A) and mucus viscoelasticity (B) for freshly excised *Rana catesbiana* prior to and after six successive nebulizations of frog Ringer.

## REFERENCES

1. Stewart WC. Weight-carrying capacity and excitability of excised ciliated epithelium. Am J Physiol 1948; 152:1–5.
2. Sadé J, Eliezer N, Silberberg A, Nevo AC. The role of mucus in transport by cilia. Am Rev Respir Dis 1970; 102:48–52.

3. King M, Gilboa A, Meyer FA, Silberberg A. On the transport of mucus and its rheologic simulants in ciliated systems. Am Rev Respir Dis 1974; 110: 740–745.
4. Rubin BK, Ramirez O, King M. Mucus-depleted frog palate as a model for the study of mucociliary clearance. J Appl Physiol 1990; 69:424–429.
5. Festa E, Macchione M, Paiva PSO, Saldiva PHN, King M. Effects of aerosolized amiloride on mucociliary transport velocity and transepithelial potential difference in isolated frog palate. J Aerosol Med 1995; 8:167–176.
6. Festa E, Guimarães E, Macchione M, Saldiva PHN, King M. Acute effects of uridine 5′-triphosphate on mucociliary clearance in isolated frog palate. J Aerosol Med 1997; 10:25–39.
7. Festa E, Saldiva PHN, King M. Role of exosurf on mucociliary velocity, transepithelial potential difference and rheology in isolated frog palate (abstr). Eur Respir J 1995; 8:297s.
8. Festa E, Feng W, Saldiva P, King M. Dose-response study with benzamil on mucociliary velocity, transepithelial potential difference and rheological properties on frog palate (abstr). Eur Respir J 1996; 9:414s.
9. Festa E, Feng W, Saldiva P, King M. The freshly excised frog palate: a model for studying ion transportation. XII Intl CF Congr, Jerusalem, June 1996.
10. King M, Rubin BK. Rheology of airway mucus: relationship with clearance function. In: Takashima T, Shimura S, eds. Airway Secretion: Physiological Bases for the Control of Mucus Hypersecretion (Lung Biology in Health and Disease Series). New York: Marcel Dekker, 1994:283–314.
11. Sadé J, Meyer FA, King M, Silberberg A. Clearance of middle ear effusions by the mucociliary system. Acta Otolaryngol 1975; 79:277–283.
12. Dulfano MJ, Adler KB. Physical properties of sputum. VII. Rheologic properties and mucociliary transport. Am Rev Respir Dis 1975; 112:341–347.
13. Puchelle E, Girard F, Zahm JM. Rhéologie des sécrétions bronchiques et transport muco-ciliaire. Bull Eur Physiopathol Respir 1976; 12:771–779.
14. King M. Is cystic fibrosis mucus abnormal? Pediatr Res 1981; 15:120–122.
15. Shih CK, Litt M, Khan MA, Wolf DP. Effect of nondialyzable solids concentration and viscoelasticity on ciliary transport of tracheal mucus. Am Rev Respir Dis 1975; 115:989–995.
16. King M, Viires N. Effect of methacholine chloride on rheology and transport of canine tracheal mucus. J Appl Physiol 1979; 47:26–31.
17. King M. Interrelation between mechanical properties of mucus and mucociliary transport: Effect of pharmacologic interventions. Biorheology 1979; 16: 57–68.
18. Meyer FA, Gelman RA. Mucociliary transference rate and mucus viscoelasticity: Dependence on dynamic storage and loss modulus. Am Rev Respir Dis 1979; 120:553–557.
19. King M. Relationship between mucus viscoelasticity and ciliary transport in guaran gel/frog palate model system. Biorheology 1980; 17:249–254.
20. Puchelle E, Zahm JM, Polu JM. Drug effects on viscoelasticity of mucus. Eur J Respir Dis 1980; 61:195–208.

21. King M. Role of mucus viscoelasticity in cough clearance. Biorheology 1987; 24:589–597.
22. Rubin BK, Ramirez O, Zayas JG, Finegan B, King M. Respiratory mucus from asymptomatic smokers is better hydrated and more easily cleared by mucociliary action. Am Rev Respir Dis 1992; 145:545–547.
23. Rubin BK, Ramirez O, King M. The role of mucus rheology and transport in neonatal respiratory distress syndrome and the effect of surfactant therapy. Chest 1992; 101:1080–1085.
24. Tomkiewicz RP, App EM, De Sanctis GT, Coffiner M, Maes P, Rubin BK, King M. A comparison of a new mucolytic N-acetylcysteine L-lysinate with N-acetylcysteine: Airway epithelial function and mucus changes in dog. Pulmon Pharmacol 1995; 8:259–265.
25. Saldiva PHN, King M, Delmonte VLC, Macchioni M, Parada MAC, Daliberto MC, Sakae RS, Criado PMP, Silveira PLP, Zin WA, Böhm GM. Respiratory alterations due to urban air pollution: an experimental study in rats. Environ Res 1992; 57:19–33.
26. Lorenzi G, Böhm GM, Guimarã ET, Costa Vaz MA, King M, Saldiva PHN. Correlation between rheologic properties and *in vitro* ciliary transport of rat nasal mucus. Biorheology 1992; 29:433–440.
27. Curtis LN, Misch DW. Regulation of mucociliary transport: Response of mucous cells to stimulation by amiloride applied to the frog palate in vitro. Am Rev Respir Dis 1984; 129:816–822.
28. Braga PC. In vivo observation and counting methods for ciliary motion. In: Braga PC and Allegra L, eds. Methods in Bronchial Mucology. New York: Raven Press, 1988:269–276.
29. King M, Kelly S, Cosio M. Alteration of airway reactivity by mucus. Respiration Physiol 1985; 62:47–59.
30. Rubin BK, Ramirez O, Zayas JG, Finegan B, King M. Collection and analysis of respiratory mucus from individuals without lung disease. Am Rev Respir Dis 1990; 141:1040–1043.
31. King M. Magnetic microrheometer. In: Braga PC, Allegra L, eds. Methods in Bronchial Mucology. New York: Raven Press, 1988:73–83.
32. Puchelle E, Zahm JM, Duvivier C. Spinnability of bronchial mucus. Relationship with viscoelasticity and mucus transport properties. Biorheology 1983; 20:239–249.

# 21
# Airway Mucociliary Epithelium Injury and Repair

**Edith Puchelle, Anne-Laure Herard, and Jean-Marie Zahm**
*Unité INSERM, Reims, France*

## I. INTRODUCTION

The airways are unique in that they consist of a very large epithelial surface that is continuously exposed to the environment. The source of noxious gases and agents are numerous, varying from low to high acute or chronic exposure. Due to its special structure, the airway surface epithelium is more easily injured than any other epithelium, such as the skin, since the epithelium is normally not multilayered but simply composed of a pseudostratified respiratory type epithelium. In contrast to the skin, which is mainly protected by its histological impermeable multilayered cell structure, the airway epithelium uses specialized protective cells like the secretory cells, which continuously provide a renewable protective mucus layer to neutralize and avoid the noxious agents that might interact with the underlying epithelial cells.

This airway mucus closely interacts with the ciliary motion by means of the ciliated cells to, under normal conditions, continuously remove exogenous inhaled particles. This airway epithelium protection system, described as the mucociliary clearance mechanism or mucociliary escalator, continuously sweeps trapped inhaled materials out of the airways.

## II. HISTOLOGICAL STRUCTURE AND MAIN FUNCTIONS OF THE HUMAN RESPIRATORY SURFACE EPITHELIUM

### A. Structure

The surface airway epithelium acts as a first defense barrier against a variety of noxious agents such as pollutants and bacterial, viral, or chemical agents that can rapidly alter the airway epithelium defense functions. In the upper and lower proximal airways, the surface epithelium is pseudostratified (i.e., all the cells are attached to the basal membrane and only part of them reach the airway lumen). A large number of different cell types have been described in the airways as able to participate in the airway response to environmental factors [1,2]. Among the epithelial lining cells, which vary in their morphological phenotype according to their location and distribution in the upper or lower airways, five types of epithelial cells may play a major role in the defense function of the airway epithelium. These cell types include basal cells, ciliated cells, mucous (or goblet) cells, neuroendocrine cells, and undifferentiated cells or indeterminated cells that are in the process of differentiation. The terminal and the peripheral bronchioles are lined by a single layer of columnar cells mainly composed of ciliated and nonciliated secretory cells also called Clara cells. These Clara cells may also be identified at the tracheal level of murine species such as mice or rats.

We will mainly focus on the structure and functions of the cell phenotypes that are directly involved in the efficiency of the mucociliary system: the ciliated and mucous cells.

#### Ciliated Cells

Ciliated cells are terminally differentiated cells that do not divide. Since they are particularly susceptible to injury from inhaled irritants, they are very easily exfoliated and represent the first target cells to noxious agents. Apart from their major role in the propulsion of airway mucus and mucociliary transport, these ciliated cells are characterized by several junctional complexes that provide mechanical cohesion and probably also play an active role in transmitting signals that control cell growth and differentiation.

The ciliated cells represent 50–80% of the surface epithelial cells, they exhibit a columnar shape, and they extend along the surface epithelium to the airway lumen. Most of them are attached to the basement membrane but they can also be attached to basal cells by desmosomes [3]. The ciliated cells are attached to the adjacent ciliated cells by different junctional complexes as described below.

*Tight Junctions.* Tight junctions, also called zonula occludens (ZO), are situated at the apical side of the polarized surface epithelial cells [4]. The characteristic structure of these tight junctions, found using freeze-fracture technique, corresponds to a series of interwoven intramembrane strands [5]. These tight junctions, which are composed of different proteins such as $ZO_1$ and $ZO_2$, cingulin, and occludin, control the epithelial paracellular permeability to ions, water, and macromolecules. Selective permeability of the tight junctions to these molecules, together with active translocation of ions, results in the generation of a transepithelial potential difference. These tight junctions also confine specialized areas of the cell membrane allowing separation of apical from baso-lateral domains. In fact, the tight junctions are one of the intercellular junctions of epithelial cells that play two major and distinct biological functions: one is a selective barrier function and the other is the fence function in regard to plasma membrane constituents.

*Zonula Adherens.* Zonula adherens are identified at the apex of the ciliated cells and link the actin filament network between adjacent cells.

*Desmosomes.* Desmosomes are mainly identified at the lower lateral face of the cells and connect with the intermediate filament network of adjacent basal cells. These desmosomes are the major cell surface attachment sites for intermediate filaments at cell–cell contact. These junctional complexes, which belong to the new division of the cadherin family, are cell adhesion molecules known as desmosomal cadherins. They are very important for the regulation of junction assembly and specific attachment of intermediate filaments via adapter proteins [6].

## B. Main Functions of the Ciliated Cells

Ciliated cells are classically considered as playing a major role in the transport of particles by means of active ciliary beating. In fact, a number of other roles are assigned to these cells such as providing mechanical stability to the epithelium, aiding epithelial cross-talk, and regulating the water content at the mucus and airway epithelial surface.

The active transport across airway epithelium is secondary to active transepithelial transport, which plays a critical role in mucociliary clearance by regulating the depth of the periciliary or sol layer in which the cilia beat as well as the hydration of the gel phase of the mucus that lies on the tips of cilia [7]. In human airways, there is an active ion-transport process corresponding to an amiloride-sensitive absorption of $Na^+$ that regulates the water absorption, particularly in distal airways, and $Cl^-$

secretion, which promotes the secretion of fluid into the airway lumen [8]. The active secretion of $Cl^-$ is mainly regulated by the CFTR protein (Cystic Fibrosis Transmembrane Conductance Regulator). This membrane glycoprotein of $M_r = 170{,}000$ is restricted to the cytoplasm in nondifferentiated airway epithelial cells and to the apical domain of the ciliated cells in differentiated airway epithelium [9,10].

There is a great deal of evidence that this CFTR protein, identified in the apical plasma membrane of the ciliated cells, participates in a number of chloride and nonchloride transporting functions such as transport of nucleotides, water, and other ions. It may also regulate the functioning of other cellular components and may even represent a receptor for bacteria [11].

It is very difficult to study the role of ciliated cells in mucociliary clearance without analyzing in a dynamic and nondestructive way the interaction between cilia and the two phases: the sol phase, which is mainly controlled by the ciliated cells, and the gel phase produced by the mucous cells.

## Mucous Cells and Structure of the Mucociliary System

Airway secretions are produced by surface epithelial mucous cells, also called goblet cells, and by submucosal glandular mucous and serous cells that are present in the lamina propria. The secretions from goblet cells are discharged into the airway lumen during the exocytosis process.

The current concept of the mucus secretory process is that the secretory granules in the mucous membrane become discontinuous during the fusion of adjacent granules and that the content of the granules is exocytosed when the limiting membrane of the granules join and fuse with the surface of the apical membrane of the secretory cell [12]. In fact, according to Verdugo [13] and to our own experiences [14] using three-dimensional imaging of the exocytosis process after quick-freezing fixation (−260°C), it can be hypothesized that the mucus granules are discharged as individual granules (Figure 1). The mucins contained in the secretory granules undergo a massive swelling only after release from the cell. This gel expansion mechanism appears to be dependent on the hydration and osmolality of the environmental milieu [15], which in turn is controlled by an active ion and water transport largely determined by the ciliated cells. There is therefore a close interdependence between ciliated and mucous cells. The rapid freezing of the respiratory epithelium is a very useful technique that allows preservation of the sol and gel phase of mucus in native conditions. It perfectly preserves the integrity of the mucus blanket and induces an

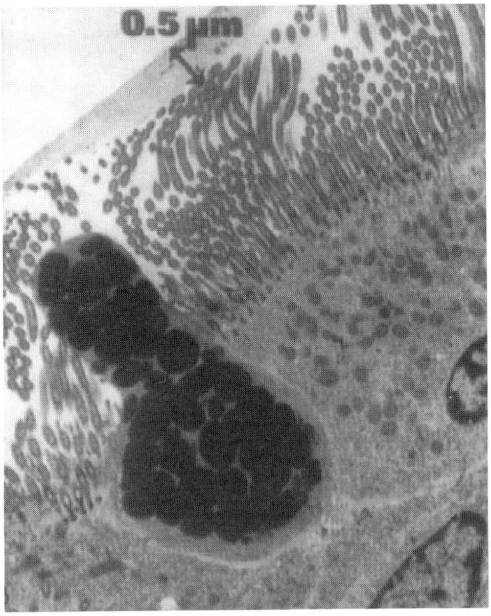


**Figure 1** Transmission electron microscopy of the mucociliary epithelium after rapid freezing and cryosubstitution allowing the gel mucus to be perfectly preserved. The two phases—gel (upper layer) and sol (periciliary) mucus—are clearly visible.

instantaneous interruption of the ciliary movement. A gradient of concentration of macromolecules can be observed at the ciliary level: few macromolecules appear to be randomly distributed in the so-called periciliary layer, whereas at the upper part of the cilia the macromolecules appear more dense and more aligned and constitute the electron-dense gel mucus layer. At the extreme apical part of the cilia, a capping structure with a crown of small bristles is seen that penetrates the upper gel mucus layer to depth of 0.5 μm, lifts it up, and pushes the mucus forward (Figure 2). The mucus gel layer appears as a continuous blanket characterized by a thickness ranging between 0.5 and 3.5 μm. The lowest values of mucus thickness are generally observed when the apical part of the cilia penetrates the gel [16].

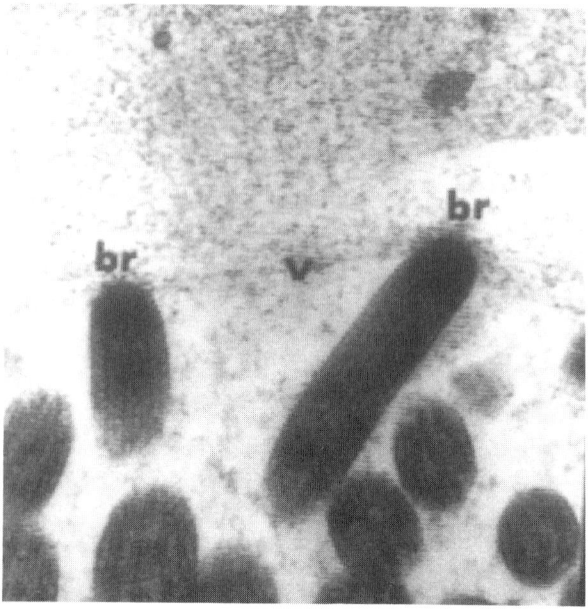


**Figure 2** Transmission electron microscopy of the mucociliary epithelium after rapid freezing and cryosubstitution. The tips of the cilia penetrate the upper gel layer during the active stroke of the ciliary beating.

## III. EFFICIENT MUCOCILIARY CLEARANCE DEPENDS ON OPTIMAL PHYSICAL AND RHEOLOGICAL PROPERTIES OF MUCUS

The efficacy of the mucociliary system depends not only on the number of active cilia, but also on ciliary coordination and speed at the tips of the cilia and on the rheological properties of the mucus. Hyperviscosity of the mucus leads to a decrease in ciliary beat frequency and is associated with a decrease in mucociliary transport [16]. The depth of the sol phase, which is determined by active ion transport and passive permeability through the respiratory epithelium, also plays a major role in the mucociliary clearance.

If the depth of the sol phase is less than the length of the cilia, the ciliary beating is ineffective. It has been shown in the rat that a decrease in the sol phase is accompanied by the arrest of ciliary beating [17]. By contrast, if the sol phase increases significantly, the cilia are unable to penetrate into the gel phase during active beating, and the ciliary beating becomes ineffective.

Other factors such as gravity and the airflow-mucus interaction during respiration may also affect mucociliary transport. The role of gravity is considered as negligible in healthy subjects [18]. However, if the depth of the periciliary layer is greater than 10 μm or if its viscosity increases, the influence of gravity then becomes important. Airflow seems to have an effect on mucociliary clearance only in the first-generation bronchi. During coughing, the interaction of the airflow with the mucosa provides a greater displacement of the mucus when the depth of the gel phase is increased [7]. It has also been shown that the mucus is necessary to ensure the transport of particles. In the absence of mucus, mucociliary transport is reduced or even abolished. The transport of particles may be restored after the deposition of a mucus substance or of substances with rheological properties similar to those of mucus.

## IV. IN VIVO ROLES OF MUCUS AND CILIA IN AIRWAY MUCOCILIARY CLEARANCE

In healthy subjects, large interindividual variations in nasal and airway mucociliary clearance have been reported. These variations can be attributed to differences in ciliary activity and/or in mucus properties. Proctor et al. [19], Puchelle et al. [20], and Lioté et al. [21] have found, either by measuring time of nasal transport of saccharin or using a radionucleotide technique, that in a population of healthy subjects, about 80% are fast movers whereas 20% are slow movers. Moreover, large intraindividual changes can be observed between repeated measurements performed several weeks apart. In a group of 20 healthy subjects, after measuring nasal transport time with saccharine, we collected mucus with a curved curette and collected ciliated cells by brushing the nasal mucosa. The transport rate capacity of the mucus was evaluated using the depleted frog palate and the ciliary beat frequency of the ciliated cells was analyzed using a photometric technique. It was of interest to note that the ex vivo transport rate capacity of the nasal mucus was significantly lower in the slow movers than in the fast movers. A negative and significant correlation was observed between the ex vivo nasal mucus transport rate and the in vivo nasal mucociliary transport rate. The faster the ex vivo nasal mucus transport rate, the shorter the in vivo nasal mucociliary transport time [21].

On the other hand no significant correlation could be observed between the ciliary beat frequency of the nasal ciliated cells and the in vivo mucociliary clearance. Moreover, we could also demonstrate that the ex vivo transport rate of the nasal mucus was significantly and positively correlated to the daily mean external temperature and negatively correlated to the $SO_2$ content of ambient air and to the unburned particle content of the air. In other words, these results suggest that physical properties of mucus (degree of hydration, osmolality, rheological properties) appear to play an essential role in the in vivo nasal mucociliary function and that environmental factors are likely to modify these physical and rheological properties.

## V. DOES THE SURFACE EPITHELIUM RESTORE ITS INTEGRITY AND REDIFFERENTIATE INTO A CILIATED RESPIRATORY EPITHELIUM AFTER INJURY?

### A. Epithelium Integrity Restoration after Injury

Inhalation of a variety of infectious as well as noninfectious agents such as ambient and occupational air pollutants are known to result in the loss of epithelial barrier integrity and impairment of mucociliary function.

It is well known that microbial infection perturbs junctional permeability. In an animal model, Azgahri et al. [22,23] have shown that elastase from Pseudomonas aeruginosa increases the alveolar epithelium permeability to albumin and that tight junctions are the primary site of enhanced permeability. These authors have also recently shown that elastase and exotoxin A, Pseudomonas aeruginosa exoproducts, have different but complementary actions that diminish barrier functions: elastase appears to increase alveolar epithelial permeability by rapidly damaging tight junction-associated proteins ($ZO_1$ and $ZO_2$), reducing the transepithelial resistance and increasing the mannitol epithelial permeability by a factor of 6 within 16 h. This means that brief contact with Pseudomonas elastase opens the junctions sufficiently to allow the bacteria to cross the epithelial barrier. Once the bacteria penetrate the epithelial lining, the secreted virulence factors like exotoxin A may render the cells unable to repair the damaged tight junctions and therefore facilitate the bacterial infection. It has also been shown by Stutts et al. [24] that the bacterial toxins of Pseudomonas aeruginosa are able to modify epithelial functions, either by altering the junctional barrier or by reducing the transcellular flux of chloride. It also has been shown by Phipps et al. [25] that pneumococcus infection increases the absorption of sodium and water. All these data suggest that infection may be associated with a marked dehydration of the periciliary sol layer and of the mucus with a resulting increased viscosity.

Apart from bacterial exoproducts, noninfectious agents such as $SO_2$, $NO_2$, and $O_3$ may also seriously injure the airway epithelium. The severity of the injury depends on the physical properties of the gas, its solubility, its concentration, the duration of the exposure, and the type of exposed cells. The ciliated cells are the most sensitive cells to most of the air pollutants. Desquamation of ciliated cells and increased paracellular permeability are observed within a short period of oxidant gas injury [26]. Whatever the source of injury, the response of the epithelium to injury can be expressed through a succession of cellular lesions varying from the loss of the junctional barrier to complete epithelial desquamation [27].

The reepithelialization of the surface airway epithelium after a wound has been analyzed by Zahm et al. [28] and Hérard et al. [29]. Cell spreading and migration are important steps in this repair mechanism allowing the reepithelialization of the denuded epithelium. Extracellular matrix proteins like fibronectin and the corresponding cell surface receptor, the integrin $\alpha_5\beta_1$, are overexpressed during the repair process [30]. Much of the fibronectin detected in injured lung is of plasma origin but is also locally synthesized by mesenchymal, endothelial, and epithelial cells. During repair, both synthesis and degradation of the extracellular matrix proteins occurs, allowing the epithelial cells to migrate. We have recently shown [29] that the complete restoration of junctional integrity is achieved only 1 to 2 days after wound closure. During this period, the airway epithelium may remain exposed to the noxious effect of airway environment in vivo owing to oxidants, proteases, and osmolarity imbalance, which are all capable of modifying tight-junction formation [31] and may therefore prevent the complete restoration of the epithelial barrier. Such a situation may be observed in vivo during chronic airway inflammation such as in asthma, chronic bronchitis, or cystic fibrosis [32]. We can also hypothesize that even after wound closure, the impairment of epithelial integrity may favor bacterial migration through the epithelial surface across intercellular spaces. After injury and disruption of the tight junctions, receptors might be exposed on basolateral membranes as well as on basal cells and the basement membrane, leading to the preferential adhesion of bacteria and viruses [33]. This mechanism could partly explain the adherence of Pseudomonas aeruginosa to the repairing airway epithelium. In addition, secretion of bacterial exoproducts is responsible for enhancing injury, thus preventing the repair process [34]. Mechanisms of bacterial adherence during airway epithelium repair need better understood in order to prevent infections in patients after non-infectious epithelial injury.

### B. Redifferentiation of the Mucociliary Epithelium after Injury

It is well known that the main target cells of inhaled pollutants, chemical agents, and bacterial exoproducts are the ciliated cells. The injury may either induce a loss of cilia or even a complete desquamation of ciliated cells and consequently a severe impairment of mucociliary function. In vivo, in different types of animal models, the epithelial regeneration begins 2–5 days after injury and is completed within 2–4 weeks. As the ciliated cells are not able to divide, their progenitor cells may be from different sources.

Of the main three cell types that constitute the tracheal epithelium (basal, secretory, and ciliated cells), only basal cells and secretory cells divide. The still-accepted dogma is that the basal cells of the airway epithelium are progenitors for secretory and ciliated cells, but evidence has accumulated in favor of multiple cell origin of the ciliated cells. According to McDowell et al. [35], during regeneration of hamster tracheal epithelium, following diverse types of injury, secretory cells play a major progenitor role. These authors also demonstrated in culture that ciliated cells were absent from surface epithelial cell cultures at day 5 but present at day 7. Ultramicroscopic analysis gave evidence that these ciliated cells had been generated in the cultures from nonciliated progenitors called preciliated immature cells, which contained mucous granules and clusters of basal bodies in the apical cytoplasm. Such preciliated cells arise in large numbers during human tracheobronchial epithelial development [36] as well as epithelium regeneration in the adult after injury. In fact, there are close similarities between normal human mucociliary development in the fetus and epithelial regeneration following injury in the adult. In each case, after a rapid proliferation of cells initially undifferentiated, there is a subsequent increase in the number of secretory cells followed by an increase in ciliated cells and a reciprocal decrease in secretory cells. Transitional cell phenotypes such as preciliated cells are observed. As suggested by Gordon and Lane [37], it is probable that ciliated cells arise partly from basal or directly from cycling cells but that post-mitotic redifferentiation of secretory cells may represent an important source of ciliated cells in regenerating tracheal epithelium.

It is also highly probable that the ionic environment as well as autocrine or paracrine factors such as EGF secreted by epithelial cells may directly control the differentiation of the ciliated cells [38]. Moreover, it has been shown that certain air-liquid culture conditions are essential for ciliary differentiation. In contrast, when airway cells are grown immersed in culture medium, attached to either tissue culture plastic or some type of biomatrix, they tend to lose their normal differentiated mucociliary features. An optimal concentration of EGF (0.5 to 25 ng) results in a much more uniform columnar mucociliary epithelium whereas a higher concentration may cause hyperplasia, multilayering, and squamous metaplasia. Regulation of mucociliary differentiation is also largely determined by cell-extracellular matrix molecules. We have earlier shown that in serum-free defined culture medium the percentage of ciliated cells and the induction of ciliogenesis was significantly dependent on the substrate characteristics. Collagen I, prepared as a thick gel, was optimal for ciliary differentiation [39]. Recent data from Davenport and Nettesheim [40] confirms that the mechanophysical properties of the substrate are important in modulating ciliated cell differentiation. Their data suggests that plating cells on type I collagen gel enhances cell attachment and accelerates the onset of secretory differentiation, and that type I collagen gel seems to be more critical for ciliated cell than for secretory cell differentiation.

Cell-matrix culture conditions may also directly modulate ciliated cell differentiation. We have shown [41] that freshly isolated surface respiratory epithelial cells cultured in a three-dimensional (3-D) collagen gel lattice are capable of forming branching structures with tubules similar to that described during fetal development of the human lung.

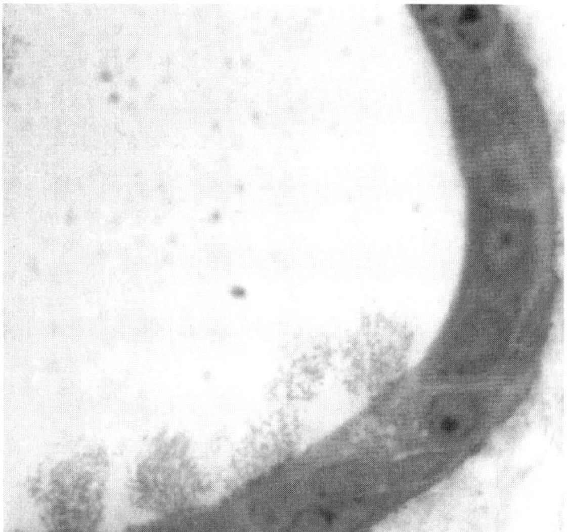

**Figure 3** Light microscopy of human respiratory surface epithelial cells cultured in a collagen I 3-D lattice. Well-differentiated basal cells, ciliated cells and secretory cells are identified.

Cross sections of these tubules examined by light and transmission electron microscopy showed that the lumen formed by the surface epithelial cells in 3-D culture was covered by cuboidal-shaped polarized cells with numerous cilia actively beating with a mean ciliary beat frequency ($9.1 \pm 1.1$ Hz) very similar to that identified in human nasal and tracheal cilia (Figure 3). Interestingly, we also observed active ciliogenesis in preciliated cells that exhibited both secretory granules and centrioles migrating toward the apex of the cells.

The process of tubule formation and ciliary differentiation in a 3-D lattice may result from several factors including (a) the physical properties of the lattice collagen I gel, which possesses a loose and flexible structure that permits changes in cell shape and (b) expression of extracellular matrix components like type-IV collagen and laminin, which interact with the cell surface integrin receptors. In fact, when the cells are embedded in a 3-D lattice, they first loose their initial polarity, the extracellular matrix molecules begin to organize on the basal side of the tubules, and a progressive reappearance of apical polarity is observed when the interacytoplasmic cavity and subsequent tubule lumen appear.

It is likely that the ability of the epithelial cells surrounding the newly formed lumen to regulate their ion and liquid environment favors ciliogenesis. We can also hypothesize that the secretion of growth factors by the embedded cells may modulate the synthesis of extracellular matrix proteins as well of that of matrix-degrading proteases.

## REFERENCES

1. Breeze RG, Wheeldon EB. The cells of the pulmonary airways. State of the Art. Am Rev Respir Dis 1977; 11:705–777.
2. Danel C. Morphologic characteristics of human airway structures. Diversity and unity. In: Chretien J, Dusser D, eds. Environmental Impact on the Airways. From Injury to Repair. Vol. Series Lung Biology in Health and Disease, 1996:19–41.
3. Evans MJ, Cox RA, Shami SG, Plopper GC. Junctional adhesion mechanisms in airways basal cells. Am J Respir Cell Mol Biol 1989; 3:341–347.
4. Carlson JL, Collier AM, Gambling TM, Leigh MW, Boucher RC, Hu SC, Boat TF. Development, organization and function of tight junctional complexes in the tracheal epithelium of infant ferrets. Am Rev Respir Dis 1988; 138:666–674.
5. Godfrey RWA, Severs NJ, Jeffery PK, A comparison between the epithelial tight junction morphology of human extrapulmonary bronchi and rat trachea. Eur Respir J 1994; 7:1409–1415.

6. Green K, Jones JCR. Desmosomes and hemidesmosomes: structure and formation of molecular components. FASEB J 1996; 10:871–881.
7. Sleigh MA, Blake JR, Liron N. The propulsion of mucus by cilia. Am Rev Respir Dis 1988; 137:726–741.
8. Knowles M, Murray G, Shallas J, Askin F, Ranga K, Gatzy J, Boucher R. Bioelectric properties and ion flow across excised human bronchi. J Appl Physiol 1984; 56:868–877.
9. Puchelle E, Gaillard D, Ploton D, Hinnrasky J, Fuchey C, Boutterin MC, Jacquot J, Dreyer D, Pavirani A, Dalemans W. Differential localization of the cystic fibrosis transmembrane conductance regulator in normal and cystic fibrosis airway epithelium. Am J Respir Cell Mol 1992; 7:485–491.
10. Brézillon S, Dupuit F, Hinnrasky J, Marchand V, Kälin N, Tümmler B, Puchelle E. Decreased expression of the CFTR protein in non cystic fibrosis patients. Lab Invest 1995; 72:191–200.
11. Biwersi J, Verkman AS. Functions of CFTR other than as a plasma membrane chloride channel. In: Dodge JA, Brock DJH, Widdicombe JH, Cystic Fibrosis. Current Topics. Vol. 2 eds. 1994:155–171.
12. Kim SK, Nasjletti CE, Han SS. The secretion process in mucous and serous secretory cells of the rat sublingual gland. J Ultrastruct Res 1972; 38:371.
13. Verdugo P. Goblet cells secretion and mucogenesis. Annu Rev Physiol 1990; 52:157–176.
14. Puchelle E, Beorchia A, Menager M, Zahm JM, Ploton D. Three-dimensional imaging of the mucus secretory process in the cryofixed frog respiratory epithelium. Biol Cell 1991; 72:159–166.
15. Widdicombe JH, Widdicombe JG. Regulation of human airway surface luqid. Respiration Physiology 1995; 99:3–12.
16. Puchelle E, Zahm JM, Quemada D. Rheological properties controlling mucociliary frequency and respiratory mucus transport. Biorheology 1987; 24: 557–563.
17. Melville GN, Iravani J. Factors affecting ciliary beat frequency in the intrapulmonary airways of rats. Can J Physiol Pharmacol 1975; 53:1122–1128.
18. Pavia D. Lung mucociliary clearance. In: Clarke SW, Pavia D, eds. Aerosols and the Lung: Clinical and Experimental Aspects. London.
19. Proctor D. The mucociliary system. In: Proctor D, Andersen I, eds. The Nose Upper Airway Physiology and the Atmospheric Environment. Amsterdam: Elsevier Biomedical Press, 1982:245:2.
20. Puchelle E, Aug F, Zahm JM, Bertrand A. Comparison of nasal and bronchial mucociliary clearance in young non smokers. Clin Sci 1982; 62:13–16.
21. Liote H, Zahm JM, Pierrot D, Puchelle E. Role of mucus and cilia in nasal mucociliary clearance in healthy subjects. Am Rev Respir Dis 1989; 140: 132–136.
22. Azghani AO, Kondepudi AY, Johnson AR. Interaction of *Pseudomonas aeruginosa* with human lung fibroblasts: role of bacterial elastase. Am J Respir Cell Mol Biol 1992; 6:652–657.
23. Azghani AO. *Pseudomonas aeruginosa* and epithelial permeability: Role of

virulence and factors elastase and exotoxin A. Am J Respir Cell Mol Biol 1996; 15:132–140.
24. Stutts JM, Schwals JH, Chen MG, Knowles MR, Boucher RC. Effects of *Pseudomonas aeruginosa* on bronchial epithelial ion transport. Am Rev Respir Dis 1986; 134:17–21.
25. Phipps RJ, Torealba PJ, Lauredo IT. Bacterial pneumonia stimulates macromolecule secretion and ion water fluxes in sheep trachea. J Appl Physiol 1987; 67:2388–2397.
26. Samet JM, Cheng PW. The role of airway mucus in pulmonary toxicology. Environ Health Perspect 1994; 102:89–103.
27. Puchelle E, Zahm JM. Repair processes of the airway epithelium. In: Chretien, Dusser, eds. Lung Biology in Health and Disease. Environmental Impact on the Airways. From Injury to Repair. Vol. 1995:157–182.
28. Zahm JM, Chevillard M, Puchelle E. Wound repair of human surface respiratory epithelium. Am J Respir Cell Mol Biol 1991; 5:242–248.
29. Herard Al, Zahm JM, Pierrot D, Hinnrasky J, Fuchey C, Puchelle E. Epithelial barrier integrity during in vitro wound repair of the airway epithelium. Am J Respir Cell Mol Biol 1996; 15:624–632.
30. Herard Al, Pierrot D, Hinnrasky J, Kaplan H, Sheppard D, Puchelle E, Zahm JM. Fibronectin and its $\alpha 5\beta_1$ integrin receptor are involved in the wound repair process of airway epithelium. Am J Physiol 1996; 271: L726–733.
31. Lewis Sa, Berg JR, Kleine TJ. Modulation of epithelial permeability by extracellular molecules. Physiol Rev 1995; 75:561–589.
32. Carson JL, Collier AM, Gambling TM, Knowles MR, Boucher RC. Ultrastructure of airway epithelial cell membranes among patients with cystic fibrosis. Hum Pathol 1990; 21:640–647.
33. de Bentzmann S, Plotkowski C, Puchelle E. Receptors in the *Pseudomonas aeruginosa* adherence to injured and repairing airway epithelium. Am J Respir Crit Care Med 1996; 154:S155–162.
34. Ishihara H, Honda I, Shimura S, Sasaki H, Takishima T. Role of chronic *Pseudomonas aeruginosa* infection in airway mucosa permeability. Chest 1991; 100:1607–1613.
35. McDowell EM, Ben T, Newkirk C, Chang S, de Luca L. Differentiation of tracheal mucociliary epithelium in primary cell culture recapitulates normal fetal development and regeneration following injury in hamsters. Am J Pathol 1987; 129:511–522.
36. Gaillard D, Lallemand A, Petit A, Puchelle E. In vivo ciliogenesis in human fetal tracheal epithelium. Am J Anat 1989; 185:415–428.
37. Gordon RE, Lane BP. Wound repair in rat tracheal epithelium. Division of G1 and G2-arrested cells, following injury. Lab Invest 1989; 42:616–624.
38. Zahm JM, Pierrot D, Puchelle E. Epidermal growth factor promotes the wound repair of human respiratory epithelium. Wound Rep Reg 1993; 1: 175–180.

39. Chevillard M, Hinnrasky J, Zahm JM, Plotkowski MC, Puchelle E. Proliferation, differentiation and ciliary beating of human respiratory ciliated cells in primary culture. Cell and Tissue Research 1991; 264:49–55.
40. Davenport EA and Nettesheim P. Regulation of mucociliary differentiation of rat tracheal epithelial cells by type I collagen gel substratum. Am J Respir Cell Mol Biol 1996; 14:19–26.
41. Benali R, Tournier JM, Chevillard M, Zahm JM, Klossek JM, Hinnrasky J, Gaillard D, Maquart FX, Puchelle E. Tubule formation by human surface respiratory epithelial cells cultured in a three-dimensional collagen lattice. Am J Physiol 1993; 264:L183–192.

# 22

# Septal Deformities Can Influence the Mucociliary Transport Impairment of the Middle Turbinate Mucosa

**Ranko Mladina**
*University Hospital Šalata, Zagreb, Croatia*

**Damir Miličić and Davorin Djanić**
*General Hospital Slavonski Brod, Slavonski Brod, Croatia*

## I. INTRODUCTION

Almost twenty years ago it was clearly shown by Ninderer [1] that septal deformities can cause metaplastic changes of the nasal mucosa owing to high-speed inspiratory and expiratory air currents, which on sniffling or snorting may reach hurricane speeds. This statement was confirmed later on by the fact that there is considerable influence of shapes and dimensions on velocity fields in the nasal cavity [2]. The airflow is considered to be incompressible, steady, and laminar [3]. This is in accordance with the finding of Hess [4] who stated that under static conditions laminar flow without turbulence can be seen in all sections of the nasal cavities. Some authors, however, determined airflows also to be moderately turbulent [5], depending on airflow velocity. According to these data septal deformities also can cause variations in airflow velocities: the narrow places in the nasal cavity give rise to increased velocity and vice versa.

Recent investigations [6] have shown that the type-3 septal deformity as classified by Mladina [7] most commonly produces obvious squamous cell metaplasia and subsequent impairment of mucociliary transport of the mucosa of both medial and lateral surfaces of the middle turbinate.

The goal of the current study was to analyze histological changes of the middle turbinate mucosa in cases of septal deformities of various types.

## II. METHODS AND MATERIALS

Thirty-seven pairs of cadaveric middle turbinates were used in the study. None of the subjects had a history of chronic nasal or sinus disease. Thirty-one of them had one of the septal deformities according to Mladina's classification [7] whereas six showed an almost straight septum and served as a control group.

Almost the whole middle turbinate was removed using microscissors and the FESS technique. The cut was always done from anterior to posterior, parallel to the anterior cranial base, starting from the highest point of the middle turbinate insertion to the lateral nasal wall. After removal

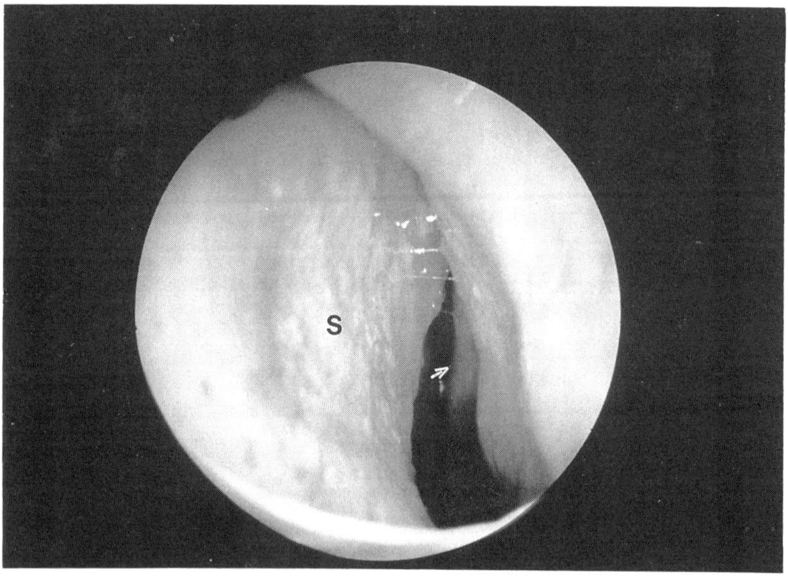


**Figure 1** Rhinoscopic appearance of a left-sided type 3 septal deformity. The almost vertical ridge at the junction of quadrangular and perpendicular lamina is located in the very close vicinity of the head of the middle turbinate. S = septum, white arrow = middle turbinate hidden by the deformity.

the medial surface of the turbinate was marked using a small 4-0 silk stich, so that during the preparation of the sample for microscopic analysis it was possible to orientate the turbinate in the proper position. After the removal of the turbinate, the middle meatus was examined throughout by an endoscope to evaluate whether any morphological changes of the regional mucosa were present. Every turbinate was then cut in three coro-

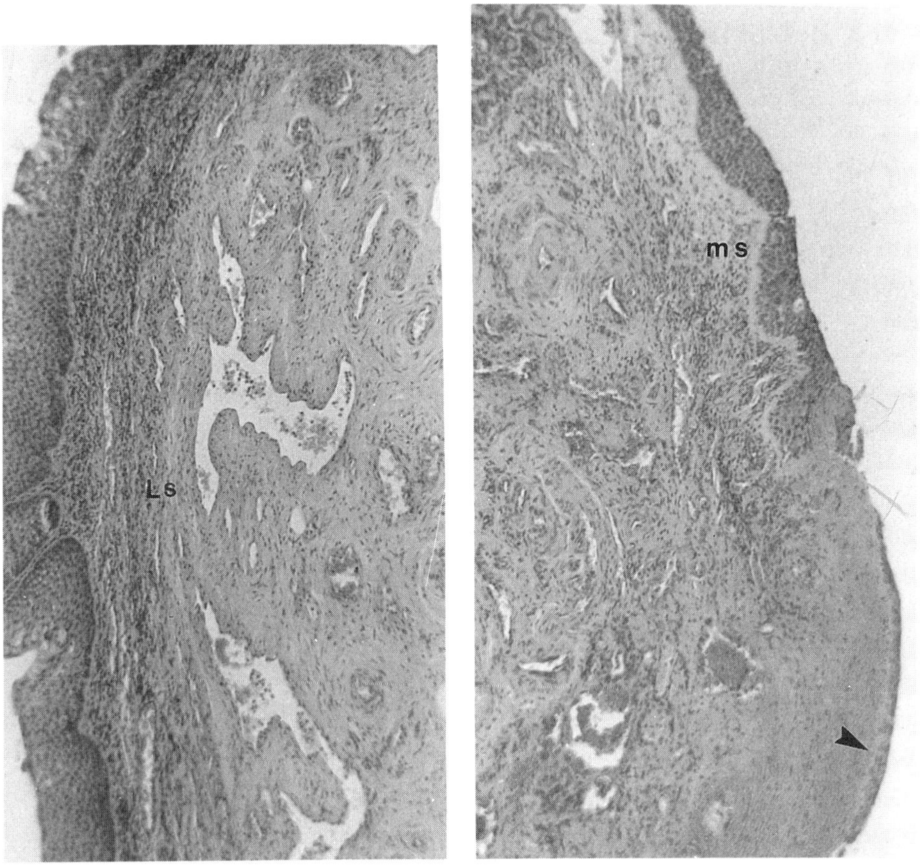


a b

**Figure 2** (a) Histological appearance of the frontal section through the anterior third of the right middle turbinate in right-sided type 3 septal deformity. Ls = lateral surface, ms = medial surface. (b) Medial surface shows scarce epithelial layer, which in the lower part seems to be totally wiped out (black arrow).

nal, i.e. frontal cuts, which were made in the anterior, middle, and posterior third of the turbinate. Cutting the turbinates in frontal sections enabled us to observe the lateral and medial surface of the turbinate at the same time.

The histological analysis was performed by means of light microscopy. The samples were stained in a standard manner using the May-Grunwald-Giemsa technique. The presence or absence of squamous cell metaplasia was noted.

## III. RESULTS

Various types of septal deformities were found: the most frequent was type 3 (Figure 1), which was found in 17 (54.8%) of the subjects. Type 5 was found in 8 (25.8%), whereas types 7, 2, and 4 were seen less frequently. No endoscopically visible morphological mucosal changes were found in any middle meatus examined.

Histological changes were found in all of the 17 subjects with the type-3 septal deformity. Complete squamous cell metaplasia with no trace of ciliated cells (lateral and medial surfaces) was seen in both left and right middle turbinate (Figure 2a,b). In the other types of septal deformities only one subject showed incomplete changes of the medial mucosal surface of the ipsilateral middle turbinate. The subjects with straight septums showed normal respiratory mucosa (Figure 3a,b).

## IV. DISCUSSION

The interest of clinicians in the histological changes of the nasal mucosa and their relation to inspiratory and expiratory air-stream speed through the nose has a long history. More than twenty years ago it was shown that nasal mucosa can be affected long-term by inspiratory and expiratory air currents, which on sniffling or snorting may reach hurricane speed [1].

Furthermore, it was shown by means of laser anemometry [2] that shapes and dimensions have a considerable influence on velocity fields in the nasal cavity and that the air flow was streamlined or made more uniform by the turbinates. The same authors showed that the airflow velocity in the nose was greater near the septum and that the flow generally was greater in the lower half of the nasal cavity. The airflow is considered to be incompressible, steady, and laminar [3]. This is in accordance with the finding of Hess [4] who stated that under static conditions laminar flow without turbulence can be seen in all sections of the nasal cavities.

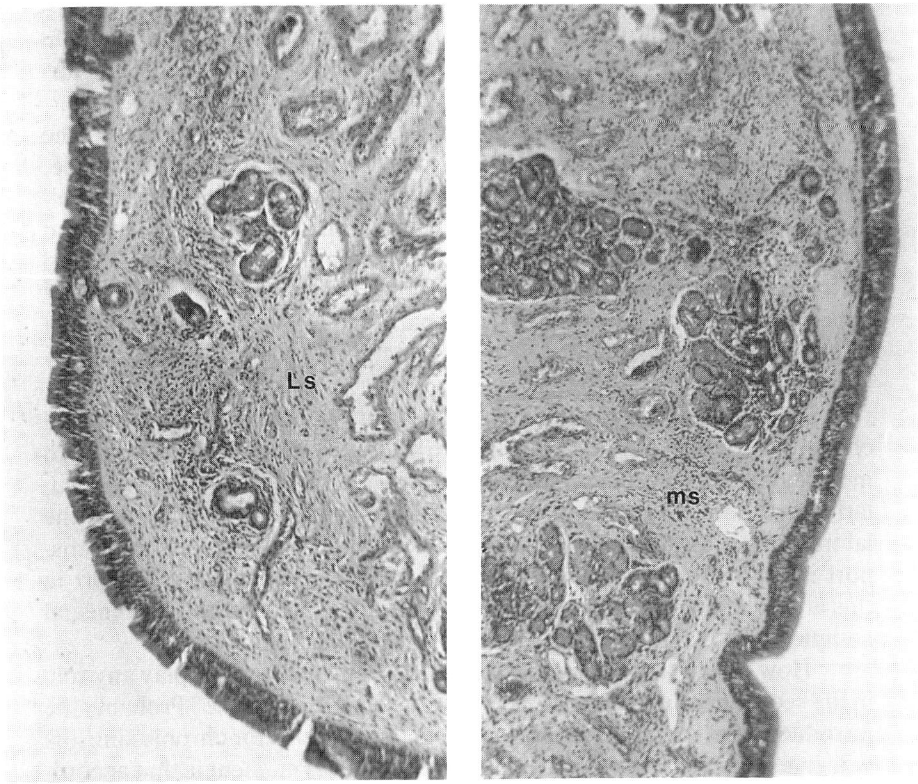


a b

**Figure 3** Histological appearance of the frontal cut through the anterior third of the right middle turbinate in straight septum. Normal respiratory epithelium can be seen both on the (a) lateral and (b) medial surface. Ls = lateral surface, ms = medial surface.

Some authors, however, have determined that airflows may be moderately turbulent [5]. This was found to be dependent on the airflow velocity.

However, no data on the possible impairment by this mechanism of mucociliary transport of the respiratory mucosa are available in the literature. It is, of course, clear that destroyed respiratory epithelium can not support any mucociliary activity. Therefore any factor causing changes in the histomorphology of the nasal mucosa must be considered in detail.

# 23

# Aspects of Mucociliary Flow and Mechanics in Nasal Cavity and Sinuses

**Elina Toskala***
*Tampere University Hospital, Tampere, Finland*

## I. INTRODUCTION

The epithelium of the upper airways forms the first line of defence against toxic and infectious agents in the inhaled air. The healthy mucous membrane has many important tasks: a transport function, a secretory function, a phagocytic function, and an immunological function. The most effective and important of these to the mucosal defence mechanism is the removal function carried out by the mucus blanket and ciliated cells [1,2]. When mucociliary function is disturbed, the impaired mucociliary function is usually associated with long-term respiratory diseases, nasal infections, sinusitis, and otitis media [3]. Primary ciliary dyskinesia (PCD), with absence of dynein arms, is an inherited disorder and characterised by inappropriate motility of the cilia [4].

## II. RESPIRATORY EPITHELIUM

The typical respiratory tract epithelium consists of columnar pseudostratified ciliated cells. This type of epithelium covers the middle and posterior thirds of the nasal cavity and the thickness of epithelium is about 2.5 μm. Nonciliated columnar cells are found on the anterior part of the nasal septum, as well as on the middle and lower turbinate. The human nose epithelium consists of basal cells and intermediate cells, goblet cells, and ciliated and nonciliated columnar cells.

* *Current affiliation:* Bowman-Gray School of Medicine, Winston-Salem, North Carolina

On the epithelial cell surface, the cilia are motile projections that keep the film of superficial mucus moving. Cilia are 5–7 μm long and about 0.2 μm wide. Cilia have nine peripheral microtubule doublets and two central tubules. Adjacent doublets are connected by nexin links. Peripheral tubules have inner and outer dynein arms. Dynein arms are morphological equivalents of ATP-splitting proteins and can be regarded as the actual energy source of the cilia [1,4]. A cross-section from a cilium shows most important ultrastructural details of cilia, which are microtubules A and B, radial spokes, spoke heads, central sheet, two central tubules, inner and outer dynein arms, and nexin links [1,5].

## III. PATHOLOGY OF RESPIRATORY CILIA

Loss of ciliated cells and cilia and ultrastructural abnormalities in cilia impair mucociliary function [2,3]. Primary ciliary dyskinesia (PCD) with absence of dynein arms is the best known of all these conditions. Many kinds of ultrastructural abnormalities have been published during the last decades. Ciliary disorientation is reported in PCD and in chronic sinusitis [3,6]. Tubular anomalies, radial spoke and nexin link defects have also been reported in PCD. Compound cilia are seen in PCD but also in healthy individuals.

## IV. MUCOCILIARY TRANSPORT

Mucociliary transport in the respiratory tract is necessary for health and normal function of the tissue, particularly in resistance to respiratory infection. Effective mucociliary function is based on a proper functional relationship between three components: mucus, the moving cilia, and the periciliary fluid. The mucus blanket lining the major airways is approximately 5 μm thick and biphasic in organization. The mucus blanket is divided into the more luminal gel phase (mucus layer) and the abluminal sol phase (serous layer), the latter in direct contact with the apical cytoplasm of the epithelial cells. Water is the main component (98%) of the mucous secretation the rest being mucous and serum-like glycoproteins, lipids, and inorganic salts produced by submucosal glands and goblet cells. The cilia of the respiratory tract beat in the low-viscosity periciliary layer that fills the interciliary space [7].

The motility of the cilia is characterised by a rapid forward stroke (effective stroke) and a slow return stroke (recovery stroke). The movements of the cilia are synchronized, producing metachronal waves that pass over the epithelial surface, thereby moving the film of mucus in one direction. In maxillary sinuses this direction is from the bottom of the sinus towards the normal ostium (even after fenestration in the inferior meatus) and the nasal transit is towards the nasopharynx. The ethmoid and sphenoid sinuses empty directly into the ostium. In frontal sinuses, transit has a lateral stream to the ostium from the lateral part of the floor and a descending stream in the lateral part of the frontonasal duct.

## V. FACTORS AFFECTING MUCOCILIARY FLOW

Mucociliary function can be disturbed by viral and bacterial infections, because of cilia toxic endotoxins produced by bacteria and because of epithelial metaplasia and loss of cilia and ciliated cells. Also abnormalities in ciliary morphology, PCD, certain air pollutants and dust, and systemic dehydration disturb normal ciliary function.

In acute infection, focal fibrosis and edema appear. In chronic inflammatory processes, the infiltration of mononuclear cells, luminal dilation of the glands exhibiting zymogen granule depletion, and increased thickness of the whole mucosal layer is seen [8].

Septal deformities, nasal polyps, bullous medial meatus, and other anatomical variations can narrow the nasal cavity causing decreased air flow in the narrow nostril and on the other hand increased air flow in the wider side of the nose. Increased air flow in the nostril may cause the drying of the epithelium and may increase mucus viscosity and impair mucociliary flow. Increased air flow will also cause epithelial metaplasia and decrease the amount of ciliated cells. When the nasal cavity or the middle meatus is so narrow that the medial wall is in contact with mucosa of the lateral wall, the function of cilia is disturbed and the mucus transit is mechanically blocked. The obstruction of the sinus ostium will cause changes in the sinus atmosphere and a low pH, which impairs ciliary function as well.

## REFERENCES


1. Satir P, Sleigh MA. The physiology of cilia and mucociliary interactions. Annu Rev Physiol 1990; 52:137–155.
2. Ohashi Y, Nakai Y. Functional and morphological pathology of chronic sinusitis mucous membrane. Acta Otolaryngol Suppl 1983; 397:11–48.
3. Nuutinen J, Toskala E, Saano V, Joki S. Ciliary beating frequency in sinusitis. Arch Otolaryngol Head Neck Surg 1993; 119:645–647.

4. Afzelius BA. A human syndrome caused by immotile cilia. Science 1976; 66: 317–319.
5. Rautiainen M, Collan Y, Nuutinen J. Kärjä J. Ultrastructure of human respiratory cilia: A study based on serial sections. Ultrastruct Pathol 1984; 6:331–339.
6. Toskala E, Rautiainen M, Nuutinen J. Scanning and transmission electron microscopic findings in cilia from human nasal turbinate and sinus mucosa following respiratory infection. Eur Arch Otorhinolaryngol 1994; 251:76–79.
7. Sleigh MA, Blake JR, Liron N. The propulsion of mucus by cilia. Am Rev Respir Dis 1988; 137:726–741.
8. Westrin KM, Stierna P, Carlsöö B, Hellström S. Mucosal fine structure in experimental sinusitis. Ann Otol Rhinol Laryngol 1993; 102:639–645.

# 24
# Clearance Function of the Eustachian Tube: What We Know, What We Don't Know

**Charles D. Bluestone and Cuneyt M. Alper**
*Children's Hospital of Pittsburgh, and University of Pittsburgh School of Medicine, Pittsburgh, Pennsylvania*

## I. INTRODUCTION

The eustachian tube is part of a system of contiguous organs including the nose, palate, nasopharynx, middle ear, and mastoid air cells. The tube has at least three physiologic functions with respect to the middle ear: (a) pressure regulation (ventilation) of the middle ear (b) protection from nasopharyngeal sound pressure and secretions, and (c) clearance of secretions produced within the middle ear (or foreign material) into the nasopharynx. Clearance of secretions from the middle ear is provided by the mucociliary system of the tube and some areas of the middle-ear mucous membrane. Also involved in clearance is the pumping action of the eustachian tube during closing. Clearance can be affected by many factors such as viruses, bacteria (and their toxins), irradiation, tobacco smoke, leukotienes $C_4$ and $D_4$, prostaglandin $E_2$, negative middle-ear pressure, and presence of a middle-ear effusion. Most investigators consider impairment of clearance function to be related to failure to resolve middle-ear effusions as opposed to being the primary cause of the disease. However, patients who have primary ciliary dyskinesia have chronic middle-ear effusions. This is what we know, but there is a great *deal* we do not know about the clearance function of the healthy tube and the role tubal clearance plays in the pathogenesis of otitis media and related conditions.

## II. NORMAL EUSTACHIAN TUBE FUNCTION

The eustachian tube is part of a system of contiguous organs including the nose, nasopharynx, middle ear, and mastoid gas cells. The eustachian tube is not just a tube but is an organ consisting of a lumen with its mucosa, cartilage, surrounding soft tissue, peritubal muscles (i.e., tensor veli palatini, levator veli palatini, salpingopharyngeus, and tensor tympani), and its superior bony support, the sphenoid sulcus. The structure of the tube is closely associated with function.

In addition to clearance of secretions produced within the middle ear into the nasopharynx, the eustachian tube has two other physiologic functions with respect to the middle ear: pressure regulation (i.e., ventilatory function) of the middle ear to equilibrate gas pressure in the middle ear with ambient pressure and protection from nasopharyngeal foreign secretions and sound pressure.

### A. Pressure Regulation

Regulation of pressure within the middle ear is the most important function of the eustachian tube, since hearing is optimum when middle-ear gas pressure is relatively the same as ambient pressure in the external auditory canal, i.e., tympanic membrane and middle ear compliance is physiologic. When the tube is functioning normally, intermittent active opening of the eustachian tube, due only to contraction of the tensor veli palatini muscle during swallowing, maintains nearly ambient pressures in the middle ear [1–5]. Opening of the eustachian tube permits gas exchange and equalization of pressures between the environment and the middle ear. Under normal conditions, fluctuations in ambient pressure are bidirectional (i.e., either to or from the middle ear), relatively small in magnitude, and not readily appreciated [6]. These fluctuations reflect the rise and fall in barometric pressures associated with changing weather conditions and elevation, or both. However, the changes in middle-ear pressure show mass directionality, can achieve appreciable magnitudes, and can result in pathologic changes. A major reason for these conditions is that the middle ear is a relatively rigid (i.e., noncollapsible) gas pocket that is surrounded by mucous membrane in which gases are exchanged between the middle-ear space and the mucosa. Differential pressure exceeds 54 mm Hg between the middle ear space at atmospheric pressure and the microcirculation in the mucuos membrane. This represents a diffusion-driven gradient from the middle ear cavity to the mucosa that can produce an underpressure (relative to ambient pressure) in the middle ear of more than 600 mm Hg during equilibration.

## B. Protective Function

The eustachian tube normally protects the middle ear due to its anatomic structure, which depends also on an intact middle ear and mastoid so that a middle-ear and mastoid gas cushion is present. The protective function has been assessed by radiographic techniques [7–9], a modification of a tubal patency test described by Wittenborg and Neuhauser [10]. In these studies, radiopaque material was instilled into the nose and nasopharynx of children who had otitis media and compared to those who were otologically normal. In the physiologic state, contrast material entered the nasopharyngeal end of the eustachian tube during swallowing activity, but did not enter the middle ear. The dye did reflux into the middle ear in some patients who had middle-ear disease, especially during closed-nose swallowing. At rest, the normal eustachian tube is collapsed and the tubal lumen is closed, thus preventing liquid from entering the nasopharyngeal end of the tube. During swallowing, when the proximal end (i.e., cartilaginous portion) opens, liquid can enter this part of the tube, but does not gain entrance into the middle ear, due to the narrow midportion of the tube, the isthmus, and the gas pocket in the middle ear and mastoid.

In addition to the anatomic aspects of the eustachian tube protective function related to the middle ear, respiratory epithelium of the tubal lumen provides local immunologic defense, as well as mucociliary defense, i.e., clearance.

## C. Clearance Function

Clearance of secretions from the middle ear is provided by two mechanisms: (a) the mucociliary system of the eustachian tube and some areas of the middle-ear mucous membrane and (2) muscular clearance or pumping function of the eustachian tube when the tube closes. Before describing these two functions, the histology of the mucous membrane of the system needs to be reviewed.

## D. Mucosa of Eustachian Tube–Middle-Ear–Mastoid

The mucous membrane of the middle ear and mastoid is continuous with that of the nasopharynx via the eustachian tube and is characterized as respiratory epithelium. This membrane covers all structures within the middle ear, including the ossicles, vessels, and nerves. Examination of cells of the mucous membrane within the tympanic cavity reveals a gradual change from tall, columnar cells with interspersed goblet cells to shorter cuboid cells at the posterior portion of the promontory and aditus-ad-antrum and mastoid cells.

Structural differentiation of the mucosal lining of the lumen of the eustachian tube is evident: mucous glands predominate at the nasopharyngeal orifice and there is a graded change to a mixture of goblet, columnar, and ciliated cells near the tympanum [11]. The products of these glands include active components [12,13]. Matsune and coworkers [14] found that the density of goblet cells was significantly lower in the roof of the eustachian tube than in the floor of the tube and was highest in the mid-cartilaginous area. From the same laboratory, Sando and colleagues [15] have postulated from their extensive examination of temporal bone specimens that the superior portion of the tubal lumen is associated with pressure regulation of the middle ear and the inferior portion is responsible for clearance, whereas both parts are involved with protection of the middle ear. These investigators also have shown that the folds in the mucosa of the lumen of the tube are most pronounced in infants and children as compared to adults; the folding affords greater surface area for the tubal lumen, which in turn provides greater abundance of ciliated cells, which should enhance clearance function, [16]. The mucociliary defense system within the tube begins in fetal life and is well established soon after birth [17].

## E. Mucociliary Clearance

Mucociliary clearance has been investigated for the past 50 years utilizing a variety of methods (Table 1). Clearance has been studied by instilling radiopaque material into the middle ear of children whose tympanic membrane was not intact, or when, or when the material entered the middle ear (intact tympanic membrane) from the nasopharynx, [8,9] by radioisotopic methods (e.g., technetium 99m) [18,19], with sequential contrast computerized tomography [20], and following insertion of foreign material into the middle ear of animal models [21,22]. Following placement of foreign material in humans with perforations of the tympanic membrane, Sade [23] reported that the anterior half to two-thirds of the middle ear cavity had the most active clearance properties. Such material will flow toward the middle-ear portion of the eustachian tube and out the tube. This movement is related to ciliary activity that occurs in the eustachian tube and parts of the middle ear; these ciliated cells in the middle ear are increasingly more active as their location becomes more distal to the opening of the eustachian tube [24].

## F. Muscular Clearance

In a series of elegant experiments employing cineradiography by Honjo [25–27], the eustachian tube was shown to pump liquid out of the middle

**Table 1** Methods Used To Assess Mucociliary Clearance Function of Middle-Ear–Eustachian Tube in Humans and Animals

| Method | Reference |
| --- | --- |
| Radiographic | Welin, 1947 |
| | Compere, 1960 |
| | Wittenborg and Neuhauser, 1963 |
| Fluorescein | Rogers et al., 1962 |
| Fluoroscopic with contrast material | Bluestone, 1971 |
| Methylene blue | Baucr, 1975 |
| Saccharin | Elbord and Larsen, 1976 |
| Foreign bodies | Albiin et al., 1983 |
| | Sade, 1987 |
| Radioisotopic | LaFaye et al., 1974 |
| | Karja and Nuutinin, 1987 |
| | Takeuchi et al., 1990 |
| Sequential CT with contrast material | Niwa et al., 1990 |

ear in both animal models and humans. These investigators also studied both the mucociliary and muscular clearance in the cat using various viscosities of colored liquid and showed that when the volume of middle-ear liquid was small, the fluid was cleared by the mucociliary system, whereas when the volume of liquid was large and of low viscosity it was cleared by muscular activity. Highly viscous fluid was cleared by both ciliary and muscular clearance. Also, clearance time was affected by viscosity in the mastoid, but not the middle ear, and clearance was more effective in the tympanic cavity than the mastoid [28].

## G. Surface Tension Factors

Surface tension factors have been identified that are related to normal eustachian tube function. Birkin and Brookler [29] were the first to isolate surface-tension–lowering substances from washings of eustachian tubes of dogs. They postulated that these substances could act to enhance eustachian tube functions, similar to surfactant in the lung. Rapport and colleagues [30] described a similar substance and demonstrated the effect of washing out the eustachian tube on the opening pressure in the experimental animal; others also have demonstrated a surfactant-like phospholipid in the middle ear and eustachian tube of animals and humans [31–33]. In a recent study in gerbils, Fornadley and Burns [34] produced middle-ear effusions by injecting killed *S. pneumoniae* into the middle-ear through the tympanic membrane, which increased the opening pressure of the eustachian tube. When the investigators introduced exogenous surfactant, the opening pressure dropped. Some investigators have described the similarity between lung surfactant and surface-active substances in the eustachian tube [35].

From these studies, it is apparent that the clearance function of the eustachian-tube–middle-ear system is important in maintaining a healthy middle ear. Since otitis media is so common in humans, especially infants and young children, efficient removal of middle-ear effusions must depend, to a large extent, on this function.

## III. ABNORMAL EUSTACHIAN TUBE FUNCTION

All three functions of the eustachian tube can be abnormal, which in turn can be related to the pathogenesis of otitis media: (a) impairment of pressure regulation function of the tube, (b) loss of protective function of the eustachian tube-middle ear, and (c) impairment of clearance function of the tube and middle ear. Abnormality of one of these functions may not be associated with an abnormality of another, such as abnormal pressure regulation and clearance [20]. Pathologic changes in other parts of the system, such as nasal obstruction or palatal dysfunction, may also adversely influence eustachian tube-middle ear function [36].

### A. Impairment of Pressure Regulation

Impairment of the regulation of pressure within the middle ear can be caused by either: (a) anatomic obstruction or (b) failure of the opening mechanism (i.e., functional obstruction) of the eustachian tube. The site of anatomic (i.e., mechanical) obstruction in the tube can be either intraluminal, periluminal, or peritubal. Anatomic obstruction of the lumen or disease within the periluminal tissues (i.e., intrinsic obstruction) can be due to inflammation secondary to viral or bacterial infection [37–39] or allergy [40]. Peritubal obstruction (i.e., extrinsic obstruction) could be the result of compression caused by a tumor or possibly an adenoid mass [9,41].

Failure of the opening mechanism may be due to either persistent collapse of the eustachian tube due to increased tubal compliance (i.e., lack of stiffness, floppiness), an inefficient active opening mechanism, or both. This has also been termed *functional obstruction* of the eustachian tube—the tube is not anatomically (i.e., mechanically) obstructed, but it is functionally obstructed—which was first described by in infants with unrepaired palatal clefts who had chronic middle-ear effusion Bluestone [7]. Failure of the opening mechanism of the eustachian tube is also common in infants and younger children without cleft palate or history of middle-ear disease [42,43], but more common in those children with middle-ear disease [8,44–46].

Possible explanations for the opening failure of the tube in infants and young children may be related to either (a) the cartilage support, since the amount of cartilage of the eustachian tube is less in the infant than in older children and adults [47] and the cell density of the cartilage decreases with advancing age [48] or (b) the marked age differences in the craniofacial base. The angle of the tube in the child is different from that of the adult. In the adult, the tube is approximately 45 degrees related to the horizontal plane, but in infants this inclination is only 10 degrees [49]. This difference in angulation most likely has an effect on the function of the active opening mechanism (i.e., tensor veli palatini muscle contraction).

## B. Loss of Protective Function

The loss of protection of the middle ear can occur when (a) the eustachian tube has abnormal patency, (b) the tube is relatively short, (c) abnormal gas pressure develops at either end of the tube, or (d) there is a nonintact middle ear and mastoid, e.g., perforation of the tympanic membrane, which results in a loss of the middle-ear gas pocket or cushion.

The eustachian tube can be abnormally patent, and when extreme, open at rest (i.e., patulous tube). Lesser degrees of abnormal patency result in a semipatulous eustachian tube that is closed at rest but has low resistance in comparison with the normal tube. A patulous eustachian tube may permit unwanted secretions from the nasopharynx to gain access, i.e., reflux, into the middle ear. Certain populations, who are at high risk of developing otitis media, have been found to have abnormally patent eustachian tubes, such as Native Americans [50], and patients who have Down's syndrome [51].

Probably one of the most important differences in the structure of the eustachian tube between infants and young children versus older children and adults is the length of the tube; the tube is shorter in children below the age of seven [47]. Young children with cleft palate have eustachian tubes that are statistically shorter than in age-matched controls below the age of six years; also the tube is shorter in children with Down's syndrome [47]. The shorter the tube, the more likely secretions can reflux into the middle ear.

Increased negative middle-ear pressure—as the result of either obstruction of the eustachian tube, secondary to either anatomic obstruction, or failure of active opening, or both—may result in aspiration of nasopharyngeal secretions into the middle ear. At the other end of the system, increased positive nasopharyngeal pressures can result in insufflation of secretions into the middle ear, which may occur during blowing the nose, crying in the infant, or when nasal obstruction is present, i.e., Toynbee Phenomenon [42,52,53]. Rapid alterations in ambient pressures, which can occur during diving, airplane flying, and when patients receive hypobaric pressure treatments, may also result in aspiration or insufflation of nasopharyngeal secretions.

When the tympanic membrane is not intact, such as when a perforation or tympanostomy tube is present, the physiologic middle-ear gas pocket is absent, which can allow secretions from the nasopharynx to reflux into the middle ear [8,44].

## C. Impairment of Clearance Function

Clearance can be affected by several conditions that can occur in the eustachian tube and middle ear (Table 2). Ohashi and colleagues [54] in studies conducted in guinea pigs demonstrated that bacteria, their toxins, and irradiation can impair ciliary function. Irradiation of middle-ear mucosa was also found to impair clearance in the rat [55]. Also, Park and associates [56] demonstrated that the influenza A virus alters ciliary activity and dye transport function in the eustachian tube of the chinchilla. The presence of middle-ear effusion can also impede ciliary activity. Takeuchi and associates [19] utilizing human serum albumin labeled technetium 99*m* in humans with and without middle-ear effusion, demonstrated that clearance was significantly lower in ears with effusion as compared to those without. Ohashi and colleagues [57] also demonstrated that human middle-ear effusions impair middle-ear mucociliary clearance in guinea pigs; clearance was less in mucoid than serous effusions. Silberberg pointed out the gradual change in the physicochemical characteristics of mucus glycoprotein lead to the impairment of the clearance mechanism [58]. In addition, inflammatory mediators associated with middle-ear effusions may be one of the factors related to impairment of ciliary activity [59,60[. Indeed, Ganbo and associates [61] found that leukotrienes $C_4$ and $D_4$ inhibited mucociliary clearance of the eustachian tube in guinea pigs, whereas prostaglandin $E_4$ promoted it. Prostaglandin $E_2$, which is in high concentrations in middle-ear effusions, has also been implicated in impairment of clearance [62].

Just the presence of middle-ear negative pressure can impair clearance, as demonstrated in the cat by Takahashi and colleagues [63]. It appeared to these investigators that the clearance of liquid in the middle ear into the ustachian actually enhanced negative pressure.

Smoking may also be implicated in impairment of clearance function. Ciliary beat frequency was evaluated by a computerized photometric technique in patients with chronic suppurative otitis media and found to be impaired, as compared to controls, and more decreased in those patients who smoked than those who did not [64[. These investigators, using similar methods, also have reported that otitis media with effusion in patients who smoke had significantly less ciliary beat frequency than those who did not; children who had middle-ear effusions and who were exposed to smoke in the home, i.e., passive smoking, also were found to have similar results, as compared to those children with effusions who were not exposed [65]. The effect of smoking on the mucous membrane of the upper respiratory tract may be related to cotitine, the main metabolite of nicotine [66].

Also, tubal pumping action is most likely ineffective when the opening mechanism is inadequate, and this function has been demonstrated to be impaired when negative pressure was present within the middle ear [67,68]. Clearance by mucociliary activity is impeded when there is mucous membrane inflammation, i.e., obstruction of the middle-ear portion of the eustachian tube [8,9,20].

Another important aspect of impairment of clearance of liquid from the middle ear is that when an effusion is present it is trapped in a closed space. To better understand this phenomenon the entire eustachian tube-middle ear system has been likened to a flask with a long, narrow neck in which the mouth of the flask represents the nasopharyngeal end; the narrow neck, the isthmus; and bulbous portion, the middle ear and mastoid air cell system [8,69]. Liquid flow from the middle ear into the nasopharynx can be demonstrated by inverting the flask of the model (Figure 1). In this case the liquid trapped in the bulbous portion of the flask does not flow out of the vessel because of the relative negative pressure that develops inside the chamber. However, if a hole is made in the vessel, the liquid drains out of the flask because the suction is broken.

Clinically, these conditions occur in cases of middle-ear effusion; pressure is relieved by spontaneous rupture of the tympanic membrane or by myringotomy. This phenomenon is most likely the reason that the effusion rapidly clears the middle ear following tympanostomy tube insertion, irrespective of whether the effusion is aspirated or not at the time of surgery. Immediate restoration of ciliary function as has been previously suggested [23] appears not to be the explanation.

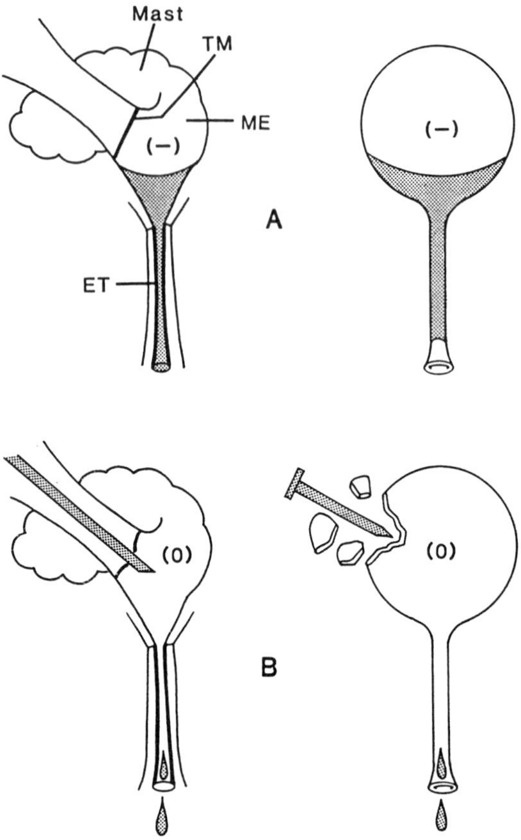


**Figure 1** Fluid flow from an inverted flask as a model of the eustachian tube–middle ear system in which the mouth of the flask represents the nasopharyngeal end of the eustachian tube, the neck is the cartilaginous portion of the tube, and the bulbous portion represents the middle ear and mastoid gas cells. (A) Fluid trapped in bulbous portion of flask with relative negative pressure in the chamber. (B) Effect of perforation of the chamber.

## IV. ROLE OF CLEARANCE FUNCTION IN PATHOGENESIS OF OTITIS MEDIA

Most investigators consider impairment of clearance function to be related to failure to resolve middle-ear effusions as opposed to being the primary cause of the disease [18,70]. However, patients who have primary ciliary dyskinesia (PCD) of their respiratory tract mucous membrane have been observed to have chronic middle-ear effusion [71,73]. But PCD is rare and persistent middle-ear effusion is common, especially in infants and young children. In fact, a group of children who had middle-ear effusion was evaluated for an underlying ciliary dyskinesia, and none were found to have this disorder [74].

Whereas mucociliary clearance impairment plays a role in the pathogenesis of otitis media, the disease is related to, and influenced by various factors (Figure 2). Mucociliary clearance impairment may be secondary to acute otitis media, the presence of middle-ear effusion, viral upper respiratory infections, eustachian tube dysfunction, or allergy. The complex interaction between the factors in the pathogenesis of otitis media may be dissected to better understand what we know and what we don't know. The presence of inflammatory mediators in acute otitis media has been shown to reduce mucociliary activity. Mucociliary impairment may be the intermediary factor in the chronicity of the effusion after an acute otitis media. Viral upper respiratory infections may cause an acute otitis media by eustachian tube dysfunction or direct invasion of the middle ear [75]. Viral upper respiratory infections alter the mucociliary clearance of both the eustachian tube and nose. Where the former have a direct effect, altering the nasal mucociliary clearance may predispose to otitis media by stagnation of mucus in the eustachian tube orifice. Even though the role of allergy in the middle ear or eustachian tube is not clear, nasal allergy is known to alter the mucociliary clearance rate leading to stagnation of secretions around the tubal orifice similar to the effect of viral infections. Eustachian tube dysfunction has been shown to impair mucociliary clearance due to inflammation, negative pressure, or decrease in the pumping function (Table 2).

From the above review, it appears that the role of mucociliary clearance in the pathogenesis of otitis media is the result of complex interaction of multiple factors. When a middle-ear effusion develops, with or without high negative middle-ear pressure, mucociliary and muscular clearance is further impaired, and since the anatomy of the system also impedes drainage, persistence of the effusion follows.

## V. TREATMENT AND CLEARANCE

Mucoactive drugs may have a beneficial effect on clearance of otitis media with effusion. Carbocisteine has been shown to be effective for otitis media with effusion in rabbits [76,77], and *S*-carboxymethylcysteine in chinchillas [78]. A recent meta-analysis of the effect of *S*-carboxymethyl-

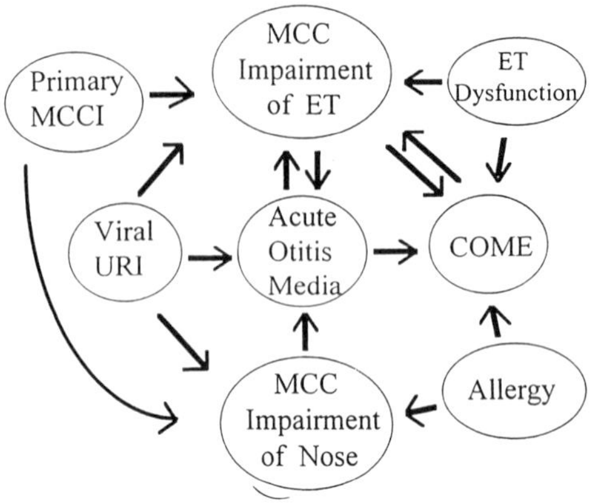


**Figure 2** Possible mechanisms of the relationship between mucocilary clearance and middle ear disease. MCCI = mucociliary clearance impairment, ET = eustachian tube, URI = upper respiratory infection, COME = chronic otitis media with effusion.

**Table 2** Factors Attributed to Impairment of Middle-Ear–Eustachian Tube Clearance in Humans and Animals

| Factor | Reference |
|---|---|
| Bacteria (and toxins) | Ohashi et al., 1989 |
| Irradiation | |
| Virus (influenza A) | Park et al., 1993 |
| Middle-ear effusion | Takeuchi et al., 1990 |
| | Ohashi et al., 1995 |
| Inflammatory mediators | |
| Prostaglandin $E_2$ | Herman et al., 1994 |
| Leukotrienes $C_4$, $D_4$ | Ganbo et al., 1995 |
| Middle-ear negative pressure | Nozoe et al., 1984 |
| | Takahashi et al., 1992 |
| Smoking | Aguis et al., 1995 |
| Inflammatory obstruction of osseous eustachian tube | Bluestone et al., 1972 |
| | Niwa et al., 1990 |
| Anatomy of system (trapping effusion in closed space) | Bluestone et al., 1972 |

cysteine or its lysine salt, or both, evaluated 10 clinical trials and found that of the 355 subjects included, 52.2% on active treatment were considered to be improved and did not require surgery, whereas only 38.7% on placebo were clinically improved, i.e., a 13.5% difference (odds ratio of 2.31, C.I. 1.28–4.20, $p < 0.01$) [79]. However, the authors failed to reach a definite conclusion because cure by tympanogram was of borderline signifance (odds ratio of 2.25, C.I. 0.97–5.22, $p = 0.058$) and because of heterogeneity and methodological problems in the trials.

## VI. WHAT WE DON'T KNOW AND FUTURE RESEARCH

From this review it is apparent that we know a great deal concerning the physiological and pathophysiological aspects of clearance function of the eustachian tube and about the role clearance plays in the pathogenesis of otitis media. Yet more must be learned about normal and abnormal clearance to better understand the mechanisms related to persistence of middle-ear effusions due to impaired clearance. An area of research neglected in the past, that may shed light on pathogenesis is to determine why patients with cystic fibrosis rarely develop chronic middle-ear effusions, even though other parts of the respiratory tract, such as the sinuses, are commonly involved by chronic inflammation. Related to possible medical treatment that involves clearance, there is an urgent need to conduct large randomized, placebo-controlled, clinical trials to evaluate the safety and efficacy of mucolytic agents, especially *S*-carboxymethylcysteine or its lysine salt, or both, since these agents may be beneficial for patients with chronic otitis media with effusion. Effective medical treatment of this disease could reduce the need for the commonly employed surgical interventions.

## REFERENCES


1. Cantekin EI, Doyle WJ, Reichert TJ, Phillips DC, Bluestone CD. Dilation of the eustachian tube by electrical stimulation of the mandibular nerve. Ann Otol Rhinol Laryngol 1979; 88:40–51.
2. Honjo I, Okazaki N, Kumazawa T. Experimental study of the eustachian tube function with regard to its related muscles. Acta Otolaryngol (Stockh) 1979; 87:84.
3. Rich AR. A physiological study of the eustachian tube and its related muscles. Bull Johns Hopkins Hosp 1920; 31:206.
4. Casselbrant ML, Cantekin EI, Dirkmaat DC, Doyle WJ, Bluestone CD. Experimental paralysis of tensor veli palatini muscle. Acta Otolaryngol (Stockh) 1988; 106:178.
5. Alper CM, Tabari R, Seroky JT, Doyle WJ. Magnetic resonance imaging of

the development of middle-ear effusion secondary to experimental paralysis of tensor veli palatini muscle. Otolaryngol Head Neck Surg 1994; 111:122.
6. Doyle WJ, Seroky JT. Middle ear gas exchange in rhesus monkeys. Ann Otol Rhinol Laryngol 1994; 103(8):636–645.
7. Bluestone CD. Eustachian tube obstruction in the infant with cleft palate. Ann Otol Rhinol Laryngol 1971; 80:1.
8. Bluestone CD, Paradise JL, Beery QC. Physiology of the eustachian tube in the pathogenesis and management of middle ear effusions. Laryngoscope 1972; 82:1654–1670.
9. Bluestone CD, Wittel RA, Paradise JL, et al. Eustachian tube function as related to adenoidectomy for otitis media. TAAOO 1972; 76:1325.
10. Wittenborg MH, Neuhauser EB. Simple roentgenographic demonstration of eustachian tubes and abnormalities. Am J Roentgenol Radium Ther Nucl Med 1963; 89:1194.
11. Tos M. Anatomy and histology of the middle ear. Clin Rev Allergy 1984; 2: 267.
12. Lim DJ. Functional morphology of the tubotympanum. An overview. Acta Otolaryngol (Stockh) 1984; Suppl 414:13–18.
13. Svane-Knudsen V, Larsen HF, Brask T. Secretory otitis media: a question of surface activity in the eustachian tube. Acta Otolaryngol (Stockh) 1988; 105:114.
14. Matsune S, Sando I, Takahashi H. Comparative study of elastic at the hinge portion of eustachian tube cartilage in normal and cleft palate individuals. Ann Otol Rhinol Laryngol 1992; 101:163.
15. Sando I, Takahashi H, Matsune S, Aoki H. Localization of function in the eustachian tube: a hypothesis. Ann Otol Rhinol Laryngol 1994; 103(4 pt 1): 311–314.
16. Sudo M, Sando I. Developmental change in folding of the human eustachian tube. Acta Otolaryngol (Stock) 1996; 116(2):307–311.
17. Park K, Ueno K, Lim DJ. Developmental anatomy of the eustachian tube and middle era in mice. Am J Otolaryngol 1992; 13:93.
18. Karja J, Nuutinen J. Mucociliary clearance in secretory otitis media. In: Sade J, ed. The Eustachian Tube. Amsterdam/Berkley: Kugler Publications, 1987:181–184.
19. Takeuchi K, Majima Y, Hirata K, Hattori M, Sakakura Y. Quantitation of tubotympanal mucociliary clearance in otitis media with effusion. Ann Otol Rhinol Laryngol 1990; 99:211–214.
20. Niwa H, Takahashi M, Yanagita N, Naganawa S. Evaluation of clearance function of the eustachian tube by sequential contrast CT. Acta Otolaryngol (Stockh) 1990; 471:43–50.
21. Albiin N, Hellstrom S, Stenfors LE. Clearance of effusion material from the attic space—an experimental study in the rat. Int J Pediatr Otorhinolaryngol 1983; 5:1.
22. Stenfors LE, Hellstrom S, Albiin N. Middle ear clearance in eustachian tube function: physiology and role in otitis media. Ann Otol Rhinol Laryngol 1985; 94:30.

23. Sade J. The middle ear mucociliary system. In: Sade J, ed. The Eustachian Tube. Amsterdam/Berkeley: Kugler Publications, 1987:159–180.
24. Ohashi Y, Nakai Y, Koshimo H, et al. Ciliary activity in the in vitro tubotympanum. Arch Otorhinolaryngol 1986; 243:317.
25. Honjo I, Okazawa N, Ushiro K, Kumazawa T. Cineradiographic analysis of eustachian tube function. Ann Otol Rhinol Laryngol 1980; 89(3):276–278.
26. Honjo I. Experimental study of the pumping function of the eustachian tube. Acta Otolaryngol (Stockh) 1981; 91:85.
27. Honjo I, Hayashi M, Ito S, et al. Pumping and clearance function of the eustachian tube. Am J Otolaryngol 1985; 6:241.
28. Honjo I, Hayashi M, Shinsuke I, Takahashi H. Pumping and clearance function of the eustachian tube. Am J Otolaryngol 1985; 6:241–244.
29. Birkin EA, Brookler KH. Surface tension lowering substance of the canine eustachian tube. Ann Otol Rhinol Laryngol 1972; 81:268.
30. Rapport PN, Lim DJ, Weiss HJ. Surface active agent in eustachian tube function. Arch Otolaryngol 1975; 101:305.
31. Hagen WE. Surface tension lowering substance in eustachian tube function. Laryngoscope 1977; 87:1033.
32. White P. Effect of exogenous surfactant on eustachian tube function in the rat. Am J Otolaryngol 1989; 10:301.
33. Karchev T, Watanabe N, Fujiyoshi T, et al. Surfactant-producing epithelium in the dorsal part of the cartilaginous eustachian tube of mice. Acta Otolaryngol (Stockh) 1994; 114:64.
34. Fornadley JA, Burns JK. The effect of surfactant on eustachian tube function in a gerbil model of otitis media with effusion. Otolaryngol Head Neck Surg 1994; 110(1):110.
35. Mira E, Benazzo M, Manzoni M, Lligona L, Salmona M, Tacconi MT. Effect of ambroxil on rabbit eustachian tube surfactant. ORL 1992; 54:130–132.
36. Bluestone CD. Pathogenesis of otitis media: role of eustachian tube. Pediatr Infect Dis J 1996; 15:281–291.
37. Bluestone CD, Cantekin EI, Beery QC. Effect of inflammation on the ventilatory function of the eustachian tube. Laryngoscope 1977; 87:493.
38. McBride TP, Doyle WJ, Hayden FG, Gwaltney JM. Alterations of the eustachian tube, middle ear, and nose in rhinovirus infection. Arch Otolaryngol 1989; 115:1054.
39. Buchman CA, Doyle WJ, Skoner D, et al. Otologic manifestations of experimental rhinovirus infection. Laryngoscope 1994; 104(10):1295–1299.
40. Friedman RA, Doyle WJ, Casselbrant ML, et al. Immunologic-mediated eustachian tube obstruction: a double-blind crossover study. J Allergy Clin Immunol 1983; 71:442.
41. Bluestone CD, Cantekin EI, Beery QC. Certain effects of adenoidectomy on eustachian tube ventilatory function. Laryngoscope 1975; 85:113.
42. Bluestone CD, Beery QC, Andrus WS. Mechanics of the eustachian tube as it influences susceptibility to and persistence of middle ear effusions in children. Ann Otol Rhinol Laryngol 1974; 83:27.
43. Bylander A, Tjernstrom O, Ivarsson A. Pressure opening and closing func-

tions of the eustachian tube by inflation and deflation in children and adults with normal ears. Acta Otolaryngol (Stockh) 1983; 96:255.
44. Bluestone CD, Cantekin EI, Beery QC, et al. Function of the eustachian tube related to surgical management of acquired aural cholesteatoma in children. Laryngoscope 1978; 88:1155.
45. Stenstrom C, Bylander-Groth A, Ingvarsson L. Eustachian tube function in otitis-prone and healthy children. Int J Pediatr Otorhinolaryngol 1991; 21: 127.
46. Iwano T, Hamada E, Kinoshita T, et al. Passive opening pressure of the eustachian tube. In: Lim DJ, Bluestone CD, Klein JO, Nelson JD, Ogra PL, eds. Proceedings of the Fifth International Symposium on Recent Advances in Otitis Media. Burlington, Ontario: Decker Periodicals, 1993:76–78.
47. Sadler-Kimes D, Siegel MI, Todhunter JS. Age-related morphologic differences in the components of the eustachian tube/middle ear system. Ann Otol Rhinol Laryngol 1989; 98:854–858.
48. Yamaguchi N, Sando I, Hashida Y, Takahashi H, Matsune S. Histologic study of eustachian tube cartilage with and without congenital anomalies: a preliminary study. Ann Otol Rhinol Laryngol 1990; 99:984–987.
49. Proctor B. Embryology and anatomy of the eustachian tube. Arch Otolaryngol 1967; 86:503–514.
50. Beery QC, Doyle WJ, Cantekin EI, et al. Eustachian tube function in an American Indian population. Ann Otol Rhinol Laryngol 1980; 89:28.
51. White BL, Doyle WJ, Bluestone CD. Eustachian tube function in infants and children with Down's syndrome. In: Lim DJ, Bluestone CD, Klein JO, Nelson JD, eds. Proceedings of the Third International Symposium on Recent Advances in Otitis Media with Effusion. Toronto: BC Decker, 1984: 62–66.
52. Jorgensen F, Holmquist J. Toynbee phenomenon and middle ear disease. Am J Otolaryngol 1984; 4:291.
53. Thompson AC, Crowther JA. Effect of nasal packing on eustachian tube function. J Laryngol Otol 1991; 105:539.
54. Ohashi Y, Nakai Y, Furuya H, Esaki Y, Ikeoka H, Kato S, Kato M. Mucociliary diseases of the middle ear during experimental otitis media with effusion induced by bacterial endotoxin. Ann Otol Rhinol Laryngol 1989; 98:479–484.
55. Chao WY, Leung HWC. Effects of irradiation on the rat middle ear mucosa. Eur Arch Otorhinolaryngol 1995; 252:244–248.
56. Park K, Bakaletz LO, Coticchia JM, Lim DJ. Effect of influenza A virus on ciliary activity and dye transport function in the chinchilla eustachian tube. Ann Otol Rhinol Laryngol 1993; 102:551.
57. Ohashi Y, Nakai Y, Ohno Y, Sugiura Y, Okamoto H. Effects of human middle ear effusions on the mucociliary system of the tubotympanum in the guinea pig. Eur Arch Otorhinolaryngol 1995; 252:35–41.
58. Silberberg A. Physiochemical characteristics of epithelial mucus and mucus glycoprotein. In: Sadé J, ed. Secretory Otitis Media and Its Sequelae. New York: Churchill Livingstone, 1979:285-.
59. Sadé J, Eliezer N, Silberberg A, Nevo AC. The role of mucus in transport by cilia. Am Rev Respir Dis 1970; 102:48–52.

60. Tos M. Pathogenesis and pathology of chronic secretory otitis media. Ann Otol Rhinol Laryngol 1980; 89(suppl 68):91–97.
61. Ganbo Y, Hisamatsu K, Shimomura S, Nakajima T, Inoue H, Murakami Y. Inhibition of mucociliary clearance of the eustachian tube by leukotrienes C4 and D4. Ann Otol Rhin Laryngol 1995; 104:231–236.
62. Herman P, Yen P-T, Tu T-Y, Loiseau A, Cassingena R, Grodet A, Friedlander G, Huy P. Pathophysiology of middle ear epithelium: a new role for prostaglandin E2. Am J Otolaryngol 1994; 15(4):258–266.
63. Takahashi H, Honjo I, Hayashi M, Fujita A. Clearance function of eustachian tube and negative middle ear pressure. Ann Otol Rhinol Laryngol 1992; 101:759.
64. Aguis AM, Wake M, Pahor AL, Smallman LA. Nasal and middle ear ciliary beat frequency in chronic suppurative otitis media. Clin Otolaryngol 1995; 20:470–474.
65. Aguis AM, Wake M, Pahor AL, Smallman A. The effects of in vitro cotitine on nasal ciliary beat frequency. Clin Otolarygol 1995; 20:465–469.
66. Wake M, McCullough DeE, Binnington JD. Effect of nasogastric tubes on eustachian tube function. J Laryngol Otol 1990; 104:17.
67. Nozoe T, Okazaki N, Koda Y, Ayani N, Kumazawa T. Fluid clearance of the eustachian tube. In: Lim DJ, Bluestone CD, Klein JO, Nelson JD, eds. Proceedings of the Third International Symposium on Recent Advances in Otitis Media with Effusion. Toronto: BC Decker, 1984:66–68.
68. Hayashi M, Takahashi H, Sato H, Honjo I. Clearance of the eustachian tube under negative middle ear pressure. Am J Otolaryngol 1986; 4:399–401.
69. Bluestone CD, Klein JO. Otitis Media in Infants and Children. 2nd ed. Philadelphia: WB Saunders, 1995:145–240.
70. Mogi G, Kawauchi H, Kurono Y. Tubal dysfunction or infection? Role of bacterial infection and immune response. In: Mogi G, ed. Recent Advances in Otitis Media: Proceedings of the Second Extraordinary International Symposium on Recent Advances in Otitis Media. New York: Kugler Publications, 1993:73–77.
71. Shikowitz MJ, Ilardi CF, Gero M. Immotile cilia syndrome associated with otitis media with effusion: A case report. In: Lim DJ, Bluestone CD, Klein JO, Nelson JD, eds. Proceedings of the Fourth International Symposium on Recent Advances in Otitis Media. Philadelphia: BC Decker, 1988:304–307.
72. Bent JP, Smith RJH. Intraoperative diagnosis of primary ciliary dyskenesia. Otolaryngol Head Neck Surg 1997; 116(1):64–67.
73. Teknos YN, Metson R, Chasse T, Balercia G, Dickerson GR. New developments in the diagnosis of Kartener's syndrome. Otolaryngol Head Neck Surg 1997; 116(1):68–74.
74. Wake M, Smallman LA. Ciliary beat frequency of nasal and middle ear mucosa in children with otitis media with effusion. Clin Otolaryngol 1992; 17:155–157.
75. Alper CM, Doyle WJ, Skoner DP, Buchman CA, Seroky JT, Gwaltney JM, Cohen SA. Prechallenge antibodies: Moderators of infection rate, signs, and symptoms in adults experimentally challenged with rhinovirus type 39. Laryngoscope 1996, 106(10):1298–1305.

76. Ohashi Y, Nakai Y, Ikeoka H, Koshimo H, Kihara S, Maruoka K. Effects of carbocisteine on the experimental otitis media with effusion caused by long-term exposure to SO2. J Otolaryngol Jpn 1985; 88:1051–1055.
77. Ohashi Y, Nakai Y, Koshimo H, Esaki Y. Mucosal pathology of experimental otitis media with effusion due to nitrogen dioxide exposure and the effects of carbocisteine on pathological middle ear mucosa. J Otolaryngol Jpn 1988; 91:71–87.
78. Hori F, Kawauchi H, Mogi G. Effect of S-carboxymethylcysteine on the clearance of middle ear effusion: an experimental study. Ann Otol Rhinol Laryngol 1994; 193:567–575.
79. Pignataro O, Pignataro L, Gallus G, Calori G, Cordaro I. Otitis media with effusion and S-carboxymethylcysteine and/or its lysene salt: a critical review. Int J Pediatr Otorhinolaryngol 1996; 35:231–241.

# 25

# Clinical Significance of Tubotympanal Mucociliary Clearance in Otitis Media with Effusion

**Kazuhiko Takeuchi, Yuichi Majima, Keisuke Hirata,***
**and Yasuo Sakakura**
*Mie University School of Medicine, Mie, Japan*

## I. INTRODUCTION

There are a variety of clinical courses in pediatric otitis media with effusion. Some children recover from effusions without any particular intervention and others have persistent effusions despite frequent myringotomies. Middle ear effusions also vary from one patient to another. In this paper, I intend to clarify the relationship between the physical property of effusions and prognosis of the disease by answering the following three specific questions. (a) Is there a difference in tubotympanal clearance between ears with effusions and those without effusions? (b) Is the rheological property of effusions related to tubotympanal clearance? (c) Is the rheological property of effusions related to the prognosis of the disease?

## II. TUBOTYMPANAL CLEARANCE

First we measured mucociliary clearance of the tubotympanum in 52 ears of 34 children with otitis media with effusion. In 36 of the ears examined effusions were present at the time the measurement was done. A group of 16 ears without effusions consisted of the patients who had drainage tubes inserted or those who did not have effusions when myringotomy

**Current affiliation*: Yamamoto Hospital, Kuwana, Japan

was performed. Myringotomy was performed and middle ear effusion, if present, was aspirated. Then we instilled 0.01 ml of sterile human serum albumin labeled with technetium 99*m* into the middle ear cavity and monitored the tracer by a gamma camera *placed* above the head of the patient. Sequential scintigrams of the radioactive tracer were taken at intervals of 30 seconds by the gamma camera. The radioactivity of the tympanic cavity was quantified in every sequential scintigram. The decay expressed percentage of the initial radioactivity was plotted against time. In most of the cases tested, there was a plateau in the clearance curve, indicating that it took time for the tracer to start moving into the eustachian tube. This plateau was followed by a rapid fall, which appeared exponential.

We analyzed the clearance curves using three parameters. First the clearance rate was defined as a decrease in the percentage of radioactivity in the tympanic cavity in 15 minutes. The clearance rate indicates the overall mucociliary function of the tubotympanum. Second, the time of plateau was defined as the length of time until the onset of the rapid fall of radioactivity. This is indicative of mucociliary clearance function of the tympanic cavity and of the tympanic orifice of the eustachian tube. Third, in order to express the slope of the curve, we noted the length of time taken for the radioactivity to be reduced by half (the half period). The half period was calculated as follows: the radioisotopic counts after the onset of transport were plotted on semilogarithmic paper with time on the abscissa. A straight line was drawn to give the best fit through the slope. The half period [T(1/2)] was calculated by the following equation

$$T(1/2) = \log_{10} 2/m$$

where $m$ is the slope of the straight line.

As a result, the ears with effusions had a significantly lower clearance rate than those without effusions ($p<0.05$). The ears with effusions had significantly longer time of plateau than those without ($p<0.05$). No significant difference was observed in the half period between the two groups. However, the half period was examined only for ears that transported human serum albumin within fifteen minutes. Thus, the ears that did not transport the albumin were excluded from the analysis of the third parameter.

## III. RHEOLOGICAL PROPERTY OF EFFUSIONS AND TUBOTYMPANAL CLEARANCE

Next I compared the clearance rate with the viscosity of middle ear effusions measured by an oscillating sphere magnetic rheometer [1]. The elastic modulus (G′) and the dynamic viscosity (η′) of the effusions were measured within 4 hours after the collection of samples. The G′ is indicative of elastic behavior and the η′ is a measure of viscous behavior. All the measurements were made at 25°C and at a frequency of 1 Hz. The viscoelasticity was measured for 33 ears with effusion. The G′ at 1 Hz ranged from 1 to 27,000 dyne/cm$^2$, and the η′ at 1 Hz ranged from 0.2 to 1790 poise. No significant correlation was observed between the G′ of the effusions and the clearance rate. On the other hand, significant negative correlation was observed for values of viscosity above 2 poise. The value of 2 poise was determined because the optimal η′ value for mucociliary clearance rate was about 2 poise, and there is a positive correlation below an η′ of 2 poise and a negative correlation above an η′ of 2 poise in vitro [2]. Thus increased viscosity is one of the causes of reduced tubotympanal mucociliary clearance rate.

## IV. RHEOLOGICAL PROPERTY OF EFFUSIONS AND PROGNOSIS OF THE DISEASE

Next we examined the impact of rheological properties of middle ear effusions on the prognosis of the disease. We selected 93 untreated ears of 69 children suffering from otitis media with effusion who had effusions at the first visit. We treated the patients with myringotomy every 1 or 2 months. We gave them no antibiotics of long duration, mucolytic agents, or steroids. In order to evaluate the prognosis, we assessed the duration of middle-ear fluid retention by tympanometry. We performed serial tympanometry every 1 to 2 months. We calculated the length of time to note a change in the tympanogram from type B or C2 to type A or C1.

The G′ at 1 Hz in 93 freshly collected effusions ranged from 1.92 to 5890 dyne/cm$^2$. The η′ at 1 Hz ranged from 0.156 to 460 poise. The distribution of both G′ and η′ was divided into four groups (1, 2, 3, and 4) in relation to mucociliary transport. As effective mucociliary transport is obtained at a G′ of about 20 dyne/cm$^2$ and an η′ of about 2 poise in vitro [2]; the G′ of group 2 was determined as ranging from 6.31 to 63.1 dyne/cm$^2$. This range is equal to 20 dyne/cm$^2$ $\pm$ 0.5 log units, where the mucociliary transport works best. The G′ of group 1 was less than 6.31 dyne/cm$^2$, in which cases the effusions appear serous. Effusions with a G′ of more than 63.1 were further divided into groups 3 and 4. The G′ of group 3 ranged from 63.1 to 1000 dyne/cm$^2$. The G′ of group 4 was more than 1000 dyne/cm$^2$, in which case the effusions appear gluey. The distribution of η′ was similarly divided into four groups.

We arbitrarily defined a good prognosis as fluid retention for less than four months. As regards elasticity versus duration of fluid retention, the distribution of good prognoses (fluid retention<4 months) was significantly higher in group 2 than group 1 ($p<.05$), group 3 ($p<.05$), and group 4 ($p<.01$), respectively. As regards viscosity versus duration of fluid retention, a statistical difference in good prognosis was observed between group 2 and group 4 ($p<.05$). But the number of good prognoses in group 2 was not significantly different when compared with that in group 1 or group 3. In other words, ears with effusions suboptimal for mucociliary transport had poorer prognosis.

## V. SUMMARY

In summary, (a) the mucociliary clearance in the tubotympanum was decreased in otitis media with effusion, (b) significant negative correlation was observed between the dynamic viscosity of the effusions and the tubotympanal clearance rate, and (c) the ears with effusions whose rheological properties are suboptimal for mucociliary transport had poorer prognosis.

## REFERENCES

1. Hirata K. Dynamic viscoelasticity of nasal mucus from children with chronic sinusitis. Mie Med J 1985; 84:205–219.
2. Majima Y, Sakakura Y, Matsubara T, et al. Rheological properties of middle ear effusions from children with otitis media with effusion. Ann Otol Rhinol Laryngol 1986; 95(suppl 124):1–4.

# 26

# Effects of Urban Air Pollution on Mucociliary Clearance

**Mariangela Macchione, György Miklós Böhm, and Paulo Hilário N. Saldiva**

*University of São Paulo School of Medicine, São Paulo, Brazil*

## I. INTRODUCTION

The contamination of the atmosphere of large urban centers has been thought to induce adverse health effects in the exposed population. In fact, epidemiological studies indicate that increased pollution levels are associated with increased morbidity and mortality [1–3]. The consistency of the mortality studies performed by different groups in different locations reinforced the association between pollution and adverse health effects and led to the establishment of air quality standards that try to keep pollution levels within the limits necessary to preserve human health. The bulk of epidemiological evidence relating pollution to health impairment clearly indicates the necessity of studying the mechanisms responsible for the injury promoted by airborne contaminants. The mucociliary clearance function is a natural target for air pollutants, and mucociliary dysfunction is likely to be one of the pathophysiologic ways to explain pollution-dependent injury.

The proper evaluation of the effect of pollution on mucociliary epithelium presents significant experimental problems. In addition to the difficulties in determining the dose and depth of penetration of the pollutant into the pulmonary parenchyma [4–6], there is the problem of characterizing the pollutant to be evaluated [7]. In the real world of streets, pollution sources simultaneously emit different types of pollutants that interact among themselves through complex reactions dependent on the degree of photochemical input and wind dispersion. Field studies make possible the development of experiments in more realistic conditions, however, they fail in controlling for exposure [8–11]. Thus, although field experiments are more efficient in detecting the effects of combinations of pollutants, they fail in ascribing to a single agent the responsibility for any observable effect.

## II. PATHOGENETIC MECHANISMS FOR POLLUTION-DEPENDENT MUCOCILIARY DYSFUNCTION

Considering the different families of pollutants present in the urban locale, it is difficult to postulate the mechanisms responsible for the impairment of mucociliary clearance. It is, however, possible for different pathogenetic pathways to account for the natural course of mucociliary impairment after exposure to air pollution.

### A. Oxidative Stress to Ciliated Cells

This stress due to different pollutants may promote oxidative injury of the ciliated cells. In addition, inflammatory cells recruited after pollution-mediated inflammation could increase the oxidative burst in the mucociliary environment. Lipid peroxidation of ciliary membranes may lead to structural and functional abnormalities, thus impairing the efficiency of ciliary transport of the mucus blanket [12,13]. Membrane lesions may alter the hydric balance between intra- and extracellular environments, promoting an overload to the ATP-dependent pumps located in cell membranes.

### B. Changes of Epithelial Permeability

Since the efficiency of mucociliary transport is dependent on the adequate mechanical coupling between cilia and mucus, the depth of the sol layer is critical to mucociliary transport: an excess of liquid would promote uncoupling between mucus and cilia, whereas dehydration would cause mechanical inefficiency by making cilia interact with mucus during their recovery movement. Changes in epithelial cells represent an early event after inhalation injury and may play a significant role in mucociliary abnormalities [14,15].

## C. Impairment of Cellular Respiration and Cilia Energetics

This is important because ciliated cells require a great amount of energy to maintain ciliary beating. Air pollutants may directly interfere with the mechanisms of capturing and/or utilizing the $O_2$ necessary for cellular respiration, as in the case of CO [16]. The ciliary stroke is driven by ATPases located in the ciliary apparatus that provide the energy necessary for ciliary motion. The activity of ATPases is impaired after exposure to oxidative injury [17,18], thus providing a biochemical basis for ciliary dysfunction.

## D. Imbalance of Protein Turnover

Interference with the mechanisms that regulate the balance between removal and restitution of proteins represents a classical example of cellular injury in different pathological processes. Damage to cell protein structures is expected to occur as a consequence either of direct action of pollutants or of proteases delivered by inflammatory cells [19]. These changes may speed the turnover of the cellular components of the mucociliary system, thus implying that a great amount of new protein structure has to be assembled. Any increase of this ratio—removal/restitution—favor the development of mucociliary dysfunction.

## E. Abnormal Cellular Restitution

Increased epithelial turnover is commonly observed after injury, as a consequence of increased cellular removal necrosis or apoptosis. Inhaled pollutants are known to alter the structural integrity of the airway epithelium. Depending on the injury, changes involving secretory cell hypertrophy and hyperplasia or apparent transdifferentiation of epithelial cells may occur [9,10,20]. This alteration in epithelial morphology results in fewer ciliated cells, more mucus produced by the epithelium, and changes in the physical properties (elasticity and viscosity) of mucus, which further impair its clearance. Air pollution may promote DNA damage either by oxidative stress or by the induction of stable DNA adducts [21]. These changes were thought to be responsible for the increased urban/rural gradient observed for lung carcinomas [22]. One may postulate that DNA alterations may modify the rate of cell turnover of airways, thus interfering with mucociliary clearance by the presence of abnormal cells lining the airways, as in the case of epithelial dysplasia.

## III. ACUTE AND SUBACUTE STUDIES

Most of the knowledge about the effects of pollutants on mucociliary clearance is derived from acute studies. These investigations usually report the behaviour of mucociliary clearance within a short period of time after exposure to a single air pollutants. Table 1 presents some selected papers reporting short-term effects of air pollution on mucociliary clearance, including both human and animal studies.

The references cited in Table 1 indicate that the acute effects of pollution on mucociliary clearance were studied for isolated or simple combinations of air pollutants. Clearly, further studies are necessary to provide information on the possible interaction of pollutants and, consequently, to know whether synergistic phenomena are involved in the pathogenesis of pollution-dependent injury.

## IV. CHRONIC STUDIES

The effects of the prolonged exposure to air pollution deserve a great amount of attention because most of the people living in large urban centers are continously and chronically exposed to low levels of pollutants. However, from the experimental point of view, the simulation of these chronic exposures in the laboratory is difficult. First, it is virtually impossible to carry out prospective chronic human studies. It is necessary to adequately control any confounding variables, such as occupational exposures and tobacco smoke exposure, before attributing to air pollution the observed alterations. On the other hand, it is quite difficult to reproduce the urban pollution scenario in chronic-exposure animal experiments. Despite the aforementioned limitations, the coherence of the results reported by different groups indicates that there is solid evidence that chronic exposure to air pollutants, within the observed ambient levels, is able to promote significant changes in airways.

Based on necropsy data, some authors concluded that individuals living in polluted areas develop inflammatory changes of the airways and lung parenchyma, with changes of airway cellular components and mucus hypersecretion [23–25]. These human findings were reinforced by chronic-exposure animal experimental studies focusing on the effects of ozone in monkeys [26] and sulfur dioxide in dogs [27].

More recently, new approaches have been employed in the study of the chronic effects of air pollution on the mucociliary system by housing animals in or collecting animals from regions with different pollution backgrounds. Our laboratory used this approach in different exposure proto-

**Table 1** Selected References on Acute and Subacute Effects of Air Pollution on Mucociliary Components

| Pollutant | Concentration | Species | Airway level | Exposure protocol | Effects | Reference |
|---|---|---|---|---|---|---|
| $SO_2$ | 1, 5, 25 ppm | Human | Nasal | 7 to 8 hours during 3 days | Decrease of mucus flow | 29 |
| $O_3$ | 0.5, 1, 2 ppm | Sheep | Trachea Bronchi | 2 hours | No changes | 30 |
| $O_3$, or aerosol contained a mixture of $O_3$ + sulfates | 3.5 $mg/m^3$ of aerosol sulfate and 0.8 ppm $O_3$ | Rat | Trachea Bronchi Lung | 4 hours | ↓ Clearance after $O_3$ | 31 |
| Aerosol of $H_2SO_4$ | 0.5 and 1.0 $mg/m^3$ | Dogs | Trachea | 1 hour | ↓ Clearance | 32 |
| Aerosol of $H_2SO_4$ | 110 to 980 $\mu g/m^3$ | Human | Trachea Bronchi | 1 hour | ↓ Clearance | 33 |
| $O_3$ and $SO_2$ | 0.3 ($O_3$) and 3 ($SO_2$) ppm | Sheep | Trachea | 5 hours during 3 days | ↓ Clearance | 34 |
| $O_3$ | 0.5 ppm | Sheep | Trachea | 4 hours/day during 6 weeks | Mucus hypersecretion | 35 |
| $O_3$ | 0.1; 0.25 and 0.6 ppm | Rabbit | Trachea Bronchi | 2 hours/day for 14 days | ↓ Clearance | 36 |
| $O_3$ | 0.2 and 0.4 ppm | Human | Trachea Bronchi | 2 hours | ↑ Clearance | 37 |
| $NO_2$ and $H_2SO_4$ | 0.3 or 1 ppm for $NO_2$ and 0.5 $mg/m^3$ of $H_2SO_4$ | Rabbit | Trachea Bronchi | 2 hours/day during 14 days | $NO_2$—no changes $H_2SO_4$—↓ clearance | 38 |
| $O_3$ | 1.0 ppm | Sheep | Trachea | 2 or 5 hour/day during 4 days | ↓ Clearance | 39 |
| $O_3$ | 0.4 ppm | Human | Trachea Bronchi | 1 hour | No changes | 40 |

cols, and the results obtained were quite consistent in reporting that air pollution interferes with mucociliary clearance. Animals living in São Paulo, downtown, presented significant evidence of mucus hypersecretion and/or change in the mucus biochemical profile [8–10]. These changes were associated with the presence of rheological alterations, leading to an increased mechanical impedance of mucus, thus causing an overload on the mucus transport system [8–9]. In addition, loss of cilia and other ciliary abnormalities were observed in pollution-exposed animals. The same findings were reported by other investigators [28], indicating that chronic exposure to air pollution is responsible for abnormalities of the mucociliary system. A list of selected studies relating chronic exposure to air pollution with alterations of mucociliary system components is presented in Table 2.

## V. FINAL REMARKS

Although there is a great amount of evidence indicating that air pollutants interfere with mucociliary functioning there are some points that deserve further clarification. An example is the discrepancy between toxicological and epidemiological studies focusing on the adverse health effects of diesel particles. As a general rule, the levels of inhalable particles are statistically associated with acute morbidity and mortality according to epidemiological studies. Toxicological investigations, however, do not coraborate to those findings. One of the reasons for this discrepancy may be synergistic interaction among different pollutants not controlled for in the epidemiological studies. Another point the presence of highly susceptible population groups such as those individuals with chronic bronchitis or cardiopulmonary disease that respond to low levels of pollutant particles. Thus, employing diseased animals in toxicological studies could provide useful information about the effects of air pollution on human health. In this scenario much more work is necessary to close the present gap between human epidemiological and animal-experimental studies. Taking into account this problem, our laboratory has performed some in vitro studies aiming to verify the effects of urban inhalable particles suspended in the atmosphere of São Paulo on mucociliary transport by means of exposing frog palates to near-ambient concentrations of this pollutant. In fact, inhalable particles do not represent a single pollutant, but are characterized by different components aggregated or adsorbed to carbon soot. Measurements were done during 120 minutes after immersion in increasing concentrations (50 to 10,000 $\mu g/m^3$) of $PM_{10}$. In these experiments, urban levels of $PM_{10}$ were able to induce significant alterations of mucociliary transport,

**Table 2** Selected References on Chronic and Effects of Air Pollution on Mucociliary Components

| Pollutant | Concentration | Species | Airway level | Exposure protocol | Effects | Reference |
|---|---|---|---|---|---|---|
| $O_3$ | 0.8 ppm | Rat | Trachea | 8 hours/day; 1 to 90 days | Mucus hypersecretion | 41 |
| $NO_2$ | 30 ppm | Hamster | Bronchioles | 22 hours/day during 7 days/week for 5 months | Morphologic abnormalities | 42 |
| $SO_2$ | 15 and 50 ppm | Dog | Bronchi | 5 to 11 months | Morphologic abnormalities | 27 |
| Diesel exhaust | 0.35 to 7 mg/m$^3$ | Rat | Trachea | 6, 12, 18, 24 months | ↓ Clearance. | 43 |
| $O_3$ | 0.15 ppm | Monkey | Nasal cavity | 8 hours/day during 6 or 90 days | Mucus hypersecretion | 26 |
| $O_3$ | 0.15 or 0.3 ppm | Monkey | Nasal cavity | 8 hours/day during 6 or 90 days | Inflammation and mucus hypersecretion. | 44 |
| Urban air | Variable | Rat | Nasal cavity and bronchi | 6 months | Hypersecretion and ↓ clearance. | 8 |
| Urban air | Variable | Rat | Trachea Bronchi | 6 months | Hypersecretion, ciliary abnormalities and ↓ clearance | 9 |
| Urban air | Variable | Rat | Nasal cavity | 6 months | Mucus hypersecretion. | 10 |
| Urban air | Variable | Several | Trachea | 5 and 12 months | Ciliary abnormalities | 28 |
| $NO_2$ | 0.5 or 10 ppm | Ferret | Trachea Bronchi | 4 hours/day, 5 days/week for 8 or 15 weeks | ↓ Clearance. | 45 |
| Urban air | Variable | Rabbit | Bronchi | 40 days | Morphologic abnormalities. | 46 |

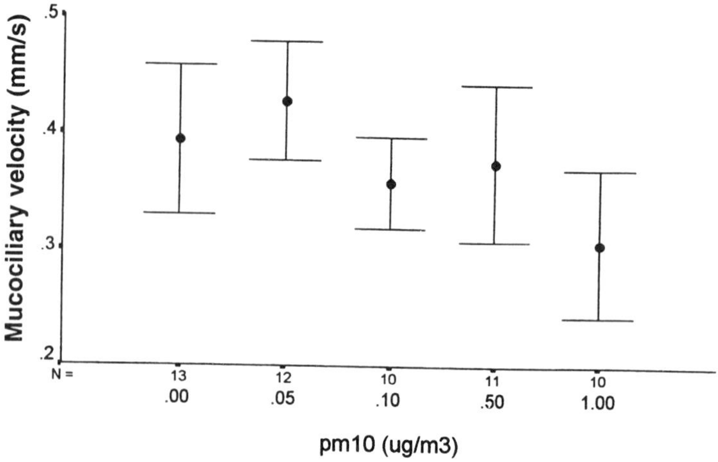


**Figure 1** Mucociliary velocity measured in frog palates immersed during 120 minutes in different concentrations of inhalable particles sampled in São Paulo, Brazil.

presenting a biphasic pattern: stimulation for lower concentrations, about 100 μg/m$^3$, and depression from this point and above (Fig. 1). These findings provide some basis for the epidemiological observations of acute effects of air pollution on health.

## ACKNOWLEDGMENTS

This paper was made possible by the Brazilian grants from FAPESP, CNPq, and LIM-HCFMUSP

## REFERENCES

1. Firket J. Fog along the Meuse Valley. Trans Faraday Soc 1936; 32: 1192–1197.
2. Shenk HH, Heimann H, Clayton GD, Gafafer WM, Wexler H. Air pollution in Donora, PA: epidemiology of the unusual smog episode of October 1948, Prelim Rep Public Health Bull 303. Public Health Serv Washington, DC; 1949.

3. Logan WPD. Mortality in the London fog incident, 1952. Lancet 1953; 1: 336–339.
4. Muir DCE, Davies CN. The deposition of 0.5 micron diameter aerosols in the lungs of man. Ann Occup Hyg 1957; 10:161–174.
5. Hatch TF, Gross P. Pulmonary deposition and retention of inhaled aerosols. New York: Academic Press, 1964.
6. Clarke SW, Pavia D, Demetri P. Section E: Defense mechanisms and immunology, deposition and clearance. In: Murray, JK, Nadel JA, eds. Textbook of Respiratory Medicine. Philadelphia: WB Saunders, 1988:313–331.
7. Pavia D, Bateman JRM, Clarke SW. Deposition and clearance of inhaled particles. Bull Europ Physiopath Resp 1980; 16:335–366.
8. Böhm GM, Saldiva PHN, Pasqualucci CAG, Massad E, Martins MA, Zin WA, Cardoso WV, Criado PMP, Komatsuzaki M, Sakae RS, Negri EM, Lemos M, Capelozzi VM, Crestana C, Silva R. Biological effects of air pollution in São Paulo and Cubatão. Environ Res 1989; 49:208–216.
9. Saldiva PHN, King M, Delmonte VLC, Macchione M, Parada MAC, Daliberto ML, Sakae RS, Criado PMP, Silveira PLP, Zin WA, Böhm GM. Respiratory alterations due to urban air pollution: an experimental study in rats. Environ Res 1992; 57:19–33.
10. Lemos M, Linchtenfels AJFC, Amaro E, Macchione M, Martins MA, King M, Böhm GM, Saldiva PHN. Quantitative pathology of nasal passages in rats exposed to urban levels of air pollution. Environ Res 1994; 66:87–95.
11. Pereira PM, Saldiva PHN, Sakae RS, Böhm GM, Martins MA. Urban levels of air pollution increase lung responsiveness in rats. Environ Res 1995; 69: 96–101.
12. Kobayashi K, Salathé M, Pratt MM, Cartagena NJ, Soloni F, Seybold ZV, Wanner A. Mechanism of hydrogen peroxide-induced inhibition of sheep airway cilia. Am J Respir Cell Mol Biol 1992; 6:667–673.
13. Feldman C, Anderson R, Kanthakumar K, Vargas A, Cole PJ, Wilson R. Oxidant-mediated ciliary dysfunction in human respiratory epithelium. Free Rad Biol Med 1994; 17:1–10.
14. Liedtke CM. Regulation of chloride transport in epithelia. Annu Rev Physiol 1989; 51:143–160.
15. Boucher RC. Human airway ion transport. Part one. Am J. Respir Crit Care Med 1994; 150:271–281.
16. Morgan JE, Blair DF, Chan SI. The reactivity of pulsed cytochrome c oxodase toward carbon monoxide. J Inorg Biochem 1985; 23(3–4):295–302.
17. Wright DT, Cohn LA, Li H, Fischer B, Li CM, Adler KB. Interactions of oxygen radicals with airway epithelium. Environ Health Perspect 1994; 102(suppl 10):85–90.
18. Harris ZL, Ridge KM, Gonzalez-Flecha B, Gottlieb L, Zucker A, Sznajder JI. Hyperbaric oxygenation upregulates rat lung Na,K-ATP-ase. Eur Respir J 1996; 9:472–477.
19. Committee of Environmental and Occupational Health Assembly of the American Thoracic Society: Bascon R, Bromberg, PA, Hill C, Costa DA,

Devlin R, Dockery DW, Frampton MW, Lambert W, Samet JM, Speizer FE, Utell M. Health effects of outdoor air pollution. Am J Respir Crit Care Med 1996; 153:3–50.
20. Basbaum CB, Jany B, Finkbeiner WE. The serous cell. Annu Rev Physiol 1990; 52:97–113.
21. Health Effects Institute Diesel Working Group. Diesel exhaust: a critical analysis of emissions, exposure and health effects. In: Health Effects Institute Special Report, 294, 1995.
22. Doll R. Atmospheric pollution and lung cancer. Environ Health Perpect 1978; 22:23–31.
23. Ryder RC, Dunnil MS, Anderson JA. A quantitative study of bronchial mucous gland volume, emphysema and smoking in a necropsy population. J Path 1971; 104:59–71.
24. Scott KWM. An autopsy study of bronchial mucous gland hypertrophy in Glasgow. Am Rev Resp Dis 1973; 107:239–245.
25. Sherwin RP, Richters V. Centriacinar region disease in the lungs of young adults. A preliminary report in: Tropospheric Ozone and Environment In: Beglund RL, Lawson PR, Mckee RJ, eds. Air and Waste Management Associations; Pittsburg, Pennsylvania, 1991:178.
26. Harkema JR, Plopper CG, Hyde DM, George JAS, Wilson DW, Dungworth DL. Response of the macaque nasal epithelium to ambient levels of ozone: a morphologic and morphometric study of the transitional and respiratory epithelium. Am J Pathol 1987; 128:129–144.
27. Scanlon PD, Seltzer J, Ingram RH Jr, Reid L, Drazen JM. Chronic exposure to sulfur dioxide. Physiological and histologic evaluation of dogs exposed to 50 or 15 ppm. Am Rev Respir Dis 1987; 135:831–839.
28. Gorriz A, Llacuna S, Durfort M, Nadal J. A study of the ciliar tracheal epithelium on passerine birds and small mammals subjected to air pollution: ultrastructural study. Arch Environ Contam Toxicol 1994; 27(1):137–142.
29. Andersen I, Camner P, Jensen PL, Phillipson K, Proctor DF. Nasal clearance in monozygotic twins. Am Rev Resp Dis 1974; 110:301–301.
30. Abraham WM, Januszkiewicz AJ, Mingle M, Welker M, Wanner A, Sackner MA. Sensitivity of bronchoprovocation and tracheal mucous velocity in detecting airway responses to $O_3$. J Appl Physiol: Respirat Environ Exercise Physiol 1980; 48(5):789–793.
31. Phalen RF, Kenoyer JL, Crocker TT, McClure TR. Effects of sulfate aerosols in combination with ozone on elimination of tracer particles inhaled by rats. J Toxicol Environ Health 1980; 6:797–810.
32. Wolff RK, Muggenburg BA, Silbaugh SA. Effect of 0.3 and 0.9 μm sulfuric acid aerosols on tracheal mucous clearance in beagle dogs. Am Rev Respir Dis 1981; 123:291–294.
33. Leikauf G, Yeates DB, Wales KA, Spektor D, Albert RE, Lippmann M. Effects of sulfuric acid aerosol on respiratory mechanics and mucociliary particle clearance in healthy nonsmoking adults. Am Ind Hyg Assoc J 1981; 42:273–282.

34. Abraham WH, Sielczak MW, Delehunt JC, Marchette B, Wanner A. Impairment of tracheal mucociliary clearance but not ciliary beat frequency by a combination of low level ozone and sulfur dioxide in sheep. Eur J Dis 1986; 68:114–120.
35. Phipps RJ, Denas SM, Sielczak MW, Wanner A. Effects of 0.5 ppm ozone on glycoprotein secretion, ion and water fluxes in sheep trachea. J Appl Physiol 1986; 60(3):918–927.
36. Schlesinger RB, Driscoll KE. Mucociliary clearance from the lungs of rabbits following single and intermittent exposures to ozone. J Toxicol Environ Health 1987; 20:125–134.
37. Foster MW, Costa DL, Langenback EG. Ozone exposure alters tracheobronchial mucociliary function in humans. J Appl Physiol 1987; 63(3):996–1002.
38. Schlesinger RB, Driscoll KE, Vollmuth TA. Effect of repeated exposures to nitrogen dioxide and sulfuric acid mist alone or in combination on mucociliary clearance from the lungs of rabbits. Environ Res 1987; 44(2):294–301.
39. Allegra L, Moavero NE, Rampoldi C. Ozone–induced impairment of mucociliary transport and its prevention with N–acetylcysteine. Am J Med 1991; 91:67s–71s.
40. Gerrity TR, Bennett WD, Kehrl H, DeWitt P. Mucociliary clearance of inhaled particles measured at 2 h after ozone exposure in humans. J Appl Physiol 1993; 74(6):2984–2989.
41. Last JA, Jennings MD, Schwartz LW, Cross CE. Glycoprotein secretion by tracheal explants cultured from rats exposed to ozone. 1977; 116:695–703.
42. Heller RF, Gordon RE. Chronic effects of nitrogen dioxide on cilia in hamster bronchioles. Exp Lung Res 1986; 10(2):137–152.
43. Wolff RK, Henderson RF, Snipes MB, Griffith WC, Mauderly JL, Cuddihy RG, McClellan RO. Alterations in particle accumulation and clearance in lungs of rats chronocally exposed to diesel exhaust. Fundam Appl Toxicol 1987; 9(1):154–166.
44. Harkema JR, Plopper CG, Hyde DM, St-George JA, Dungworth DL. Effects of an ambient level of ozone on primate nasal epithelial mucosubstances. Quantitative histochemistry. Am J Pathol 1987; 127(1):90–96.
45. Rasmussen RE, Mannix RC, Oldham MJ, Phalen RF. Effects of nitrogen dioxide on respiratory tract clearance in the ferret. J Toxicol Environ Health 1994; 41(1):109–120.
46. Gulisano M, Pacini P, Marceddu S, Orlandini GE. Scanning electron microscopic evaluation of the alterations induced by polluted air in the rabbit bronchial epithelium. Anat Anz 1995; 177(2):125–131.

# 27

# Improved Mucociliary Clearance Following Nebulization with Hypertonic Saline in Adults with Cystic Fibrosis

**Michael Robinson, Ariane Hemming, Jeff A. Regnis, Peter Bye, and George Bautovich**
*Royal Prince Alfred Hospital, Sydney, New South Wales, Australia*

**Dale L. Bailey**
*Hammersmith Hospital, London, United Kingdom*

**Malcolm King and Wei Feng**
*University of Alberta, Edmonton, Alberta, Canada*

## I. INTRODUCTION

In cystic fibrosis (CF), mutations in the CF gene result in aberrant ion transport across the airway epithelium. Although considerable controversy currently exists on the exact ionic composition of the sol phase of the CF airway surface fluid [1], expectorated sputum from patients with CF is known to be low in both sodium and chloride [2,3]. These changes are believed to alter the rheologic properties of the mucus and lead to a decrease in mucociliary clearance (MCC) [4]. This in turn is thought to play a pivotal role in the development of the subsequent vicious cycle of airway obstruction, infection, and ongoing lung damage that is characteristic of this disease. A major goal of therapy in CF is the effective removal of thick tenacious secretions.

Pavia and colleagues [5] have shown a twofold increase in MCC in patients with chronic bronchitis following the inhalation of a 7.0% hypertonic saline (HS) solution compared with isotonic saline (IS). Inhalation of 14.4% HS has been shown to significantly increase MCC in healthy subjects and asthmatics, both groups having normal baseline clearance [6]. Hypertonic saline has also been used in many centers to induce sputum production in a variety of respiratory diseases. This paper reports the effect of various concentrations of aerosolized hypertonic saline in improving mucociliary clearance in patients with cystic fibrosis.

## II. STUDY PROTOCOL

A schema of the procedures performed on each of the study days is shown in Figure 1. After inhalation of a standardized dose of bronchodilator, the patients inhaled the radioaerosol and baseline clearance was collected for 30 minutes. The patients then inhaled the intervention solution prior to being repositioned under the gamma camera for the postintervention clearance measurement for a further 30–60 minutes. Spirometry was performed immediately prior to the intervention, 5 minutes after the end of the intervention, and at the end of the study.

### A. Measurement of Mucociliary Clearance

Mucociliary clearance was assessed using the methods previously described by our group [4]. The patient was connected to a closed breathing

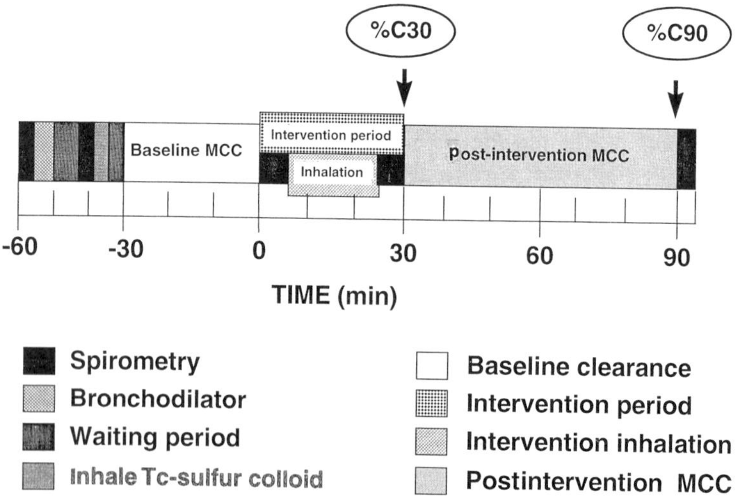


**Figure 1** Timeline detailing the protocol used in both studies.

circuit and inhaled a calibrated amount of radioisotope ($^{99m}$Tc-sulfur colloid). The radioaerosol was generated using an Acorn nebulizer, selected for its large particle size (mass medium aerodynamic diameter [MMAD] = 4.8 μm, geometric standard deviation [GSD] = 1.6). The patient was requested to follow a preprogrammed breathing pattern that featured a fast inspiratory flow (mean inspiratory flow = 56.0 L/min) and a small tidal volume (0.45L) [7]. This promoted deposition of the radioaerosol predominantly in the ciliated airways. The system provides a breath-by-breath analysis of the patient's breathing pattern.

Following deposition of the radioaerosol, patients were placed supine on the scan bed. Data were acquired using a large field of view gamma camera fitted with a low-energy all-purpose collimator (Phillips Diagnost Tomo, Hamburg, Germany). Simultaneous emission and transmission scans were acquired using a scanning line source [8]. The transmission images were subsequently used to define the lung contour and regions of interest. Alternating twenty-second anterior and posterior images were collected every minute (allowing 20 sec for camera rotation) and a dynamic geometrically meaned data set was subsequently generated [9].

## B. Data Analysis

Mucociliary clearance was defined as the rate of removal of radioactivity from the right lung. The geometric mean data set was used to generate the activity versus time curves that were then fitted with an exponential function using a least-squares model. Curves were independently fitted to the baseline and postintervention data sets and then linked. The amount of activity cleared from the lung during the baseline measurement, intervention period (percent clearance at 30 minutes, %C30), and postintervention period (%C90) was then calculated. To remove the influence of any variation in baseline clearance between individuals and within individuals between study days, the amount of activity remaining at the end of the baseline acquisition was standardized to 100% and the curves refitted to the transformed data set.

## C. Delivery System for Hypertonic Saline

All intervention solutions were delivered via a modified Omron-NE-U06 ultrasonic nebulizer (Omron Corporation, Tokyo, Japan). This nebulizer has a high volume output (1.5 mL/minute) and a small dead volume (0.5 mL). The MMAD of this delivery system was 3.7 μm and 73% of the particles were in the respirable range (Malven Mastersizer-X, Malvern, UK). The nebulizer was charged with 7.0 mL of the trial solution on all days. Solutions were inhaled using a tidal breathing pattern to "dryness"

**Table 1** Anthropometric and Lung Function Data for the Patients in Both Studies

| | Study 1 ($n$ = 9) | Study 2 ($n$ = 10) |
|---|---|---|
| Age (year) | 22.6 ± 3.1 (19–28) | 22.1 ± 3.8 (18–27) |
| Gender | 8M/1F | 7M/3F |
| BMI (kg $m^{-2}$) | 21.4 ± 2.6 (17.7–24.8) | 20.6 ± 2.3 (16.2–22.8) |
| $FEV_1$ (% pred) | 55.4 ± 30.6 (27–112) | 52.0 ± 6.7 (30.9–83.8) |
| FVC (% pred) | 72.3 ± 23.6 (44–118) | 71.2 ± 6.6 (57.0–84.1) |
| FEF 25–75 | 36.3 ± 37.1 (9.0–104) | 29.7 ± 2.7 (23.1–52.7) |

Values are mean ± SD. Ranges are shown in parentheses.

(mean nebulization time = 9.2 ± 3.1 min, mean volume leaving delivery system = 4.6 ± 1.3 mL).

### D. Patient Group

All patients were studied in stable clinical condition and baseline medications were not altered throughout the study period. None of the patients were taking rhDNase. Patients were instructed not to take any bronchodilators or perform physiotherapy on the morning of each of the study days. Anthropemteric and lung function data for both studies is shown in Table 1.

## III. STUDY 1: EFFECT OF HYPERTONIC SALINE AND COUGH ON MUCOCILIARY CLEARANCE IN PATIENTS WITH CYSTIC FIBROSIS

### A. Intervention Solutions

An additional aim of this study was to determine whether any observed changes in MCC following the inhalation of the hypertonic saline solution could be accounted for by the coughing induced by this solution. Thus, three interventions were used.

1. Isotonic saline (IS): 7.0 mL of 0.9% NaCl.

2. Hypertonic Saline (HS): 7.0 mL of a 7.0% solution of NaCl.
3. Voluntary cough (COUGH): the number of cough maneuvers performed was equal to or slightly greater than the maximum number of coughs recorded during the intervention period on the hypertonic saline day

## B. Results

There was no significant variation in baseline characteristics (lung function, radioaerosol deposition, or baseline clearance, $p = 0.5$) between study days that would have accounted for the observed changes in postintervention clearance. Mean baseline clearance was impaired in all patients.

Mean values for percent clearance at 30 minutes (%C30) and 60 minutes (%C60) are shown in Table 2. Mucociliary clearance was enhanced by the inhalation of hypertonic saline. The percent cleared at 30 and 60 minutes was significantly increased with HS (%C30 = 26.5%, %C60 = 29.4%) compared with both IS (%C30 = 14.7%, %C60 = 17.5%) and COUGH (%C30 = 18.0%, %C60 = 19.5%, $p < 0.01$) [10].

The improvement in MCC seen with HS could not be entirely accounted for by the induced coughing as the total number of coughs (both spontaneous and intervention induced) recorded during the intervention period and postintervention MCC measurement was greatest on the cough clearance day (IS 36.5 ± 9.6, HS 86.8 ± 20.5, COUGH 132.6 ± 19.8, $p = 0.0001$).

There was a mean fall in forced expiratory volume in one second ($FEV_1$) of 8.4 ± 1.7% following the inhalation of the HS solution. However this fall was not statistically different ($p = 0.4$) from the falls seen

**Table 2** Mean (± SEM) Values for Mucociliary Clearance at the Designated Time Points for Study 1

| | %C30 | %C60 |
|---|---|---|
| IS (0.9%) | 14.7 ± 3.8 | 17.5 ± 4.5 |
| Cough | 18.0 ± 4.5 | 19.5 ± 4.7 |
| HS (7%) | 26.5 ± 5.7[a] | 29.4 ± 6.4[a] |

[a]Significantly different from the isotonic saline and cough days ($p < 0.01$).
%C30 and %C60 = amount cleared from the lung at 30 and 60 min postintervention respectively.

with IS (3.7 ± 1.8%) and COUGH (1.9 ± 4.3%). Upon questioning, none of the patients reported any symptoms such as "chest tightness," "wheeze," or "difficulty in breathing" with any of the interventions.

## IV. STUDY 2: EFFECT OF INCREASING DOSES OF HYPERTONIC SALINE ON MUCOCILIARY CLEARANCE IN PATIENTS WITH CYSTIC FIBROSIS

### A. Intervention Solutions

Different doses of HS were achieved by varying the concentration of NaCl in the nebulizer bowl, while keeping all other delivery parameters constant. The solutions used were

1. 7 mL × 3% HS
2. 7 mL × 7% HS
3. 7 mL × 12% HS
4. CONTROL—a combination of isotonic (0.9%) saline and voluntary cough. On this day we aimed to have the patients voluntarily cough a number of times that equaled or exceeded the number of coughs on any of the other study days. The coughs were evenly dispersed throughout the inhalation period.

### B. Results

There was no significant difference in baseline characteristics (lung function, radioaerosol deposition, or baseline mucociliary clearance, $p = 0.1$) between study days. Baseline clearance at 30 min was significantly impaired in 9 of the 10 patients (8 ± 6% compared with normal subjects in our laboratory 20 ± 13%, $p < 0.01$).

Mean values for percent clearance of radioisotope at 30 min (%C30) and 90 min (%C90) postintervention are shown in Table 3. Figure 2 shows the standardized mean activity–time curves. There was a significant increase in the amount of activity cleared from the right lung at 30 and 90 minutes with *all* concentrations of HS compared with control ($p = 0.002$). In addition there was a significant increase in %C90 for 12% HS solution compared with 3% HS solution ($p = 0.001$).

The change seen in %C90 for all concentrations of HS was not solely due to that seen at %C30. That is, there was an ongoing effect beyond the immediate inhalation period ($p = 0.01$). The actual number of patients

**Table 3** Mean (± SEM) Values for Percent Clearance Immediately Postintervention (%C30) and an Hour Later (%C90) in Study 2

| | %C30 | %C90 |
|---|---|---|
| Control | 8.7 ± 1.7 | 12.7 ± 1.4 |
| 3% HS | 13.5 ± 2.6[a] | 19.7 ± 3.1[a] |
| 7% HS | 15.8 ± 3.0[a] | 23.8 ± 4.0[a] |
| 12% HS | 17.3 ± 2.7[a] | 26.0 ± 3.1[a,b] |

[a]Significantly different from the control day.
[b]Significantly different from 3% HS, (p = 0.002).

that achieved the highest clearance on 12% HS ($n$ = 6) was significantly more than expected ($n$ = 2.5) under the null hypothesis of each intervention having an equal probability (One group Chi square test, p = 0.0001).

The observed changes in MCC were not entirely due to the cough induced by the intervention solution as the number of coughs recorded

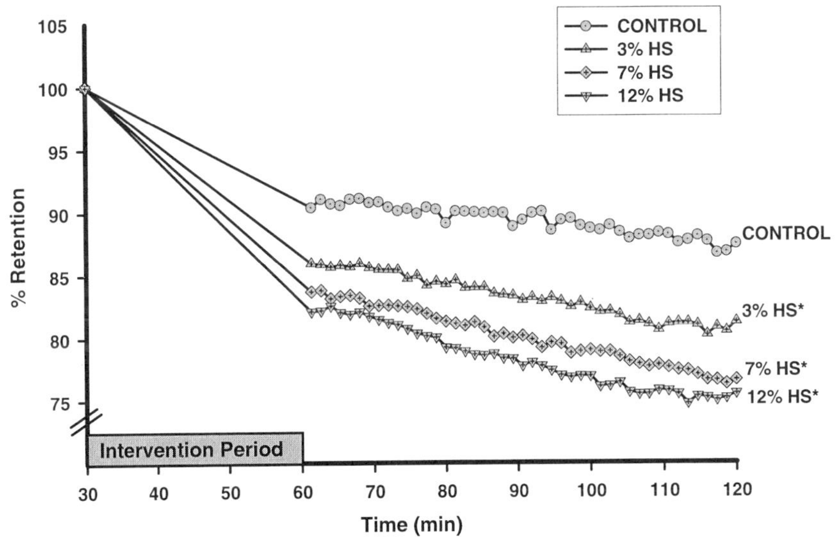


**Figure 2** Standardized mean clearance curves for the whole right lung. Curves marked with an asterisk (*) are significantly different from the control day.

**Table 4** Mean (± SEM) Number of Coughs Recorded During the Measurement of Baseline Clearance, the Intervention Period, and the Measurement of Postintervention Clearance in Study 2

| | Baseline MCC | Intervention period | Postintervention MCC |
|---|---|---|---|
| Control | 3.8 ± 1.3 | 133.0 ± 22.9 | 8.9 ± 3.2 |
| 3% HS | 4.8 ± 1.7 | 78.2 ± 15.5[a] | 6.6 ± 2.4 |
| 7% HS | 3.6 ± 1.4 | 105.1 ± 20.9[a] | 6.5 ± 2.9 |
| 12% HS | 3.3 ± 1.5 | 123.5 ± 24.5[a,b] | 12.0 ± 3.5 |

[a]Significantly different from control day.
[b]Significantly different from 3.0% HS ($p = 0.0001$).

on the control day was greater than on any of the other intervention days (Table 4).

There was a small mean percentage fall in $FEV_1$ immediately following inhalation of 7% HS, 12% HS, and the control (Table 5). However, this was not statistically or clinically significant ($p = 0.5$). This fall in $FEV_1$ had been reversed by the end of the study (i.e., an hour later) resulting in a small (nonsignificant) increase in $FEV_1$ with all the HS solutions compared with the preintervention values.

**Table 5** Mean (± SEM) Percent Change in $FEV_1$ from Baseline to End Intervention (30 min) and at the End of the Subsequent Clearance Measurement (90 min) in Study 2

| | Δ at 30 min | Δ at 90 min |
|---|---|---|
| Control | −1.5 ± 4.8 | −3.4 ± 4.6 |
| 3% HS | 0.6 ± 9.5 | 5.4 ± 10.4 |
| 7% HS | −3.7 ± 8.5 | 3.6 ± 8.2 |
| 12% HS | −4.9 ± 5.8 | 2.8 ± 7.8 |

There was no significant change in $FEV_1$ at either time point ($p = 0.2$).

## V. DISCUSSION

These studies show that mucociliary clearance is enhanced by the inhalation of hypertonic saline in patients with cystic fibrosis. The improvement occurred with all concentrations of HS examined and the effect appears to be dose dependent. As both isotonic saline [6] and cough [11] had been shown previously to improve lung clearance in normals, the enhanced clearance seen in these studies with HS was in addition to any increase that is likely to have occurred from the inhalation of IS and/or voluntary cough. On the basis of our previous cross-sectional study [4] showing values of MCC for normal subjects and CF patients stratified according to their lung function, we consider a 10% increase in MCC to be clinically significant. This was achieved by intervention with 7% and 12% HS.

The improvement in MCC with HS was seen in patients with a wide range of lung function and was independent of disease severity. The improved clearance could not be entirely accounted for by the cough induced by these solutions, as the maximum number of coughs was recorded on the control days and clearance on these days was significantly less than with HS.

### A. Hypertonic Saline, the Airway Surface Fluid, Infection, and Inflammation

The small volume and relative inaccessibility of the airway surface fluid (ASF) make it difficult to directly measure the ionic composition without perturbing the system. Quinton and colleagues [12] were the first to show that cystic fibrosis was characterized by abnormal chloride permeability across exocrine epithelia. Boucher et al. [13] have reported that CF epithelia also demonstrate a secondary hyperabsorption of sodium ions. Thus the accepted paradigm for CF airway disease for many years involved impaired secretion of $Cl^-$ ions into the airway lumen and the hyperabsorption of $Na^+$ ions in the opposite direction. Water was thought to follow the ions passively and lead to dehydration of airway secretions. This in turn resulted in a decrease in mucociliary clearance, airway obstruction, and colonization with a variety of pathogenic bacteria. Mizgerd et al. [14] have shown that the low sodium environment that is typical of CF sputum reduces the ability of human peripheral neutrophils to phagocytose and kill *Pseudomonas aeruginosa* in vitro. By restoring the sodium concentration to normal, neutrophil phagocytosis and killing was returned to normal. This linked the ion abnormality to the CF airway's propensity towards infection.

A number of strategies have been developed to restore an hypo-osmolar cystic fibrosis ASF milieu to normal. These include the sodium channel blockers amiloride [3,15] and benzamil and the chloride secretagogues uridine triphosphate (UTP) and adenosine triphosphate (ATP) [16]. Within this paradigm, it seems logical that nebulization of hypertonic saline may go some way to restoring the ionic imbalance in the ASF. Due to osmotic regulation, this change is likely to be transient and this transience may account for the relatively short effect of HS seen in this and other studies [6,17].

Recently more sophisticated techniques have been used to measure the ion composition of the ASF [1,18]. These studies indicate that the ASF in patients with CF may in fact be hyperosmolar with respect to normal ASF. This is in keeping with the hypothesis of Quinton [19] that the airway epithelium, like the sweat duct, primarily acts to regulate the composition (rather than volume) of the ASF by absorbing $Cl^-$ from the airway lumen. Failure to absorb $Cl^-$ by a dysfunctional cystic fibrosis transmembrane regulator protein (CFTR) results in an increase in both $Na^+$ and $Cl^-$ concentration in the ASF and hyperosmolar dehydration of airway secretions [1].

Smith [20] and Goldman [18] have recently been able to demonstrate a possible link between an hyperosmolar ASF and infection in the CF airway. The high salt content of the CF airway surface liquid deactivates a low-molecular weight peptide that is secreted onto the airway surface by the epithelium. This peptide is known as human β-defensin-1 and is believed to be essential in the early elimination of bacteria, fungi, and viruses. Once this primary line of defense has been violated a more sustained immune response is evoked (by the release of cytokines). This initiates the migration of large numbers of neutrophils and macrophages into the airway lumen. Here they release a number of proteases, elastases, and other degradative enzymes that are responsible for much of the lung damage and ongoing inflammation. Deoxyribonucleic acid and F-actin are also released from the degenerative neutrophils and are known to significantly increase the viscosity of CF respiratory secretions, thus sustaining the vicious cycle.

The reported differences in ion concentrations between expectorated sputum and samples obtained by microsampling of the ASF via a bronchoscope suggest that ions and water may not move freely between the gel and sol layers but are in a Donnan equilibrium and that changes in one may not necessarily be reflected in the other. The complex nature of the mucus macromolecule means that it is capable of trapping ions and water within its three-dimensional structure. In the CF airways the volume of mucus is likely to be far greater than that of the ASF and may take up much of the additional $Na^+$ and $Cl^-$ before they reach the ASF where the defensins are located.

Any effect of HS on defensins is likely to be more important early in the disease process and may lie in delaying the onset of colonization. Once suppurative lung disease is established, the beneficial effect of hypertonic saline in debulking the airways of copious amounts of viscous secretions and the pathogens and degradative enzymes contained therein is likely to be more significant. It is also likely that any effect HS has on the ion concentration of the ASF, and hence on defensin deactivation, would soon be counteracted by the rapid influx of water into the airway lumen along its osmotic gradient [21].

## B. Hypertonic Saline and Mucus Rheology

The hypertonic saline may have achieved its effect, at least in part, by promoting hydration of airway secretions. Water is believed to move into the airway lumen in response to the transient osmotic gradient created following inhalation of hypertonic saline [22]. The movement of water into the airway lumen may better hydrate airway secretions, improving their viscoelastic properties. It is also possible that the HS solution induced the release of prostaglandin $E_2$ ($PGE_2$) from the airway epithelium [23]. The $PGE_2$ is known to increase ciliary beat frequency [24]. The combination of improved rheologic properties and increased ciliary beat frequency would have promoted clearance of airway secretions.

The hypertonic saline also could have achieved its effect through improvement of the rheological properties of the mucus independent of hydration. The mucin macromolecules in respiratory secretions carry fixed negative and positive charges. In cystic fibrosis, there is a great excess of negative charges, primarily due to increased oligosaccharide sulfation [25]. The repulsion between these fixed negative charges along the length of the macromolecule will result in CF mucin being more extended in aqueous solution. The addition of mobile, low molecular weight ions will assist in reducing the repulsion between fixed charges, resulting in a more compact conformation. In solutions sufficiently concentrated in mucin macromolecules to produce significant entanglements and gelation, which is the case in airway mucus, the reduced effective size will result in reduced numbers of entanglements. This would increase the deformability of the mucus and lead to more effective clearance.

The effect of salt on gel strength is well known in synthetic polyelectrolyte solutions [26]. Lutz and colleagues [27] compared the viscoelastic properties of canine mucin gels prepared in 0.15 *M* NaCl and in salt-free medium. The lack of salt increased the gel modulus profoundly, particularly at low mucin concentrations. McCullagh and coworkers [28] have reported that human tracheobronchial mucins from CF patients exhibit increasing gel strength as the solvent is changed from 6.0 *M* guanidinium HCl to 0.1 *M* NaCl to distilled water. Other investigators have observed that added salt inhibits the aggregation of tracheobronchial mucins [29,30].

Scheffner and coworkers [31] have shown that the addition of 3% sodium chloride to homogenized sputum markedly reduces its viscosity. They suggested that sodium salts have a specific effect of reducing the viscosity of deoxyribonucleic acid (DNA) solution and that 2 M (12%) sodium chloride is as effective in this regard as DNase. There is also evidence that DNA complexes with mucoprotein to make it unavailable to the natural proteases in purulent sputum. Hypertonic saline acts, as does DNase, to dissociate the DNA from the mucoprotein, thereby making the latter susceptible to natural proteolytic enzyme digestion. Lieberman [32] demonstrated that this effect of saline increases with elevations of ionic strength. By increasing the ionic strength to 0.167 with sodium chloride, intrinsic sputum protease activity was enhanced by as much as 370%.

Hypertonic saline is known to have a favorable effect on mucus rheology. Dasgupta et al. [33] have examined the effect of 3% HS compared with normal saline on mucus viscoelasticity in vitro. They reported that HS reduced spinnability and sputum rigidity. In addition they showed that HS improved mucus clearability more than rhDNase (25 μg/mL). Ziment [34] has postulated that hypertonic saline breaks the ionic bonds within the mucin gel, thereby reducing the effective degree of crosslinking and entanglements, and lowering the viscosity and elasticity.

Wills et al. [17] have studied the transportability of sputum from cystic fibrosis patients using a mucus-depleted bovine tracheal model. They measured the intrinsic transportability of bovine mucus and compared this with the transportability of expectorated sputum. The relative transportability of CF sputum was reduced at baseline but improved from $33 \pm 3\%$ to $57 \pm 5\%$ ($p = 0.001$) following overnight incubation with solid NaCl (0.5% w/w). In their ex vivo model, ciliary beat frequency was constant over a wide range of ionic concentrations. Thus, they postulated that the observed increase in transportability was due solely to changes in sputum rheology caused by an increase in osmolarity.

### C. Hypertonic Saline and Airway Narrowing

In these studies, some minor and transient airway narrowing was seen in the patients following application of nebulized hypertonic saline despite premedication with a bronchodilator. On questioning, none of the patients reported any symptoms such as chest tightness, wheeze or difficulty in breathing. It is possible that the mobilization of secretions into the large airways may have been partially responsible for the fall in $FEV_1$. The combination of the inhaled fluid volume and the movement of water into the airway lumen along its osmotic gradient may also have contributed to the fall in $FEV_1$ following inhalation of the HS. The former may explain why a small decrease in $FEV_1$ was also seen following the inhalation of the isotonic saline solution in both studies.

Hypertonic saline is known to be a bronchoprovactive agent. Rodwell and colleagues [35] have reported the effect of hyperosmolar (10%) saline challenge in 23 cystic fibrosis patients. All patients were selected because they carried a concurrent clinical diagnosis of asthma. In this study, 30% of the patients showed a greater than 15% fall in $FEV_1$ at the end of the challenge. A second group of patients (40%) showed a transient decrease in $FEV_1$ with a partial spontaneous recovery before the challenge finished. A third group (30%) showed spontaneous recovery during the challenge with the $FEV_1$ recorded at the end of the challenge significantly higher (4.5%) than immediately prior to the challenge. All groups had a further significant increase in lung function after administration of a bronchodilator at the completion of the challenge. It is expected that the percentages of CF patients showing a fall in $FEV_1$ and the magnitude of such a fall would be considerably less in the general CF population.

### D. Future Role of Hypertonic Saline in the Treatment of Patients with CF

The improved clearance seen with HS in our studies appears to last for approximately 60 minutes. In a clinical setting, it is likely that nebulization of HS would be limited to twice daily. It is not known what long-term impact this more effective, but somewhat transitory, removal of purulent secretions would have on morbidity and mortality. However, Eng and coworkers [36] in a parallel, controlled study have examined the effect of a two-week course of twice daily 10 mL $\times$ 6% HS on lung function in a group of 52 CF patients. They found a 15.0 $\pm$ 16.0% improvement in $FEV_1$ in the HS group compared with a 2.8 $\pm$ 13.1% improvement with normal saline ($p = 0.004$). Using a VAS, the patients in the HS group also had a subjective improvement in the effectiveness of chest physiotherapy, quality of sleep, and exercise tolerance. There was no difference in reported adverse events between the two groups and the authors state that treatment with 6% HS was well tolerated.

## VI. SUMMARY

In conclusion, there is now considerable preliminary evidence for a beneficial effect of HS in cystic fibrosis. This comes from radioaerosol studies of mucociliary clearance, in vitro studies of mucus rheology and transportability, and a short-term clinical trial. Further studies are required to fully exclude any potential undesirable effects on airway inflammation or unfavorable effects on host defense mechanisms. Long-term studies of clinical safety and efficacy are also required. The role of HS may prove to be short term in enhancing clearance of secretions alone or in combination with physiotherapy. However, the possibility remains that HS could be of long-term clinical benefit in reducing morbidity and mortality. This needs to be assessed in long-term clinical trials.

## ACKNOWLEDGMENTS

This work is supported by the National Health and Medical Research Council of Australia and the Australian Cystic Fibrosis Research Trust.

## REFERENCES

1. Joris L, Dab I, Quinton PM. Elemental composition of human airway surface fluid in healthy and diseased airways. Am Rev Respir Dis 1993; 148: 1633–1637.
2. Potter JL, Matthews LW, Spector S, Lemm J. Studies on pulmonary secretions II. Osmolality and the ionic environment of pulmonary secretions from patients with cystic fibrosis, bronchiectasis and laryngectomy. Am Rev Respir Dis 1967; 96:83–87.
3. Tomkiewicz RP, App EM, Zayas JG, Ramirez O, Church N, Boucher RC, Knowles MR, King M. Amiloride inhalation therapy in cystic fibrosis. Influence on ion content, hydration, and rheology of sputum. Am Rev Respir Dis 1993; 148:1002–1007.
4. Regnis JA, Robinson M, Bailey DL, Cook P, Hooper P, Chan H-K, Gonda I, Bautovich GJ, Bye PTP. Mucociliary clearance in patients with cystic fibrosis and in normal subjects. Am J Resp Crit Care Med 1994; 150:66–71.
5. Pavia D, Thomson ML, Clarke SW. Enhanced clearance of secretions from the human lung after the administration of hypertonic saline aerosol. Am Rev Respir Dis 1978; 117:199–203.
6. Daviskas E, Anderson SD, Gonda I, Eberl S, Meikle S, Seale JP, Bautovich GI. Inhalation of hypertonic saline aerosol enhances mucociliary clearance in asthmatic and healthy subjects. Eur Respir J 1996; 9:725–732.

7. Phipps PR, Gonda I, Anderson SD. Apparatus for the Control of Breathing Patterns During Aerosol Inhalation. J Aerosol Med 1992; 5:155–169.
8. Bailey DL, Robinson M, Meikle SR, Bye PTP. Simultaneous emission and transmission measurements as an adjunct to dynamic planar gamma camera studies. European Journal of Nuclear Medicine 1996; 23:326–331.
9. Bailey DL, Fulton RR, Jackson CB, Hutton BE. Dynamic geometric mean studies using a single headed rotrating gamma camera. J Nuc Med 1989; 30: 1865–1869.
10. Robinson M, Regnis JA, Bailey DL, King M, Bautovich GJ, Bye PT. Effect of hypertonic saline, amiloride, and cough on mucociliary clearance in patients with cystic fibrosis. Am J Respir Crit Care Med 1996; 153:1503–1509.
11. Bennett WD, Foster WM, Chapman WF. Cough-enhanced mucus clearance in the normal lung. J Appl Physiol 1990; 69:1670–1675.
12. Quinton PM, Bijman J. Higher bioelectric potentials due to decreased chloride absorption in the sweat glands of patients with cystic fibrosis. N Engl J Med 1983; 308:1185–1189.
13. Boucher RC, Stutts MJ, Knowles MR, Cantley L, Gatzy JT. $Na^+$ transport in cystic fibrosis respiratory epithelia: abnormal basal rate and response to adenylate cyclase activation. J Clin Invest 1986; 78:1245–1252.
14. Mizgerd JP, Kobzik L, Warner AE, Brain JD. Effects of sodium concentration on human neutrophil bactericidal functions. Am J Physiol 1995; 269: L388–393.
15. Knowles MR, Church NL, Waltner WE, Yankaskas JR, Gilligan P, King M, Ewards LJ, Helms RW, Boucher RC. A pilot study of aerosolized amiloride for the treatment of lung disease in cystic fibrosis. N Engl J Med 1989; 322:1189–1194.
16. Bennett WD, Olivier KN, Zeman KL, Hohnecker KW, Boucher RC, Knowles MR. Effect of Uridine 5′ Triphosphate plus amiloride on mucociliary clearance in adult cystic fibrosis. Am J Respir Crit Care Med 1996; 153: 1796–1801.
17. Wills PJ, Hall RL, Chan WM, Cole PJ. Sodium chloride increases the ciliary transportability of cystic fibrosis and bronchiectasis sputum on the mucus-depleted bovine trachea. J Clin Invest 1997; 99:9–13.
18. Goldman MJ, Anderson GM, Stolzenberg ED, Kari UP, Zasloff M, Wilson JM. Human β-defensin-1 is a salt-sensitive antibiotic in lung that is inactivated in cystic fibrosis. Cell 1997; 88:553–60.
19. Quinton PM. Cystic fibrosis: a disease in electrolyte transport. FASEB J 1990; 4:2709–2717.
20. Smith JJ, Travis SM, Greenberg EP, Welsh MJ. Cystic fibrosis airway epithelia fail to kill bacteria because of abnormal airway surface fluid. Cell 1996; 85:229–236.
21. Folkesson HG, Kheradmand F, Matthay MA. The effect of salt water on alveolar epithelial barrier function. Am J Respir Crit Care Med 1994; 150: 1555–1563.
22. Boucher RC. Human airway ion transport. Part two. Am J Respir Crit Care Med 1994; 150:581–593.

23. Assouline G, Leibson V, Danon A. Stimulation of prostaglandin output from rat stomach by hypertonic solutions. Eur J Pharmacol 1977; 44:271–273.
24. Wanner A, Sielczak M, Mella JF, Abraham WM. Ciliary responsiveness in allergic and nonallergic airways. J Appl Physiol 1986; 60:1967–1971.
25. Shankar V, Naziruddin B, de la Rocha SR, Sachdev GP. Evidence of hydrophobic domains in human respiratory mucins: effect of sodium chloride on hydrophobic binding properties. Biochemistry 1990; 29:5856–5864.
26. King M, Eisenberg A. Viscoelastic properties of polyelectrolyte solutions at high concentrations. Am Chem Soc Org Coatings Plast Chem 1972; 32: 327–335.
27. Lutz RJ, Litt M, Chakrin LW. Physical–chemical factors in mucus rheology. In: Gabelnick HL, Litt M, eds. Rheology of Biological Systems. Springfield, IL: Charles C Thomas, 1973; 158–194.
28. McCullagh CM, Jamieson AM, Blackwell J, Gupta R. Viscoelastic properties of human tracheobronchial mucin in aqueous solution. Biopolymers 1995; 35:149–159.
29. Chace KV, Naziruddin B, Desai VC, Flux M, Sachdev GP. Physical properties of purified human respiratory mucus glycoproteins: effects of sodium chloride concentration on the aggregation properties and shape. Exp Lung Res 1989; 15:721–737.
30. Slayter HS, Lamblin G, Treut AL, Galabert C, Houdret N, Degand P, Roussel P. Complex structure of human bronchial mucus glycoprotein. Eur J Biochem 1984; 142:209–218.
31. Scheffner AL, Medler EM, Jacobs LW, Sarett HP. The in vitro reduction in viscosity of human tracheobronchial secretions by acetylcysteine. Am Rev Respir Dis 1964; 90:721–729.
32. Lieberman I, Kurnick NB. Influence of deoxyribonucleic acid content on the proteolysis of sputum and pus. Nature (Lond.) 1962; 196:988–990.
33. Dasgupta B, Tomkiewicz RP, Brown NE, King M. Combined effects of hypertonic saline and rhDNase on cystic fibrosis sputum in vitro. Ped Pulmonol 1995; 20:A201:236.
34. Ziment I. Hypertonic solutions, urea and ascorbic acid. In: Eds. Braga PC, Allergra L, Drugs in Bronchial Mucology. New York: Raven Press, 1989; 137–143, 300–303.
35. Rodwell LT, Anderson SD. Airway responsiveness to hyperosmolar saline challenge in cystic fibrosis—a pilot study. Pediatric Pulmonology 1996; 21: 282–289.
36. Eng PA, Morton J, Douglass JA, Riedler J, Wilson J, Robertson CF. Short-term efficacy of ultrasonically nebulized hypertonic saline in cystic fibrosis. Ped Pulmonol 1996; 21:77–83.

# 28

# Ciliary Transportability of Sputum Is Governed by Its Osmolality

**Peter John Wills and W. M. Chan**
*National Heart and Lung Institute, London, United Kingdom*

**Roderick L. Hall**
*Bayer plc, Stoke Poges, United Kingdom*

**Peter J. Cole**
*National Heart and Lung Institute, and Royal Brompton Hospital, London, United Kingdom*

## I. INTRODUCTION

In health, the airways below the larynx contain only microscopic quantities of mucus and are virtually sterile. However, in several common and serious diseases mucus accumulates and tends to become infected. Clearance of pulmonary mucus is delayed in acute infections, chronic bronchitis, bronchiectasis of various causes including cystic fibrosis, and asthma in exacerbation [1]. When this situation becomes chronic, the host inflammatory response fails to eliminate the pathogen and in fact may be largely responsible for the lung damage that ensues [2].

Chest physiotherapy techniques, particularly the forced expiratory maneuver, are effective at helping to clear the retained secretions, and antibiotics and anti-inflammatory agents have important roles. However, the more fundamental problem of mucostasis remains practically untreatable pharmacologically despite the plethora of cough remedies available.

## II. PROPERTIES OF AIRWAY MUCUS

We investigated the properties of airway mucus that govern its transportability by cilia using our newly described model, the mucus-depleted bovine trachea [3]. This has the same scope as the mucus-depleted frog palate system, but uses a mammalian respiratory epithelium rather than an amphibian digestive one. Tracheas obtained from the slaughterhouse were dissected into pieces 8 cm × 3 cm and depleted of their endogenous mucus by prolonged incubation at 37°C exposed to the air and by repeated passage of a large quantity of bovine tracheal mucus. When a state of mucus depletion was achieved, the tissue still transported mucus but failed to transport 1 mm particles of aluminium foil, which were readily transported by the fresh mucus-rich trachea. The transportability of applied mucus on such a preparation will therefore reflect the properties of the applied mucus. A sample of approximately 50 microlitres was applied to the caudal end of the explant and its progress recorded with the naked eye, a watch, and a millimeter scale. At least six measurements were taken for each sample at 15 sec intervals after a steady rate was reached. Bovine tracheal mucus was always transported faster than sputum samples, at 20–30 mm/minute, and was used as internal standard. The transport rate of other mucus samples was expressed as an index, which was the transport rate of the sample expressed as a percentage of that of bovine tracheal mucus. Thus a sample with a transportability index (TI) of 25 was transported at a quarter of the rate of bovine tracheal mucus. It was not necessary to incorporate any nutrient in the medium; the epithelial surface was kept moist with isotonic phosphate-buffered saline (PBS) and the cilia usually remained active throughout the working day. Addition of nutrients to the medium encouraged bacterial growth without prolonging the life of the tracheal explants. Special measures to humidify the chamber were also found not to be necessary, provided that the epithelial surface was rinsed with PBS at regular intervals to maintain its hydration.

The major constituents of mucus in molar terms are water and salt, and in this series of studies we investigated the effect on transportability of altering the osmolality of the mucus. The loose nature of mucus gels allows changes to the solute and solvent composition to be made easily.

In a previous study [4] we examined the effect of incubating sputum samples from patients with cystic fibrosis (CF) and non-CF bronchiectasis in PBS of different osmolalities. Untreated sputum had a TI of 26; sputum incubated in <100 mosM PBS had a TI of <4, but sputum incubated in 300–500 mosM PBS had a TI of 49. The transportability of bovine tracheal mucus was similarly responsive to salinity. The mean osmolality (measured by freezing point depression) of 12 CF sputa was 262 mosmoles/

kg, that of 12 non-CF bronchiectasis sputa was 240. Therefore the osmolality of sputum appears to be suboptimal for ciliary transport.

The osmolality of 12 sputum samples (5 CF, 7 non-CF) was increased by inspissation at room temperature under a current of air for up to 2 hours until half the original weight was lost; this caused the transportability of the gel to increase from 28 (sem 4) to 57 (sem 4), p 0.004 by the Wilcoxon signed rank test. In this experiment, all the products of infection and inflammation are concentrated in the inspissated gel, yet the ciliary transportability increased. This emphasizes the importance of osmolality in determining the ciliary transportability of sputum on the bovine trachea model and brings into question the generally held assumption that sputum is retained because it is poorly hydrated.

The effect of the nonelectrolyte mannitol on the ciliary transportability of sputum was also investigated. In this experiment, mannitol in solid form was added to aliquots of sputum (approx 0.5 mL) to a final concentration of 0.1 *M*. Another aliquot of the same sputum sample was untreated. After gentle mixing, both samples were incubated overnight at 4°C and the ciliary transportability measured the following day. Again, this experimental design allowed no egress of infective or inflammatory material from the gel. Sputum samples from 14 patients (8 CF, 6 non-CF) were treated in this way; the mean TI of the untreated sputa was 29 (sem 3), that of the aliquots to which mannitol had been added was 67 (sem 4), p 0.001 by the Wilcoxon test. Therefore, the nonionized mannitol when added to sputum increases its ciliary transportability at least as well as does sodium chloride in experiments of similar design [4]. Daviskas and colleagues have shown in this congress that administration of dry powder mannitol also increases mucociliary clearance in vivo.

In all the above experiments, no difference was detected between CF and non-CF sputa in their response to altered osmolality.

## III. SUMMARY

Therefore, expectorated sputum was transported submaximally on the mucus-depleted bovine trachea. Increasing the osmolality of the mucus in various ways, including incubating in excess buffered saline, evaporating off water, or adding mannitol caused it to be more rapidly transported. In most cases, at least a doubling of its transportability could be achieved. This suggests the possibility that increasing the osmolality of mucus retained in the airways might aid its expulsion.

## REFERENCES


1. Kobayashi K, Wanner A. Mucociliary clearance and ciliary activity. In: Chung KF, Barnes PJ, eds. Pharmacology of the Respiratory Tract: Experimental and Clinical Research. New York: Dekker, 1995: 621–653.
2. Cole P, Wilson R. Host–microbial relationships in respiratory infection. Chest 1989; 95:217S–221S.
3. Wills PJ, Garcia Suarez MJ, Rutman A, Wilson R, Cole PJ. The ciliary transportability of sputum is slow on the mucus-depleted bovine trachea. Am J Respir Crit Care Med 1995; 151:1255–1258.
4. Wills PJ, Hall RL, Chan WM, Cole PJ. Sodium chloride increases the ciliary transportability of cystic fibrosis and bronchiectasis sputum on the mucus-depleted bovine trachea. J Clin Invest 1997; 99:9–13.

# 29

# Mucociliary Clearance in the Guinea Pig is Increased by Ouabain (i.v.) and by Hypertonic Saline (14.4%) Aerosol

**B. Ben Newton and Roderick L. Hall**
*Bayer plc, Slough, United Kingdom*

## I. INTRODUCTION

In health, airway hygiene is maintained by an efficient mucociliary clearance (MCC) system. In disease MCC appears to fail, leading to mucus retention and accumulation. The resulting mucostasis is a clinical feature of airway diseases such as chronic bronchitis (CB), cystic fibrosis (CF), and bronchiectasis (BE). Therapeutic strategies to improve the impaired clearance should therefore be of clinical benefit.

The underlying reasons for the retention of lung mucus remain unclear. However, mechanisms involved in mucus transport can be studied in vitro using the mammalian ciliated epithelium of the mucus depleted bovine trachea (MDBT). Using the MDBT, CB, CF, and BE sputa were transported poorly [1]. This poor transportability may be linked to low electrolyte/osmolyte content [2]. Indeed, diseased sputum is known to have low electrolyte content relative to plasma [3,4,5] (Figure 1).

Clinical studies have concluded that MCC can be significantly increased in both health and disease following inhaled hypertonic saline or amiloride [6,7]. These agents would appear to act by increasing the ionic content of airway surface liquid (ASL). In the MDBT improved transportability of disease sputum was achieved both by directly applied sodium chloride (NaCl) or by inspissation to increase osmolality [2]. However a functional link between ASL ion composition and mucus clearance requircs further evaluation in vivo.

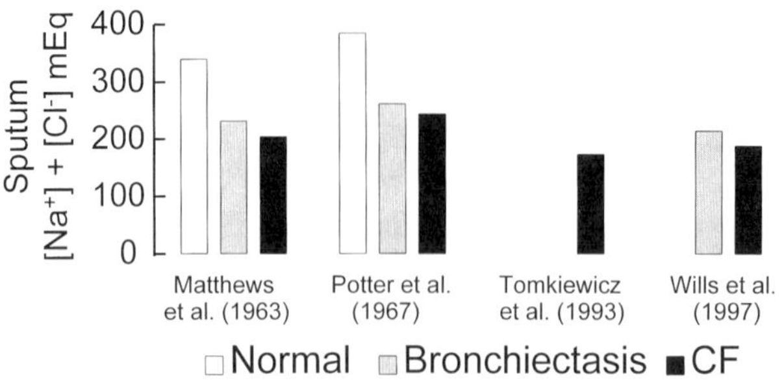


**Figure 1** The reduced electrolyte content of sputum samples obtained from normal, bronchiectatis and cystic fibrosis patients.

In our design of an in vivo model of MCC we had two main objectives. The first was to develop a model capable of reproducible and repeat measurements in a small animal. The second was to functionally investigate a link between MCC and ASL ion content and osmolality. To do this we have measured tracheal mucus velocity (TMV) in the guinea-pig. We have used a technique originally described by Morley and Sanjar [8] and refined more recently by Read et al. [9].

## II. METHODOLOGY

Naive male animals were anesthetized with Hypnorm and Hypnovel given intramuscularly. A 3-cm length of the cervical trachea was exposed by blunt dissection. A single hole was made with a sharp-pointed electrocautery needle in the distal portion of the exposed trachea. The animal was then placed on a thermostatically controlled operating table. A 25 μL microsyringe loaded with the $^{32}$P-labeled marker was lowered into the trachea until the tip just touched the posterior tracheal mucosa (Figure 2). The wound site was then covered with polythene film and the entry hole made airtight with cyanoacrylate glue and paper tissue.

A lead-collimated miniature β-particle detector probe was placed 15 mm proximally, anterior and externally to the trachea. The β-particles emitted from the injected marker were detected by the probe and relayed to a Harwell instruments interface (AEA Technology, Oxfordshire, UK).

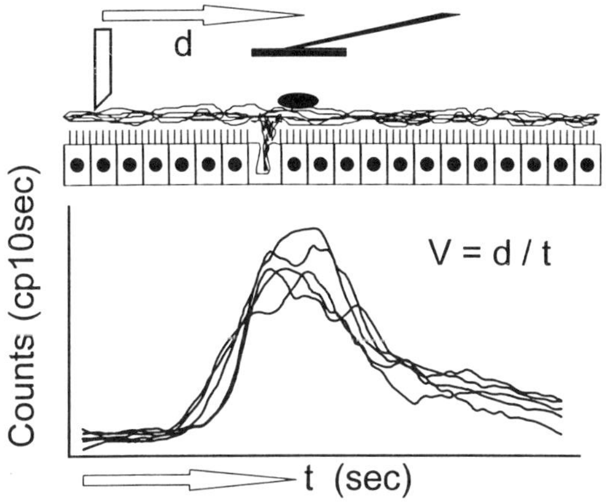


**Figure 2** Diagram to show the physical arrangement of microsyringe needle and β-probe. The lower panel shows the counts detected by the β-probe as the $^{32}P$-labeled *Saccharomyces cerevisiae* are transported along the tracheal mucociliary layer.

Events were recorded as the sum of the counts in a 10 second period by a Windows-based data acquisition system (EMMS, Hampshire, UK).

The first 0.7 μL aliquot of the $^{32}P$-labeled marker was then injected onto the tracheal mucosa 20 minutes after induction of anesthesia. Once in contact with the mucociliary layer the marker was transported by MCC and approached the β-particle detector probe. Counts peaked and then decayed as the marker bolus was transported beneath and then beyond the probe. As the distance (d) between the microsyringe needle and the probe was known and the time (t) to peak was measured, the velocity was calculated (Figure 2). Since the residual count decreased back to near the background level, successive measurements were made every 15 minutes in each animal.

The first study was carried out to compare the velocity using $^{32}P$-labeled *S. cerevisiae* with that obtained using $^{32}P$-labeled erythrocytes (Figure 3). The median plateau TMV measured using $^{32}P$-labeled erythrocytes was between $6.3 \pm 0.6$ and $9.9 \pm 3.0$ mm.min$^{-1}$ (median ± semi-interquartile range; $n = 6$). Using $^{32}P$-labeled *S. cerevisiae*, median TMV over the same time points was between $5.4 \pm 1.5$ and $8.2 \pm 1.5$ mm. min$^{-1}$ ($n = 6$). The TMV using *S. cerevisiae* gave comparable values to erythrocytes. *S. cerevisiae* provide a more osmotically robust radio-

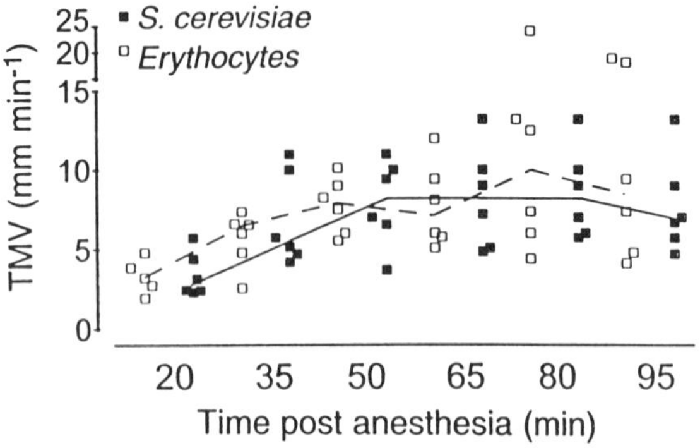


**Figure 3** Comparison of tracheal mucus velocity (TMV) measurements obtained using $^{32}$P-labeled *Saccharomyces cerevisiae* or $^{32}$P-labeled erythrocytes in the spontaneously breathing anesthetized guinea pig.

labeled marker than do erythrocytes and so were used in all subsequent TMV studies. Using this procedure reproducible and repeatable measurements can be taken.

## III. AGENT STUDIES

To begin to investigate a functional link between MCC and epithelial ion transport we used two inhibitors: (a) Ouabain, the selective $Na^+$, $K^+$-ATPase inhibitor, and (b) Bumetanide, the $Na^+$-$K^+$-$2Cl^-$ cotransport inhibitor. In a further study we modified directly the ASL composition by the administration of hypertonic saline by aerosol inhalation.

Ouabain (10 or 30 μg.kg$^{-1}$) or saline (0.9%) was administered intravenously and TMV was measured after 5, 20, 35, 50, and 65 min. A significant increase in TMV was observed at 35 min after 30 μg.kg$^{-1}$ ouabain injection. The TMV was 5.1 ± 1.2 mm.min$^{-1}$ ($n$ = 7) in saline injected animals compared to 7.8 ± 2.3 mm.min$^{-1}$ ($n$ = 8) in animals injected with 30 μg.kg$^{-1}$ ouabain ($p \leq 0.05$; median ± semi-interquartile range; Mann–Whitney U-test) (Figure 4). This is a 53% increase over the vehicle control values. These results are consistent with TMV data in the anesthetized cat [10], in which acetylouabain (25–50 μg.kg$^{-1}$ i.v.) significantly increased mucus flow 20 minutes after administration.

Bumetanide (1 mg.kg$^{-1}$) or 33% PEG 400 in phosphate-buffered saline (0.9%) (vehicle) was administered intravenously. The TMV was measured after 5, 20, 35, 50, and 65 min. No significant change in TMV was

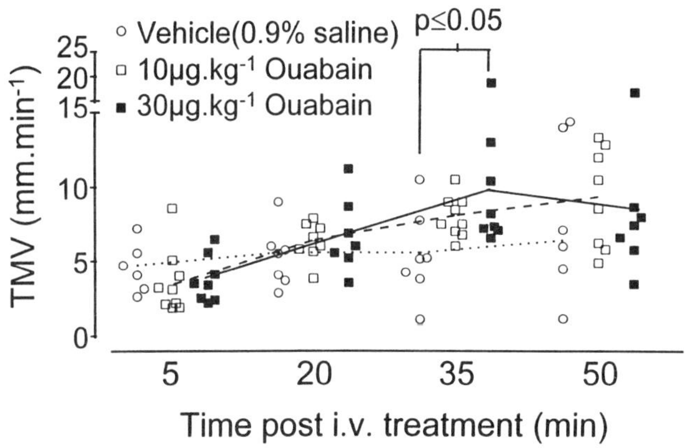


**Figure 4** The effect of intravenous ouabain (10 or 30 μg.kg$^{-1}$) on tracheal mucus velocity (TMV) in the spontaneously breathing anesthetized guinea pig.

observed at any time point following bumetanide treatment. The TMV was 3.6 ± 1.6 mm.min$^{-1}$ ($n$ = 10) in vehicle-injected animals compared to 3.1 ± 0.8 mm.min$^{-1}$ ($n$ = 12) in animals injected with bumetanide 5 min post injection. Later, at 35 min post injection, TMV was 6.8 ± 1.2 mm.min$^{-1}$ ($n$ = 10) in vehicle injected animals compared to 6.9 ± 1.5 mm.min$^{-1}$ ($n$ = 10) in animals injected with bumetanide (Figure 5).

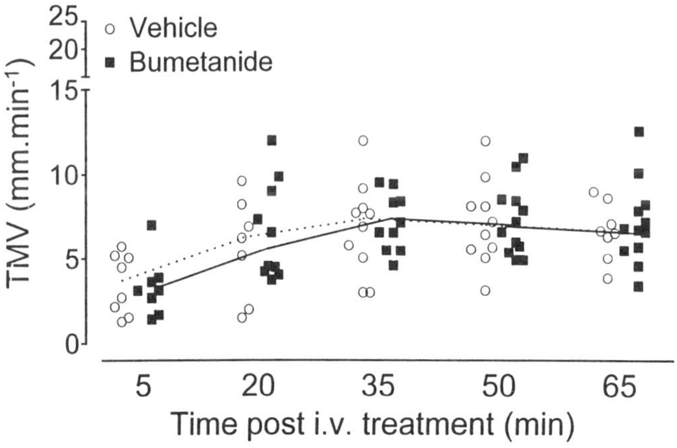


**Figure 5** The effect of intravenous bumetanide (1 mg.kg$^{-1}$) on tracheal mucus velocity (TMV) in the spontaneously breathing anesthetized guinea pig.

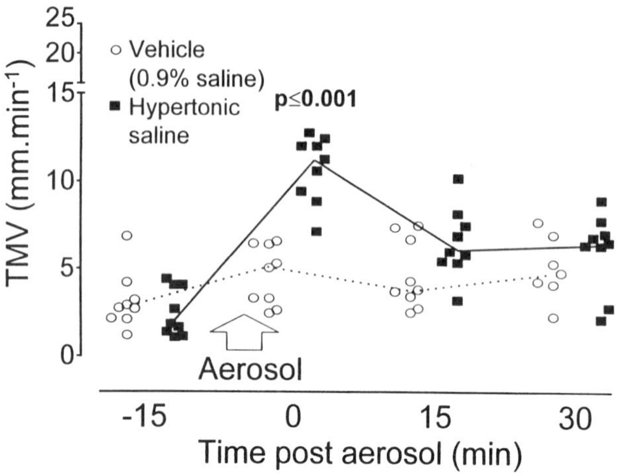


**Figure 6** The effect of 5-min aerosol of hypertonic saline (14.4%) on tracheal mucus velocity (TMV) in the spontaneously breathing anesthetized guinea pig.

A 5-min aerosol of hypertonic saline (14.4%) or isotonic saline (0.9%) was administered, and TMV was monitored at 1, 15, and 30 min following termination of the aerosol. A significant increase in TMV was only observed 1 min after hypertonic saline aerosol. The TMV was 5.1 ± 1.0 $mm.min^{-1}$ ($n = 9$) in 0.9% saline-exposed animals compared to 11.3 ± 1.3 $mm.min^{-1}$ ($n = 9$; $p \leq 0.001$) in hypertonic saline-exposed animals (Figure 6). This is a 122% increase in TMV over the vehicle values. At 15 and 30 min there were no significant differences in TMV between the two treatment groups.

## IV. MECHANISM STUDIES

To investigate the basis of the increased TMV seen following ouabain and hypertonic saline administration, we studied two possible mechanisms: (a) a sensory afferent-driven neuronal mechanism and (b) ciliary beat frequency (CBF) stimulation.

In the rat, 14.4% saline aerosol caused a highly significant plasma protein extravasation (PPE) in the trachea and bronchi that was prevented by pretreatment with capsaicin [11]. Capsaicin causes depletion of neuropeptides, e.g. substance-P, from sensory nerve terminals. This results in abolition of neuropeptide-based responses upon subsequent sensory nerve stimulation [12]. Neuropeptides are released from the terminals of sensory fibres following activation of irritant receptors and subsequent antidromic stimulation [13]. Since release of neuropeptides can cause PPE, then measurement of PPE serves as a rapid and convenient means of studying sensory afferent activation.

Guinea pigs ($n=6$–8) were anesthetized and then pretreated with Evan's blue dye (25 mg $kg^{-1}$ i.v.). A 5-min aerosol of hypertonic saline (14.4%) or isotonic saline (0.9%) was administered 20 min later. Animals were killed at 1 and 15 min following termination of the aerosol and PPE was measured in the trachea and bronchi [14]. No significant PPE was elicited following hypertonic saline inhalation suggesting that there was no significant release of neuropeptides (Figure 7). Neuropeptides are therefore unlikely to be involved in the increase in TMV seen in response to hypertonic saline.

Guinea pig CBF was measured on the epithelial cells of tracheal rings at 37°C using a photometric technique [15,16]. We investigated the effect of suspension medium osmolality on CBF by increasing the concentration of NaCl in Krebs buffer or by diluting the medium with water (Figure 8). There were no significant differences in CBF measured over the range 140–480 mOsm at 1 and 15 min after exposure to the medium. Maximum median CBF was 12.3 ± 0.9 Hz at 280 mOsm after 1 min and minimum median CBF was 11.0 ± 0.6 Hz at 477 mOsm after 15 min ($n=8$

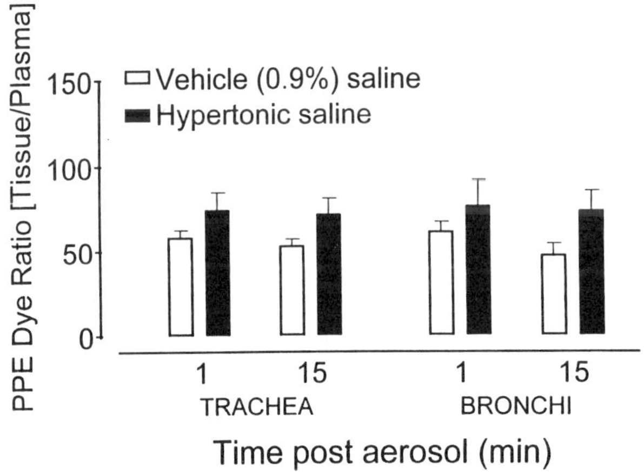


**Figure 7** The effect of 5-min aerosol of hypertonic saline (14.4%) on plasma protein extravasation (PPE) in the spontaneously breathing anesthetized guinea pig.

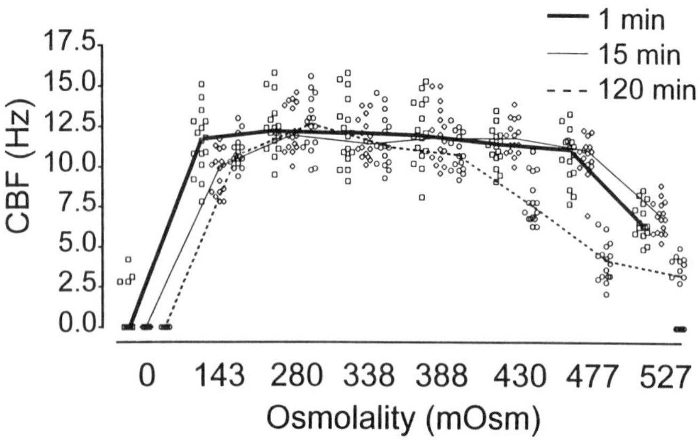


**Figure 8** The relationship between suspension medium osmolality and cilia beat frequency (CBF) of guinea pig tracheal epithelial cells.

in duplicate). There was a significant reduction in CBF at 1 and 15 min in 527 mOsm medium. The CBF in 527 mOsm medium was 6.4 ± 0.8 Hz at 1 min ($n = 8$ in duplicate) and 6.9 ± 0.9 Hz at 15 min ($n = 8$ in duplicate) compared to 11.0 ± 1.0 Hz at 1 min ($p \leq 0.001$; $n = 8$ in duplicate) and 11.0 ± 0.6 Hz at 15 min ($n = 8$ in duplicate; $p \leq 0.001$) in medium of 477 mOsm.

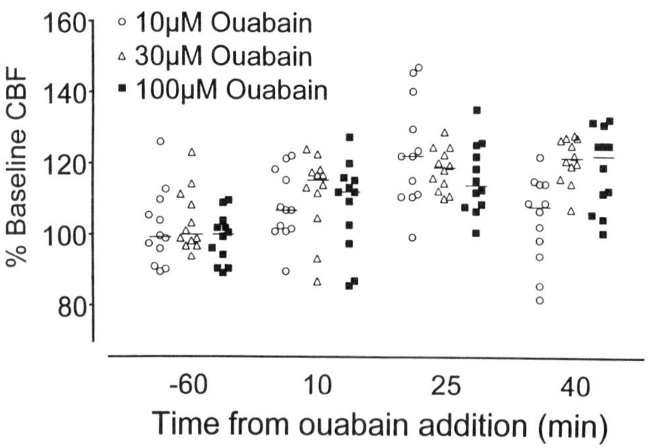


**Figure 9** The effect of ouabain (10–100 $\mu M$) on cilia beat frequency (CBF) of guinea pig tracheal epithelial cells.

At 120 min there were no significant differences in CBF measured over the range 140–390 mOsm. Maximum median CBF was 12.6 ± 1.3 Hz at 280 mOsm ($n=8$ in duplicate) and the minimum CBF was 10.7 ± 1.1 Hz at 388 mOsm ($n=8$ in duplicate). At 430 to 527 mOsm CBF declined. In water, ciliostasis resulted at all time points. These data accord well with the values obtained by Luc and Dulfano [17] on human bronchial explants and more recently by Wills et al. [2] on bovine cilia. Although it is difficult to compare the effect of nebulized hypertonic saline to this in vitro technique, a predominantly CBF-driven increase in TMV seems unlikely to account for the TMV increases seen in vivo.

In contrast, ouabain in this preparation caused a 20% increase after 15- and 30-min incubation, but only at concentrations above 30 μM. These concentrations are far higher than could be achieved in vivo following 30 $\mu g.kg^{-1}$ administration (Figure 9), and again CBF increases are unlikely to explain the increases in TMV achieved with ouabain.

## V. SUMMARY

Ouabain (30 $\mu g.kg^{-1}$ i.v.) significantly increased TMV by 53% at 35 min. A five-minute aerosol of hypertonic saline (14.4%) significantly increased TMV by 122%. Collectively these data support the hypothesis that there is a relationship between the ion content or osmolality of the ASL and mucociliary transport rate in the guinea-pig trachea.

PPE was not increased following inhalation of hypertonic saline. This suggests that neuropeptides are unlikely to be involved in the increase in TMV seen in response to hypertonic saline. The CBF of tracheal epithelial cells in vitro was constant up to 2 h over the range 140–390 mOsm. This indicates that an increase in beat frequency is unlikely to account for the increase in TMV in vivo.

## REFERENCES


1. Wills PJ, Garcia-Suarez MJ, Rutman A, Wilson R, Cole PJ. The ciliary transportability of sputum is slow on the mucus-depleted bovine trachea Am J Respir Crit Care Med 1995; 15:1255–1258.
2. Wills PJ, Hall RL, Chan W, Cole PJ. Sodium chloride increases the ciliary transportability of cystic fibrosis and bronchiectasis sputum on the mucus-depleted bovine trachea. J Clin Invest 1997; 99:9–13.
3. Matthews LW, Spector S, Lemm J, Potter JL. Studies on pulmonary secretions I: The overall chemical composition of pulmonary secretions from pa-

tients with cystic fibrosis, bronchiectasis, and laryngectomy. Am Rev Respir Dis 1963; 88:199–204.
4. Potter JL, Matthews LW, Spector S, Lemm J. Studies on Pulmonary secretions II: Osmolality and the ionic environment of pulmonary secretions from patients with cystic fibrosis, bronchiectasis and laryngectomy. Am Rev Respir Dis 1967; 96:83–88.
5. Tomkiewicz RP, App EM, Zayas JG, Ramirez O, Church N, Boucher RC, Knowles MR, King M. Amiloride inhalation therapy in cystic fibrosis: Influence of ion content, hydration, and rheology of sputum. Am Rev Respir Dis 1993; 148:1002–1007.
6. Pavia D, Thomson ML, Clarke SW. Enhanced clearance of secretions from the human lung after the administration of hypertonic saline aerosol. Am Rev Respir Dis 1978; 117:199–203.
7. App EM, King M, Helfesrieder R, Kohler D, Matthys H. Acute and long-term amiloride inhalation in cystic fibrosis lung disease. Am Rev Respir Dis 1990; 141:605–612.
8. Morley J, Sanjar S. Use of a solid state beta-particle detector for measurement of mucociliary transport in the trachea of small laboratory animals (abstr). Proc Physiol Soc 1983; 2:16–17P.
9. Read RC, Roberts P, Munro N, Rutman A, Hastie A, Shryock T, Hall R, McDonald-Gibson W, Lund V, Taylor G, Cole PJ, Wilson R. Effect of *Pseudomonas aeruginosa rhamnolipids* on mucociliary transport and ciliary beating. J Appl Physiol 1992; 72(6):2271–2277.
10. Laurenzi GA, Yin S. Studies of mucus flow in the mammalian respiratory tract I. The beneficial effects of acetylouabain on respiratory tract mucus flow. Am Rev Respir Dis 1971; 103:800–807.
11. Umeno E, McDonald DM, Nadel JA. Hypertonic saline increases vascular permeability in the rat trachea by producing neurogenic inflammation. J Clin Invest 1990; 85:1905–1908.
12. Dray A. Mechanism of action of capsaicin-like molecules on sensory neurons. Life Sci 1992; 51(23):1759–1765.
13. Piedimonte G, Bertrand C, Geppetti P, Snider RM, Desai MC, Nadel JA. A new $NK_1$ receptor antagonist (CP-99,994) prevents the increase in tracheal vascular permeability produced by hypertonic saline. J Pharmacol Exp Ther 1993; 266:270–273.
14. Evans TW, Chung KF, Rogers DF, Barnes PJ. Effect of platelet activating factor on airway vascular permeability: possible mechanisms. J Appl Physiol 1987; 63:479–484.
15. Yager J, Chen T-M, Dulfano MJ. Measurement of frequency of cilia beats of human respiratory epithelium. Chest 1978; 73:627–633.
16. Rutland J, Cole PJ. Non-invasive sampling of nasal cilia for measurement of beat frequency and study of ultrastructure. Lancet 1980; 9:564–565.
17. Luk CK, Dulfano MJ. Effect of pH, viscosity and ionic-strength changes on ciliary beating frequency of human bronchial explants. Clin Sci 1983; 64: 449–451.

# 30

# Interventions that Increase Mucociliary Clearance in Asthmatic and Healthy Subjects

**Evangelia Daviskas and Sandra D. Anderson**

*Royal Prince Alfred Hospital, Sydney, New South Wales, Australia*

## I. INTRODUCTION

Mucociliary clearance (MCC) is a physiological response of the respiratory tract to clear locally produced debris, excessive secretions, or unwanted inhaled particles. For normal MCC to occur it is necessary that the epithelial cells are intact, the ciliary activity and the rheology of mucus is normal, and also that the depth and chemical composition of the periciliary fluid layer is optimal. Physiological, pathological, and pharmacological factors can cause changes in any of these, resulting in increased or decreased mucus transport. Of great concern are the pathological situations that affect the epithelial cells and normal hydration of the airways thereby resulting in abnormal MCC. Under normal circumstances the hydration of the airways is controlled locally by the ion transport across the epithelial cells.

Asthma is an inflammatory disease and asthmatic subjects have been characterized as having loss of epithelial cells and mucus hypersecretion. Studies of MCC in asthmatic subjects during exacerbation of their asthma or during remission showed that they had abnormal MCC compared with the MCC of the healthy subjects [1,2]. Recent work [3,4,5] has shown that subjects with stable asymptomatic asthma have good baseline MCC in the small and large airways. This can be attributed to optimal asthma treatment that leads to the repair of the epithelial cells, reduction of inflam-

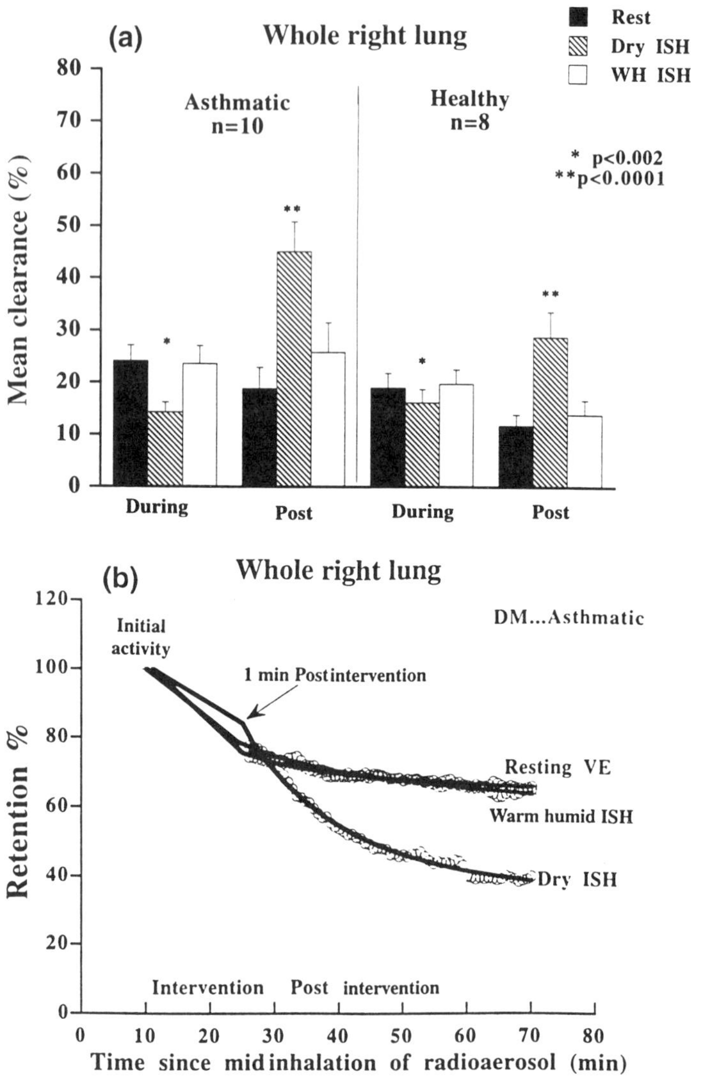


**Figure 1** (a) Mean % clearance (± SEM) during and post isocapnic hyperventilation (ISH) with dry and warm humid (WH) air and nasal resting breathing at the same time intervals in the whole right lung in asthmatic and healthy subjects. Mucociliary clearance decreased during and increased after ISH with dry air when compared to ISH with warm humid air and nasal resting breathing. The magnitude of the changes in MCC in response to ISH with dry air is greater in the asthmatic compared to the healthy subjects. (b) Example of the % retention curves in the whole right lung in an asthmatic subject on three study days: (1) nasal resting

matory cells, and decrease in mucus hypersecretion. However, an asthma attack triggered by hyperpnea with dry air may be a result of acute dehydration of the airways. During hyperpnea with dry air, the demand to humidify large volumes of air is high and asthmatic subjects may have an abnormally low water transport when the demand is high. It is thought that during hyperpnea with dry air, the rate of evaporative water loss exceeds the rate of water return resulting in a reduction of the depth and subsequent hyperosmolarity of the periciliary fluid layer [6]. The mechanism whereby the evaporative water loss leads to airway narrowing is thought to relate to its dehydrating and osmotic effects [6,7,8]. Dehydration of the airways can impair MCC by reducing the depth of the periciliary fluid layer and by altering the viscoelastic properties of mucus that affect its rheology.

We have investigated the effect of airway drying and hyperosmolarity on MCC in asthmatic and healthy subjects using isocapnic hyperventilation (with dry and with warm humid air) and inhalation of hyperosmotic agents (hypertonic saline aerosol and dry powder mannitol). These interventions were found to result in an increase in mucociliary clearance both in asthmatic and healthy subjects [3,4,9] and this paper presents a short review of these findings. These findings, which have implications for the mechanism of hyperventilation-induced asthma, may also have therapeutic implications for patients with abnormal mucociliary clearance as suggested by other workers using hypertonic saline [10,11].

## II. MEASUREMENT OF MUCOCILIARY CLEARANCE

Mucociliary clearance was measured using a radioaerosol technique ($^{99m}$Tc-sulfur colloid, MMAD 6–7 μm). The radioaerosol was delivered with a controlled breathing pattern [12] to maximize deposition in the conducting airways. Lung images were obtained using a gamma camera. An anterior and a posterior emission image was collected 10 minutes after the inhalation of the radioaerosol in order to define the initial deposition in terms of distribution and intensity. As soon as the intervention finished, successive anterior and posterior emission images were collected for 45 minutes. Only the right lung was analysed, and it was divided into three

---

breathing, (2) isocapnic hyperventilation (ISH) with dry air, and (3) ISH with warm humid air. The % retained activity was plotted vs time since mid inhalation of the radioaerosol. This figure demonstrates that mucociliary clearance (initial – % retained activity) decreases during ISH and increases post-ISH with dry air.

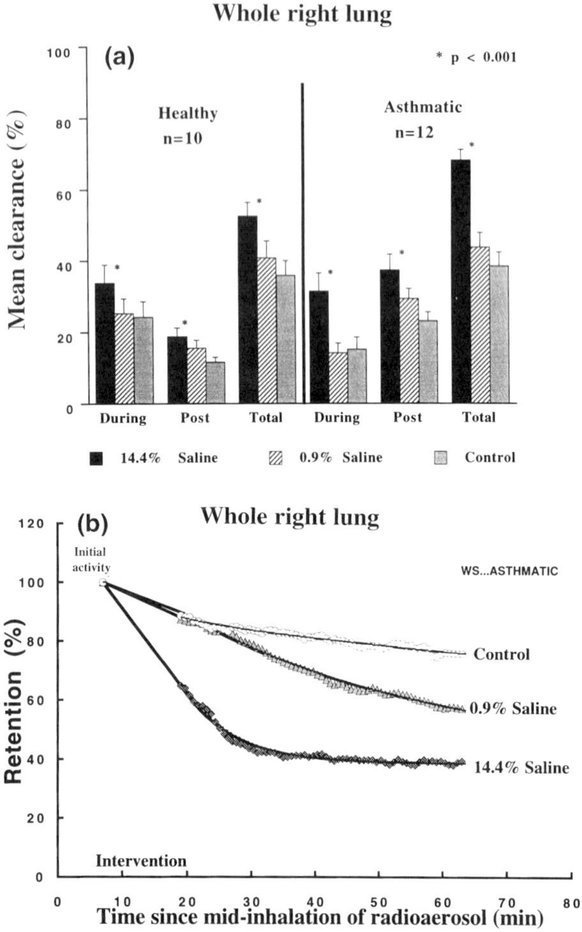


**Figure 2** (a) Mean % clearance (±SEM) during, post intervention, and total in 1 h in the whole right lung in asthmatic and healthy subjects, on the three study days. This figure demonstrates the significant increase in the mean clearance in response to inhalation of 14.4% saline, compared to 0.9% saline and nasal resting breathing (control) over the same time interval. The asthmatic subjects compared to the healthy subjects had a greater increase in clearance after 1 h in response to 14.4% saline ($p < 0.02$). Both asthmatic and healthy subjects had similar clearance ($p > 0.6$) after 1 h with 0.9% saline and control (less than 4% difference). (b) Example of the % retention curves in the whole right lung in an asthmatic subject on the three study days as defined by the intervention: (1) control (nasal resting breathing), (2) 14.4% saline; and (3) 0.9% saline. This figure demonstrates the increase in the mucociliary clearance (initial activity −% retained activity) in response to inhalation of 14.4% saline compared to 0.9% saline and control. It

regions of interest: central, intermediate, and peripheral [13]. Details of the methods used are described elsewhere [3,4].

## III. EFFECTS OF INTERVENTIONS ON MUCOCILIARY CLEARANCE

### A. Isocapnic Hyperventilation (ISH) with Dry Air and with Warm Humid Air

Isocapnic hyperventilation with dry or warm humid air was performed for 6–8 minutes continuously, at ventilation rates of (mean ± SD) 70 ± 23 and 83 ± 15 L/min in asthmatic and healthy subjects respectively. The ISH with dry air caused a mean fall in forced expiratory volume in one second ($FEV_1$) of 29% in the asthmatic subjects, which was abolished when the air was warm and humid.

The MCC was found to decrease during ISH with dry air and increase immediately after ISH with dry air when compared to the MCC with warm humid air and resting nasal breathing (Figure 1) both in asthmatic and healthy subjects [3]. However, there were differences in the magnitude and location of these responses in the asthmatic and healthy subjects. The changes in MCC were present in all lung regions in the asthmatic subjects, while only in the central region in the healthy subjects. The reduction in MCC during ISH with dry air is consistent with the view that the airways dehydrate and the findings suggest that asthmatic airways may dehydrate more than the healthy airways. The increase in MCC after ISH with dry air is most likely related to the possibility that dry air leads to transient hyperosmolarity of the airway fluid, especially since ISH with warm humid air did not have any effect on MCC (Figure 1). The presence of inflammatory mediators in the asthmatic airways that are released in response to ISH [14] and hyperosmolarity [15] may contribute to the more marked and sustained increase in MCC compared to healthy subjects. The released mediators, apart from stimulating the ciliary beat frequency, also stimulate chloride secretion and thus may facilitate recovery of the periciliary fluid layer in the asthmatic subjects.

The increase in MCC after ISH with dry air was delayed in the presence of frusemide (36 mg) for about ten minutes in the asthmatic but not in the healthy subjects [16]. Frusemide, an inhibitor of the $Na^+/K^+/Cl^-$ cotransporter at the basolateral membrane, may affect the recovery

---

also demonstrates that the increase in the clearance started during the inhalation of the 14.4% saline and that the clearance reached its maximum in a relatively short time.

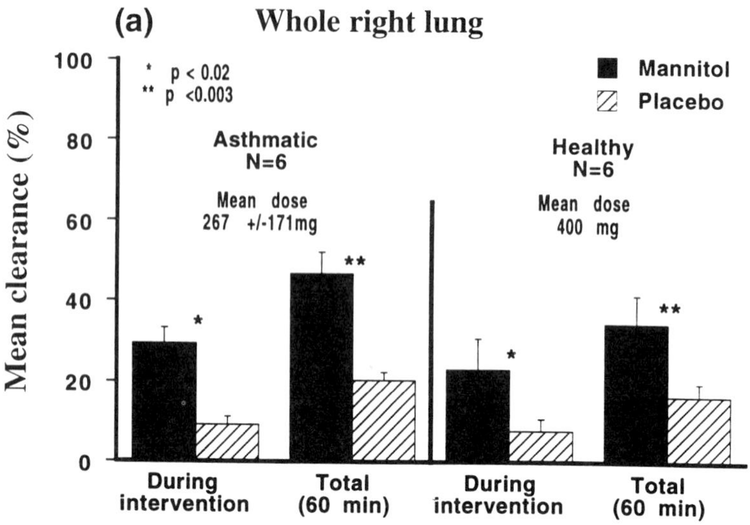


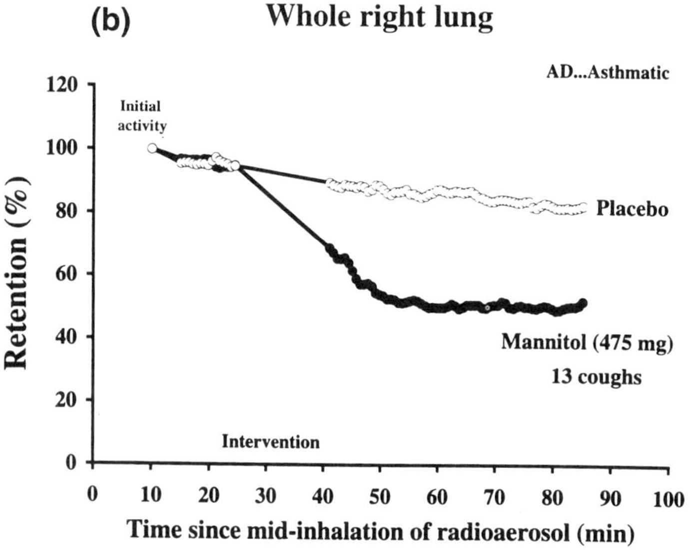


**Figure 3** (a) Mean % clearance (±SEM) in response to inhaling dry powder mannitol or placebo, during and total in 1 h after the start of the inhalation, in the whole right lung in asthmatic and healthy subjects. Inhalation of dry powder mannitol from capsules, using a dry powder inhaler, caused a significant increase in mucocilary clearance compared to placebo both in asthmatic and healthy subjects. The placebo involved the same number of inhalations through the dry powder inhaler, having an empty capsule, and the same number of coughs. (b) Example

of the water at the airway lumen in the asthmatic subjects by a mechanism that is not clear but may relate to inhibition of chloride secretion [17] and possibly to the inhibition of mediators [18].

The hypothesis that ISH with dry air increases MCC by transiently increasing the osmolarity of the airway fluid is supported by the findings of effect of hypertonic saline on MCC.

## B. Inhalation of Hypertonic Saline Aerosol

Inhalation of hypertonic saline aerosol (14.4%) compared to 0.9% saline and the control (no intervention) caused an immediate increase in MCC in asthmatic and healthy subjects [4] (Figure 2). The magnitude and location of this response was different in the two groups, as with ISH with dry air. The increase in MCC was greater and present in all regions in the whole right lung in the asthmatic subjects, despite the fact that they inhaled less hypertonic saline than the healthy subjects. The mean volume of the ultrasonically nebulized hypertonic saline was $2.2 \pm 1.2$ mL for asthmatics and $3.2 \pm 0.7$ mL for the healthy subjects, delivered over a maximum period of 6 minutes. Most of the increase in MCC in response to hypertonic saline occured while inhaling the aerosol and the MCC rate remained increased for up to 20 minutes after inhalation of the aerosol.

These findings of the effect of hypertonic saline on MCC suggest that hypertonic saline was very effective in increasing MCC in both the large and small airways in the asthmatic subjects and the large airways in the healthy subjects.

## C. Inhalation of Dry Powder Mannitol

Having established that an increase in the osmolarity of the airway fluid increases MCC, we further investigated the effect of inhaling a dry powder of mannitol on MCC in asthmatic subjects and healthy subjects [9]. Dry powder preparation of mannitol BP was inhaled from capsules using a dry powder inhaler, with 67% of the particles in the respirable range (below 7 μm). Mannitol is a naturally occurring sugar alcohol that is not absorbed by the gastrointestinal tract and is not metabolized to any appreciable extent when injected. The advantage of using dry powder mannitol is that

---

of the % retention curves in an asthmatic subject in response to inhaling dry powder mannitol (475 mg) or placebo. The number of coughs induced by the inhalation of mannitol was small; however, it was reproduced on the placebo day. This figure demonstrates the immediate marked increase in mucociliary clearance in response to inhaling dry powder mannitol compared to placebo.

it is a stable substance that is a powerful hyperosmotic stimulus in the lower airways and can be delivered using simple nonexpensive inhaler devices [19]. Inhalation of dry powder mannitol caused a marked increase in MCC immediately, both in asthmatic and healthy subjects (Figure 3). The mean inhaled dose was $267 \pm 171$ mg in the asthmatic subjects and 400 mg in the healthy subjects. Inhalation of dry powder mannitol ($267 \pm 171$ mg) caused a fall in $FEV_1$ of $21.8 \pm 2.7\%$ in asthmatic subjects. The mean actual increase in MCC in the whole right lung for the 60 minute period immediately after the start of inhalation of mannitol was 26.4 ($\pm 11.9\%$) and 18.1 ($\pm 4.9$) % compared to placebo, in asthmatic and healthy subjects respectively. The increase in MCC in response to inhaling an aerosol of dry powder mannitol is similar to the increase observed in response to hypertonic saline [4]. The increase in MCC following inhalation of dry powder mannitol is not related to induced cough, as the number of cough was low ($13 \pm 10$ and $8 \pm 11$ in asthmatic and healthy subjects respectively) and was reproduced on the placebo day.

## IV. DISCUSSION

Continuous hyperpnea with dry air of short duration, inhalation of hypertonic saline, and inhalation of dry powder mannitol increase mucociliary clearance in asthmatic and healthy subjects, most likely via some common mechanism linked to indirect effects of increased osmolarity. Exposure of cells to a hyperosmotic environment causes several effects on the cells including mechanical and electrical changes that trigger a series of intracellular biochemical events. Water moves out of the cells due to the osmotic stimulus causing the cells to shrink. Shrinking of the cells is followed by a regulatory volume increase (RVI) that initiates biochemical events leading to an increase in intracellular calcium $[Ca^{2+}]_i$ [20]. An increase in intracellular calcium $[Ca^{2+}]_i$, has been shown to increase the ciliary activity [21,22,23]. Hyperosmolarity also triggers release of mediators from epithelial cells (e.g., prostaglandin $E_2$) [24] and mast cells (e.g., histamine) [15] and possibly neuropeptides (e.g., supstance P) [25] from sensory nerves that have been shown to increase ciliary beat frequency [26,27,28]. Histamine has been demonstrated to increase MCC both in asthmatic and healthy subjects [29,30]. Histamine was also found to increase $[Ca^{2+}]_i$ [31] and it is possible that other mediators may do so as well. Therefore, the increase in mucociliary clearance in response to hyperosmolarity is most likely due to an increase in ciliary activity modulated by intracellular biochemical mechanisms with changes in the intracellular calcium playing a key role.

These interventions, which transiently increased the osmolarity of the airway surface fluid, increased mucociliary clearance above the normal baseline value and revealed some if not all of the reserve capacity of the mucociliary system. An increase in the osmolarity of the airway surface fluid either with hypertonic saline or mannitol provides three major advantages for MCC: (a) the osmotic gradient that it creates can improve the hydration of the airways immediately by favoring the movement of water towards the airway lumen, (b) it can stimulate the ciliary activity to increase MCC, and (c) it can change the rheological properties of mucus [32]. This can be of great benefit to patients with abnormal mucociliary clearance as a result of having abnormal hydration of the airways and abnormal rheology of the mucus. The superiority of the use of dry powder mannitol over hypertonic saline as a therapeutic tool needs to be investigated further in patients who require a therapeutic intervention to improve their mucociliary clearance.

In conclusion, continuous hyperpnea with dry air of short duration, inhalation of hypertonic saline, and inhalation of dry powder mannitol increase mucociliary clearance in asthmatic and healthy subjects. This increase occurs most likely via some common mechanism involving an increase in intracellular calcium of the ciliated epithelial cells of the lower airways.

## ACKNOWLEDGMENTS

The authors would like to thank Mr. S. Eberl, Dr. D. Bailey, Dr. S. Meikle, Mr. R. Fulton, Dr. G. Bautovich and the technical staff of the Department of Nuclear Medicine for their help. The authors would also like to thank Dr. I. Gonda from Aradigm Corporation, Hayward, USA and Dr. H-K. Chan from the Department of Pharmacy and Professor P. Seale from the Department of Pharmacology of University of Sydney for their help.

The studies were supported by a grant from the National and Medical Research Council of Australia.

## REFERENCES

1. Messina MS, O'Riordan TG, Smaldone GC. Changes in mucociliary clearance during acute exacerbations of asthma. Am Rev Respir Dis 1991; 143: 993–997.
2. Pavia D, Bateman JRM, Shehan NF, Agnew JE, Clarke SW. Tracheobronchial mucociliary clearance in asthma: impairment during remission. Thorax 1985; 40:171–175.

3. Daviskas E, Anderson SD, Gonda I, Chan K, Cook P, Fulton R. Changes in mucociliary clearance during and after isocapnic hyperventilation in asthmatic and healthy subjects. Eur Respir J 1995; 8:742–751.
4. Daviskas E, Anderson SD, Gonda I, Eberl S, Meikle S, Seale JP, Bautovich G. Inhalation of hypertonic saline aerososl enhances mucociliary clearance in asthmatic and healthy subjects. Eur Respir J 1996; 9:725–732.
5. Svartengren K, Phillipson K, Svartengren M, Anderson M, Camner P. Tracheobronchial deposition and clearance in small airways in asthmatic subjects. Eur Respir J 1996; 9:1123–1129.
6. Anderson SD. Is there a unifying hypothesis for exercise induced asthma? J Allergy Clin Immunol 1984; 73:660–665.
7. Anderson SD, Daviskas E, Smith CM. Exercise-induced asthma: a difference in opinion regarding the stimulus. Allergy Proc 1989; 10:215–226.
8. Anderson SD, Daviskas E. The airway microvasculature and exercise induced asthma. Thorax 1992; 47:748–752.
9. Daviskas E, Anderson SD, Brannan J, Chan HK, Eberl S, Bautovich G. Inhalation of dry powder mannitol increases mucociliary clearance (MCC). Eur J Respir Dis 1997, in press.
10. Pavia D, Thomson ML, Clarke SW. Enhanced clearance of secretions from the human lung after the administration of hypertonic saline aerosol. Am Rev Respir Dis 1978; 117:199–203.
11. Robinson M, Regnis JA, Bailey DL, King M, Bautovich GJ, Bye PTP. Effect of hypertonic saline, amiloride and cough on mucociliary clearance in patients with cystic fibrosis. Am J Respir Crit Care Med 1996; 153:1503–1509.
12. Phipps PR, Gonda I, Anderson SD. Apparatus for the control of breathing patterns during aerosol inhalation. J Aerosol Med 1992; 5:155–170.
13. Phipps PR, Gonda I, Bailey DL, Borham P, Bautovich G, Anderson SD. Comparisons of planar and tomographic gamma scintigraphy to measure the penetration index of inhaled aerosols. Am Rev Respir Dis 1989; 139: 1516–1523.
14. Pliss LB, Ingenito EP, Ingram RH, JR, and Pichurko B. Assessment of bronchoalveolar cell and mediator response to isocapnic hyperpnea in asthma. Am Rev Resp Dis 1990; 142:73–78.
15. Silber G, Proud D, Warner J, Naclerio R, Kagey-Sobotka A, Lichtenstein L, Eggleston P. In vivo release of inflammatory mediators by hyperosmolar solutions. Am Rev Respir Dis 1988; 137:606–612.
16. Daviskas E, Anderson SD, Gonda I, D Bailey, G Bautovich, JP Seale. Mucociliary clearance during and after isocapnic hyperventilation with dry air in the presence of frusemide. Eur Respir J 1996; 9:716–724.
17. Widdicombe JH, Nathason IT, Highland E. Effects of loop diuretics on ion transport by dog tracheal epithelium. Am J Physiol 1983; 245:C388–C396.
18. Anderson SD, Wei HE, Temple DM. Inhibition by furosemide of inflammatory mediators from lung fragments. N Engl J Med 1991; 324:131.
19. Anderson SD, Brannan JB, Spring J, Spalding N, Rodwell LT, Chan HK, Gonda I, Walsh A, Clark A. A new method for bronchial provocation testing

in asthmatic subjects using a dry powder of mannitol. Am J Resp Crit Care Med 1997;156:758–765.
20. Eveloff JL and Warnock DG. Activation of ion transport systems during cell volume regulation. Am J Physiol (Renal Fluid Electrolyte Physiol 21), 1987; 252:F1–F10.
21. Villalon M, Hinds TR and Verdugo P. Stimulus–response coupling in mammalian ciliated cells. Demonstration of two mechanisms of control for cytosolic [$Ca^{2+}$]. Biophys J 1989; 56:1255–1258.
22. Lansley AB, Sanderson MJ, Dirksen EP. Control of the beat cycle of respiratory tract by $Ca^{2+}$ and cAMP. Am J Physiol 1992; 263:L232–L242.
23. Korngreen A and Priel Z. Simultaneous measurement of ciliary beating and intracellular calcium. Biophys J 1994; 67:377–380.
24. Assouline G, Leibson V, Danon A. Stimulation of prostaglandin output from rat stomach by hypertonic solutions. Eur J Pharmacol 1977; 44:271–273.
25. Jongejan RC, de Jongste JC, Raatgee RC, Stijnen T, Bonta IL, Kerrebijn KF. Effect of hyperosmolarity on human isolated central airway. Br J Pharmacol 1991; 102:931–937.
26. Wanner A, Maurer D, Abraham WM, Szepfalusi Z, Sielczak M. Effects of chemical mediators of anaphylaxis on ciliary function. J Allergy Clin Immunol 1983; 72:663–667.
27. Wanner A, Sielczak M, Mella JF, Abraham WM. Ciliary responsiveness in allergic and nonallergic airways. J Appl Physiol 1986; 60(6):1967–1971.
28. Wong LB, Miller IF, Yeates D. Pathways of substance P stimulation of canine tracheal ciliary beat frequency. J Appl Physiol 1991; 70(1):267–273.
29. Mussatto DJ, Garrard CS, Lourenco RV. The effect of inhaled histamine on human tracheal mucus velocity and bronchial mucociliary clearance. Am Rev Respir Dis 1988; 138:775–779.
30. Garrard CS, Mussatto DJ, Lourenco RV. Lung mucociliary transport in asymptomatic asthma: Effects of inhaled histamine. J Lab Clin Med 1989; 113:190–195.
31. Rugolo M, Barzanti F, Gruenert DC and Hrelia S. Histamine activates phospholipase C in human airway epithelial cells via ester-sensitive pathway. Am J Physiol (Lung Cell Mol Physiol 15) 1996; 271:L665–L671.
32. Wills P, Hall R, Chan W-m, Cole PJ. Sodium chloride increases the ciliary transportability of cystic fibrosis and bronchiectasis sputum on the mucus-depleted bovine trachea. J Clin Invest 1997; 99:9–13.

# 31

# Mucociliary and Cough Clearance: Role of Ion Transport and the $P_2Y_2$ Receptor-Mediated System

**Michael R. Knowles, Peadar G. Noone, William D. Bennett, and Richard C. Boucher**
*The University of North Carolina at Chapel Hill, Chapel Hill, North Carolina*

## I. INTRODUCTION

Mucociliary and cough clearance are critical features of pulmonary host defense, and the net effect of ion transport, mucin secretion, and ciliary beat frequency determine the efficiency of these defense mechanisms [1]. Under normal circumstances, the response of each of the individual elements is key to this defense mechanism, and these responses are regulated, in part, by the nucleotide ($P_2Y_2$) receptor system [2–6]. Cystic fibrosis (CF) is a recessive genetic disorder that reflects mutations in the cystic fibrosis transmembrane regulator (CFTR) gene [7]. Cystic Fibrosis is associated with defects in ion transport and the abnormal airway secretions, and chronic bacterial infection in CF demonstrates the adverse impact of defective ion transport on host defense. Cystic Fibrosis also offers the possibility for novel therapy that targets the defective ion transport. Patients with primary ciliary dyskinesia (PCD) have chronic airway disease that reflects mutations that cause ciliary dysfunction [8–11]. These patients highlight the importance of normal ciliary clearance to airway host defense. Because these patients have no mucociliary clearance [8], they are dependent on cough as their primary mechanism for airway clearance. Drugs that can modify ion transport, mucin secretions, and/or ciliary beat frequency may be useful in the treatment of airway diseases such as those associated with abnormal secretions and defective airway clearance. This chapter provides an overview of such therapeutic approaches in the treatment of CF and PCD.

## II. NORMAL AIRWAYS

### A. Normal Airway Epithelia

The epithelium defends against inhaled noxious agents and infectious particles by the clearance of a mucus layer by cilia and by regulation of the volume and composition of airway surface liquid by active ion transport across surface epithelia [1,5,6]. Regulation of airway surface liquid depth and composition is a complex function involving the genesis of liquid at distal airway and/or alveolar units and the absorbtion of surface liquid that is moved from a large cross-sectional surface area (distal sites) to a much smaller surface area (four thousand-fold reduction in cross-sectional surface area in proximal conducting airways). To assist in volume absorption, the surface epithelium of conducting airways exhibits active $Na^+$ transport through amiloride-sensitive $Na^+$ channels on the apical surface. The accompanying anion ($Cl^-$) flow may occur through transcellular and/or paracellular pathways [12,13]. Normal airway epithelia have at least two $Cl^-$ conductances that are available in the apical membrane; the cAMP-regulated (CFTR) $Cl^-$ conductance and the $Ca^{2+}$-activated alternative $Cl^-$ conductance ($Cl^-{}_a$) [5,6]. Activation of either of these $Cl^-$ conductances may result in $Cl^-$ (liquid) movement in either the secretory or the absorptive direction, depending on the net electrochemical driving force.

### B. Role of $P_2Y_2$ in Normal Airways

The $P_2Y_2$ (purinoceptor) receptor system has recently been recognized as having a role in the regulation of at least three elements of airway defense, i.e., ion transport, mucus secretion, and ciliary function. The $P_2Y_2$ receptor is a member of an emerging family of receptors that interact with extracellular triphosphate nucleotides (adenosine triphosphate, ATP, and uridine triphosphate, UTP) and their metabolic breakdown products (ADP, AMP, and adenosine and UDP, UMP, and uridine) [14]. The human $P_2Y_2$ receptor has been cloned [15] and has the structure of a traditional G-protein linked receptor, i.e., seven membrane spanning segments in the protein. The nucleotides ATP and UTP are equipotent at the $P_2Y_2$ receptor, and activation of the receptor is associated with increases in the concentration of intracellular calcium via the inositol phosphate pathway [16–18]. In the airway epithelium, extracellular nucleotides activate the alternative $Cl^-$ conductance, induce goblet cell degranulation (i.e., mucin secretion), and stimulate ciliary beat frequency [3,16,17]. Thus, the $P_2Y_2$ receptor coordinates these airway mucosal defense responses. As it is present in airway surface liquid at submicromolar concentrations [19] and is released by mechanostimulation [20], ATP may be the endogenous nucleotide to activate these systems.

## III. CYSTIC FIBROSIS

### A. Disease Pathogenesis

Cystic fibrosis lung disease reflects mutations in CFTR, which normally functions as a cAMP-regulated $Cl^-$ channel [7]. The normal CFTR protein also has an inhibitory influence on apical membrane $Na^+$ transport, and mutated CFTR in CF is associated with accelerated $Na^+$ transport [21,22]. The traditional hypothesis for the pathogenesis of CF airway disease suggests that accelerated $Na^+$ transport in conducting airways is associated with excessive volume absorption and results in less-than-optimal hydration of mucin-laden airway secretions. These thick airway secretions have abnormal rheological properties, poor mucociliary and cough clearance, and predispose to chronic bacterial infection with Staphylococcus aureus and Pseudomonas aeruginosa [23]. An alternative hypothesis has recently been advanced that suggests that airways resemble sweat ducts, i.e., $Cl^-$ cannot be absorbed across CF airway epithelium [24,25]. This alternative hypothesis projects an increase in the concentration of $Cl^-$ in the airway lumen, which inhibits the activity of antimicrobial factors on the airway surface. There is little direct evidence to confirm an increase in the $Cl^-$ concentration of CF patients as compared to normal subjects [26], but this question is the focus of ongoing research.

### B. Abnormal $Na^+$ Transport

Early studies of therapy targeting abnormal ion transport in CF involved amiloride, a $Na^+$ conductance inhibitor. Initial studies indicated that amiloride inhibited excessive $Na^+$ transport [6,27] and improved mucociliary clearance [28]. A blinded cross-over, placebo-controlled trial with a small number of adults with CF suggested that the rate of loss of lung function (forced vital capacity, FVC and forced expired volume in 1 sec., $FEV_1$) was slowed during the six months of treatment with aerosolized amiloride as compared to inhaled vehicle alone [29]. However, the results of subsequent acute and chronic studies have been variable. Several acute or short-term studies of treatment with aerosolized amiloride showed evidence of potential benefit, i.e., it may be useful to shorten the length of time to achieve a peak FVC during intravenous antibiotic cleanup [30] or that more sputum may be expectorated after aerosolized amiloride as compared to control [31], but other studies failed to demonstrate an acute beneficial effect on mucociliary clearance [32,33]. Moreover, three studies testing the effect of chronic (greater than three months) effect of aerosolized-amiloride treatment have failed to confirm the initial report that pulmonary function was improved by this approach [34–36]. Each of these three studies differed in aspects of the study design as compared to the initial chronic study, which may explain, at least in part, the different outcomes. However, the two larger studies [35,36] suggest that using aerosolized amiloride as a single agent may not be beneficial when added to standard therapy. The lack of clear benefit from aerosolized amiloride may reflect an incorrect hypothesis regarding the role of abnormal $Na^+$ transport in the pathogenesis of lung disease in CF. Alternatively, dosing and the duration of effect of amiloride may be inadequate, and the use of $Na^+$ conductance inhibitors may require agents that are more potent, specific, and/or long-lasting in effect. Benzamil is one such option [37]; if it is largely metabolized to amiloride, the safety profile might be advantageous. Moreover, the use of $Na^+$ conductance inhibitors may be best envisioned as a preventive therapy, which would be most effective in a young-patient population, before the development of severe airways disease and bronchiectasis. Finally, it is possible that a $Na^+$ conductance inhibitor might amplify the potential benefit of $Cl^-$ conductance agonists such as UTP (see below) [33].

### C. Defective $Cl^-$ Transport

Recent studies have focused on the abnormal $Cl^-$ conductance in CF, which reflects mutations in the CFTR $Cl^-$ channel protein. Although beta-agonists and other agents that act via cAMP-mediated mechanisms fail to activate a $Cl^-$ conductance in CF, the triphosphate nucleotides (ATP and UTP) activate a non-CFTR alternative $Cl^-$ conductance ($Cl^-_a$) [2,17]. The concept that activation of an alternative $Cl^-$ conductance may protect against airway disease in CF patients is supported by the lack of lung disease in CF knockout mice, which possess a large endogenous $Ca^{2+}$-activated $Cl^-_a$ in the airway epithelia, as compared to a much smaller $Cl^-_a$ in human airways [38]. Initial studies of treatment with aerosolized nucleotides focused on UTP, because metabolic products from ATP

(ADP, AMP, adenosine) induce bronchoconstriction in some patients with airways disease [39] whereas uridine does not.

Initial studies testing the effects of aerosolized UTP on mucociliary clearance in normal volunteers, as indexed by external scanning, produced striking results [4]. There was a 2½-fold increase in the initial rate of mucociliary clearance in these subjects, which reflected the net effect of activation of the $Cl^-$ conductance, mucin release (goblet cell degranulation), and stimulation of ciliary beat frequency. Studies with aerosolized UTP in CF patients focused on the peripheral small airways which are the sites of initial pathogenesis of CF lung disease [7,23]. In adult CF patients with established lung disease, the rate of peripheral mucociliary clearance at baseline was approximately half that seen in normal subjects [33]. The administration of aerosolized UTP in conjunction with amiloride improved the rate of peripheral mucociliary clearance in these CF patients to approximate the baseline rate of clearance in normal subjects. The mechanism of this effect is presumably the same as that seen in normal individuals, i.e., the activation of an apical membrane $Cl^-$ conductance, goblet cell degranulation, and stimulation of ciliary beat frequency. It should be noted that if the alternative hypothesis for the pathogenesis of CF airway disease is correct, i.e., there is an increased concentration of $Cl^-$ in the airway lumen that adversely inhibits antimicrobial defenses, the activation of an apical membrane $Cl^-$ conductance by UTP would be beneficial. Specifically, activation of a $Cl^-$ conductance by UTP would allow excessive $Cl^-$ on the airway surface to be drained into the airway epithelia if a high concentration of $Cl^-$ is present. Studies are ongoing using aerosolized UTP in CF patients focusing on defining the maximally tolerated and safe dose in CF patients down to the age of 5 years.

## IV. PRIMARY CILIARY DYSKINESIA

### A. Disease Pathogenesis

The airway disease that develops in patients with PCD reflects mutations of proteins important for normal ciliary structure/function and these lead to defective mucociliary clearance [9]. Patients with PCD have no baseline mucociliary clearance [8] which results in chronic bacterial infection and bronchiectasis. Therefore, patients with PCD rely on cough as the mechanism for clearance of the airways. Patients with PCD have, in general, milder disease than CF patients which implies that cough clearance is more effective in PCD than CF or that other pathogenic pathways are present in CF.

### B. UTP and Cough Clearance

Because patients with PCD have little (or no) baseline mucociliary clearance and defective cilia that are unable to respond to the stimulatory effects of UTP, we sought to test whether aerosolized UTP might improve cough clearance in these patients. Specifically, we hypothesized that the net effect of activating a $Cl^-$ conductance and stimulating goblet cell degranulation would change the rheological properties of airway secretions in PCD patients such that cough clearance would be more effective. The study employed external gamma scanning to monitor the rate of clearance of isotopically labeled iron oxide particles with a series of voluntary coughs interspersed over the first 60 minutes of the procedure. The initial results from 12 patients with PCD were quite striking, i.e., cough clearance was improved over the initial 60 minutes, and the effect lasted for the duration of monitoring (2 hours) [40]. Thus, it appears that aerosolized UTP has a beneficial effect on the cough-sensitive rheological properties of airway secretions in patients with PCD. Direct measures of pertinent rheologic properties are being tested in sputum samples expectorated by these patients.

## V. SUMMARY

Mucociliary and cough clearance are important features of the normal pulmonary host defense. Patients with genetic diseases associated with defective ion transport (i.e., cystic fibrosis) and abnormal ciliary function (i.e., primary ciliary dyskinesia) are human experiments of nature that demonstrate lung diseases reflect dysfunction of elements that are critical to mucociliary and cough clearance. Definition of the abnormal elements associated with disease in these two genetic disorders provides the opportunity for therapeutic intervention using agents that target ion transport, mucin secretion, and ciliary beat frequency. The $P_2Y_2$ receptor-mediated system plays a key role for coordinating biological actions that are critical for mucociliary and cough clearance. Agents that act through this system (e.g. UTP) may provide therapeutic benefit for diseases with abnormal airway secretions and/or defective airway clearance mechanisms.

## ACKNOWLEDGMENTS

Supported by NIH HL000046, Inspire Pharmaceuticals, Inc., NIH P01 HL32322. Acknowledgment with a thousand thanks to Beth Godwin and Lisa Brown.

## REFERENCES


1. Wanner A, Salathe M, O'Riordan TG. Mucociliary clearance in the airways State of the Art. Am J Respir Crit Care Med 1996; 154:1868–1902.
2. Knowles MR, Clarke LL, Boucher RC. Activation by extracellular nucleotides of chloride secretion in the airway epithelia of patients with cystic fibrosis. N Engl J Med 1991; 325:533–538.
3. Lethem MI, Dowell ML, Van Scott M, Yankaskas JR, Egan T, Boucher RC, Davis CW. Nucleotide regulation of goblet cells in human airway epithelial explants: normal exocytosis in cystic fibrosis. Am J Respir Cell Mol Biol 1993; 9:315–322.
4. Olivier KN, Bennett WD, Hohneker KW, Zemen KL, Edwards LJ, Boucher RC, Knowles MR. Acute safety and effects on mucociliary clearance of aerosolized uridine 5′-triphosphate ± amiloride in normal human adults. Am J Respir Crit Care Med 1996; 154:217–223.
5. Boucher RC. Human airway ion transport (Part 1). Am J Respir Crit Care Med 1994; 150:271–281.
6. Boucher RC. Human airway ion transport (Part 2). Am J Respir Crit Care Med 1994; 150:581–593.
7. Welsh MJ, Tsui L, Boat TF, Beaudet AL. Cystic fibrosis. In: The Metabolic and Molecular Bases of Inherited Disease. Scriver CR, Beaudet AL, Sly WS, Valle D, Stanbury JB, Wyngaarden JB, Fredrickson DS, ed. New York. McGraw-Hill, Inc. 1995; 17:3799–3876.
8. Camner P, Mossberg B, Afzelius BA. Measurements of tracheobronchial clearance in patients with immotile-cilia syndrome and its value in differential diagnosis. Eur J Respir Dis Suppl 1983; 64 (Suppl. 127):57–63.
9. Greenstone M, Rutman A, Dewar A, MacKay I, Cole PJ. Primary ciliary dyskinesia: cytological and clinical features. Q J Med New Series 1988; 67: 405–430.
10. Eliasson R, Mossberg B, Camnen P, Afzelius BA. The immotile–cilia syndrome. A congenital ciliary abnormality as an etiologic factor in chronic airway infections and male sterility. N Engl J Med 1977; 297:1–6.
11. Mossberg B, Afzelius BA, Eliasson R, Camner P. On the pathogenesis of obstructive lung disease. A study on the immotile–cilia syndrome. Scand J Respir Dis 1978; 59:55–65.
12. Willumsen NJ, Boucher RC. Shunt resistance and ion permeabilities in normal and cystic fibrosis airway epithelium. Am J Physiol 1989; 256: C1054–C1063.
13. Willumsen NJ, Davis CW, Boucher RC. Intracellular $Cl^-$ activity and cellular $Cl^-$ pathways in cultured human airway epithelium. Am J Physiol 1989; 256:C1033–C1044.
14. Fredholm BB, Abbracchio MP, Burnstock G, Daly JW, Harden TK, Jacobson KA, Leff P, Williams M. Nomenclature and classification of purinoceptors: a report from the IUPHAR subcommittee. Pharmacol Rev 1994; 46: 143–156.
15. Parr CE, Sullivan DM, Paradiso AM, Lazarowski ER, Burch LH, Olsen JC, Erb L, Weisman GA, Boucher RC, Turner JT. Cloning and expression of a

human $P_{2U}$ nucleotide receptor, a target for cystic fibrosis pharmacotherapy. Proc Natl Acad Sci USA 1994; 91:3275–3279.
16. Mason SJ, Paradiso AM, Boucher RC. Regulation of transepithelial ion transport and intracellular calcium by extracellular adenosine triphosphate in human normal and cystic fibrosis airway epithelium. Br J Pharmacol 1991; 103:1649–1656.
17. Clarke LL, Boucher RC. Chloride secretory response to extracellular ATP in normal and cystic fibrosis nasal epithelia. Am J Physiol 1992; 263: C348–C356.
18. Brown HA, Lazarowski ER, Boucher RC, Harden TK. Evidence that UTP and ATP regulate phospholipase C through a common extracellular 5′-nucleotide receptor in human airway epithelial cells. Mol Pharmacol 1991; 40: 648–655.
19. Donaldson SH, Stutts MJ, Boucher RC, Knowles MR. Adenosine triphosphate levels in nasal surface liquid. Am J Respir Crit Care Med 1996; 153: A854 (abstr).
20. Grygorczyk R, Hanrahan JW. CFTR-independent ATP release from epithelial cells triggered by mechanical stimuli. Am J Physiol 1997; 272: C1058–C1066.
21. Stutts MJ, Canessa CM, Olsen JC, Hamrick M, Cohn JA, Rossier BC, Boucher RC. CFTR as a cAMP-dependent regulator of sodium channels. Science 1995; 269:847–850.
22. Boucher RC, Stutts MJ, Knowles MR, Cantley L, Gatzy JT. $Na^+$ transport in cystic fibrosis respiratory epithelia. Abnormal basal rate and response to adenylate cyclase activation. J Clin Invest 1986; 78:1245–1252.
23. Davis PB. Pathophysiology of the lung disease in cystic fibrosis. In: Davis PB, ed. Cystic Fibrosis Lung Biology in Health and Disease. Vol. 64. New York: Marcel Dekker, Inc. 1993; 93–218.
24. Smith JJ, Travis SM, Greenberg EP, Welsh MJ. Cystic fibrosis airway epithelia fail to kill bacteria because of abnormal airway surface fluid. Cell 1996; 85:1–20.
25. Quinton PM. Viscosity versus composition in airway pathology [editorial]. Am J Respir Crit Care Med 1994; 149:6–7.
26. Smith JJ, Travis SM, Greenberg EP, Welsh MJ. Erratum. Cell 1996; 87:
27. Knowles MR, Gatzy J, Boucher R. Increased bioelectric potential difference across respiratory epithelia in cystic fibrosis. N Engl J Med 1981; 305: 1489–1495.
28. Kohler D, E App, Schmitz-Schumann M, Wuertemberger G, Matthys H. Inhalation of amiloride improves the mucociliary and the cough clearance in patients with cystic fibroses. Eur J Respir Dis 1986; 69(Suppl. 146):319–326.
29. Knowles MR, Church NL, Waltner WE, Yankaskas JR, Gilligan PH, King M, Edwards LJ, Helms RW, Boucher RC. A pilot study of aerosolized amiloride for the treatment of lung disease in cystic fibrosis. N Engl J Med 1990; 322:1189–1194.
30. Bowler IM, Kelman B, Worthington D, Littlewood JM, Watson A, Conway SP, Smye SW, James SL, Sheldon TA. Nebulized amiloride in respiratory

exacerbations of cystic fibrosis: a randomised controlled trial. Arch Dis Child 1995; 73:427–430.

31. Hofmann T, Senier I, Bittner P, Huels G, Schwandt HJ, Lindemann H. Aerosolized amiloride: Dose-effect on nasal bioelectric properties, pharmacokinetics, and effect on sputum expectoration in patients with cystic fibrosis. J Aerosol Med 1997; 10:147–158.
32. Robinson M, Regnis JA, Bailey DL, King M, Bautovich GJ, Bye PT. Effect of hypertonic saline, amiloride, and cough on mucociliary clearance in patients with cystic fibrosis. Am J Respir Crit Care Med 1996; 153:1503–1509.
33. Bennett WD, Olivier KN, Zeman KL, Hohncker KW, Boucher RC, Knowles MR. Effect of uridine 5′-triphosphate plus amiloride on mucociliary clearance in adult cystic fibrosis. Am J Respir Crit Care Med 1996; 153:1796–1801.
34. Graham A, Hasani A, Alton EWFW, Martin GP, Marriott C, Hodson ME, Clarke SW, Geddes DM. No added benefit from nebulized amiloride in patients with cystic fibrosis. Eur Respir J 1993; 6:1243–1248.
35. Pons G, Marchand MC, Foucard C, Chaumet-riffaud P, d'Athis P, Lenoir G, Navarro J, Sautejeau A. French multicentre randomized double–blind placebo–controlled trial on nebulized amiloride in cystic fibrosis patients. Pediatr Pulmonol Suppl 1995; 12:193(abstr).
36. Church NL, Burroughs SM, Wisniewski ME, Anderson WH. AML302/303 Clinical Investigators. The effect of amiloride on the decline of pulmonary function in cystic fibrosis patients 10 years of age and older. Pediatr Pulmonol Suppl 1996; 13:279–280 (abstr).
37. Hofmann T, Ziersch A, Senier I, Blank U, Weber WM, Huels G, Lindemann H. Benzamil and amiloride in CF nasal epithelium: Time course of the effects on nasal potential difference *in vivo* Pediatr Pulmonol Suppl 1996; 13:280 (abstr).
38. Clarke LL, Grubb BR, Gabriel SE, Smithies O, Koller BH, Boucher RC. Defective epithelial chloride transport in a gene targeted mouse model of cystic fibrosis. Science 1992; 257:1125–1128.
39. Holgate ST, Mann JS, Church MK, Cushley MJ. Mechanisms and significance of adenosine–induced bronchoconstriction in asthma. Allergy 1987; 42:481–484.
40. Noone PG, Bennett WD, Zeman KL, Regnis JA, Carson JL, Boucher RC, Knowles MR. Effects on cough clearance of aerosolized uridine-5′-triphosphate ± amiloride in patients with primary ciliary dyskinesia. Am J Respir Crit Care Med 1996; 153:A530 (abstr).

# 32

# Surface Properties of Respiratory Secretions: Relationship to Mucus Transport

**Bruce K. Rubin***
*St. Louis University School of Medicine, and Cardinal Glennon Children's Hospital, St. Louis, Missouri*

## I. INTRODUCTION

The clearance of airway secretions depends upon an efficient mucus–epithelial interaction in which the physical properties of mucus as well as properly functioning cilia play a role [1]. Normally, mucus is cleared to the proximal airways by cilia beating in contact with the mucous layer [2] and this is dependent, in part, on the rheology of the secretions. When there is hypersecretion or ciliary damage, secretions are cleared by cough, which is dependent upon both air flow and the interaction between the sputum and the epithelium [3].

Viscoelasticity is directly involved in the flow and deformation-dependent transport of secretions, especially mucociliary transport [4]. Surface properties influence the interface interaction between mucus and the epithelium. The viscoelastic properties of vegetable gum mucus simulants have been shown to influence both mucociliary (MCTR) and cough transportability (CTR) [5]. However true respiratory secretions contain cellular debris, proteins, and lipids. This may explain the poor correlation between cough transportability and the viscoelastic properties of sputum.

Abhesive agents like surfactants can have a profound effect on the interaction between mucus and cilia or sputum and the underlying epithe-

**Current affiliation*: Bowman Gray School of Medicine, Winston-Salem, North Carolina lium. When applied directly to the frog palate, surfactant increases mucociliary transport [6]. This may be due to lubrication of the mucus–cilia interface preventing ciliary entanglement in the mucus gel [7].

When applied to the airway of anesthetized dogs, surfactant increases tracheal mucus transport velocity nearly 400% over control values

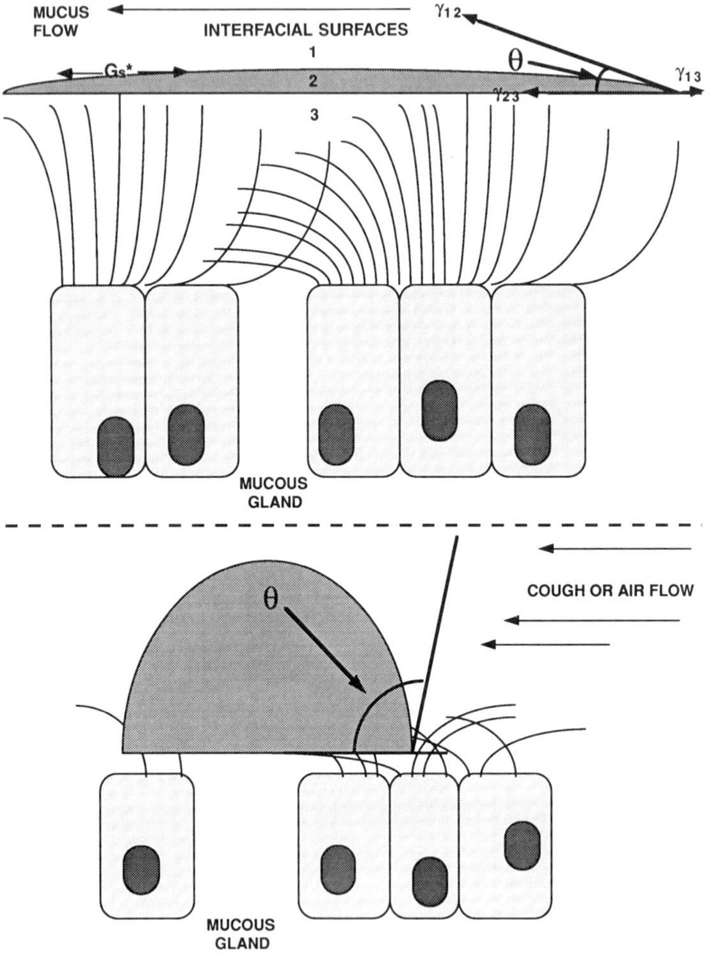


**Figure 1** Schematic showing the surface interaction between airway secretions and the ciliated epithelium in health (top) and with chronic airway inflammation and hypersecretion (bottom). Interfacial (surface) tension is $\gamma$ and wettability is measured by the contact angle $\theta$. Surface mechanical impedance, $G_s$* can be measured by the rolling ball technique.

[8] and improves the clearability of mucus in newborns with respiratory distress syndrome [9]. A surfactant vehicle can also increase the airway distribution of technetium or pentamidine when instilled intratracheally and it can improve the uniformity of aerosol distribution [10].

Surfactant is also produced in the airway and this bronchial surfactant is thought to be important for normal mucociliary interaction. Schürch and colleagues demonstrated that latex particles less than or equal to 6 μm in diameter deposited in hamster lungs were submerged in the subphase of the alveolar lining layer and became completely coated with an osmiophilic film. Similar results were obtained for particles deposited in the conducting airways. For small particles (less than 100 μm) the surface tension force was several orders of magnitude greater than forces related to gravity [11].

## II. MATERIALS AND METHODS

We wished to study the viscoelastic and surface properties of secretions from patients with different airway diseases and how these related to sputum transportability in vitro. We collected mucus from subjects with cystic fibrosis (CF), chronic bronchitis (CB), fatal asthma, and subjects without lung disease. Sputum samples were produced by expectoration during pulmonary function testing by 18 patients with stable CF lung disease (age 14–45 years), and 30 stable patients with CB all over 16 years of age. Airway secretions were collected at the time of autopsy (within 18 hours of death) from subjects who died of asthma (fatal asthma), or subjects without lung disease normals, as part of the Canadian Prairie Provinces Fatal Asthma Study. Mucociliary transportability (MCTR) was measured by the frog palate assay [12] and cough transportability (CTR) in the simulated cough machine [13]. Loss modulus (viscosity: G″) and storage modulus (elasticity: G′) were measured by magnetic microrheometry at 1 and 100 rad/sec [5]. Surface properties, schematically diagrammed in Figure 1, were measured as detailed below.

### A. Surface Mechanical Impedance, $G_s$*, by the Rolling Ball Technique

The rolling ball technique is used in industry to measure the adhesiveness of glue strips. The magnetic microrheometer set up and computer measurement software were modified to make rolling ball $G_s$* measurements on small mucus specimens. The steel microprobe is placed on the surface of the mucus layer without using a paraffin oil cover. The magnetic force needed to roll the sphere across the surface of the sample is used to calculate $G_s$* at 1 rad/s. This measures mechanical (frictional) adhesive forces.

### B. Sputum Wettability (Contact Angle θ)

A glass microscope slide is cleaned with chromic sulfuric acid and rinsed with deionized water then stored in absolute ethanol to maintain dehydration before use. A 20 μL volume of secretion is dropped onto the surface from 3–5 mm and a stabilization time of one minute is timed before capturing the image of the drop in a video processing system [14].

### C. Interfacial Tension γ by Distraction (Du Noüy Ring Method)

A specially instrumented platinum–iridium ring, a metal that is completely wettable, is pulled from the mucus sample at a distraction velocity of 10 mm/sec until separation is achieved. The force of separation is measured by a strain gauge connected to the ring. This gives the interfacial (surface) tension in dyne/cm. We use a semiautomated tensiometer for these measurements (Fisher Tensiomat Model #21, Fischer Scientific, Pittsburgh, PA). We have calibrated a platinum–iridium ring of circumference 1.7145 ± 0.0381 cm so that with a flat chamber we need only a volume of 40 μL to accurately measure interfacial tension [15].

The work of adhesion ($W_{ad}$) is calculated from the contact angle and interfacial tension as follows:

$$W_{ad} = \gamma (1 + \cos \theta)$$

and tenacity or work of distraction is the product of cohesiveness (in cm) and $W_{ad}$ (dyne/cm).

## III. RESULTS AND DISCUSSION

We found that viscosity was significantly higher ($p < 0.05$) in CF and CB secretions compared with normal and asthma mucus. Elasticity and cohesiveness (spinnability) were highest in fatal asthma secretions. Interfacial tension and adhesive work were greater than normal in all diseases, but significantly so only in CF (Table 1).

As previously demonstrated, CF mucus was not more viscous than secretions from subjects with CB [16]. However, CF sputum was significantly more adhesive and this probably explains the poor cough transport-

**Table 1** The Physical and Transport Properties of Secretions from Patients with Cystic Fibrosis, Chronic Bronchitis, Stable Asthma, Fatal Asthma, and Patients Without Lung Disease

| | CF | CB | Asthma | Fatal asthma | Normal |
|---|---|---|---|---|---|
| $\gamma$ (mN/m) | 81.1 ± 3.4 | 72.1 ± 3.7 | 73.0 ± 8.3 | 69.6 ± 8.2 | 65.1 ± 3.0 |
| $\theta$ glass (degrees) | 45.5 ± 2.5 | 31.9 ± 2.4 | 42.6 ± 5.0 | 56.9 ± 4.5 | 42.6 ± 5.2 |
| Cohesiveness (mm) | 27.4 ± 1.5 | 31.3 ± 1.5 | 21.2 ± 3.9 | 34.1 ± 4.6 | 26.6 ± 2.1 |
| G″ 1 rad/sec | 109.9 ± 18.7 | 111.2 ± 16.8 | 18.6 ± 5.2 | 37.4 ± 10.7 | 30.4 ± 6.2 |
| G′ 1 rad/sec | 446.2 ± 84.9 | 550.6 ± 91.5 | 162.4 ± 59.6 | 444.4 ± 86.4 | 161.7 ± 32.3 |
| MCTR | 0.71 ± 0.04 | 0.65 ± 0.05 | 0.95 ± 0.04 | 0.71 ± 0.1 | 0.84 ± 0.09 |
| CTR (mm) | 15.1 ± 1.3 | 21.5 ± 1.4 | 30.5 ± 11.5 | 28.5 ± 9.5 | 31.2 ± 2.6 |
| $W_{ad}$ (mN/m) | 137.1 ± 28.2* | 128.3 ± 22.9 | 107.8 ± 24.5 | — | 110.9 ± 7.0 |

All results are given as means ± 1 standard deviation.
*$p < 0.05$ when compared to normal.

ability of CF sputum in comparison with other diseases. The work of adhesion of sputum inversely correlated with the ability to clear these secretions in the cough machine.

Mucociliary transportability was equally compromised in all diseases, but MCTR did not correlate well either with viscoelasticity or surface properties. The lack of a correlation between $W_{ad}$ and MCTR may be due to a more complex dependency of mucociliary clearability on viscoelasticity.

The $W_{ad}$ is a characteristic of two substances, each with unique surface tensions, that share an interface. It is the work that is done to overcome the interfacial tension and effect a separation of these two substances. When describing the $W_{ad}$ for respiratory secretions we must specify the interface that is being evaluated. The ideal will be to describe the mucus/epithelium interface. Differences in the interface may arise in either component of the system. This emphasizes the need to develop ways of studying mucus as part of a mucus/epithelium system in which the characteristics of both the mucus and the epithelium can be determined.

Puchelle and colleagues have shown the importance of phospholipids in determining the high interfacial tension of CF sputum [17]. The phospholipid composition of CF sputum is different from that of sputum from patients with CB or bronchiectasis. These differences in sputum phospholipid composition were related to the contact angle between the sputum sample and a glass slide in air.

The high interfacial tension and abnormal surface properties of sputum suggest that aerosolized surfactant could be an effective mucokinetic therapy for patients with CF or CB. We have conducted multicenter, randomized, placebo-controlled trials of aerosolized surfactant for CF [18] or CB [19]. In the latter study, 66 patients were randomized to surfactant and 21 to saline treatment. Patient demographics between groups were similar at baseline. In patients who received 607.5 milligrams dipalmitoyl phosphatidyl choline (DPPC)/day of surfactant for two weeks, prebronchodilator forced expiratory volume in one second ($FEV_1$) increased from $1.22 \pm 0.08$ L at day 0 to $1.33 \pm 0.09$ L at day 21 [$p < 0.05$]), an improvement of 11.4 percent; postbronchodilator $FEV_1$ improved 10.4 percent by days 14 and 21 ($p < 0.05$); and the ratio of residual volume to total lung capacity, a measure of thoracic gas trapping, decreased 6.22 percent by day 21 ($p < 0.05$). In the surfactant groups, there was a dose-dependent increase of in vitro sputum mucociliary clearability.

The results of these Phase 2 studies strongly suggest that aerosolized surfactant can improve pulmonary function and reduce thoracic gas trapping in patients with chronic inflammatory, hypersecretory lung disease.

## ACKNOWLEDGMENT

Supported by a grant from the US Cystic Fibrosis Foundation (Rubin 96PO).

## REFERENCES

1. Lucas AM, Douglas LC. Principles underlying ciliary activity in the respiratory tract. Arch Otolaryngol 1934; 20:528–541.
2. Sleigh MA, Blake JR, Liron N. The propulsion of mucus by cilia. Am Rev Resp Dis 1988; 137:726–741.
3. King M, Zahm JM, Pierrot D, Vaquez-Girod S, Puchelle E. The role of mucus gel viscosity, spinnability, and adhesive properties in clearance by simulated cough. Biorheology 1989; 26:737–745.
4. Puchelle E, Zahm JM, Girard F, Bertrand A, Polu JM, Aug F, Sadoul P. Mucociliary transport in vivo and in vitro. Relations to sputum properties in chronic bronchitis. Eur J Respir Dis 1980; 61:254–264.
5. King M, Rubin BK. Mucus rheology: relationship with transport. In: Takishima, T., ed. Airway Secretion: Physiological Bases for the Control of Mucus Hypersecretion. New York: Marcel Dekker, Inc. 1994 pp. 283–314.
6. Allegra L, Bossi R, Braga PC. Influence of surfactant on mucociliary transport. Prog Resp Dis 1985; 19:441–460.
7. Morgenroth K, Bolz J. Morphological features of the interaction between mucus and surfactant on the bronchial mucosa. Respiration 1985; 47: 225–231.
8. De Sanctis GT, Tomkiewicz RP, Rubin BK, Schürch S, King M. Exogenous surfactant enhances mucociliary clearance in the anaesthetized dog. Eur Respir J 1994; 7:1616–1621.
9. Rubin BK, Ramirez O, King M. Mucus rheology and transport in neonatal respiratory distress syndrome and the effect of surfactant therapy. Chest 1992; 101:1080–1085.
10. Kharasch VS, Sweeney TD, Fredberg J, Lehr J, Damokosh AI, Avery ME, Brain JD. Pulmonary surfactant as a vehicle for intratracheal delivery of technetium sulfur colloid and pentamidine in hamster lungs. Am Rev Respir Dis 1991; 144:909–913.
11. Schürch S, Gehr P, Im Hof V, Geiser M, Green F. Surfactant displaces particles toward the epithelium in airways and alveoli. Respir Physiol 1990; 80:17–32.
12. Rubin BK, Ramirez O, King M. Mucus-depleted frog palate as a model for the study of mucociliary clearance. J Appl Physiol 1990; 69:424–429.
13. Agarwal M, King M, Rubin BK, Shukla JB. Mucus transport in a miniaturized simulated cough machine: effect of constriction and serous layer simulant. Biorheology 1989; 26:977–988.

14. Vaquez-Girod S, Zahm J-M, Dionisius JP, Pierrot D, Puchelle E. Automatic measurement of the wetting properties of biological fluids. Innov Tech Biol Med 1988; 9:402–412.
15. Albers GM, Tomkiewicz RP, May MK, Ramirez OE, Rubin BK. Ring distraction technique for measuring the surface tension of sputum and relationship of the work of adhesion to clearability. J Appl Physiol 1996; 81: 2690–2695.
16. King M. Is cystic fibrosis mucus abnormal? Pediatr Res 1981; 15:120–122.
17. Girod S, Galabert C, Lecuire A, Zahm JM, Puchelle E. Phospholipid composition and surface-active properties of tracheobronchial secretions from patients with cystic fibrosis and chronic obstructive pulmonary diseases. Pediatr Pulmonol 1992; 13:22–27.
18. Rubin BK, Ramirez OE, Smith E, Barrett J. Early results of clinical trials of Exosurf® for the therapy of cystic fibrosis lung disease. Pediatr Pulmonol 1995; 12:234.
19. Anzueto A, Jubran A, Ohar J, Piquette C, Colice G, Pattishall E, Barrett J, Engle M, Dettenmeier P, Epperson K, Mazur M, Rubin B. Effects of aerosolized Exosurf® in stable chronic bronchitis. J Am Med Assoc 1997, in press.

# 33

# Cellular Components of Airway Secretions: Recruitment and Regulation

**Stephen Rennard**
*University of Nebraska Medical Center, Omaha, Nebraska*

## I. INTRODUCTION

The components of sputum include not only the secreted products of airway epithelial cells and glands but also a population of cells. These cells include both inflammatory cells (macrophages, neutrophils, eosinophils, and lymphocytes) as well as desquamated epithelial cells. The recruitment and activation of inflammatory cells into the lumen of the airways undoubtedly plays an important role in host defense. The secreted products of these cells, however, can alter lung function by affecting mucus release and mucociliary clearance and by damaging the airway epithelium. Finally, lysis of cells within the airway lumen can release both cytoplasmic proteins and DNA. Under some circumstances, these can become important constituents altering the physical properties of mucus.

## II. INFLAMMATORY CELLS AND DESQUAMATED EPITHELIAL CELLS

Sputum obtained from the lower respiratory tract of normal volunteers contains mostly macrophages but also includes up to 25–35% neutrophils and smaller percentages of eosinophils and epithelial cells [1,2]. Sputum from smokers contains increased numbers of neutrophils, and these numbers are increased further in patients with chronic bronchitis [2–4]. Neutrophil numbers and the percentage of neutrophils increases further with the exacerbations of bronchitis [5]. This review will provide an overview of the mechanisms and consequences of neutrophil entrainment into airway secretions, with particular emphasis on cigarette smokers and patients with chronic bronchitis.

Neutrophils present within the airway lumen obviously must originate in the bone marrow. Once in the airway lumen, however, neutrophils can alter lung function. Damage of the airway epithelium may result from the action of neutrophil proteases including neutrophil elastase [6,7] as well as from other neutrophil secretions including other proteases, oxidants, and toxic peptides such as the defensins [8,9]. In this regard, Snider and colleagues have demonstrated that neutrophil elastase instilled into the airways can result in goblet cell metaplasia [10,11]. Neutrophil elastase, moreover, is a potent mucus secretagogue [12,13]. Through the release of elastase, therefore, neutrophils present within the airway lumen could contribute both acutely and chronically to altered airway secretions. The observation that the elastase content of neutrophils can vary raises the possibility that some individuals may be predisposed to develop altered secretions through this mechanism [14]. Certainly peripheral blood neutrophil counts are variable. Importantly, smokers have increased peripheral blood neutrophil counts suggesting they may have a larger reservoir of neutrophils to recruit into the airway in response to a given stimulus [15,16]. This may provide one mechanism by which smokers have more exuberant inflammation in their airways than nonsmokers. The mechanisms that lead to this neutrophilia in smokers are incompletely defined. However, the observation that airway epithelial cells can make growth factors including GMCSF and GCSF suggests this as at least one possibility [17].

The migration of neutrophils from the intravascular space into tissues is a highly regulated process that involves several distinct steps (see References 8 and 9 for reviews). First, neutrophils bind relatively loosely to endothelial cell surfaces through selectins. This results in slowing of neutrophil transit through postcapillary venules and permits binding of CD-11, 18 integrins on neutrophil surfaces, with the counter ligand ICAM-1 on the surface of endothelial cells. This results in a much firmer adhesion between neutrophils and endothelial cells. Neutrophils then bind to PECAM and undergo diapedesis between endothelial cells, thus migrating into tissue spaces. This process can be controlled by regulating the expression of the various ligands on the surface of the neutrophil as well as by regulating the expression of the appropriate counter ligands on the surface of the endothelial cell. The expression of all of these surface adhesion molecules can be modulated by complex interactions among cytokines and inflammatory mediators. Interestingly, there may be specific sites in the airway through which cells migrate into the lumen [18].

It is also possible for neutrophil transit through microvessels to be delayed as a result of alterations in neutrophil deformability [9]. In this regard, it is important to note that neutrophils are slightly larger than pulmonary capillaries, and neutrophil deformation, therefore, is essential for neutrophil transit. A decrease in neutrophil deformability has been reported in smokers [19] and, consistent with this, a delay in neutrophil transit through the lungs has also been noted in smokers [20]. This retention of neutrophils in the lung in smokers may contribute to lung damage by release of neutrophil mediators onto the endothelial surface. Alternatively, neutrophils retarded in lung capillaries may serve as a reservoir for cells to be recruited into the air space of the lung. Such a mechanism may play an important role in pulmonary circulation and has been suggested to play a pathogenic role in acute respiratory distress syndrome (ARDS) [21,22]. Whether a similar process takes place in the bronchial circulation and contributes to neutrophil recruitment into airway secretions is unknown at present.

## III. CHEMOATTRACTANTS

Once neutrophils migrate from vessels into the tissue space, it is likely that they migrate through tissues in response to chemical gradients of specific chemotaxins. In this regard, a number of neutrophil chemoattractants have been described including several that may play a role in cigarette smokers [23].

First, cigarette smoke can activate the alternate complement pathway by converting properdin factor B to its activated form [24]. This can, therefore, convert factor 3 to its active form, which in turn can generate the neutrophil chemoattractant C5a from C5. The chemotactic potency of C5a is regulated by the presence of the co-chemotaxin GCG [25–28]. Binding of GCG by the chemotactic factor inactivator (CFI) can, therefore, serve as a mechanism to downregulate the chemotactic potency of C5a. Cigarette smoke can inactivate CFI, thus eliminating a mechanism for attenuating the chemotactic potency of C5a in the presence of GCG [29]. Cigarette smoke, therefore, can both generate a complement-derived neutrophil chemotaxin and block at least one of the mechanisms that serves to downregulate this activity.

Cigarette smoke can also activate macrophages to release neutrophil chemoattractants. In this regard, Hunninghake and colleagues have demonstrated that macrophages obtained from the lower respiratory tract of cigarette smokers spontaneously release a low molecular weight lipid chemoattractant with the properties of eicosanoid [30]. Macrophages from normal individuals, in contrast, do not release such a factor, although they can be stimulated by cigarette smoke to do so. While the identity of macrophage derived neutrophil chemoattractants is incompletely defined, it is likely that both LTB4 and HETEs may play a role.

Airway epithelial cells can also produce neutrophil chemoattractants. In this regard, Shoji et al. demonstrated that cigarette smoke stimulates the release of both a low molecular weight lipid neutrophil chemoattractant and a higher molecular weight peptide chemoattractant from bovine bronchial epithelial cells [31]. Mio and colleagues have extended these observations demonstrating that human bronchial epithelial cells release interleukin 8 (IL-8) in response to stimulation with cigarette smoke [32]. Consistent with this, bronchoalveolar lavage fluid from smokers was demonstrated to contain increased amounts of interleukin 8 when compared to bronchoalveolar lavage fluid from nonsmokers. The interleukin 8 levels, moreover, were higher in lavage fluid recovered from the proximal airways than in fluid recovered from the alveoli, which suggests that airway epithelial cells may be an important source of this mediator. The airway epithelial cell, moreover, has the capacity to release a complex set of cytokines including several that can modulate endothelial cell expression of ICAM-1 [33]. Thus, it seems reasonable to suggest that airway epithelial cells can regulate neutrophil recruitment both through the release of chemoattractants and by modulating their transendothelial migration.

Airway epithelial cells may also play an important role in regulating neutrophil activation within the airway lumen. For example, IL-8 is a potent neutrophil activating factor [34]. In this regard, supernatant from cultured airway epithelial cells has been demonstrated to augment the ability of neutrophils to phagocytose and kill opsonized bacteria [35]. Thus, epithelial cells may play a role not only in recruiting neutrophils into airways, but also in regulating their behavior.

## IV. CLEARING AIRWAY OF NEUTROPHILS

Once neutrophils are recruited into the airway, they eventually must be cleared. If they undergo lysis, they can release intracellular contents into the air space. Of these, two may contribute to pathogenic airway secretions. Actin can polymerize within the airway making secretions highly viscous [36], and DNA can also contribute to airway viscosity by virtue of its extended polymeric structure [37]. Importantly, neutrophils within the airway need not lyse and contribute to pathologic secretions. It is likely that many cells can be cleared by the mucociliary escalator prior to lysis. Other cells may undergo apoptosis together with phagocytosis by macrophages. This process would result in clearance of neutrophils without release of intracellular contents [38]. In this regard, neutrophils appear to be committed to apoptosis both through the expression of cell surface fas and through the production of fas ligand(s). It is, therefore, of interest that airway epithelial cell derived cytokines can prolong neutrophil survival [17], and nicotine can inhibit neutrophil apoptosis [39]. It is possible, therefore, that the presence of such substances in the airway milieu could lead to delayed apoptosis associated with increased neutrophil necrosis in release of intracellular contents into airway secretions.

## V. SUMMARY

This brief review has attempted to provide an overview of the mechanisms by which neutrophils can be recruited into airway secretions and by which the activity of these cells can be regulated. The various steps in this process represent potential opportunities for therapeutic intervention. It is likely, moreover, that similar mechanisms will also govern the recruitment of other inflammatory cells into airway secretions. Definition of the cytokines and adhesion molecules responsible for the selective recruitment and activation of various inflammatory cells is an area of active research and will likely provide important insights into pathogenesis of airway diseases. Understanding the cellular components of airway secretions will be important in understanding the role of airway secretions in health and disease.

## REFERENCES

1. Fahy JV, Liu J, Wong H, Boushey HA. Cellular and biochemical analysis of induced sputum from asthmatic and from healthy subjects. Am Rev Respir Dis 1993; 147:1126–1131.
2. Floreani AA, Rennard SI, Heires A, Buchalter SE, Thompson AB. Induction of sputum by hypertonic saline does not cause airway inflammation. Am J Respir Crit Care Med 1995; 151:A704.
3. Pizzichini E, Pizzichini MM, Efthimiadis A, Evans S, Morris MM. Indices of airway inflammation in induced sputum; reproducibility and validity of cell and fluid-phase measurements. Am J Respir Crit Care Med 1996; 154: 308–317.

4. Maestrelli P, Saetta M, Di-Stefano A, Calcagni PG, Turato G, Ruggieri MP, Roggeri A, Mapp CE, Fabbri LM. Comparison of leukocyte counts in sputum, bronchial biopsies and bronchoalveolar lavage. Am J Respir Crit Care Med 1995; 152:1926–1931.
5. Medici TC, Chodosh S. Sputum cell dynamics in bacterial exacerbations of chronic bronchial disease. Arch Intern Med 1972; 129:597–603.
6. Rickard KA, Taylor J, Rennard SI. Observations on development of resistance to detachment of cultured bovine epithelial cells in response to protease treatment. Am J Respir Cell Molec Biol 1992; 6:414–420.
7. Rennard SI, Rickard K, Beckmann JD, Huerta G, Koyama S, Robbins RA, Romberger D, Spurzem J, Thompson AB. Protease injury in airways disease. Annals of New York Academy of Sciences 1991; 624:278–285.
8. Holland SM, Gallin JI. Neutrophils. In: Crystal RG, West JB, eds. The Lung: Scientific Foundations. Philadelphia: Lippincott-Raven, 1997:877–890.
9. Hogg JC. Neutrophil traffic. In: Crystal RG, West JB, eds. The Lung: Scientific Foundations. Philadelphia: Lippincott-Raven, 1997:891–904.
10. Christensen TG, Korthy AL, Snider GL, Hayes JA. Irreversible bronchial goblet cell metaplasia in hamsters with elastase-induced panacinar emphysema. J Clin Invest 1977; 59:397–404.
11. Snider GL, Lucey EC, Christensen TG, Stone PJ, Calore JD, Catanese A, Franzblau C. Emphysema and bronchial secretory cell metaplasia induced in hamsters by human neutrophil products. Am Rev Respir Dis 1984; 129: 155–160.
12. Niles RM, Christensen TG, Breuer R. Serine proteases stimulate mucous glycoprotein release from hamster tracheal ring organ culture. J Lab Clin Med 1986; 459–497.
13. Sommerhoff CP, Nadel JA, Basbaum CB, Caughey GH. Neutrophil elastase and cathepsin G stimulate secretion from cultured bovine airway gland serous cells. J Clin Invest 1990; 85:682–689.
14. Sachs CW, Christensen RH, Pratt PC, Lynn WS. Neutrophil elastase activity and superoxide production are diminished in neutrophils of alcoholics. Am Rev Respir Dis 1990; 141:1249–1255.
15. Corre F, Lellouch J, Schwartz D. Smoking and leucocyte-counts. Lancet 1971; 2:632–634.
16. Howell RW. Smoking habits and laboratory tests. Lancet 1970; 2:152.
17. Cos G, Gauldie J, Jordana M. Bronchial epithelial cell-derived cytokines (G-CSF and GM-CSF) promote the survival of peripheral blood neutrophils in vitro. Am J Respir Cell Mol Biol 1992; 7:507–513.
18. Hulbert WC, Walker DC, Hogg JC. The site of leukocyte migration through the tracheal mucosa in the guinea pig. Am Rev Respir Dis 1981; 124:310–316.
19. Selby C, Drost E, Wraith PK, MacNee W. In vivo neutrophil sequestration within lungs of humans is determined by in vitro "filterability." J Apply Physiol 1991; 71:1996–2003.
20. MacNee W, Wiggs B, Belzberg AS, Hogg JC. The effect of cigarette smoking on neutrophil kinetics in human lungs. New Engl J Med 1989; 321:924–928.

21. Weiland JE, Davis WB, Holter JF, Mohammed JR, Dorinsky PM, Gadek JE. Lung neutrophils in the adult respiratory distress syndrome. Clinical and pathophysiologic significance. Am Rev Respir Dis 1986; 133:218–225.
22. Wortel CH, Doerschuk CM. Neutrophils and neutrophil–endothelial cell adhesion in adult respiratory distress syndrome. New Horiz 1993; 1:631–637.
23. Robbins RA, Thompson AB, Koyama S, Metcalf JP, Rickard KA, Spurzem JR, Rennard SI. Cigarette smoking and neutrophil migration. J Immunol Res 1990; 2:178–188.
24. Robbins RA, Nelson KJ, Gossman GL, Koyama S, Rennard SI. Complement activation by cigarette smoke. Am J Physiol 1991; 260:L254–L259.
25. Bullard DC, Qin L, Lorenzo I, Quinlin WM, Doyle NA, Bosse R. P–selectin/ICAM-1 double mutant mice: acute emigration of neutrophils into the peritoneum is completely absent but is normal into pulmonary alveoli. J Clin Invest 1995; 95:1782–1788.
26. Robbins RA, Hamel FG. Chemotactic factor inactivator interaction with Gc-globulin (vitamin D-binding protein). A mechanism of modulating the chemotactic activity of C5a. J Immunol 1990; 2371–2376.
27. Metcalf JP, Thompson AB, Gossman GL, Nelson KJ, Koyame S, Rennard SI, Robbins RA. GcGlobulin function as a co-chemotaxin in the lower respiratory tract. A potential mechanism for lung neutrophil recruitment in cigarette smokers. Am Rev Respir Dis 1991; 143:844–849.
28. Perez HD. Gc globulin (vitamin D-binding protein) increases binding of low concentrations of C5a des Arg to human polymorphonuclear leukocytes: an explanation for its cochemotaxin activity. Inflammation 1994; 18:215–220.
29. Robbins RA, Gossman GL, Nelson KJ, Koyama S, Thompson AB, Rennard SI. Inactivation of chemotactic factor inactivator by cigarette smoke. A potential mechanism of modulating neutrophil recruitment to the lung. Am Rev Respir Dis 1990; 142:763–768.
30. Hunninghake GW, Crystal RG. Cigarette smoking and lung destruction: accumulation of neutrophils in the lungs of cigarette smokers. Am Rev Respir Dis 1983; 128:833–838.
31. Shoji S, Erti RF, Koyama S, Robbins R, Leikauf G, Von Essen S, Rennard SI. Cigarette smoke stimulates release of neutrophil chemotactic activity from cultured bovine bronchial epithelial cells. Clin Sci 1995; 88:337–344.
32. Mio T, Romberger DJ, Thompson AB, Robbins RA, Heires A, Rennard SI. Cigarette smoke induces interleukin-8 release from human bronchial epithelial cells. Am J Respir Crit Care Med 1997; 155:1770–1776.
33. Shelhamer JH, Levine SJ, Wu T, Jacoby DB, Kaliner MA, Rennard SI. Airway inflammation. NIH conference. Ann Intern Med 1995; 123:288–304.
34. Strieter RM, Kunkel SL. Chemokines. In: Crystal RG, West JB, eds. The Lung: Scientific Foundations, Philadelphia: Lippincott-Raven, 1997; 1: 155–186.
35. Ertl R, Oddera S, Sacco O, Rossi G, Rennard SI. Airway epithelial cells potentiate neutrophil phagocytosis and killing of bacteria. Am J Respir Crit Care Med 1994; 149:A118.

36. Vasconcellos CA, Allen PG, Wohl ME, Drazen JM, Janmey PA, Stossel TP. Reduction in viscosity of cystic fibrosis sputum in vitro by gelsolin. Science 1994; 263:969–971.
37. Shak S, Capon DJ, Hellmiss R, Marsters SA, Baker CL. Recombinant human DNase I reduces the viscosity of cystic fibrosis sputum. Proc Natl Acad Sci USA 1990; 87:9188–9192.
38. Hale AJ, Smith CA, Sutherland LC, Stoneman VE, Longthorne VL, Culhane AC, Williams GT. Apoptosis: molecular regulation of cell death. Eur J Biochem 1996; 236:1–26.
39. Aoshiba K, Nagai A, Yasui S, Konno K. Nicotine prolongs neutrophil survival by suppressing apoptosis. J Lab Clin Med 1996; 127:186–194.

# 34

# DNA and Actin Filament Ultrastructure in Cystic Fibrosis Sputum

**Robert P. Tomkiewicz,* Chikako Kishioka,† and John Freeman**
*St. Louis University School of Medicine, St. Louis, Missouri*

**Bruce K. Rubin‡,§**
*St. Louis University School of Medicine, and Cardinal Glennon Children's Hospital, St. Louis, Missouri*

## I. INTRODUCTION

The rheology of airway secretions strongly influences mucociliary clearance measured in vitro using the mucus depleted frog palate [1,2]. The viscoelastic properties of normal mucus are believed to depend in part upon the concentration, molecular weight, and entanglements between mucin polymers. In concentrated polymer solutions consisting of a single species, bulk viscosity is proportional to the 3.4-power of the molecular weight of the monomer-forming monomolecular linear polymers, e.g., $G'' \sim MW^{3.4}$ [3,4]. Purulent airway secretions from patients with cystic fibrosis (CF) lung disease contains several species of linear polymers including mucin, high molecular weight neutrophil-derived DNA [5,6], and filamentous actin (F-actin) [7]. When CF sputum is treated with rhDNase [8] or gelsolin, an actin-severing peptide [7], there is a reduction in sputum dynamic viscosity.

Laser scanning confocal microscopy (LSCM) allows the detailed morphologic assessment of tissue sections by capturing sequential image

*Current affiliations:*
* Private practice
† Mie University, Mie, Japan
‡ Bowman Gray School of Medicine, Winston-Salem, North Carolina
§ To whom correspondence should be addressed

data from different tissue planes. Image analysis software can then be used to evaluate the structure of these digitized images. These image analysis techniques could potentially be used for specimens, such as sputum, that have a three-dimensional polymeric structure.

The purpose of this study was to evaluate the relationship between the rheology of CF sputum and the length and density of DNA and F-actin polymers in the same sputum samples using LSCM. We also evaluated the polymeric relationship between the DNA and F-actin polymer networks. We then treated these sputum specimens with rhDNase (Pulmozyme, Genentech, South San Francisco, CA) to assess the relationship between changes in sputum filament morphology and sputum rheology after enzymatic digestion of sputum DNA.

## II. MATERIALS AND METHODS

### A. Sputum Collection and Sample Preparation

Sputum was collected from 9 patients (4 females and 5 males; ages 14–25 years) with CF lung disease. None of the patients was receiving mucoactive medications. The samples were collected in mid morning by spontaneous expectoration over 30 minutes during pulmonary function testing. Sputum was visually separated from saliva and specimens were stored in O-ring containers at −70°C until analyzed.

Sputum samples, 30 μL in volume, were placed onto glass microscope slides partitioned into two lanes using strips of laboratory time tape and double-stick tape to a total thickness of approximately 70 μm. After this, 7.5 μL of buffered saline with (in lane 1) and without (in lane 2) 30 mcg/mL of Pulmozyme was layered onto the sample with 1 μL each of the fluorochromes YOYO-1 and rhodamine-phalloidin added to both lanes for dual visualization of DNA and F-actin. A glass cover slip was placed over the sample and onto the double-stick tape. This was incubated for 30 minutes at room temperature. The slide was then inverted onto the stage of the LSCM.

### B. Image Analysis

The samples were scanned using the argon/krypton laser setting of the Zeiss 410 LSCM (Carl Zeiss LSCM Version 3.92 Beta, Oberkochen, Germany). Fluorescence images were transformed by pseudo color so DNA appeared green, F-actin appeared red, and merged images of colocalized fields of DNA and F-actin (YOYO-1 and rhodamine fluorescence within 30 nm of each other) appeared yellow. Images were analyzed for the intensity of fluorescence, distribution of the DNA and F-actin filaments, filament number, and filament length using the KS400 image analysis software.

For the purpose of geometric analysis of filament length and number, the red and green fluorescence images were converted into a gray scale with pixel intensity scaled from 0 to 255 for Marr analysis [9]. This is an edge analysis by region defining the contours and delineating boundaries of shapes at discrete intensity changes. On the transformed images, filament length and size distribution were calculated. The shortest filaments were less than 30 μm in length and the longest were 120–210 μm with the maximal calculated length limited by the diagonal of the viewing field. Statistical analysis was completed using the LSCM computer software and the StatView 4.5 statistical package (Abacus Concepts, Palo Alto, CA).

### C. Physical Properties of Secretions

#### Sputum Rheology

In the magnetic microrheometer a 100 μm steel ball was positioned in a 3–5 μL sample of sputum, and the motion of this sphere under the influence of an electromagnet was used to determine the sputum viscoelasticity. The magnified image of the steel ball was projected onto a pair of photocells whose output was amplified and transmitted to an oscilloscope. By plotting the displacement of the ball against the magnetic driving force the dynamic loss modulus (viscosity, $G''$) and storage modulus (elasticity, $G'$) were measured at 1 and 100 rad/sec [2].

#### Sputum Cohesiveness (Spinnability)

Cohesiveness is the internal force of attraction of like molecules within a substance, and for mucus polymers, is measured as the length of the thread formed under the influence of low amplitude deformation [10]. A 30-μL mucus sample is stretched at a velocity of 10 mm/sec to form a thread. An electric signal conducted through the mucus is interrupted at the point where the thread is broken and the maximum length of this thread is measured and reported in millimeters.

## III. RESULTS

### A. Images of DNA and F-Actin in CF Sputum

All sputum samples demonstrated staining for both DNA and F-actin although the amount of staining was quite variable between different specimens. Figure 1 shows the appearance of these filaments after gray-scale

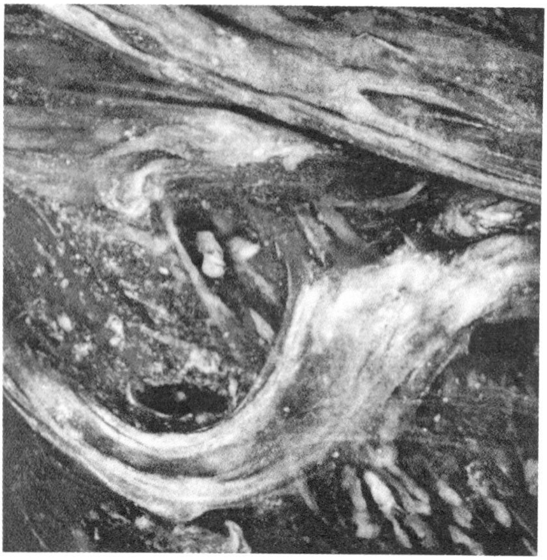

**Figure 1** Gray scale transformation of an LSCM image of CF sputum abundant in long filaments of DNA stained with YOYO-1.

transformation in a specimen containing abundant filaments. In sputum containing less DNA, there was also less F-actin, and staining was punctate and scattered throughout the field. There was a high correlation ($r = 0.93$, $p < 0.0001$) between the rhodamine and YOYO-1 intensity distribution data reflecting colocalization of DNA and actin in all sputum samples, regardless of the amount and intensity of DNA visualized. Colocalization analysis of staining proximity demonstrated that nearly all F-actin was found within 30 nm of DNA filaments confirming that these two species copolymerized in CF sputum.

## B. Relationship Between Sputum Ultrastructure and Rheology

Dynamic sputum loss modulus, or viscosity $G''$ at 1 rad/sec, directly correlated with the percent of long filaments ($r = 0.81$, $p = 0.01$) as shown in Figure 2. There was a weaker relationship between the percentage of long filaments and storage modulus, elasticity $G'$ at 1 rad/sec ($r = 0.49$, $p = 0.18$).

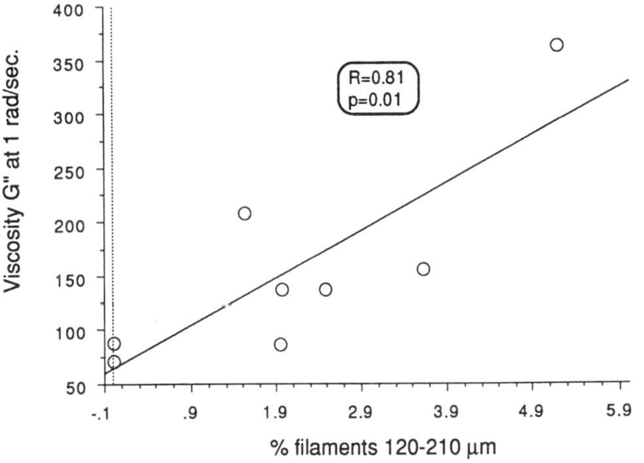


**Figure 2** Sputum viscosity G″ correlated with percentage of long filaments (120–210 μm).

There was an inverse correlation between sputum cohesiveness and the percent of short filaments ($r = -0.69$, $p = 0.06$) as shown in Figure 3. This indicated that the greater the number of small filaments the lower the sputum cohesiveness.

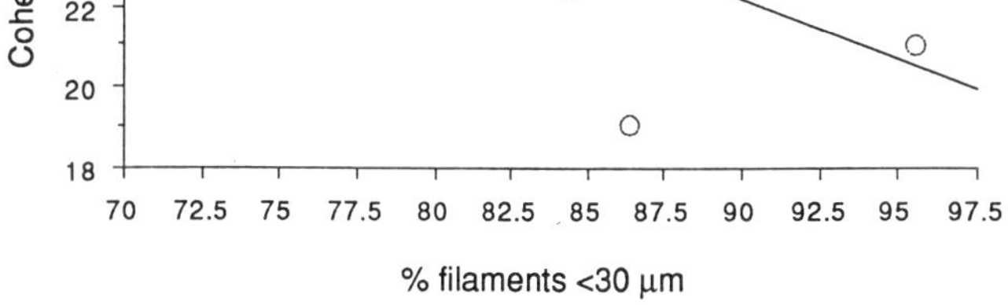


**Figure 3** Sputum cohesiveness inversely correlated with the percentage of short filaments (<30 μm in length).

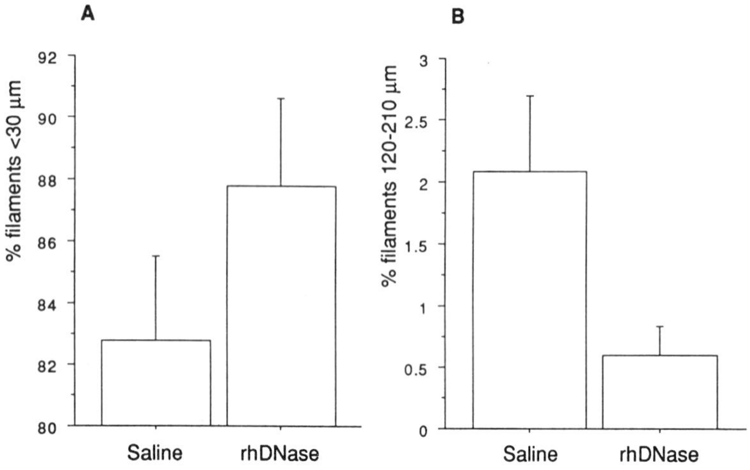


**Figure 4** Filament length shift after Pulmozyme treatment. (A) Pulmozyme increased the percentage of short filaments (<30 μm), and (B) Pulmozyme decreased the percentage long filaments (120–210 μm).

### C. Effects of Pulmozyme on Sputum Ultrastructure

The LSCM images of sputum treated in vitro with buffered saline solution showed distinct green fibers of DNA as well as yellow colocalized fields of DNA and actin. Sputum treated in vitro with Pulmozyme showed only short filaments and punctate fluorescence for DNA and F-actin. Quantitative analysis of filament length distribution performed on the converted images showed that there was a shift in the percentage of filaments in each length subgroup after Pulmozyme treatment. Pulmozyme increased the number of short filaments (<30 μm) while decreasing the number of longest filaments (120–210 μm) as shown in Figure 4. In this study, Pulmozyme significantly increased the ratio of smaller to longer filaments ($n = 7$, $p = 0.004$).

## IV. DISCUSSION

The LSCM technique allows dynamic visualization of the ultrastructure of tissue in three dimensions by sequential acquisition of two-dimensional (planar) data through successive depths of tissue with computer reconstruction of these planes into a three dimensional image. This is similar to tomographic imaging. These digitized images are ideal for spatial and temporal image analysis of regions of interest. This work is the first published report using this technique to quantitatively evaluate the structure of neutrophil-derived actin and DNA filaments in CF sputum.

Polymers of DNA and F-actin have been visualized in CF sputum using fluorescent microscopy and it has been postulated that actin contributes to DNA filament formation [11]. These studies confirm this hypothesis by clearly documenting consistent colocalization of these filaments. Although DNase treatment of sputum disrupted both DNA and F-actin polymers, this alone does not necessarily indicate that these networks colocalize. The DNase enzyme not only shortens DNA filaments, it is also an actin-binding peptide and this would shift the actin polymerization–depolymerization equilibrium toward depolymerization [12].

The rheology of airway secretions depends upon a three-dimensional network of gel-forming polymers [4,5]. Thus we would expect to find a relationship between mucus viscoelasticity and filament length and density. Although this relationship has been suggested by others both for mucin polymers [3] and for sputum DNA [5] this study actually demonstrates that the length and density of DNA filaments contribute to sputum dynamic viscosity.

It has been shown that DNase reduces the length of DNA macromolecules both in vitro [8] and in vivo [13]. In these studies, we have documented that changes in sputum rheology paralleled changes in sputum DNA filament morphology after in vitro treatment with DNase. Pulmozyme decreased the number of the longest filaments while increasing the number of shorter ones. This decrease in the percentage of longest filaments was paralleled by a decrease in viscosity and to a lesser degree in elasticity.

Cohesiveness is the internal force of attraction between similar molecules in a fluid or gel. Because polymerization is one of the strongest attractive forces between like molecules, depolymerization of macromolecules that contribute to cohesive forces should have a profound effect on sputum cohesiveness. After CF sputum was treated with Pulmozyme we indeed saw an increase in the number of smaller DNA filaments, which is a direct measurement of DNA depolymerization. This depolymerization was strongly correlated with a reduction in sputum cohesiveness. We have previously suggested that the length of DNA macromolecules in CF sputum was more closely related to sputum cohesiveness (spinnability) than to viscoelasticity [14]. This is of significant physiologic interest because sputum cohesiveness has a far more powerful influence on the in vitro cough transportability of secretions than does viscoelasticity [15].

The LSCM method appears to be a highly precise way of evaluating the morphology of the macromolecular polymers that contribute to the physical properties of airway secretions. This tool will enable us to better understand the pathogenesis of airway diseases associated with mucus hypersecretion and to better evaluate potential therapies for these diseases.

## ACKNOWLEDGMENTS

This work was supported by grants from the Cystic Fibrosis Foundation (Rubin 96PO) and Alpha-1 Biomedicals, Inc. Dr. Tomkiewicz was an American Lung Association-Gosselin Pediatric Asthma Research Associate.

## REFERENCES

1. Denton R, Forsman W, Hwang SH, Litt M, Miller CE. Viscoelasticity of mucus: its role in ciliary transport of pulmonary secretions. Am Rev Respir Dis 1968; 98:380–391.
2. King M, Rubin BK. Mucus rheology, relationship with transport. In: Takishima T, Ed. Airway secretion: Physiological Basis for the Control of Mucus Hypersecretion. Marcel, New York: Dekker, Inc., 1994:283–314.
3. Lutz RJ, Litt M, Charkin LW. Physical-chemical factors in mucus rheology. In: Gabelnik HL, Litt M, Eds. Rheology of Biological Systems. Springfield, IL: Charles C. Thomas 1973:158–194.
4. Ferry JD. Molecular theory for undiluted polymers and concentrated solutions: networks and entanglements. In: Viscoelastic Properties of Polymers, 2nd Ed. New York: Wiley, 1970:268–270.
5. Matthews LW, Spector S, Lemm J, Potter JL. Studies on pulmonary secretions. I. The overall composition of pulmonary secretions from patients with cystic fibrosis, bronchiectasis and laryngectomy. Am Rev Respir Dis 1963; 88:199–204.
6. Lethem MI, James SL, Marriott C, Burke JF. The origin of DNA associated with mucous glycoproteins in cystic fibrosis sputum. Eur Respir J 1990; 3: 19–23.
7. Vasconcellos CA, Allen PG, Wohl ME, Drazen JM, Janmey PA, Stossel TP. Reduction in viscosity of cystic fibrosis sputum in vitro by gelsolin. Science 1994; 263:969–971.
8. Shak S, Capon DJ, Hellmiss R, Marsters SA, Baker CL. Recombinant human DNase I reduces the viscosity of cystic fibrosis sputum. Proc Natl Acad Sci USA 1990; 87:9188–9192.
9. Kontron Elektronik Imaging System KS 400, Rel. 2.0. August, 1995.
10. Puchelle E, Zahm JM, Duvivier C. Spinnability of bronchial mucus: relation-

ship with viscoelasticity and mucus transport properties. Biorheology 1983; 20:239–249.
11. Sheils CA, Käs J, Travassos W, Allen PG, Janmey PA, Wohl ME, Stossel TP. Actin filaments mediate DNA fiber formation in chronic inflammatory airway disease. Am J Pathol 1996; 148:919–927.
12. Hitchock SE, Carlsson L, Lindberg U. Depolymerization of F-actin by deoxyribonuclease I. Cell 1976; 7:531–542.
13. Brandt T, Breitenstein S, von der Hardt H, Tummler B. DNA concentration and length in sputum of patients with cystic fibrosis during inhalation with rccombinant human DNase. Thorax 1995; 50:880–882.
14. Rubin BK, Ramirez O, Shak S, Toy KJ, Sinicropi D. DNA concentration of CF sputum correlates with hydration and spinnability but not with viscoelasticity. Pediatr Pulmon 1994; 10S:241.
15. Albers GM, Tomkiewicz RP, May MK, Ramirez OE, Rubin BK. Ring distraction technique for the measurement of the surface tension of sputum. J Appl Physiol 1996; 81:2690–2695.

# 35

# Is Bacterial Colonization of the Airways Relevant?

**Adam Wanner, Matthias Salathé, Kenji Kobayashi, William M. Abraham, Zsuzsanna Seybold, and Jolanta T. Jackowski†**
*University of Miami School of Medicine, Miami, Florida*

## I. INTRODUCTION

Bacteria can gain access to the bronchial tree by inhalation, aspiration from the upper airways, or translocation from the upper airways by endotracheal tubes, suction catheters, and endoscopes. Under certain circumstances, the bacteria can colonize the bronchi. Traditionally, bacterial colonization of the bronchi per se has been considered irrelevant except as a prerequisite for bacterial adhesion to and infection of the bronchial wall. However, Wilson et al. [1,2] have shown that soluble bacterial products, especially products of *Pseudomonas aeruginosa* can impair the ciliary function of the airway epithelium without airway infection. By this mechanism, *Pseudomonas aeruginosa* could counteract its own clearance from the airway.

We therefore conducted a series of studies designed to elucidate some of the mechanisms underlying the cilioinhibition caused by soluble products of *Pseudomonas aeruginosa*. All experiments were carried out using sheep. We found that the cell-free supernatant of a *Pseudomonas aeruginosa* culture and the *Pseudomonas aeruginosa* products pyocyanin and 1-hydroxyphenazine cause a concentration-dependent decrease in tracheal ciliary activity in vitro. We also found that this effect is blocked by catalase and superoxide dismutase, which implicates the generation of

† Deceased

reactive oxygen species. The bacterial product-induced cilioinhibition required the presence of luminal neutrophils. One bacterial product potentiated oxygen-radical generation by phorbol-12-myristate-13-acetate (PMA) activated neutrophils, suggesting that neutrophils have a mediating role in this process. We then examined the effects of exogenous $H_2O_2$ on ciliary beat frequency in cultured tracheal explants. At concentrations between $10^{-8}$ and $10^{-6}$ *M*, $H_2O_2$ decreased ciliary beat frequency reversibly. At higher concentrations of $H_2O_2$, incomplete recovery of ciliary beat frequency ($10^{-5}$ *M*) or irreversible ciliostasis ($10^{-4}$ *M*) were present. A protein kinase C (PKC) inhibitor abolished the cilioinhibition at low, but not at high, $H_2O_2$ concentrations. A dose-dependent cilioinhibition that was caused by PMA was blocked by the PKC inhibitor. Activated PKC phosphorylated a ciliary membrane-associated target, an effect temporally related to PKC-mediated cilioinhibition.

These observations suggested that at nonlethal concentrations, $H_2O_2$ inhibited ciliary activity reversibly through the activation of second messengers including PKC. We confirmed these in vitro observations in vivo by showing catalase-blockable impairment of tracheal mucociliary clearance by inhaled *Pseudomonas aeruginosa* supernatant. Finally, we showed that impairment of mucociliary clearance prolongs the persistence of viable bacteria in the airways. We conclude that *Pseudomonas aeruginosa* can impair airway mucosal function by interacting with luminal phagocytes, and that the mucociliary impairment favors bacterial colonization of the airway. The studies supporting this sequence of events are summarized below.

## II. EFFECT OF *PSEUDOMONAS AERUGINOSA* PRODUCTS ON CILIARY FUNCTION

This investigation was carried out to evaluate the effects of soluble products derived from *Pseudomonas aeruginosa* on the function of airway cilia and to assess the role of phagocytes and oxygen radicals in the responses [3]. Ciliary beat frequency was measured in a perfusion chamber with a microscopic technique using tracheal epithelial cells obtained from normal sheep by brush biopsy (70% epithelial cells, 18% macrophages, 11% neutrophils). Baseline ciliary beat frequency ranged between 678 and 1126 $min^{-1}$. After 20 minutes of perfusion with a cell-free supernatant of *Pseudomonas aeruginosa* culture (mucoid strain), a concentration-dependent depression of ciliary beat frequency was observed with a 58% inhibition at a 1:1 dilution. The *Pseudomonas aeruginosa* derived products pyocyanin and 1-hydroxyphenazine also decreased ciliary beat frequency in a concentration-related fashion ($10^{-7}$–$10^{-4}$ *M*). The cilioinhibitory effect of the supernatant and bacterial products were markedly attenuated after near complete removal of neutrophils from the brush preparation by centrifugation. Glucose/glucose oxidase also caused a rapid, concentration-dependent cilioinhibition or ciliostasis. Catalase blocked or attenuated the ciliary effect of the supernatant, bacterial products, and glucose/glucose oxidase. Finally, we found that pyocyanin potentiated oxygen-radical generation by PMA activated neutrophils, supporting the critical role of neutrophils in this process. Thus, bacterial products released from *Pseudomonas aeruginosa* impaired ciliary activity by a pathway that involved neutrophils and was mediated by reactive oxygen species.

## III. MECHANISM OF HYDROGEN PEROXIDE INDUCED CILIOINHIBITION

In an attempt to characterize some of the cellular mechanisms of $H_2O_2$ induced cilioinhibition, we measured ciliary beat frequency in cultured tracheal explants [4]. Addition of $H_2O_2$ ($10^{-8}$–$10^{-4}$ *M*) produced a concentration-dependent decrease in ciliary beat frequency between 11% and 100%; at each concentration, the maximal effect was reached by 20–25 min. Between $10^{-8}$ and $10^{-6}$ *M* $H_2O_2$, the decrease in ciliary beat frequency was reversible, LDH release was not significantly increased, and major morphologic lesions were not seen. At higher concentrations of $H_2O_2$, incomplete recovery of ciliary beat frequency ($10^{-5}$ *M*) or irreversible ciliostasis ($10^{-4}$ *M*) developed, and significant increases in LDH and morphologic lesions were present. Catalase and H-7, a protein kinase inhibitor, abolished the cilioinhibition produced by $H_2O_2$ at $10^{-6}$ *M* and lower concentrations but not at $10^{-5}$ *M* and higher concentrations. The PKC activator PMA caused a concentration-dependent ($10^{-11}$–$10^{-5}$ *M*), reversible decrease in ciliary beat frequency; this effect was abolished by H-7. These observations suggest that at nonlethal concentrations, $H_2O_2$ inhibits ciliary activity of airway epithelial cells reversibly through the activation of second messengers including PKC.

## IV. PKC-DEPENDENT PHOSPHORYLATION OF A CILIARY MEMBRANE PROTEIN

The purpose of this study was to examine whether PKC phosphorylates ciliary proteins and whether this phosphorylation is temporally correlated with cilioinhibition [5]. Activation by PKC decreased ciliary beat frequency on tracheal epithelium, an effect fully blockable by pretreatment with H-7. On cilia removed from these epithelial surfaces and incubated in solutions containing stimulators of PKC along with [γ-$^{32}$P]ATP or [γ-$^{35}$S]ATP, a single protein target of ciliary PKC was identified. The protein turned out to be a 37-kDa polypeptide as estimated by SDS-polyacrylamide gel electrophoresis. The PKC dependency of phosphorylation was proven by showing that Calphostin C, a specific PKC inhibitor, blocked label incorporation into the protein completely and by demonstrating that the protein was phosphorylated by purified PKC. The protein was recovered in a Triton X-100-extractable fraction of the ciliary preparation and was present in a pellet representing reconstituted membranes. This suggested that the PKC target is membrane or matrix associated. The protein did not precipitate with annexin I or II antibodies. We interpreted these results as showing that activated PKC phosphorylates a membrane- or matrix-associated ciliary target, an effect temporally related to PKC mediated cilioinhibition.

## V. MUCOCILIARY IMPAIRMENT

In this study, we wished to determine if tracheal mucus velocity is impaired by *Pseudomonas aeruginosa* supernatant and if oxygen radicals have a mediating role [6]. In conscious sheep, tracheal mucus velocity was determined before and serially following aerosol challenge with a cell free supernatant of a *Pseudomonas aeruginosa* culture or unconditioned culture medium (control). Tracheal mucus velocity decreased by 29, 35, and 25% at 0.5, 3, and 24 hours after challenge, respectively, and returned to baseline 1 week later. Control medium had no effect on tracheal mucus velocity. Aerosolized catalase blunted the effect of *Pseudomonas aeruginosa* supernatant on tracheal mucus velocity. To determine if the impairment of tracheal mucus velocity involved ciliary inhibition, tissues were mounted in a chamber and ciliary beat frequency and surface liquid velocity were measured with a microscopic method. *Pseudomonas aeruginosa* supernatant decreased both ciliary beat frequency and surface liquid velocity in a dose-dependent fashion, with a correlation between the two parameters; these effects were blocked by catalase. These results indicate that short-term exposure to soluble *Pseudomonas aeruginosa* products decreases airway mucociliary clearance in vivo in part by a pathway involving oxygen radicals, that the impairment persists for at least 24 hours but less than one week, and that cilioinhibition has a role in the *Pseudomonas aeruginosa* induced decrease in mucociliary clearance.

## VI. EFFECT OF MUCOCILIARY IMPAIRMENT ON BACTERIAL CLEARANCE IN VIVO

In this experiment, normal control sheep and sheep in which tracheal mucus velocity was decreased by approximately 50% with the intravenous administration of atropine sulfate, were given an aerosol challenge of a *Pasteurella haemolytica* suspension [7]. *Pasteurella haemolytica* is a natural respiratory pathogen in sheep. The persistence of viable bacteria was assessed by serially obtaining samples of tracheal secretions with a calibrated microbiology brush, and performing quantitative cultures (colony counts). The colony counts decreased exponentially in the control and atropine-treated sheep. However, there was a marked difference in the persistence of viable bacteria in the trachea between control and atropine-treated sheep: in the former, no viable bacteria were recoverable beyond 6 hours after bacterial challenge; while in the latter, viable bacteria persisted between 6 and 48 hours. This experiment does not establish a definitive causal link between mucociliary impairment and persistence of lower airway bacteria because atropine could effect bacterial clearance by mechanisms other than a decrease in mechanical transport by the mucociliary apparatus. However, the impairment of bacterial clearance by mucociliary impairment is certainly a possible explanation.

## VII. CONCLUSIONS

*Pseudomonas aeruginosa* is an important colonizer of the lower airways of patients with cystic fibrosis and bronchiectasis and of critically ill patients. Impairment of mucociliary clearance by soluble products released by *Pseudomonas aeruginosa* appears to be one of several possible strategies developed by this organism to promote its persistence in the lower airway. The bacterial products pyocyanin and its metabolite 1-hydroxyphenazine have a critical role in this process, and at least in sheep, oxygen radicals released from pyocyanin stimulated luminal neutrophils seem to mediate the mucociliary impairment. The in vitro and in vivo observations that were made in sheep may be relevant to human disease as the biological effects of pyocyanin and 1-hydroxyphenazine occurred at concentrations that have been measured in expectorated sputum of patients with cystic fibrosis [8].

## REFERENCES


1. Wilson R, Pitt T, Taylor G, et al. Pyocyanin and 1-hydroxyphenazine produced by *Pseudomonas aeruginosa* inhibit the beating of human respiratory cilia in vitro. J Clin Invest 1987; 79:221–229.
2. Wilson R, Roberts D, Cole P. Effect of bacterial products on human ciliary function in vitro. Thorax 1985; 40:125–131.
3. Jackowski JT, Szepfalusi Z, Wanner DA, et al. Effects of *Pseudomonas aeruginosa* derived bacterial products on tracheal ciliary function: role of oxygen radicals. Am J Physiol 1991; 261:L61–L67.
4. Kobayashi K, Salathé M, Pratt M et al. Mechanism of hydrogen peroxide induced inhibition of sheep airway cilia. Am J Respir Cell Mol Biol 1992; 6: 667–673.
5. Salathé M, Pratt MM, Wanner A. Protein kinase C-dependent phosphorylation of a ciliary membrane protein and inhibition of ciliary beating. J Cell Sci 1993; 106:1211–1220.
6. Seybold ZV, Abraham WM, Gazeroglu H, Wanner A. Impairment of mucociliary transport by *Pseudomonas aeruginosa* products: role of oxygen radicals. Am Rev Respir Dis 1992; 146:1173–1176.
7. Whiteside M, Lauredo I, Chapman GA, Ratzan KR, Abraham WM, Wanner A. Effect of atropine on tracheal mucociliary clearance and bacterial counts. Bull Europ Physiopath Resp 1984; 20:347–351.
8. Wilson R, Sykes DA, Watson D, Rutman A, Taylor GW, Cole PJ. Measurement of *Pseudomonas aeruginosa* phenazine pigments in sputum and assessment of their contribution to sputum sol toxicity for respiratory epithelium. Infect Immun 1988; 56:2515–2517.

# 36
# Mucin Synthesis Is Upregulated at the Transcriptional Level by *P. aeruginosa* LPS

**Carol B. Basbaum and Jian-Dong Li**
*University of California, San Francisco, San Francisco, California*

## I. INTRODUCTION

Although abnormalities can be detected in mucin of the cystic fibrosis lung prior to infection, it is regularly observed that airway mucus obstruction becomes particularly pronounced after the onset of bacterial infection. The sequelae of infection and obstruction bring about mortality in over 90% of CF patients (the remainder succumbing to failure of other organ systems such as the pancreas). The molecular events following bacterial infection of the cystic fibrosis lung are poorly understood and are currently refractory to therapeutic intervention.

## II. MATERIALS AND METHODS

Based on histopathological observations of overabundant (PAS-staining) mucin in cystic fibrosis airways, we hypothesized that stimuli associated with bacterial infection actively upregulate mucin synthesis. Results of in situ hybridization analysis of cystic fibrosis versus normal airways confirmed that mRNA encoding the mucin MUC 2 and MUC 5 gene products was indeed more abundant in cystic fibrosis than in normal tissues. To begin to determine whether or not bacterial products are responsible for this upregulation, we exposed normal human bronchial explants to *P.*

*aeruginosa* supernatant and monitored mucin mRNA levels. After 6 hours of exposure, we observed upregulation of MUC 2 and MUC 5 steady-state mRNA with respect to the buffer-incubated controls. Because the explants contained multiple cell types including hematopoietic inflammatory cells, however, it was unclear from the results of the explant experiments whether the *P. aeruginosa* was affecting mucin synthesis directly or through activation of mediator release from an intermediate cell type.

We therefore repeated the experiment but this time put *P. aeruginosa* supernatant in contact with isolated epithelial cells growing in culture. We observed mucin mRNA upregulation in the epithelial cells under these conditions as well, indicating that bacterial exoproducts themselves trigger mucin upregulation in epithelial cells (i.e., independent of mediators released from inflammatory cells).

Although our analysis showed that mucin synthesis was upregulated at the RNA level, we could not conclude from the initial experiments whether changes in steady-state levels of mRNA occurred as a result of changes in DNA transcription rate or RNA degradation rate. To look specifically at transcription rate, we placed a luciferase reporter gene under the control of variable length fragments of the MUC 2 promoter and compared luciferase activity in epithelial cells transfected with these constructs in the presence and absence of *P. aeruginosa* supernatant. We observed increases in activity when *P. aeruginosa* supernatant was present indicating that *P. aeruginosa* control of mucin synthesis is at least partly at the transcriptional level.

The availability of variable length promoter-luciferase constructs also permitted us to map the site of enhancers induced by *P. aeruginosa*. A strong enhancer is present at −1458/−1436 and a relatively weak one is present at −91/−73. To obtain information concerning nuclear proteins that activate these enhancers, we performed gel shift assays using DNA fragments −91/−73 and −1458/−1436 as probes. Combined with mutagenesis and supershift assays using protein-specific antibodies, we determined that the proximal enhancer is a CCAAC box that binds SP1 and/or as yet unidentified proteins. The distal enhancer is a combined NFκB and C/EBP site that binds NFκB but not C/EBP upon *P. aeruginosa* stimulation.

We also approached the question of how the *P. aeruginosa* stimulus is transmitted through the cell to activate these nuclear transcription factors. We tested the ability of a variety of chemical inhibitors to block the *P. aeruginosa*-induced transcriptional upregulation and found that both genistein, a tyrosine kinase inhibitor, and the antioxidants dimethylthiourea (DMTU) and N-acetylcysteine (NAc) inhibited the response. Additional analysis by use of both chemical inhibitors and cotransfection of luciferase-transfected cells with dominant negative mutants of components of the mitogen-activated protein kinase (MAPK) signaling cascade showed that the pathway triggered by *P. aeruginosa* activates src, Ras, Raf, MEK, and Erk 1/2. The relative roles of reactive oxygen species and MAP kinases are not as yet clear but based on several studies demonstrating the activation of MAPKs by reactive oxygen species, we hypothesize that such activation also exists in the mucin upregulation pathway.

## III. DISCUSSION

With respect to the most proximal events involved in *P. aeruginosa*-mucin activation, we have attempted to identify the active component of *P. aeruginosa* responsible for triggering mucin upregulation. Size fractionation of the bacterial supernatant revealed activity in all fractions between 30 and 200 kDa. The activity is resistant to protease and DNase degradation. A direct test of lipopolysaccharide (LPS) a major constituent of the outer membrane of gram-negative bacteria, indicated that it contained potent stimulatory activity. We further determined that it is the lipid-A portion of LPS that is required for activity. The LPS from a variety of bacteria were all effective in stimulating MUC 2 and MUC 5 upregulation. Although studies of monocytes/macrophages have shown that LPS binds lipopolysaccharide binding protein (LBP) in serum that in turn ligates the hematopoietic cell surface receptor CD14, the analogous receptor in epithelial cells is unknown. Our ongoing studies are aimed at determining the potential involvement of LBP and CD14 in mucin upregulation by LPS. Also under investigation is our finding that gram-positive bacteria (i.e., those lacking LPS) stimulate mucin synthesis at the RNA level. As LPS is not present in these bacteria, it raises the question of which components associated with these bacteria are detected by host epithelial cells to trigger the mucin response.

## IV. CONCLUSION

In conclusion, we have shown that LPS from all bacteria tested as well as supernatant from gram-positive bacteria stimulate synthesis of MUC 2 and MUC 5 by direct effects at the epithelial cell surface. This requires both oxygen radical- and MAPK-dependent signaling pathways that culminate in the activation of NFκB as well as currently unidentified transcription factors.

## ACKNOWLEDGMENTS

We thank our colleagues Marianne Gallup, Greta Glugoski, Susumu Miyata, Ph.D, Jim Gum, Ph.D., Young Kim, M.D., and Jay Nadel, M.D. Supported by NIH PHS grants HL RO1 43762 and PO 1 24136.

# 37

# Tissue Distribution of Secretory Mucins and Alterations of Mucin Gene Expression in Chronic *H. pylori* Infection of the Stomach

**Samuel B. Ho**
*University of Minnesota, and Veterans Affairs Medical Center, Minneapolis, Minnesota*

## I. INTRODUCTION

Mucins are the largest and most heavily glycosylated of the O-linked glycoproteins. They can be categorized as membrane-bound or secreted mucins. The secreted mucins may be subdivided into polymerizing and nonpolymerizing mucins. The polymerizing mucins form linkages by end-to-end disulfide binding, resulting in apparent molecular weights of $10^7$ Da and high levels of viscoelasticity. Mucin polymers are the major structural proteins of mucous gels, which coat gastrointestinal, respiratory, and reproductive epithelia [1–4]. This review will focus on recent developments in the molecular structure of mucin genes, the organ distribution of various secretory mucins, and our preliminary results concerning the relationship of mucin gene expression and chronic gastritis induced by *Helicobacter pylori* infection of the stomach.

## II. MOLECULAR BIOLOGY OF MUCIN GLYCOPROTEINS

Biochemical analysis and characterization of mucin proteins has been limited because of their large size and abundant glycosylation. Recently, much has been learned about the structure and possible functions of mucins from molecular-sequence data derived from the cloning of several human and animal mucin cDNAs. These data demonstrate that mucins share certain structural features. For example, all the mucin proteins sequenced to date contain a central region with a high proportion of threonine and/or serine glycosylation sites. These threonine- and serine-rich sequences are repeated in tandem along the length of the molecule and thus are called tandem repeats. However, the specific amino acid sequences, the lengths of these repeats, and the nonrepetitive flanking domains of these molecules are quite variable (Table 1).

Two families of mucins have been characterized. The first is a membrane-bound glycoprotein that is synthesized by most, if not all, epithelial tissues. Characterization of this mucin resulted from studies of the high molecular weight glycoproteins (previously labeled as PUM, PEM (polymorphic epithelial mucin), MAM-6, PAS-O, EMA, NPG, and DF-3) that occur in human breast milk and are highly expressed in breast and other adenocarcinomas [5–7]. Data from full-length cDNA sequencing indicate that two-thirds of this protein consists of 20 amino acid tandem repeats, which contain 25% serine or threonine glycosylation sites, followed by a transmembrane sequence and a 69 amino acid cytoplasmic tail. This protein is typically localized to the apical membrane of epithelial cells, with the tandem-repeat domain extending approximately 150 nm above the cell surface. This structure is thought to physically shield epithelial surface antigens or receptors from luminal agents.

The second family of mucins consists of the secreted or soluble mucins. Biochemical studies have demonstrated considerable heterogeneity in secretory mucins purified from both the respiratory tract and various regions of the intestine [1,2,8]. Electron microscopic studies by Carlstedt and Sheehan, working with human cervical mucin, have demonstrated that secreted mucins may form long polymers, with molecules linked end-to-end [9]. Mucin monomers are thought to be linked together by terminal disulfide bonds via cysteine residues in the terminal nonrepetitive domains, since reduction of disulfide bounds results in the loss of viscosity in mucous gels [2].

To date, eight human secretory mucin cDNAs have been cloned and partially sequenced (Table 1). Two unique mucins were isolated from a human intestinal cDNA library. The first was labeled MUC2 and contained an extended array of tandem repeats of 23 amino acids that were rich in threonine and proline [10]. The second (MUC3) encoded 17 amino acid tandem repeats rich in threonine and serine [11]. The MUC2 and MUC3 mucins can possibly be considered to correspond to the acidic and neutral mucin fractions, respectively, previously isolated from human

**Table 1** Human Mucin Genes

| Designation | Source | Type[a] | # a.a. in t.r. | vWF-like domains | Chromosomal location | Tandem repeat a.a. sequence | Reference |
|---|---|---|---|---|---|---|---|
| MUC1 | Mammary<br>Pancreatic | mem. | 20 | no | 1q21q24 | GSTAPPAHGVTSAPDTRPAP | [6,49,50] |
| MUC2 | Intestinal<br>Tracheal | sec. | 23 | yes | 11p15 | PTTTPITTTTTVTPTPTPTGTQTPYPY-<br>PITTTTTVTPTPTPT(G/S)TQT | [10,31,51] |
| MUC3 | Intestinal | sec. | 17 | no | 7 | HSTPSFTSSITTTETTS | [11] |
| MUC4 | Tracheal | sec. | 16 | ? | 3q29 | (T)SS(A)ST(GHA)T(P)L(P)VT(D)[b]<br>(other reading frame possible?) | [12] |
| MUC5 (5AC) | Tracheal<br>Gastric | sec. | 8 | yes | 11p15 | TTSTTSAP | [17,52,53] |
| MUC5B | Tracheal | sec. | 29 | yes | 11p15 | SSTPGTAHTHTEQTTTATTPTAT-<br>GTTATP[b] | [54] |
| MUC6 | Gastric | sec. | 169 | yes | 11p15 | SPFSSTGPMTATSFQTTTTYPTPS<br>HPQTTLPTHVPPFSTSLVTPSTGT<br>VITPTHAQMATSASIHSTPTGTIP<br>PPTTLKATGSTHTAPPMTPTTSG<br>TSQAHSSFSTAKTSTSLHSHTSST<br>HHPEVTPTSTTTITPNPTSTGTSTP<br>VAHTTSATSSRLPTPFTTHSPPTGS | [18] |
| MUC7 | Salivary | sec. | 23 | no | 4 | TTAAPPTPSATTPAPPSSSAPPG | [22] |
| MUC8 | Tracheal | sec. | 13–41 | ? | 12 | TSCPRPLQEGTRV<br>TSCPRPLQEGTPGSRAAHALSRR<br>GHRVHELPTSSPGGDTGF | [16] |

[a] mem. = membrane-bound, sec. = secreted
[b] 0 = imperfectly conserved

intestine using biochemical techniques. Investigators have cloned and partially sequenced several human tracheo–bronchial mucins, tentatively labeled MUC4, MUC5 (also labeled MUC5AC), MUC5B, and MUC8 [12–16]. Two mucin cDNAs have been cloned from human gastric cDNA libraries, MUC5AC and MUC6 [17]. The MUC6 gene contains the longest tandem repeat sequence yet published, with 169 amino acids [18]. Chromosomal clustering of genes may suggest functional similarity, and further sequence analysis of these separate genes will be of interest.

Of the secretory mucins, the most complete cDNA sequence data is available for MUC2 and MUC6. Gum and co-workers have shown that the carboxyl terminal domain of MUC2 contains 984 residues and can be divided into mucin-like (139 residues) and cysteine-rich (845 residues) subdomains [19]. The cysteine-rich subdomain exhibits varying degrees of sequence similarity to nontandem repeat domains of mucins from mice, rats, pigs, cows, and frogs. The amino terminal segment upstream of the tandem repeats contains a repetitive mucin-like subdomain and a second cysteine-rich region. Both cysteine-rich regions have sequence similarity with the von Willebrand factor (vWF) D-domains and most likely functions to form disulfide-linked polymers.

Toribara et al. have shown that the primary amino acid structure of MUC6 deduced from the cDNA sequence consists of approximately 5100 amino acids (a.a.) organized into three regions [20]. These include an amino terminal nontandem repeat region of 1400 a.a. having segments with considerable homology to D-domains 1, 2, and 3 of von Willebrand factor; a tandem repeat region containing 20–24 repeat units; and a 361 a.a. carboxyl terminal nontandem repeat region with a 91 a.a. cysteine-rich segment with approximately 25% amino acid homology to carboxyl terminal regions of MUC2, MUC5AC, and von Willebrand factor. Unlike MUC2, MUC5B, and MUC5AC mucins, the MUC6 carboxyl terminal region did not have a vWF D4 domain. However, the conserved spacing of the cysteine residues suggests that this region serves for monomer polymerization.

In contrast with the MUC2, 5AC, 5B, and 6 mucins, the MUC3 mucin does not appear to contain vWF-like cysteine domains. This suggests that this mucin does not undergo polymerization. Preliminary sequence data regarding the carboxyl terminal region of the mouse and human MUC3 mucin indicates the presence of a cysteine-containing region corresponding to a three-lobed EGF-like tertiary structure [21] (J. Gum, personal communication). This suggests its possible role in ligand and receptor interactions involving pathogens or intestinal mucosal surfaces. A low molecular weight mucin has been cloned from a human salivary gene library (MUC7) [22]. This mucin also does not contain vWF-like cysteine-rich regions. These mucins may represent a category of non-polymerizing or non-gel-forming mucins. Their specific functions remain to be determined.

## III. ORGAN DISTRIBUTION OF SECRETORY MUCINS

Studies using mucin-specific cDNA probes and antibodies indicate that expression of secretory mucin genes is organ and cell-type specific (Table 2). Mucin MUC2 mRNA and protein are highly expressed in goblet cells of normal jejunum, ileum, and colon. Mucin MUC3 mRNA and protein are highly expressed in columnar and goblet cells of normal jejunum, ileum, colon, and gallbladder. These levels are contrasted with very low to undetectable levels in other epithelia (bronchus, breast, stomach). The MUC4 mucin appears to be limited to respiratory, colon, and cervix tissues.

In contrast to the other mucins, the MUC5 and MUC6 mucins demonstrate a wider tissue distribution (Table 2). We and others have used in situ hybridization, RNA analysis and immunohistochemistry to quantify and localize MUC6 mucin gene expression [17,18,23,24]. These data show that MUC6 is highly expressed in gastric, gallbladder, seminal vesicle, and pancreatic ductile tissues. Strong but focal expression in basal endometrial and endocervical glands was also observed. Immunohistochemical expression of MUC6 occurs early in fetal development and was observed in Brunner's glands and pancreatic ducts at 18–19 weeks and in gastric glands at 20-weeks gestation. The tissue distribution of the MUC6 secretory mucin indicates it functions to protect epithelial tissues from a wide range of substances, specifically hydrochloric acid (stomach), bile acids (gallbladder), and proteases (seminal vesicle, cervix, and pancreas).

The finding of strong reactivity for MUC6 mucin in pancreatic ductules indicates that this mucin may represent the major cytoprotective mucin in the pancreas [24,25]. We have previously reported that MUC2 and MUC3 expression is absent in normal pancreas [26]. In addition, Balague et al. reported that MUC2, MUC5AC, and MUC5B secretory mucins were weakly expressed or undetectable in normal pancreas [27]. In contrast, cancers of the pancreas frequently demonstrate aberrant expression of MUC2-5 mucins [26,27].

The normal stomach is characterized by high levels of MUC5AC and MUC6. Immunohistochemical and in situ hybridization demonstrate that these mucins are localized to distinct cell types. The MUC5AC mucin is produced by the surface mucous cells in all areas of the stomach. In contrast, MUC6 is primarily produced by the neck mucous cells of the

**Table 2** Expression of Human Secretory Mucin Genes

| | MUC2 | MUC3 | MUC4 | MUC5AC | MUC5B | MUC6 | MUC7 | MUC8 |
|---|---|---|---|---|---|---|---|---|
| Trachea | ± | – | + + | + + | + | – | – | + + |
| Middle ear | – | ± | | + + (epithelium) | + + | + (glands) | | |
| Endocervix | – | – | + + | + + | + + | + | – | |
| Seminal vesicles | – | – | | – | | + + + | – | |
| Salivary | – | – | | – | + | – | + + + | |
| Esophagus | – | – | – | – | | – | – | |
| Stomach | – | – | – | + + + (surface mucous cells) | – | + + + (neck, antral cells) | – | |
| Duodenum | + + | + + | – | – | – | + + (Brunner's glands) | – | |
| Small intest. | + + | + + | – | – | – | – | – | |
| Colon | + + | + + | + + | – | – | – | – | |
| Gallbladder | – | + + | – | + + | + + | + + | – | |
| Pancreas | – | ± | – | – | – | + + | | |

±, +, + + = trace, moderate, or high-level expression; – = none or very low-level expression; blank = not definitively determined
References 15–17, 24–26, 29, 30, 55 and Ho et al., unpublished data.

fundus and by the antral gland-type cells of the cardia and antrum [17,23]. Ota and Katsuyama have used a dual histochemical–lectin staining method to characterize the surface mucus layer of the human stomach [28]. They demonstrated that the surface mucus layer consists of a laminated array of two types of mucin, with mucin corresponding to gland cells interspersed with mucin corresponding to surface mucous cells. In addition, the secretion of the neck cells has been shown to consist of a soluble mucus that is absent from gastric juice on fasting and is secreted in response to parasympathomimetic-vagal stimulation [29]. These data suggest that gastric mucus is composed of two layers of mucin proteins (MUC5AC and MUC6) whose expression is constitutive and regulated, respectively. Similar to stomach epithelium, immunohistochemical analysis of gallbladder epithelium indicates that the cells of the superficial epithelium synthesize different mucins (MUC3 and MUC5AC) compared to the cells found in the deeper folds, which synthesize MUC6 [30]. We speculate that the two-component multilayer structure of mucin gels found in the stomach may also be present in other epithelial tissues that express multiple gel-forming mucin proteins.

Reproductive and respiratory epithelium express a complex array of secretory-type mucins. In addition to MUC6, the glands of the endocervix express MUC5AC, MUC5B, and MUC4 secretory mucins [15,31]. Expression cloning, RNA analysis, and immunohistochemistry have revealed that bronchial and pulmonary epithelia express MUC2, MUC4, MUC5AC, MUC5B, and the recently described MUC8 [12,15,16,26,32]. Using in situ hybridization techniques, Audie et al. demonstrated MUC2, MUC4, and MUC5AC in the surface epithelium of bronchi, whereas MUC4 and MUC5B transcripts were present in the subepithelial glands [15]. Inflammatory conditions in the lung induce mucin gene expression. Exposure to inhaled irritants, neutrophils, carcinogens, and infectious agents result in increased goblet cells in the epithelium; cystic fibrosis airways are enriched in MUC2 mRNA [33]. Studies by Basbaum and coworkers have shown that exposure of rats to sulfur dioxide leads to increased mucous cells and glands and causes upregulation of specific MUC2-related mucin genes [34]. We have found that neoplastic transformation in the lung is associated with a marked change in MUC1-4 gene expression. Adenocarcinomas, especially well-differentiated tumors, exhibit increased MUC1, MUC3, and MUC4 mRNA. Lung squamous cell, adenosquamous, and large-cell carcinomas are characterized by increased MUC4 transcripts only [35]. The frequent occurrence of increased MUC4 transcripts in a variety of nonsmall-cell cancers indicates the potential importance of this type of mucin in lung cancer biology.

To date we have performed immunohistochemical studies using specimens of adult temporal bones ($n = 10$) and RNA analysis from surgical scrapings of tympanic cavity epithelium ($n = 6$) and eustacian tube epithelium obtained at autopsy ($n = 1$). Transcripts for MUC1, MUC5B, and MUC5AC were detected in tympanic cavity tissue by RT-PCR analysis. Northern blot analysis of eustacian tube epithelium reveals MUC5AC, MUC5B, and MUC6 mRNA. In focal ciliated epithelial cells in 9 of 10 specimens and in some glands MUC5AC immunoreactivity was present. Preliminary data indicate that MUC5B and MUC6 core peptide was expressed in subepithelial glands and focally within the goblet cells of the surface epithelium; MUC2 core peptide expression was generally absent in these specimens. Trace MUC3 immunoreactivity was present in focal ciliated columnar cells in 5 of 9 specimens. Mucous cells of the surface epithelium and the subepithelial glands also demonstrated differences in lectin binding with PNA, MAA, and UEA lectins. These data demonstrate that the epithelium of the middle ear may contain a unique pattern of secretory mucin gene expression. The focal nature of the staining patterns of the surface epithelium indicates that other mucin proteins may be expressed at this site, including other respiratory-type mucins or other, novel mucin proteins. The functional significance of the complex pattern of secretory mucins present in reproductive and respiratory tissues remains to be determined.

## IV. HELICOBACTER PYLORI–RELATED GASTRIC DISEASE AND ALTERATIONS OF GASTRIC MUCIN

*Helicobacter pylori* is distributed worldwide, and its prevalence varies with age, race, ethnicity, and socioeconomic status [36]. Infection with *H. pylori* has emerged as a major predisposing condition for the development of peptic ulcer disease, gastric adenocarcinoma, and primary B-cell lymphoma of the stomach. *H. pylori* infection is chronic and results in histologic changes in the stomach that may progress over decades [37].

The pathophysiology of the development of peptic ulcer disease in patients infected with *H. pylori* is unknown. Altered mucin production or function has been postulated to be one important factor [37,38]. The intimate association of *H. pylori* and gastric mucin was confirmed by the identification that *H. pylori* binds to the fucosylated $Le^b$ antigen that is found in the surface mucous cells and secreted gastric mucin [39]. Histologic studies indicate that the antrum is the most common site of infection, suggesting that the intact healthy antral morphology and neutral gastric mucin in the surface epithelial cells represents an optimal environment for infection [40]. The presence of *H. pylori* organisms highly correlates with the presence of mucin depletion and degeneration, regeneration, and ulceration of the surface gastric epithelium [40–43]. Disintegration of apical mucous and the production of epithelial pits and microerosions are the most specific markers of *H. pylori* colonization. Intracellular mucin measured by histochemical methods was reported to be decreased in surface antral cells in *H. pylori*-infected subjects compared with noninfected subjects [44]. *H. pylori* may produce a number of proteases and lipases that are thought to destroy the polymeric structure of mucin proteins and the lipid components, leading to loss of viscosity and hydrophobicity of the mucous gel [38,45].

The purpose of our recent experiments was to determine if chronic *H. pylori* infection resulted in alteration of specific MUC5AC and/or MUC6 mucins in humans. Gastric biopsies and serum were obtained from consecutive patients undergoing endoscopic procedures ($n = 41$). Infection was determined by histologic and serologic testing. Specimens from patients treated with proton pump inhibitors or with histologic evidence of intestinal metaplasia or transitional mucosa were not included in analysis of mucin protein or RNA [46].

The most significant findings from these preliminary studies is the demonstration that antral biopsies from patients infected with *H. pylori* exhibit a significant reduction in surface MUC5AC immunoreactivity, native gastric mucin immunoreactivity measured by ELISA, and MUC5AC mRNA levels measured by RNAse protection assays, compared with normal antral biopsies. These findings are in agreement with recent preliminary findings that extracts of *H. pylori* decreased mucous thickness and lowered blood flow in rats, and that a lipid extract of *H. pylori* inhibited both mucin synthesis and secretion (measured by $^{14}$C-glucosamine labeling) in a gastric cancer cell line HM02 [47,48]. In contrast, we observed an increase in MUC6 tandem repeat immunoreactivity on the luminal surface, in antral glands, and in mucous neck cells in specimens with *H. pylori* infection compared with normal specimens. This was not associated with significant changes in MUC6 mRNA levels. This suggests that the increased immunoreactivity may be secondary to changes in glycosylation of the MUC6 tandem repeat rather than due to increased synthesis of MUC6 protein. The finding that MUC5AC and MUC6 are differentially regulated is supported by recent studies describing that a phorbol ester PMA increased the expression of MUC2 and MUC5AC, but not MUC6, in HT-29 colon cancer cells [49]. These data lend support to the hypothesis that decreased MUC5AC mucin results from chronic infection related to *H. pylori* and may be a contributing factor in the pathogenesis of mucosal disease related to this chronic infection. These findings are in contrast to the mucin response to acute bacterial infection in bronchial tissues, which is characterized by increased goblet cell metaplasia and increased mucin synthesis and secretion. These differences may be due to differences in the chronicity of the infection, bacterial virulence factors, and host immune mediators.

## V. CONCLUSION

Epithelial tissues express a wide variety of secretory-type mucins. Sequence data derived from individual mucin genes give insight into the structure and the specific functions of these proteins. Epithelial tissues frequently co-express multiple secretory mucins, often by different cell types within each tissue. The presence of different gel-forming mucins in these secretions may suggest the presence of a multilayer structure of the mucous coat. Chronic infection of the gastric mucosa by *Helicobacter pylori* was found to be associated with decreased native mucin and MUC5AC immunoreactivity and mRNA levels and increased MUC6 immunoreactivity. These data suggest that the MUC5AC and MUC6 mucins are differentially regulated in the stomach and that decreased MUC5AC mucin may be a contributing factor in the pathogenesis of *H. pylori*-related peptic ulcer disease.

## ACKNOWLEDGMENTS

This work was supported by a Veterans Affairs Merit Review Award, a USPHS Grant RO1 HD33171, and the Veterans Affairs Research Service.

## REFERENCES

1. Neutra M, Forstner JF. Gastrointestinal mucus: synthesis, secretion, and function. in: Physiology of the Digestive Tract. Johnson L, Ed. New York: Raven Press 1987: 975–1009.
2. Strous GJ and Dekker J. Mucin-type glycoproteins. Crit Rev Biochem Mol Biol 1992; 27:57–92.
3. LaMont JT. Mucus: the front line of intestinal mucosal defense. Annal NY Acad Sci 1993; 190–201.
4. Rose M. Mucins: structure, function and role in pulmonary diseases. Am J Physiol 1992; 263:L413–L429.
5. Hilkens J, Buijs F, Ligtenberg M. Complexity of MAM-6, an epithelial sialomucin associated with carcinomas. Cancer Res 1989; 49:786–793.

6. Gendler SJ, Lancaster CA, Taylor-Papdimitriou J, Duhig T, Peat N, Burchell J, Pemberton L, Lalani E-N, Wilson D. Molecular cloning and expression of human tumor-associated polymorphic epithelial mucin. J Biol Chem 1990; 265:15286–15293.
7. Siddiqui J, Abe M, Hayes D, Shani E, Yunis E, Kufe D. Isolation and sequencing of a cDNA coding for the human DF3 breast carcinoma-associated antigen. Proc Natl Acad Sci USA 1988; 85:2320–2323.
8. Wesley A, Mantle M, Man D, Qureshi R, Forstner G, Forstner J. Neutral and acidic species of human intestinal mucin. Evidence for different core peptides. J Biol Chem 1985; 260:7955–7959.
9. Carlstedt I, Sheehan JK. Structure and macromolecular properties of mucus glycoproteins. Monogr Allergy 1988; 24:16–24.
10. Gum JR, Byrd JC, Hicks JW, Toribara NW, Lamport DTA, Kim YS. Molecular cloning of human intestinal mucin cDNAs. Sequence analysis and evidence for genetic polymorphism. J Biol Chem 1989; 264:6480–6487.
11. Gum JR, Hicks JW, Swallow DM, Lagace RL, Byrd JC, Lamport DTA, Siddiki B, Kim YS. Molecular cloning of cDNAs derived from a novel human intestinal mucin gene. Biochem Biophys Res Commun 1990; 171:407–415.
12. Porchet N, Cong NV, Dufosse J, Audie JP, Guyonnet-Duperat V, Gross MS, Denis C, Degand P, Bernheim A, Aubert JP. Molecular cloning and chromosomal localization of a novel human tracheo–bronchial mucin cDNA containing tandemly repeated sequences of 48 base pairs. Biochem Biophys Res Com 1991; 175:414–422.
13. Aubert JP, Porchet N, Crepin M, Duterque-Coquillaud M, Vergnes G, Mazzuca M, Debuire B, Petitprez D, Degand P. Evidence for different human tracheo–bronchial mucin peptides deduced from nucleotide cDNA sequences. Am J Respir Cell Mol Biol 1991; 5:178–185.
14. Crepin M, Porchet N, Aubert JP, Degand P. Diversity of the peptide moiety of human airway mucins. Biorheology 1990; 27:471–484.
15. Audie JP, Janin A, Porchet N, Copin MC, Gosselin B, Aubert JP. Expression of human mucin genes in respiratory, digestive, and reproductive tracts ascertained by in situ hybridization. J Histochem Cytochem 1993; 41: 1479–1485.
16. Shankar V, Gilmore MS, Elkins RC, Sachdev GP. A novel human airway mucin cDNA encodes a protein with unique tandem–repeat organization. Biochem J 1994; 300:295–298.
17. Ho SB, Roberton AM, Shekels LL, Lyftogt CT, Niehans GA, Toribara NW. Expression cloning of gastric mucin cDNA and localization of mucin gene expression. Gastroenterology 1995; 109:735–747.
18. Toribara NW, Roberton AM, Ho SB, Kuo W-L, Gum E, Hicks JW, Gum JR, Byrd JC, Siddiki B, Kim YS. Human gastric mucin; identification of a unique species by expression cloning. J Biol Chem 1993; 268:5879–5885.
19. Gum JR, Hicks JW, Toribara NW, Rothe E-M, Lagace RE, Kim YS. The human MUC2 intestinal mucin has cysteine-rich subdomains located both upstream and downstream of its central repetitive region. J Biol Chem 1992; 267:21375–21383.

20. Toribara NW, Ho SB, Gum E, Gum JR, Lau P, Kim YS. The carboxyl terminal sequence of the human secreted mucin, MUC6. Analysis of the primary amino acid sequence. J Biol Chem 1997; 272:16398–16403.
21. Hunninghake DA, Shekels LL, Lyftogt CT, Kieliszewski M, Ho SB. Expression cloning of murine colonic mucin reveals differential mucin gene expression in the mouse gastrointestinal tract. Gastroenterology 1995; 108:A291.
22. Bobek LA, Tsai H, Biesbrock AR, Levine MJ. Molecular cloning, sequence, and specificity of expression of the gene encoding the low molecular weight human salivary mucin (MUC7). J Biol Chem 1993; 268:20563–20569.
23. deBolos C, Garrido M, Real F. MUC6 apomucin shows a distinct normal tissue distribution that correlates with Lewis antigen expression in the human stomach. Gastroenterology 1995; 109:723–734.
24. Bartman AE, Aubert J-P, Buisine M-P, Niehans GA, Toribara NW, Kim YS, Kelly EJ, Crabtree JE, Ho SB. MUC6 secretory mucin gene expression in tissues exposed to endogenous mucosal damaging agents. Gastroenterology 1997; 112:A861.
25. Meier PB, Ewing SE, Evans MK, Toribara NW, Ho SB. MUC6 mucin is highly expressed in normal and neoplastic pancreatic tissue and periductular biliary glands. Gastroenterology 1996; 110:A417.
26. Ho SB, Niehans GA, Lyftogt C, Yan P-S, Cherwitz DL, Gum ET, Dahiya R, Kim YS. Heterogeneity of mucin gene expression in normal and neoplastic tissues. Cancer Res 1993; 53:641–651.
27. Balague C, Gambus G, Carrato C, Porchet N, Aubert J-P, Kim YS, Real FX. Altered expression of MUC2, MUC4, and MUC5 mucin genes in pancreas tissues and cancer cell lines. Gastroenterology 1994; 106:1054–1061.
28. Ota H, Katsuyama T. Alternating laminated array of two types of mucin in the human gastric surface mucous layer. Histochemical J 1992; 24:86–92.
29. Davenport HW, Physiology of the Digestive Tract. 4th ed. Chicago: Year Book Medical Publ, Inc. 1977:103.
30. Ho SB, Shekels LL, Toribara NW, Kim YS, Purdum PP, Cherwitz DL. Human gallbladder epithelium expresses a unique mucin gene pattern. Submitted, 1997.
31. Gipson IK, Ho SB, Spurr-Michaud SJ, Tisdale AS, Zhan Q, Torlakovic E, Pudney J, Anderson DJ, Toribara NW, Hill JA. Mucin genes expressed by human female reproductive tract epithelia. Biol Reproduction 1997; 56: 999–1011.
32. Jany BH, Gallup MW, Yan P-I, Gum JR, Kim YS, Basbaum CB. Human bronchus and intestine express the same mucin gene. J Clin Invest 1991; 87: 77–82.
33. Jany B, Basbaum CB. Modification of mucin gene expression in airway disease. Am Rev Respir Dis 1991; 144:S38–S41.
34. Basbaum C, Gallup M, Gum JR, Kim Y, Jany B. Modification of mucin gene expression in the airways of rats exposed to sulfur dioxide. Biorheology 1990; 27:485–489.
35. Nguyen PL, Niehans GA, Cherwitz DL, Kim YS, Ho SB. Membrane-bound

(MUC1) and secretory (MUC2, MUC3, and MUC4) mucin gene expression in human lung cancer. Tumor Biol 1996; 17:176–192.
36. Walsh JH, Peterson WL. The treatment of *Helicobacter pylori* infection in the management of peptic ulcer disease. N Engl J Med 1995; 333:984–991.
37. Blaser MJ, Parsonnet J. Parasitism by the "slow" bacterium, *Helicobacter pylori* leads to altered gastric homeostasis and neoplasia. J Clin Invest 1994; 94:4–8.
38. Slomiany BL, Slomiany A. Mechanism of Helicobacter pylori pathogenesis: focus on mucus. J Clin Gastroenterol 1992; 14(suppl 1):S114–S121.
39. Boren T, Falk P, Roth KA, Larson G, Normark S. Attachment of *Helicobacter pylori* to human gastric epithelium mediated by blood group antigens. Science 1993; 262:1892–1895.
40. Hansing RL, D'Amico H, Levy M, Guillan RA. Prediction of *Helicobacter pylori* in gastric specimens by inflammatory and morphological histological evaluation. Am J Gastro 1992; 87:1125–1131.
41. Group GP. Rapid identification of pyloric *Campylobacter* in Peruvians with gastritis. Dig Dis Sci 1986; 31:1089–1094.
42. Hui PK, Chan WY, Cheung PS, Chan JKC, Ng CS. Pathologic changes of gastric mucosa colonized by *Helicobacter pylori*. Hum Pathol 1992; 23: 548–556.
43. Sobala GM, Crabtree JE, Dixon MF, Schorah CJ, Taylor JD, Rathbone BJ, Heatley RV, Axon ATR. Acute *Helicobacter pylori* infection: clinical features, local and systemic immune response, gastric mucosal histology, and gastric juice ascorbic acid concentrations. Gut 1991; 32:1415–1418.
44. Kawano S, Tsujii M, Nagano T, Ogihara H, Tanimura H, Hayashi N, Ito T, Sato N, Kamada T, Tamura K, Tanaka M. Different effect of *Helicobacter pylori* on the human gastric antral and body mucosal intracellular mucin. Scand J Gastroenterol 1990; 25:997–1003.
45. Sarosiek J, Peura DA, Guerrant RL, Marshall BJ, Lasewicz W, Gabryelewicz A, McCallum RW. Mucolytic effects of Helicobacter pylori. Scand J Gastroenterol 1991; 26(Suppl)187:47–55.
46. Fesenmyer ME, Bartman AE, Shekels LL, Wiehr CL, Ewing SL, Crabtree JE, Ho SB. *Helicobacter pylori* infection alters human gastric mucin MUC5AC and MUC6 gene expression. Gastroenterology 1997; 112:A971.
47. Atuma C, Engstrand L, Holm L. *Helicobacter pylori* reduces gastric mucus secretion and mucosal blood flow in rats. Gastroenterology 1996; 110:A302.
48. Beil W, Wagner S, Enss M, Sewing K-F. A lipid extract of *Helicobacter pylori* broth inhibits mucin synthesis and secretion in the gastric mucous cell line HM02. Gastroenterology 1996; 110:A300.
49. Hong D, Forstner J, Forstner G. Differential responses to phorbol ester of mucin genes clustered on chromosome 11p15. Gastroenterology 1996; 110: A957.
50. Ligtenberg MJL, Vos HL, Gennissen AMC, Hilkens J. Episialin, a carcinoma-associated mucin, is generated by a polymorphic gene encoding splice variants with alternative amino termini. J Biol Chem 1990; 265:5573–5578.

51. Lan MS, Batra SK, Qi W-N, Metzgar RS, Hollingsworth MA. Cloning and sequencing of a human pancreatic tumor mucin cDNA. J Biol Chem 1990; 265:15294–15299.
52. Gerard C, Eddy RL, Shows TB. The core peptide of cystic fibrosis tracheal mucin contains a tandem repeat structure. Evidence for a common mucin in airway and gastrointestinal tissue. J Clin Invest 1990; 86:1921–1927.
53. Guyonnet Duperat V, Audie J-P, Debailleul V, Laine A, Buisine M-P, Galiegue-Zouitina S, Pigny P, Degand P, Aubert J-P, Porchet N. Characterization of the human mucin gene MUC5AC: a consensus cysteine-rich domain for 11p15 mucin genes? Biochem J 1995; 305:211–219.
54. Meerzaman D, Charles P, Daskal E, Polymeropoulos MH, Martin BM, Rose MC. Cloning and analysis of cDNA encoding a major airway glycoprotein, human tracheobronchial mucin (MUC5). J Biol Chem 1994; 269: 12932–12939.
55. Dufosse J, Porchet N, Aubert J-P, Duperat VV, Laine A, Van-Seuningen I, Marrakchi S, Degand P, Aubert J-P. Degenerate 87-base-pair tandem repeats create hydrophilic/hydrophobic alternating domains in human mucin peptides mapped to 11p15. Biochem J 1993; 293:329–337.

# 38

# Effect of *Pseudomonas aeruginosa* Alginate on Airway Mucus Secretion

**Chikako Kishioka* and Robert P. Tomkiewicz†**
*St. Louis University School of Medicine, St. Louis, Missouri*

**Bruce K. Rubin‡,§**
*St. Louis University School of Medicine, and Cardinal Glennon Children's Hospital, St. Louis, Missouri*

**Daniel J. Hassett**
*University of Cincinnati, Cincinnati, Ohio*

## I. INTRODUCTION

*Pseudomonas aeruginosa* chronically infects the lower airways of most older patients with cystic fibrosis (CF) [1]. Strains of mucoid *P. aeruginosa* that elaborate a viscous exopolysaccharide glycocalyx are characteristic in CF lung disease. The major component of this glycocalyx is alginate. It has been proposed that hypersecretion of mucus may be a result of infection with bacteria, especially *P. aeruginosa* [2–4], and hypersecretion may further contribute to infection and airway obstruction.

Evidence from in vitro studies suggests that some bacterial products stimulate airway mucus secretion [5–9]. We hypothesized *P. aeruginosa* alginate would stimulate tracheal mucus secretion in a dose-response fashion, and to test this hypothesis, we measured mucin and lysozyme secretion in the isolated ferret trachea. We also wished to determine if alginate-induced mucus secretion could be reduced by inhibiting the usual secretion pathways.

---

*Current affiliations:*
*Mie University, Mie, Japan
† Private practice
‡ Bowman Gray School of Medicine, Winston-Salem, North Carolina
§ To whom correspondence should be addressed

## II. MATERIALS AND METHODS

### A. Short-Term Organ Culture

As previously described [10], tracheas from adult ferrets of either sex were divided into four approximately equal segments. These segments were placed in Krebs–Henseleit solution (KHS) and agitated gently at 38°C, which is ferret body temperature. After a 2-hour recovery time, the KHS was changed and the segments were incubated for 40 minutes with only KHS (period 1). Then they were incubated for another 40 minutes with purified alginate, alginate, with proteinase inhibitor, alginate with atropine, or with KHS only (period 2). After the experiment, the KHS was collected and stored at −70°C until analyzed.

The relative contribution of mucous and serous cells to the secretion was estimated by measuring the amount of mucin and lysozyme secretion, respectively. A secretory index (S.I.) expressing the relative increase in each marker over constitutive secretion was calculated for each tracheal segment as the mucin or lysozyme concentration after 40 minutes exposure to alginate (period 2) divided by the concentration measured after 40 minutes exposure to KHS alone (period 1). We randomized the administration of agents to different segments in each experiment in order to account for anatomic variation in secretion from proximal or distal tracheal segments.

The concentration of alginate used (20 and 200 μg/ml) was based on published data [11,12]. Atropine (Elkins-Sinn, Cherry Hill, NJ) was used at a concentration known to inhibit methacholine-induced tracheal secretion [13]. A cocktail of phenylmethylsulfonyl fluoride (PMSF), leupeptin and pepstatin A (Sigma, St. Louis, MO) was used to block proteolytic activity.

### B. Mucin Analysis

We used a sandwich enzyme-linked lectin assay (ELLA) to measure mucin secretion as previously described [14]. A 96-well microtiter plate was coated with Dolichos biflorus agglutinin (DBA). After rinsing with high-salt phosphate-buffered saline (HSPBS), the plate was exposed to sample buffer. Then, it was treated with DBA conjugated with horse radish peroxidase in phosphate-buffered saline (PBS) containing 1% bovine serum albumin. Before and after this step, the plate was washed with HSPBS. Tetramethylbenzidine in citrate-acetate buffer (pH 6.0) was added to each well and incubated for 10 min. The reaction was stopped by adding 4.7 *N* $H_2SO_4$. Color development was read as the difference in absorbance at 450 and 650 nm. The amount of mucin was calculated by comparison with asialo ovine submaxillary mucin standard. The change in mucin secretion after secretagogue was expressed as percent increase above excipient control in mucin weight per trachea tissue weight (ng/g).

### C. Lysozyme Assay

Lysozyme activity was determined spectrophotometrically by measuring the initial rate of lysis of a 1.38 mg/mL *Micrococcus lysodeikticus* suspension [15]. A 0.1 mL volume of sample buffer was added to 1.4 mL of substrate in PBS buffer at pH 6.0. The change in turbidity was measured at 540 nm. One unit of enzyme activity was defined as the change in absorbance per minute equivalent to that produced by 1.0 mg of egg-white lysozyme under identical assay conditions. The change in lysozyme secretion after secretagogue was expressed as percent increase above excipient control in unit per tissue weight (g).

### D. Statistical Analysis

Statistical analysis of data was performed using the StatView 4.5 statistics package (Abacus Concepts Inc., Berkeley, CA). The data were analyzed by ANOVA to assess secretory dose response and the effect of inhibitors. Those p values $\leq 0.05$ were considered statistically significant.

## III. RESULTS

### A. Mucin Secretion

The secretory response was measured as a function of alginate concentration. There was a dose-dependent response to alginate with a 26% increase in mucin secretion with alginate at 20 μg/mL ($n = 10$) and a 111% increase ($n = 10$, $p = 0.003$) with alginate at 200 μg/mL (Figure 1). The antiprotease cocktail only had a minimal effect on alginate-induced secretion. Atropine appeared to partially block alginate-induced secretion by 41% ($n = 5$, p = not significant, [NS]).

### B. Lysozyme Secretion

Lysozyme secretion was also measured as a function of alginate concentration as shown in Figure 1. There was a 12% increase in lysozyme secretion with alginate at 20 μg/mL ($n = 10$) and a 20% increase with alginate at 200 μg/mL ($n = 10$, $p = 0.024$). Atropine partially inhibited alginate

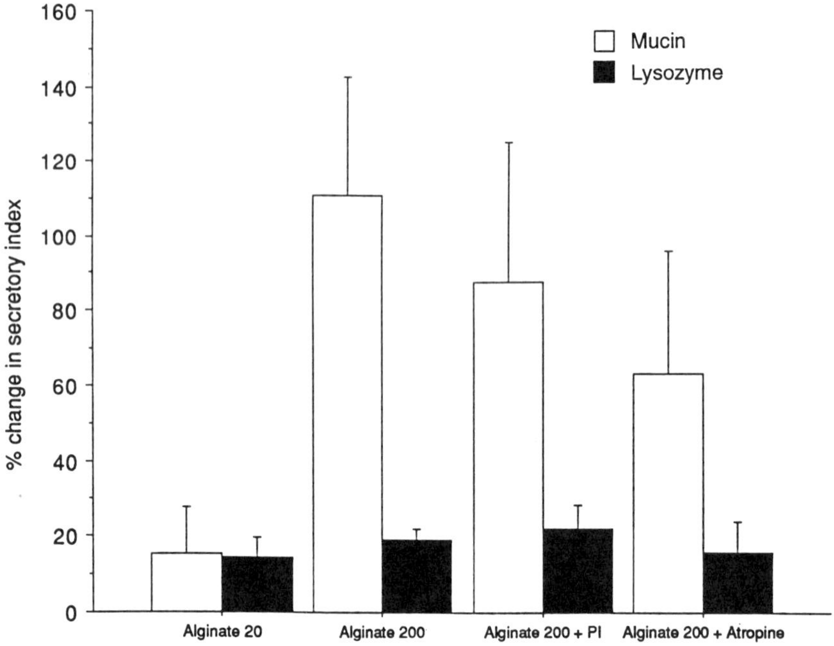


**Figure 1** Secretory response to alginate at 20 and 200 μg/mL, with and without atropine or proteinase inhibitors (PI). The percent increase in secretory index above control (constitutive secretion) is given for mucin and lysozyme secretion. The cocktail of proteinase inhibitors did not inhibit alginate-induced mucin or lysozyme secretion but atropine partially inhibited alginate-induced mucin and lysozyme secretion by 42% and 25%, respectively (p = not significant).

induced secretion by 25% ($n$ = 5, p = NS). The lysozyme secretion response was much smaller than the mucin secretory response.

## IV. DISCUSSION

The airways of most older patients with CF are chronically infected with *P. aeruginosa* [16]. *P. aeruginosa* can produce substantial amounts of an encapsulating exopolysaccharide called alginate. In most CF centers, 80% or more of the older patients harbor mucoid (alginate producing) strains of *P. aeruginosa* [17]. We have demonstrated that *P. aeruginosa* alginate stimulates mucin and lysozyme secretion in ferret tracheal segments at concentrations similar to those reported in the sputum of patients with CF [11,12].

The secretory effect of bacterial exoproduct specifically involves release of mucin and this effect is probably mediated by alginate. The mechanism by which alginate stimulates secretion is uncertain. Imaging by transmission electron microscopy of the ferret tracheal tissue showed no cell damage at the concentrations and exposure times used in this study. Because atropine appeared to partially block this stimulatory effect, the cholinergic nervous system is probably activated by alginate, contributing to the secretory effect. It is likely that alginate also acts via stimulation of other nerves or through the release of inflammatory mediators.

The alginate-stimulated secretion of mucin was greater than that of lysozyme. This suggests that the secretion induced by the alginate contains more mucin relative to lysozyme and thus greater secretion from mucous cells than from serous cells. This could produce a secretion that is more rigid (higher mechanical impedance) and might be more difficult to clear from the airway. This, in turn, might further contribute to the poor mucus clearance seen in patients with cystic fibrosis [9,18,19], increasing the risk for chronic airway infection and inflammation.

## ACKNOWLEDGMENTS

Supported by a grant from the Cystic Fibrosis Foundation (Rubin 96PO). Dr. Kishioka was an Abbott Research Associate and Dr. Tomkiewicz was a Gosselin Pediatric Asthma Research Associate.

## REFERENCES

1. Rubin BK. Cystic Fibrosis and Bronchiectasis. In: The 11th National ACCP Pulmonary Board Review. Northbrook, Illinois: American College of Chest Physicians Press, 1996:65–78.
2. Adler KB, Winn WC, Alberghini TV, Craighead JE. Stimulatory effect of *Pseudomonas aeruginosa* on mucin secretion by the respiratory epithelium. JAMA 1983; 249:1615–1617.
3. Oppenheirner EH. Similarity of the tracheobronchial mucous glands and epithelium in infants with and without cystic fibrosis. Hum Pathol 1981; 12: 36–48.
4. Sturgess J, Imrie J. Quantitative evaluation of the development of tracheal submucosal glands in infants with cystic fibrosis and control infants. Am J Pathol 1982; 106:303–311.

5. Klinger JD, Tandler B, Liedtke CM, Boat TF. Proteinases of *Pseudomonas aeruginosa* evoke mucin release by tracheal epithelium. J Clin Invest 1984; 74:1669–1678.
6. Adler KB, Hardwick DH, Craighead JE. Effect of cholera toxin on secretion of mucin by explants of guinea pig trachea. Lab Invest 1981; 45:372–377.
7. Stutts MJ, Guyman LF, Boucher RC. Interaction of *Pseudomonas aeruginosa* products with bronchial epithelia. Chest 1982; 81:13S–14S.
8. Steiger D, Hotchkiss J, Bajaj L, Harkema J, Basbaum C. Concurrent increases in the storage and release of mucin-like molecules by rat airway epithelial cells in response to bacterial endotoxin. Am J Respir Cell Mol Biol 1995; 12:307–314.
9. Somerville M, Taylor GW, Watson D, Rendell NB, Rutman A, Todd H, Davis JR, Wilson R, Cole P, Richardson PS. Release of mucus glycoconjugates by *Pseudomonas aeruginosa* rhamnolipids into feline trachea in vivo and human bronchus in vivo. Am J Respir Cell Mol Biol 1992; 6:116–122.
10. Kishioka C, Tomkiewicz RP, Hassett DJ, Ramirez OE, de Mello D, Rubin BK. *Pseudomonas aeruginosa* alginate is a potent secretagogue in the isolated ferret trachea. Am Rev Resp Crit Care Med (Submitted).
11. Mrsny RJ, Lazazzera BA, Daugherty AL, Schiller NL, Patapoff TW. Addition of a bacterial alginate lyase to purulent CF sputum in vitro can result in the disruption of alginate and modification of sputum viscoelasticity. Pulmonary Pharmacology 1994; 7:357–366.
12. Pedersen SS, Kharazmi A, Espersen F, Høiby N. *Pseudomonas aeruginosa* alginate in cystic fibrosis sputum and inflammatory response. Infect Immun 1990; 58:3363–3368.
13. Tom-Moy M, Basbaum CB, Nadel JA. Localization and release of lysozyme from ferret trachea: effects of adrenergic and cholinergic drugs. Cell and Tissue Research 1983; 228:549–562.
14. Kishioka, C, Cheng PW, Seftor REB, Lartey PA, Rubin BK. Regulation of mucin secretion in the ferret trachea. Otolaryngolog Head Neck Surg 1997; In press.
15. Konstan MW, Cheng P-W, Sherman JM, Thomassen MJ, Wood RE, Boat TF. Human lung lysozyme: sources and properties. Am Rev Resp Dis 1981; 123:120–124.
16. Hata J, Fick RB. *Pseudomonas aeruginosa* and the airway disease of cystic fibrosis. Clin Chest Med 1988; 9:679–689.
17. Pedersen SS, Høiby N, Espersen F, Koch C. Role of alginate in infection with mucoid *Pseudomonas aeruginosa* in cystic fibrosis. Thorax 1992; 47: 6–13.
18. Currie DC, Pavia D, Agnew JE, Lopez-Vidriero MT, Diamond PD, Cole PJ, Clarke SW. Impaired tracheobronchial clearance in bronchiectasis. Thorax 1987; 42:126–130.
19. Wood RE, Wanner A, Hirsh J, Farrell PM. Tracheal mucociliary transport in patients with cystic fibrosis and its stimulation by terbutaline. Am Rev Respir Dis 1975; 111:733–738.

# 39
# Analysis of Mucociliary Dysfunction

**Yasuo Sakakura and Yuichi Majima**
*Mie University School of Medicine, Mie, Japan*

## I. INTRODUCTION

Mucociliary transport (MCT) is governed by three factors: cilia, mucus, and the interaction between cilia and mucus. With normal cilia, mucus, and the interaction between them, MCT is functioning. However, if even one of these factors is impaired, mucociliary dysfunction will occur. Functional and structural impairment of these three factors must be responsible for MCT dysfunction. The cause should be analyzed in terms of these three factors. At the same time, the effect of medical or surgical intervention for mucociliary dysfunction should be evaluated in terms of these factors.

## II. MEASUREMENT OF NASAL MCT FUNCTION

Three methods have been used for the measurement of nasal MCT function. The flow of the outer mucous layer can be determined by insoluble substances, whereas the saccharin test and nasal scintigraphy may reflect the flow of both outer mucous and periciliary fluid (PCF) layers. Particle transport in the nose of patients with Sjögren's syndrome was significantly impaired compared to normal subjects. However, no significant change in the clearance of radioisotope tagged normal saline was found between the control and the patient groups. We found that there is a harmonization between flow of gel and sol layers in the normal nose. However, there is dissociation in flow in the two layers under pathologic conditions, such as Sjögren's syndrome, chronic sinusitis (CS), and administration of atropine. Care must be taken for the selection of methods and interpretation of results.

In CS, nasal particle transport rate was significantly slower and saccharin transit time was significantly longer than those of control subjects. Thus, there is a mucociliary dysfunction in both gel and sol layers in the nose of CS. Normal numbers of ciliated cells and cilia, normally oriented distribution of cilia, normal amplitude, and frequency of ciliary beats and coordination of ciliary beats are required as cilia-related factors for maintaining normal MCT function.

Ciliary beat frequency (CBF) is determined by the photoelectric method. We did not find any impairment in CBF in the nose of the patients with allergic rhinitis. There was no MCT dysfunction in the nose of perennial and seasonal allergic rhinitis during and off season.

Before computerized tomography (CT), we used contrast X-ray studies for the evaluation of the maxillary sinus pathology of CS. According to the contrast X-ray studies, we classified maxillary pathology in CS into three grades: mild, moderate, and severe. The CBF of the maxillary sinus mucosa in CS was decreased with an increase of grade of the maxillary inflammation.

## III. DISTRIBUTION OF CILIA

For the evaluation of the distribution of cilia, we have used the term "ciliated area," (CA): CA designates the area that is occupied by cilia on the surface of the epithelium. Using an image analyzer, the CA was determined and was expressed as a percentage reflecting the ratio of CA to total epithelial surface in a given area. The CA of the superolateral wall of the maxillary sinus mucosa before and after the surgery was compared. The CA, 74% after FESS was significantly higher than the 61% before. The decreased CA in chronic inflammation can be improved by the establishment of drainage and ventilation.

## IV. THE MUCUS FACTOR

The mucus factor consists of both the amount of mucus and the rheologic properties of the mucus. These must both be normal for normal function.

The nasal secretion comprises tears, condensed water from expired air, and nasal fluid that comes from nasal cavities and paranasal sinuses. The nasal fluid is composed of passively diffused blood serum and tissue fluid, actively transported ion and water, cells migrated from blood and mucosa, and nasal mucus. Nasal mucus is secreted from mucous and serous gland cells in the lamina propria and epithelial goblet cells. This nasal mucus could be classified into neutral glycoprotein (fucomucin) and acidic glycoproteins (sialomucin and sulfomucin).

The amount of nasal secretion could be increased by increased activity of any component of nasal secretion. At the same time, changes in the quality of nasal secretions could occur as a result of disease.

In CS, several changes are observed in nasal mucus and in the mucosa, in terms of histopathology and rheology. Histopathologically, the number of goblet cells is unchanged, but the number of serous and mucous cells in the lamina propria in the nasal cavity is increased. The ratio of serous to mucous secretory cells however is decreased. In the maxillary sinus of CS patients, the number of goblet cells is unchanged, and the number of serous and mucous cells is increased. In CS, the ratio of serous to mucous cells is decreased.

We investigated the histochemical properties of secretory cells of nasal and maxillary mucosa from patients with CS. This was a quantitative study by PAS-AB using three different pH values. No difference in mucous glycoprotein was observed in goblet cells of control, allergic rhinitis (AR), and CS. In normal maxillary sinus goblet cells, the neutral glycoprotein, fucomucin, was predominant. In CS, a conversion from neutral to acid mucous glycoproteins occurred. There was no difference in the mucous glycoproteins of nasal glands in the three groups. In CS, neutral glycoprotein of normal maxillary sinus glands converted to acid glycoprotein. These findings indicate that when we control mucous secretion, we have to modulate both quantity and quality of secretion.

We have used an oscillating sphere magnetic rheometer developed in this department for the determination of rheologic properties of nasal mucus. We found that there is an optimal viscoelasticity of mucus for MCT. The viscoelasticity of nasal mucus from patients with CS is significantly higher than that of controls and far from the optimal value for MCT. This indicates that high viscoelasticity of nasal secretion of CS per se is a cause of impaired MCT in this disease.

Reduced nasal MCT in CS improved after repeated antral lavages. This indicates that the mucociliary dysfunction in the nose of CS is not the cause of CS but an effect of chronic inflammation of the respiratory mucosa. Viscoelasticity of nasal mucus significantly decreased after repeated antral lavage treatments.

## V. MUCOCILIARY INTERACTION

In normal mucociliary interaction, the mucous blanket should be double-structured. Normal depth of the periciliary fluid (PCF) and continuity of PCF layer are required.

For normal mucociliary interaction this double-structured mucous blanket is essential. Mucociliary interaction is expressed by the depth and continuity of the PCF layer. In the normal mucous blanket, the depth of the PCF layer should be a little shorter than the length of cilia and only cilia should disrupt the continuity of the PCF layer.

There are two patterns in an abnormal periciliary layer. The first is a decrease of PCF. In this situation, the sticky mucus penetrates into the periciliary space. Cilia must beat in this viscous gel layer instead of a sol layer. It is obvious that CBF and amplitude in this situation are not enough for efficient MCT to take place.

The second factor that might impair mucociliary interaction is abnormal secretion from mucus-secreting cells. One example is a failure of the formation of a double-structured mucous blanket. The other is a tethering phenomenon: strands of mucus connect the mucous blanket with the epithelial surface; which disrupts the continuity of the PCF layer and impairs MCT. This phenomenon might be explained by abnormal hydration of mucous glycoprotein released from secretory cells or abnormal surfactant production.

Normal saline nebulization temporally improved mucociliary clearance in the nose of CS patients. This fact suggests that there is a decrease in PCF in these patients. Since saline nebulization did not affect the clearance in normal subjects another abnormality of the depth of the PCF layer, that is an increase of PCF, would not occur in vivo. Since the PCF layer is not closed, excess fluid will flow from both anterior and posterior edges of the PCF layer. Mucus could absorb fluid.

## VI. ERYTHROMYCIN TREATMENT

In 1984, low-dose and long-term erythromycin (EM) treatment was first reported to be very effective for patients with diffuse panbronchiolitis (DPB) in Japan. The five-year survival rates of DPB before and after the introduction of EM treatment dramatically improved from 71.0% to 93.4% after the use of EM treatment. The clinical efficacy of low-dose and long-term-treatment was only observed in 14-membered macrolide antibiotics including clarithromycin (CAM) and roxithromycin (RXM). Recently, the usefulness of this treatment has been extended to patients with chronic airway disease, and significant clinical effectiveness has been reported for patients with chronic bronchitis, bronchiectasis, bronchial asthma, CS, and otitis media with effusion. The effectiveness of this treatment is shown by a decrease in rhinorrhea, otorrhea, and sputum.

## VII. MECHANISMS OF EPITHELIAL MUCUS HYPERSECRETION

To elucidate the mechanisms of epithelial mucus hypersecretion in upper respiratory inflammation, we have produced hypertrophic and metaplastic changes of goblet cells in rat nasal epithelium by the intranasal instillation of endotoxin. Instillation 0.1 mg of endotoxin induced significant goblet cell metaplasia and mucus production in this epithelium. Intraperitoneal administration of EM inhibited endotoxin-induced mucus production in the rat nasal cavity.

Patients with CS were given the 14-membered macrolide CAM at a 400 dosage of mg/day for three months. The number of neutrophils in the nasal secretion significantly decreased 1 to 3 months after the beginning of macrolide treatment. The IL-8 concentration in the nasal secretions significantly decreased 2 to 3 months after the beginning of macrolide treatment.

Macrolide treatment was clinically effective for the patients with chronic airway disease; however, the mechanism by which macrolide antibiotics control chronic inflammation is still unclear. It has been suggested that these effects depend on an anti-inflammatory action rather than an antimicrobial one, because (a) low-dose treatment (only a half of the usual dose) resulted in a good effect, (b) long-term treatment was required, (c) this treatment was also effective for the patients with pathogens insensitive to the macrolide such as *Pseudomonas aeruginosae* and *Haemophilus influenzae* (d) clinical efficacy did not relate to the disappearance of the pathogens, and (e) a 16-membered macrolide antibiotic was not effective in this situation.

Many investigators have reported anti-inflammatory actions of 14-membered macrolide antibiotics, which include (a) inhibitory effects on neutrophil functions, (b) modulation of the function of lymphocytes and macrophages, (c) inhibition of cytokine production, (d) inhibitory effects on mucus secretion, (e) inhibition of elastase production from *Pseudomonas aeruginosae*, and (f) inhibition of bacterial biofilm formation.

Studies of erythromycin suggest that antimicrobial treatment in general is less significant for hypersecretion of airway inflammation. From my daily ENT practice, I would recommend a 14-membered macrolide for the treatment of inflammation with mucus hypersecretion.

## VIII. NASAL SURFACE EPITHELIUM

The nasal surface epithelium exhibits the capacity for net volume absorption or secretion. Under normal conditions, the surface epithelia of airways exhibit volume absorption, driven by active sodium transport. The surface epithelia also secrete liquid, driven by active chloride secretion. The balance between the sodium absorption and chloride secretion determines the water movement. Therefore, the transepithelial ion transport is important. It determines the depth of PCF and viscoelasticity of outer mucus.

We have looked at the changes of ion transport in AR and CS. In AR, sodium absorption was decreased and chloride secretion was increased. Thus, water is secreted into the lumenal side, which might contribute to watery nasal secretion observed in the early phase of nasal allergy. In CS, sodium absorption was increased, and this may cause the absorption of water from the lumenal side into the submucosal side, resulting in thick, viscous nasal secretion.

We would propose a vicious circle of self-mediated inflammation for the pathogenesis of chronic airway inflammation. Lysosomal proteases released from neutrophils recruited into the cavities enhance mucus production, which results in the increase of viscoelasticity of the effusion. Mucociliary dysfunction causes stagnation of the effusion that contains inflammatory cells and various chemical mediators including lysosomal proteases, leukotriens, prostaglandins, and cytokains. Dysfunction of MCT may trigger the mediators, which are host protective in nature in airway infection, to become host invasive.

## IX. CONCLUSION

In conclusion, MCT is governed by cilia, mucus, and their interaction. We can analyze the cause of mucociliary dysfunction and evaluate the effect of medical and surgical intervention on MCT dysfunction. Now, we must develop more effective therapeutic modalities for each impaired factor of mucociliary function: MCT dysfunction initiates a vicious circle of self-mediated inflammation that may cause deterioration of airway function.

# 40
# Effects of Mucokinetic Agents on Nasal Mucus

**Yuichi Majima**
*Mie University School of Medicine, Mie, Japan*

## I. INTRODUCTION

The human body may produce respiratory mucus with optimal rheologic properties for mucociliary transport, but during upper or lower respiratory diseases excess mucus may be too thick or too thin and both cases may behave suboptimally. In such conditions, the drugs that could correct the abnormal rheologic properties of mucus may be useful for improvement of mucociliary transport. Here, I would like to discuss the in vivo effects of several drugs on the viscosity and elasticity of respiratory mucus.

## II. CHRONIC SINUSITIS (CS) AS A MODEL FOR DRUG EXPERIMENTS

Chronic sinusitis is one of the prevalent chronic upper airway diseases. Nasal mucus is hypersecreted both from epithelial goblet cells and submucosal gland cells in the nose and the paranasal sinuses. In this disease the nasal mucociliary transport was reduced in about half of the patients. The viscosity ($\eta'$) and elasticity ($G'$) of the nasal mucus from CS are much higher than the optimal viscoelasticity for mucociliary transport. The distribution of $\eta'$ and $G'$ of nasal mucus from CS are similar to those of sputa from chronic obstructive lung diseases (COLD). Moreover, the nasal mucus is easily collected without contamination from saliva. These results indicate that CS is a good model for evaluating in vivo effects of mucokinetic drugs.

## III. EFFECTS OF SYSTEMATICALLY ADMINISTERED DRUGS ON RHEOLOGIC PROPERTIES OF NASAL MUCUS

### A. L-cysteine Ethyl Ester Hydrochloride (CEE)

In Japan CEE, one of the sulfhydryl oral mucolytics, has been used as an expectorant for the patients with obstructive lung diseases. The drug was administered orally at a dose of 300 mg/day for 4 weeks to 13 adult patients with CS and 20 paired nasal mucus samples were collected both before and 4 weeks after the administration. Significant reduction in both $G'$ and $\eta'$ were observed after CEE administration. This indicates that in vivo administration of CEE brought the viscoelastic values closer to the optimal values for mucociliary transport.

It is known that orally or intravenously administered CEE undergoes a biotransformation to *N*-acetyl-1-cysteine (NAC). As NAC is first metabolized to cysteine and increases tissue sulfhydryl groups, in vivo administration of CEE may change the conformation of protein substance or cleave the covalent bonds of nasal mucus and this may lead to the reduction in viscoelasticity.

### B. Serratiopeptidase (SER)

A proteolytic enzyme derived from strain $E_{15}$ of the genus *Serratia*, SER was given orally at a dose of 30 mg/day for 4 weeks to 11 patients with CS. Four weeks administration of SER significantly reduced $\eta'$ but did not change $G'$ of nasal mucus.

Absorbed from the intestinal mucosa into the bloodstream, the absorbed form of SER is enzymatically active. Marriott reported that in vitro application of proteolytic enzyme to bronchial mucus digested naked peptide regions of mucus glycoprotein and that the reduction in elasticity was not as great as that in viscosity. Our experience and these reports suggest that in vivo administration of SER could reduce the viscosity of nasal mucus in a fashion similar to that observed in vitro.

### C. Norfloxacin (NFLX)

An oral antibacterial agent, NFLX was administered at a dose of 600 mg/day for 2 weeks to 10 patients with CS and resulted in a significant decrease in $G'$. The $\eta'$ value did not change in this observation period.

This antibacterial agent may decrease the amount of DNA derived from damaged host and bacterial cells by its bacteriocidal effect and also may decrease the exudation of serum proteins by protecting the permeability of respiratory epithelium.

## D. Other Drugs Recently Developed

### Macrolide Antibiotics

Macrolid antibiotics possess a variety of actions other than antibacterial activity. The oral administration of these drugs reduced the amount of sputum expectorated by patients with COLD and also decreased the nasal hypersecretion in patients with CS. The oral administration of clarithromycin (CAM), a newly developed macrolide, to patients with COLD at a dose of 200 mg/day for 8 weeks increased both the ratio of wet to dry weight and G′ of the sputum. The effects of this drug on CS nasal mucus rheology have never been reported.

### Agent with a Blocked Sulfhydryl Group

*S*-carboxymethylcysteine (SCMC) is a cysteine derivative with a blocked thiol group. Inconsistent results have been obtained measuring mucus viscoelasticity. The oral administration of this drug changed neither G′ nor $\eta'$ of nasal mucus in the patients with CS; whereas oral administration of SCMC reduced both G′ and $\eta'$ of middle ear effusion in the patients with otitis media with effusion. This drug does not appear to break mucin disulfide bonds, but may act via alternative mechanisms.

### Ambroxol Hydrochloride

Ambroxol is related to bromhexine and appears to stimulate mucus secretion. The oral administration of this drug to 15 patients with CS at a dose of 45 mg/day for 4 weeks did not alter the viscoelasticity of nasal mucus.

# 41

# Pharmacologic Management of Mucus Retention Based on a Taxonomy of Mucoactive Medications

**Bruce K. Rubin***
*St. Louis University School of Medicine, St. Louis, Missouri*

## I. INTRODUCTION

Airway mucus clearance depends upon the physical properties of the mucus gel as well as interactions between mucus and the ciliated epithelium [1]. Mucus retention is usually due to a combination of hypersecretion and impaired mucociliary clearance [2]. Infected secretions, poorly cleared from the airway, induce an inflammatory response and both infection and the resulting inflammation can further damage the airway epithelium [3,4].

In order to break this cycle, vigorous efforts are made to reduce the infectious and inflammatory burden in the airway [5]. Numerous mucoactive agents with a variety of actions on the airway or secretions have been used for this purpose. Understanding the properties of airway secretions and the mechanisms of action of mucoactive therapy is important for determining appropriate clinical use of these medications [6–8].

There has been some confusion regarding the nomenclature used to classify mucoactive medications [9]. Although many authors have used "mucolytic" as a generic term for these agents, it is clear that most of these medications are thought to mobilize secretions by mechanisms other than the direct thinning of mucus. For example, while the mucociliary transportability of sputum is often improved by reducing sputum viscosity

**Current affiliation:* Bowman Gray School of Medicine, Winston-Salem, North Carolina

[10] the cough clearability of secretions is independent of sputum viscosity [11].

## A. Expectorants

Expectorants are agents that have been proposed to increase the hydration of sputum either by the direct addition of water or by stimulation of water secretion into the airway. Expectorants do not directly improve mucociliary clearance. The expectorants include water, glycerol guiacolate (guaifenesin), and the iodide-containing medications. Despite widespread use of these agents, there are no well-controlled clinical trials that support the use of expectorants as effective mucokinetic agents.

The use of hypertonic saline as an expectorant has attracted recent attention. Short-term studies suggest that the inhalation of 6% saline can increase expectoration and improve pulmonary function in patients with cystic fibrosis. This is probably due, in part, to stimulation of mucus secretion in a hyperosmolar environment. Because hypertonic saline is inexpensive and readily available, this could be an attractive therapy. However, recent concerns regarding the inactivation of airway beta-defensins in a hypertonic milieu, for example in the cystic fibrosis (CF) airway, suggest that long-term clinical trials will be needed to properly evaluate the safety and efficacy of this therapy.

## B. Mucolytics

Mucociliary clearance is dependent, in part, on the viscoelasticity of the secretions [10]. Mucolytics reduce sputum viscosity by disrupting polymer networks in the secretions: Classic mucolytic agents work through the severing of disulfide bonds, binding of calcium, depolymerizing mucopolysaccharides, and liquefying proteins. Agents containing free sulfhydryl groups reduce the disulfide bridges interconnecting the mucin molecules. These agents include *N*-acetyl-L-cysteine (acetylcysteine) and similar drugs. Oral acetylcysteine improves pulmonary function and secretion clearance in selected patients with chronic lung disease. Evidence for the efficacy of acetylcysteine by the aerosol route is less clear and there are reports that inhaled acetylcysteine (pH 2.2) can cause bronchospasm and airway inflammation. There are no data that clearly support the use of carbocysteine, mesna, or similar agents as mucoactive medications.

Sputum contains products of inflammation including neutrophil-derived DNA and filamentous actin (F-actin). The DNA and F-actin copolymerize to form a rigid network entangled with the mucin gel. Peptide mucolytics degrade these abnormal filaments. Alpha dornase (Pulmozyme, Genentech, South San Francisco, CA) has been approved for the treatment of cystic fibrosis. Other peptide mucolytics under investigation include the F-actin depolymerizing agents gelsolin (Biogen, Inc., Cambridge, MA) and thymosin β4 (Alpha-1 Biomedicals, Bethesda, MD).

## C. Mucokinetic Agents

For cough to be effective, there must be sufficient airflow to detach sputum from the epithelium and to mobilize secretions so that they can be expectorated. Mucokinetic agents improve the cough clearance of secretions without altering viscoelasticity, either by increasing air flow or by altering the sputum-epithelium interaction. By this definition, bronchodilators can be considered mucokinetic agents but only in patients who demonstrate a significant improvement in airflow with bronchodilator therapy. As beta agonists are mucus secretagogues, care must be taken to document improvement in pulmonary function before prescribing these agents. Beta agonists increase ciliary beat frequency but this does not appear to improve mucus clearance in patients with lung disease.

Mucus adheres to the cilia and epithelium. It is thought that a bronchial surfactant layer between the periciliary fluid and the mucous gel promotes spreading of the mucous layer and efficient transfer of energy from beating cilia to the mucus, which prevents entanglement of the cilia in the mucus. Some adhesiveness is probably necessary to entrap particulate matter. Decreasing sputum tenacity (the product of adhesiveness and cohesiveness) generally increases sputum cough transportability (Figure 1). Some of the expectorant activity of classic mucolytics may be attributed to their ability to exert their activity on the sputum from the outside inward. Although it would be of little benefit to reduce the viscosity of an entire

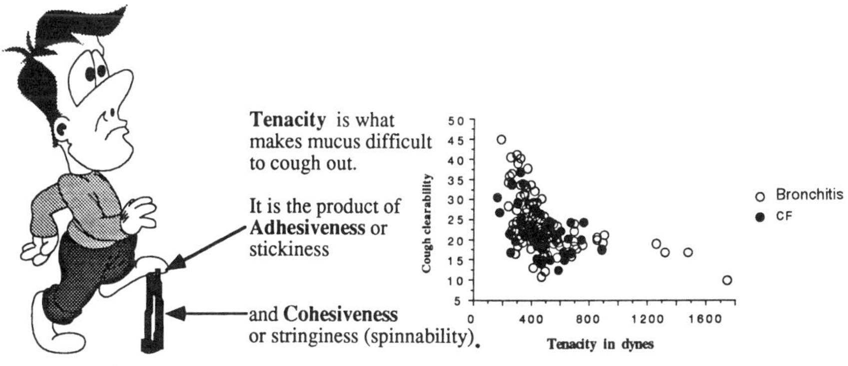


Figure 1

mucous plug, by thinning the surface it is possible to unstick the secretions from the ciliated epithelium.

As a lubricant, surfactant augments the efficient transfer of energy from the cilia to the mucous layer. In anesthetized dogs, the tracheal instillation of surfactant increased mucociliary transport velocity nearly 400% over saline instillation. Phase 2 clinical trials with an artificial surfactant in CF and chronic bronchitis patients have demonstrated a 6–12% increase in forced vital capacity (FVC), as well as a decrease in trapped thoracic gas. This improvement was associated with increased sputum mucociliary transportability on the frog palate.

According to one hypothesis, abnormal ion and water transport in CF epithelia with excessive sodium and water reabsorption may dehydrate the mucous gel layer and reduce the depth of the periciliary fluid layer in the CF airway. Therefore, blocking sodium reabsorption at the epithelium with agents like amiloride, or increasing non-CFTR chloride and water secretion at the ciliated epithelium with tricyclic nucleotides (UTP) should correct the CF defect. However there are no data demonstrating that CF secretions are dehydrated and recent studies suggest that amiloride is ineffective in promoting sputum clearance in patients with CF. UTP aerosol increases airway mucociliary clearance in persons with CF, but as tricyclic nucleotides are secretagogues, it is possible that some of the increase in clearance is due to increased mucus secretion.

### D. Mucoregulatory Agents

Mucoregulatory agents inhibit mucus production or mucus secretion. Anticholinergic medications are the best-studied agents in this class. Anticholinergics can reduce the volume of stimulated secretions without increasing viscosity. The topical anticholinergic bronchodilator, ipratropium bromide, has been shown to decrease the volume of secretions in patients with chronic bronchitis without changing mucus viscoelasticity.

Inflammation leads to mucous gland hyperplasia and many inflammatory mediators are potent secretagogues. Oral glucocorticosteroids have been used as mucoregulatory agents for many years. Indomethacin has also been administered by aerosol for the treatment of mucus hypersecretion in persons with diffuse panbronchiolitis, a chronic inflammatory airway disease seen most commonly in Japan and Korea.

Some macrolide antibiotics are also able to reduce airway mucin secretion. This action is unrelated to the antibacterial activity but may be due in part, to an anti-inflammatory action. Erythromycin and clarithromycin have been successfully used to treat mucus hypersecretion in patients with bronchorrhea, asthma, and sinusitis. Ongoing research is directed at elucidating the mechanism for this action and to identify other macrolides with similar properties.

## II. THE CLINICAL USE OF MUCOACTIVE THERAPY

The principal indication for mucoactive therapy is to reduce airway obstruction by abnormal secretions. By decreasing the volume of secretions, gas trapping is reduced and there is improved performance of the muscles of respiration. Therapy should first be directed at decreasing infection and inflammation and minimizing exposure to irritants. The use of mucoactive medications and therapy to decrease mucus production to facilitate sputum expectoration can then be considered. Chest physical therapy, with or without accessory devices, is usually prescribed along with the use of mucoactive medications.

Patients most likely to benefit from mucoactive therapy usually have a history of increased sputum expectoration and preserved airflow. With the possible exception of external chest wall oscillation, most clearance assistive devices and mucoactive medications are less beneficial in patients with severe lung disease. In general, when $FEV_1$ is less than 25% of predicted, it becomes difficult to expectorate secretions. Theoretically, when airflow is profoundly compromised, mucolytic agents could reduce sputum clearance due to the retrograde flow of airway secretions.

The effectiveness of therapy in an individual patient can be difficult to assess. When the patient feels better and there is an improvement in airflow and/or a reduction in trapped thoracic gas, benefit is clear. However, changes in $FEV_1$ poorly reflect clinical improvement with mucoactive therapy. Intuitively, one might expect that expectorated sputum volume would be a good way to assess the effectiveness of therapy but the expectorated sputum volume relates poorly, at best, to improvement in pulmonary function or the clinical status of the patient. This is partly due to limitations of measuring sputum volume because of patient reluctance to expectorate, swallowing of secretions, and salivary contamination of secretions. The sputum volume is also variable from day to day and at different times of the day. Finally, increased volumes of collected secretions could represent increased production of mucus as well as increased clearance.

Patients with gastroesophageal reflux and inability to protect their airway are at risk from postural drainage and may be at risk from the use of mucus mobilizing or expectorant therapy, particularly when the medication increases the volume of secretions produced.

Patients with acute mucus retention such as acute bronchitis or exacerbations of CF appear to be less responsive to mucoactive medications than stable patients. This may be due to decreased airflow caused both by the increase in infection and to muscular weakness in association with the pulmonary exacerbation, further reducing airflow dependent clearance mechanisms.

The scientific evaluation of secretion properties and response to therapy should facilitate the development of effective mucoactive therapy and allow us to better determine which patients are most likely to benefit from a specific therapy.

## ACKNOWLEDGMENT

Supported by a grant from the US Cystic Fibrosis Foundation (Rubin 96PO).

## REFERENCES

1. King M, Rubin BK. Mucus rheology relationship with transport. In: Takishima T, ed. Airway secretion: Physiological Bases for the Control of Mucus Hypersecretion. New York: Marcel Dekker, Inc, 1994:283–314.
2. King M, Rubin BK. Mucus physiology and pathophysiology: therapeutic aspects. In: Derenne JP, Similowski T, Whitelaw WA, Eds. Chronic Obstructive Lung Disease. New York: Marcel Dekker, Inc., 996:391–411.
3. Lange P, Nyboe J, Appleyard M, Jensen G, Schnohr P. Relation of ventilatory impairment and of chronic mucus hypersecretion to mortality from obstructive lung disease and from all causes. Thorax 1990; 45:579–585.
4. Jamal K, Cooney TP, Fleetham JA, Thurlbeck WM. Chronic bronchitis. Correlation of morphologic findings to sputum production and flow rates. Am Rev Respir Dis 1984; 129:719–722.
5. Clarke SW. Chronic bronchitis in the 1990s: up-to-date treatment. Respiration 1991; 58:43–46.
6. Wanner A. The current status of mucolytic drugs. Contemp Int Med 1991; May: 29–37.
7. Rubin BK, van der Schans CP. Management of mucus retention: evaluation of mucoactive medications. Eur Resp J. In press.
8. Dasgupta B, King M. Molecular basis for mucolytic therapy. Can Resp J 1995; 2:223–230.
9. Braga PC, Ziment I, Allegra L. Classification of agents that act on bronchial mucus. In: Braga PC, Allegra L, ed. Drugs in Bronchial Mucology. Raven Press, New York: 1989.

10. Puchelle E, Zahm JM, Girard F, Bertrand A, Polu JM, Aug F, Sadoul P. Mucociliary transport in vivo and in vitro. Relations to sputum properties in chronic bronchitis. Eur J Respir Dis 1980; 61:254–264.
11. Albers GM, Tomkiewicz RP, May MK, Ramirez OE, Rubin BK. Ring distraction technique for measuring the surface tension of sputum and relationship of the work of adhesion to clearability. J Appl Physiol 1996; 81: 2690–2695.

# 42

# Experimental Models for Mucus Drug Research

**Takeshi Miyata and Hirofumi Kai**
*Kumamoto University, Kumamoto, Japan*

## I. INTRODUCTION

Various parameters (chemical properties, physical properties, mucus production, surfactant phospholipids production, and mucociliary clearance) are considered to be important for the dynamics and the mobilization of airway secretions. Pharmacological investigation with appropriate techniques of the ability of an agent to modify these parameters can give us useful information about the mechanism of the action. However, since these parameters are strictly interconnected, it is very complicated to elucidate the precise mechanism of the action of mucoactive drugs. This means that the goal of the treatment cannot always be the modification of a single parameter, but should be more realistically aimed at a renormalization of several parameters.

On the basis of this idea, it will be taken for granted that glucocorticoids are ideal mucoactive drugs, because glucocorticoids have various pharmacological effects on the lung. However, there are few studies documenting the benefits of glucocorticoid therapy on mucociliary clearance and the excessive production of airway mucus.

We have used pigeons for the evaluation of the effect of glucocorticoids on mucociliary transport [1]. A major reason for the use of these birds is based on the findings of pharmacology, histology, and biochemistry [1–8]. The features of the birds' trachea are similar to airway inflammatory states; there are many proliferated goblet cells and submucosal glands in tracheal epithelium.

## II. MODELS FOR MUCUS DRUG RESEARCH

As shown in Figure 1, corticosterone (5.0 mg/kg) significantly increased mucociliary transport (MCT) in pigeons. Metyrapone, an adrenal 11-β steroid hydroxylase inhibitor, significantly decreased the mucociliary transport rate, and the inhibitory action was blocked by 1.0 mg/kg of corticosterone. Table 1 shows corticosterone levels in the plasma of pigeons at 30 min after administration.

Corticosterone was the main corticosteroid in the plasma of pigeons. Metyrapone prominently decreased the level of corticosterone in the plasma, and the decreased levels of corticosterone were reversed by treatment with 1.0 mg/kg corticosterone. These findings suggested that glucocorticoids modulated the mucociliary clearance, especially in diseases associated with a decreased level of endogenous corticosteroids.

As for the effects on mucus production, several reports have shown that glucocorticoids directly reduce submucosal gland secretions [9–12] without a significant alteration in the synthesis of mucins [11]. We tried to clarify the effect of dexamethasone on mucus production and mucin gene expression in a human pulmonary mucoepidermoid carcinoma cell line (NCI-H292) [13]. The NCI-H292 cells produced hyaluronidase-resistant high-molecular-weight glycoconjugates (HMWG), which elute in the void volume on Sepharose CL-4B column chromatography. Dexamethasone ($10^{-8}$–$10^{-6}$ *M*) significantly suppressed the basal production of

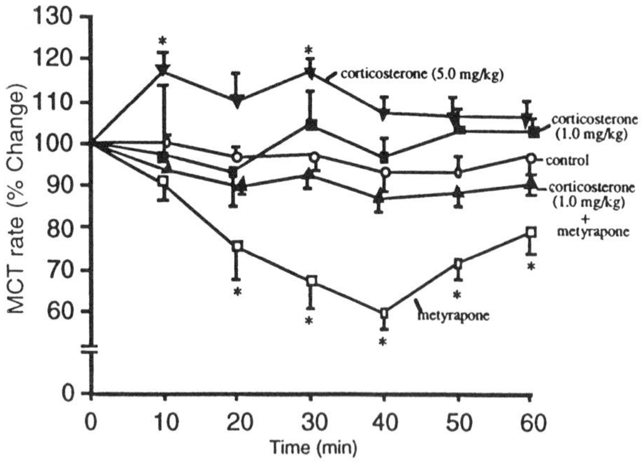


**Figure 1** Mucociliary transport in pigeons as affected by corticosterone and metyrapone. *Significantly different from each control group at $p < 0.05$.

**Table 1** Contents of Corticosterone in Pigeon Plasma

| | Vehicle | Corticosterone (mg/kg) | |
|---|---|---|---|
| | | 1.0 | 5.0 |
| Vehicle | 58.3 ± 19.3 | 213.3 ± 17.3* | 600.3 ± 112.5* |
| Metyrapone | N.D. | 372.6 ± 95.2* | 1020.5 ± 223.6* |

Each value (ng/mL plasma) represents the mean ±S.E. of 5 animals.
* Significantly different from each control group at p <0.05. N.D.: not detected

[$^3$H]glucosamine- or [$^3$H]serine-labeled HMWG in NCI-H292 cells (Figure 2).

To examine the effect on mucin gene expression, we have selected MUC-2 and MUC-5 out of eight mucin genes (MUC-1-MUC-8) [14–26]. We selected them because MUC-2 was expressed in the airways of pa-

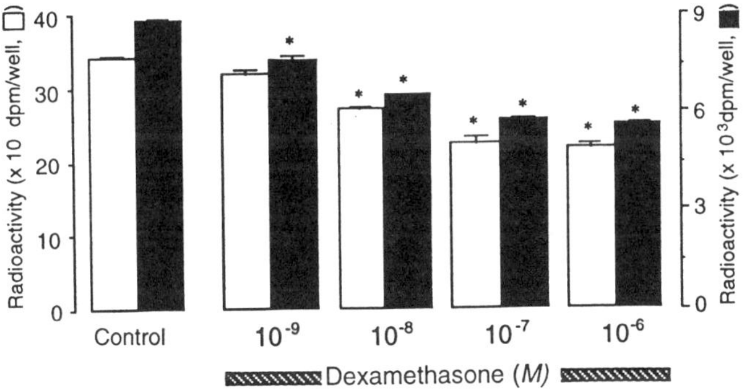


**Figure 2** Effect of dexamethasone on the produced HMWG in NCI-H292 cells. NCI-H292 cells were plated on 24-well tissue culture dishes at a density of 5 × 10$^4$ cells/well and reached to confluency five days after plating. Cultured cells were incubated with dexamethasone in the presence of the [$^3$H] glucosamine (open bars) or [$^3$H]serine (solid bars) for 24 h. After treatment with hyaluronidase, the void fraction of each sample in Sepharose CL-4B chromatography was rendered as HMWG. The produced HMWG was represented by the sum of the released HMWG and the cell-associated HMWG. Each value represents a mean + S.E. of triplicate samples from three different experiments. The difference between groups was assessed by Dunnett's test for multiple comparisons. *$p < 0.01$ vs. control.

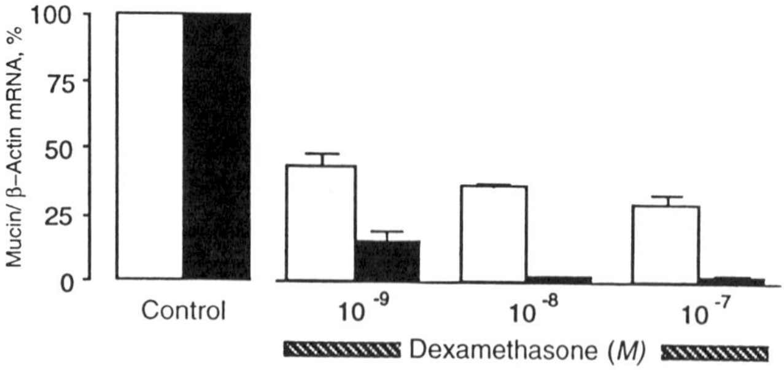


**Figure 3** Effect of dexamethasone on steady-state mRNA levels of MUC-2 and MUC-5AC mucin gene in NCI-H292 cells. Cultured cells were incubated with dexamethasone for 24 h. The panel shows the densitometric study of mRNA transcript levels of MUC-2 (open bars) and MUC-5AC (solid bars). The mRNA levels were calculated as the ratio of MUC-2 or MUC-5AC to β-actin mRNA expression and expressed as a percentage of control expression in three experiments.

tients with inflammatory airway disorders, such as cystic fibrosis and chronic bronchitis [27–30], and because MUC-5 has been cloned from the human airway [24]. In Northern blot analysis, dexamethasone ($10^{-9}$–$10^{-7}$ *M*) attenuated steady-state mRNA levels of MUC-2 and MUC-5 mucin genes (Figure 3). Thus, we concluded that dexamethasone suppressed the basal production of HMWG and decreased steady-state mRNA levels of mucin genes in airway mucus-producing cancer cells.

## III. CONCLUSION

Finally, we, at present, are interested in Chinese traditional medicines. From polypharmacological points of view, a Chinese traditional medicine can be classified as a glucocorticoid-like drug because Chinese medicines consist of many kinds of active components that have various pharmacological effects. As a future direction for research, we believe that the search for compatible actions between glucocorticoids and oriental medicines may lead us to a new world for the development of ideal mucoactive drugs with specified actions, i.e. suppression of gene expression.

## REFERENCES

1. Kai H, Yamamoto S, Takahama K, Miyata T. Influence of corticosterone on tracheal mucociliary transport in pigeons. Jpn J Pharmacol 1990; 52:496–499.
2. Miyata T, Kai H, Furusawa K, Nakamura H, Saito M, Okano Y, Takahama K. Secretomotor and mucolytic effects of mabuterol, a novel bronchodilator. Arch Int Pharmacodyn Ther 1987; 288:147–159.
3. Kai H, Saito M, Furusawa K, Oda Y, Okano Y, Takahama K, Miyata T. Protective effect of surface-active phospholipids against the acid-inducing inhibition of the tracheal mucociliary transport. Jpn J Pharmacol 1989; 49: 375–380.
4. Kubo S, Kasé Y, Miyata Y, Kito G, Uesaka I. Pharmacological studies of 1-(o-Chlorophenyl)-2-tert.-butylaminoehanol (C–78), a new bronchodilator. Arzneim-Forsch/Drug Res 1975; 25:1028–1037.
5. Inatomi N, Kuriki H, Kanno M, Kawagoe K, Nagawa Y, Miyata T, Takahama K, Kasé Y. Pharmacological studies of trans-6-hydroxy-5-hydroxymethyl-2-isopropylamino-1,2,3,4-tetrahydro-1-napht halenol hydrochloride (AA-497), a new potent bronchodilator. Arzneim-Forsch/Drug Res 1980; 30: 276–285.
6. Miyata T, Matsumoto N, Yuki H, Takahama K, Okano Y, Kai H. The effect of *N*-acetylneuraminic acid on the mucociliary transport impaired by cigarette smoke. Arch Int Pharmacodyn Ther 1988; 296:202–209.
7. Miyata T, Matsumoto N, Yuki H, Oda Y, Takahama K, Kai H. Effects of anticholinergic bronchodilators on mucociliary transport and airway secretion. Jpn J Pharmacol 1989; 51:11–15.
8. Tai S, Kai H, Isohama Y, Takahama K, Miyata T. Effect of leukotriene $D_4$ on tracheal mucociliary transport velocity in quails. Jpn J Pharmacol 1996; 70:195–197.
9. Marom Z, Shelhamer J, Alling D, Laliner M. The effect of corticosteroids on mucus glycoprotein secretion from human airways in vitro. Am Rev Respir Dis 1984; 129:62–65.
10. Lundgren JD, Hirata F, Marom Z, Logun C, Steel L, Kaliner M, Shelhamer J. Dexamethasone inhibits respiratory glycoconjugate secretion from feline airways in vitro by the induction of lipocortin (lipomodulin) synthesis. Am Rev Respir Dis 1988; 137:353–357.
11. Shimura S, Sasaki T, Ikeda K, Yamauchi K, Sasaki H, Takishima T. Direct inhibitory action of glucocorticoid on glycoconjugate secretion from airway submucosal glands. Am Rev Respir Dis 1990; 141:1044–1049.
12. Satoh M, Shimura S, Sasaki H, Ebina M, Takishima T. Dexamethasone modulation of ion-transport and fluid movement across airway epithelium. Am J Physiol 1993; 264 (Lung Cell Mol Physiol 8):L376–L381.
13. Kai H, Yoshitake K, Hisatsune A, Kido T, Isohama Y, Takahama K, Miyata T. Dexamethasone suppresses mucus production and MUC-2 and MUC-5AC gene expression by NCI–H292 cells. Am J Physiol 1996; 271 (Lung Cell Mol Physiol 15):L484–L488.

14. Aubert J-P, Porchet N, Crepin M, Duterque-Coquillaud M, Vergnes G, Mazzuca M, Debuire B, Petitprez D, Degand P. Evidence for different human tracheobronchial mucin peptides deduced from nucleotide cDNA sequences. Am J Respir Cell Mol Biol 1991; 5:178–185.
15. Bobek LA, Tsai H, Biesbrock AR, Levine MH. Molecular cloning, sequence, and specificity of expression of the gene encoding the low molecular weight human salivary mucin (MUC7). J Biol Chem 1993; 268:20563–20569.
16. Gendler SJ, Lancaster CA, Taylor-Papadimitriou J, Duhig T, Peat N, Burchell J, Pemberton L, Lalani E-N, Wilson D. Molecular cloning and expression of human tumor-associated polymorphic epithelial mucin. J Biol Chem, 1990; 265:15286–15293.
17. Gum JR, Byrd JC, Hicks JW, Toribara NW, Lamport DTA, Kim YS. Molecular cloning of human intestinal mucin cDNAs. J Biol Chem 1989; 264: 6480–6487.
18. Gum JR, Hicks JW, Swallow DM, Lagace RL, Byrd JC, Lamport DTA, Siddiki B, Kim YS. Molecular cloning of cDNAs derived from a novel human intestinal mucin gene. Biochem Biophys Res Commun, 1990; 171:407–415.
19. Gum JR Jr, Hicks JW, Toribara NW, Rothe E-M, Lagace RE, Kim YS. The human MUC2 intestinal mucin has cysteine-rich subdomains located both upstream and downstream of its central repetitive region. J Biol Chem, 1992; 267:21375–21383.
20. Gum JR Jr, Hicks JW, Toribara NW, Siddiki B, Kim YS. Molecular cloning of human intestinal mucin (MUC2) cDNA. J Biol Chem 1994; 269:2440–2446.
21. Lan MS, Batra SK, Qi W-N, Metzgar RS, Hollingsworth MA. Cloning and sequencing of a human pancreatic tumor mucin cDNA. J Biol Chem 1990; 265:15294–15299.
22. Meerzaman D, Charles P, Daskal E, Polymeropoulos MH, Martin BM, Rose MC. Cloning and analysis of cDNA encoding a major airway glycoprotein, human tracheobronchial mucin (MUC5). J Biol Chem 1994; 269: 12932–12939.
23. Porchet N, Van Cong N, Dufossse J, Audie JP, Guyonnet-Duperat V, Gross MS, Denis C, Degand P, Bernheim A, Aubert JP. Molecular cloning and chromosomal localization of a novel human tracheo-bronchial mucin cDNA containing tandemly repeated sequences of 48 base pairs. Biochem Biophys Res Commun 1991; 171:407–415.
24. Toribara NW, Gum JJR, Culhane PJ, Lagace RE, Hicks JW, Petersen GM, Kim YS. MUC-2 human small intestinal mucin gene structure. J Clin Invest 1991; 88:1005–1013.
25. Toribara NW, Roberton AM, Ho SB, Kuo W-L, Gum E, Hicks JW, Gum JR Jr, Byrd JC, Siddiki B, Kim YS. Human gastric mucin. J Biol Chem 1993; 268:5879–5885.
26. Shankar V, Gilmore MS, Elkins RC, Sachdev GP. A novel human airway mucin cDNA encodes a protein with unique tandem-repeat organization. Biochem J 1994; 300:295–298.
27. Gerard CR, Eddy L, Shows TB. The core polypeptide of cystic fibrosis

tracheal mucin contains a tandem repeat structure: evidence for a common mucin in airway and gastrointestinal tissue. J Clin Invest 1990; 86:1921–1927.
28. Gum JR Jr. Mucin genes and the proteins they encode: structure, diversity, and regulation. Am J Respir Cell Mol Biol 1992; 7:557–564.
29. Jany BH, Gallup MW, Yan PS, Gum JR, Kim YS, Basbaum CB. Human bronchus and intestine express the same mucin gene. J Clin Invest 1991; 87: 77–82.
30. Rose MC. Mucins: structure, function, and role in pulmonary disease. Am J Physiol 1992; 263 (Lung Cell Mol Physiol 6):L413–L429.

# 43

# Ciliary Beat Frequency, Ultrastructural Abnormalities, and Mucus Clearance in Primary Ciliary Dyskinesia (PCD)

**Michael T. Newhouse**
*McMaster University, and St. Joseph's Hospital, Hamilton, Ontario, Canada*

## I. INTRODUCTION

Since 1975 when Camner and Afzelius first described what they called immotile cilia syndrome [1] this rare congenital abnormality has been the subject of increasing interest. This interest is not only because what was previously called Kartagener's Syndrome, first described in 1933 as a triad of sinusitis, bronchiectasis, and situs inversus (SI) [2] (to which male infertility was latter added), appeared to have an identifiable mechanism related to a probable genetic defect but also because the absence of mucociliary transport (MCT) despite usually motile but dyskinetic cilia has taught us a great deal about other mechanisms that accomplish mucus clearance from the airways when cilia are dyskinetic, slowed, immotile, or absent (e.g. squamous metaplasia). It is perhaps not surprising that even in the complete absence of coordinated ciliary function, we are nevertheless able to clear mucus using a variety of backup mechanisms that also come into play during viral bronchitis.

These mechanisms include cough, which clears secretions mainly from major airways; gas–liquid pumping, probably the major mechanism in peripheral airways; and the surfactant gradient from peripheral to central airways described by the La Place relationship (pressure gradient is equal to 2× surface tension/radius). Other factors include the depth and rheological characteristics of the periciliary fluid, surfactant as a lubricant between mucus and periciliary fluid, and the viscoelastic properties of the mucus itself [3].

In contrast to cystic fibrosis (CF) where decreased chloride secretion into the airway lumen leads to absorption of water from the periciliary fluid and mucus layers [4] causing secondary ciliary dysfunction, in primary ciliary dyskinesia (PCD) uncoordinated and usually slowed ciliary beating results from a variety of ultrastructural defects and/or malorientation of the cilia. The ultrastructural abnormalities cause abnormal sliding and bending of ciliary microtubules while ciliary malorientation results in chaotic ciliary motion and failure of metachronal wave development. This in turn leads to impaired mucus transport, particularly in the sinuses, eustacian canal, and peripheral airways where cough clearance is poor. This results in retained upper and lower respiratory tract secretions and recurring infections in the sinuses, middle ear cavity, lower respiratory tract airways, and pulmonary parenchyma. The latter leads to bronchiectasis. Male infertility that occurs in almost all of these patients (except those with malorientation only), results from the inability of the sperm to progress due to similar ultrastructural abnormalities as in the respiratory tract cilia. The majority of these patients have outer and/or inner dynein arm deficiency, while loss of radial spokes and the peripheral doublet transposition defect are rare [5]. Interference with normal sperm function in PCD associated only with ciliary malorientation is unlikely since the progression of sperm should be unimpaired. Thus, males with this type of PCD are likely to be fertile [6,7].

Another interesting feature is SI that occurs in about 50% of PCD patients [8]. One biologically plausible hypothesis put forward by Afzelius suggests that SI occurs because the embryonal archenteron is normally ciliated, which accounts for the usual rotation of internal organs. In the absence of effective ciliary function, there should be an approximately equal chance of situs normalis and SI as is the case in PCD. Thus, in patients who present with clinical features otherwise compatible with Kartagener syndrome (KS) but without SI, PCD is nevertheless considered.

The differential diagnosis includes CF, immunodeficiency syndromes, $\alpha_1$ antitrypsin deficiency, asthma, and gastroesophageal reflux disease. Indeed the problem is made much more complex clinically by the fact that the incidence of asthma is two times greater in patients with PCD [8]. There is possibly also an increased likelihood of gastroesophageal reflux (GER) disease in patients with reversible airflow obstruction with an incidence of GER of approximately 80% in contrast to 30% in the normal population [9].

## II. LABORATORY EVALUATION OF PCD

In addition to clinical and physiological evaluation (resting spirometry, lung volumes, diffusing capacity, arterial blood gases, and exercise testing) and assuming that PCD-like syndromes have been excluded, how can the diagnosis of PCD be established with a reasonable degree of certainty? There are three components to confirming the diagnosis, namely (a) studies of MCT, (b) electron microscopy to define ciliary ultrastructure, and (c) evaluation of ciliary beat frequency and wave form.

### A. Mucociliary Transport

Mucociliary transport can be evaluated as an office procedure by means of a nasal taste transit time or by nasal or tracheobronchial MCT evaluated scintigraphically [5].

#### Taste Transit Time (TTT)

A relatively simple and inexpensive way to evaluate MCT is the saccharine or aspartame TTT that can readily be done in any physician's waiting room and takes at the most 1 hour [10]. Saccharine is 600 times and aspartame 300 times as sweet as sugar and can easily be detected when it drops from the nasopharynx onto the back of the tongue. A 1 mm diameter granule is prepared from a crushed sweetener tablet. With the patient lying comfortably on an examining table the nasal speculum is used to identify the inferior turbinate. Using a bayonet forceps the sweetener granule is advanced through the nasal speculum and placed gently onto the superior surface of the inferior turbinate about 1.5 cm inside the nares. The patient is then comfortably seated with head tilted slightly forward to prevent nasal secretions from trickling posteriorly causing false positive results. The patient should be asked to read a book or magazine without tilting the head back and without sniffing until a sensation of intense sweetness is noted. In contrast to normals, patients with PCD do not taste sweetness within 30 minutes and indeed the test may be negative even after 60 minutes. At that point the patient should be asked to sniff vigorously and swallow several times to make sure that the sweetness can be tasted. Normals will generally taste sweetness within $19 \pm 5$ minutes. We were unable to understand why one of our patients with proven PCD with inner and outer dynein arm defect (and with a sister who had Kartagener syndrome) was able to taste sweetness after approximately 15 minutes on two separate occasions. On further consideration and with the likelihood that the positive test was an artifact, we repeated the test a third time. On that occasion it was stressed that she should not tilt her head back or sniff since ongoing rhinorrhea was a major problem (as it is with virtually all PCD patients). On the third TTT there was no taste even after 60 minutes until she sniffed vigorously when sweetness could be appreciated almost at once.

### B. Scintigraphic Nasal and Tracheobronchial MCT

This has previously been described in detail [11]. Absence of nasal and/or tracheobronchial MCT does not alone confirm the diagnosis of PCD since secondary ciliary dyskinesia associated with nasal and/or tracheobronchial infection may also impair MCT [12]. However a relatively normal MCT excludes PCD [5].

### C. Evaluation of Ciliary Beat Frequency and Wave Form

After screening the patient with the TTT, ciliary beat frequency (CBF) and wave form are evaluated using nasal ciliated cells harvested by means of a small bronchoscopy brush that is inserted about 2–3 cm into the nose and directed against the lateral wall of the nasal cavity, then rapidly withdrawn. This scrapes an adequate sample of the nasal mucosa onto the brush, which is then vigorously shaken into a vial of Earl's balanced salt solution. Clumps of cells can readily be removed for microscopic examination at once or refrigerated and saved for several days if necessary. Specimens can also be readily transported in an insulated container with an ice pack. A second sample is fixed in gluteraldehyde for transmission electron microscopy (TEM).

Ciliary beat frequency and wave form can be studied using a motion analysis video camera (Sony Model No. RSC-1010, Japan) fitted to a phase contrast microscope (Leitz Dialux 22, Japan) with a magnifying capability of 600× and 1,000× (oil immersion). To visualize the ciliated cells, they are placed on a slide under a cover slip or alternatively a hanging-drop preparation can be prepared. For measurement of ciliary beat frequency, specimens are warmed on the microscope stage to 37°C. If slow motion video analysis capability is not available then the cilia can be slowed for visual analysis of wave form by cooling them, although beat frequency cannot be determined reliably.

The specimens can be evaluated for the presence of metachronal waves, specific abnormalities of wave form, or chaotic movement. In some forms of PCD the cilia may initially appear immotile at lower magnification because the light and dark oscillations produced by the metachronal waves of normally beating cilia are absent (which probably accounted for the misnomer "immotile cilia syndrome"). The abnormal beat frequency and wave form may only be evident at a magnification of 600× or more and with slow motion video analysis.

On average 200 cilia are evaluated from 10 different locations for beat frequency and wave form. After determining the CBF overall (normal 12.5 ± 1 Hz), the rate of motion of the abnormal cells is measured and the proportion of these cells is estimated. Depending on the type of ultrastructural abnormality, 75–90% of cells show abnormalities of motion in PCD. Most PCD cilia have a range of beat frequency from 6 Hz characteristic of inner and outer dynein arm deficiency to approximately 9–10.5 Hz with translocation and radial spoke defects. With ciliary malorientation as the sole abnormality, a decreased beat frequency of about 10 Hz is seen. For like cilia having a specific ultrastructural abnormality, motion is chaotic and debris or bacteria in the neighborhood of the cilia are simply bounced up and down or wafted around in small vortices, evident when viewed at 1000×. With PCD, no metachronal waves can be seen in contrast to normals where metachronicity is readily visualized at normal speed or using slow motion. In normal ciliary preparations, any debris that comes near the ciliated cells is seen to be rapidly propelled out of the microscope field [5,11,13–15].

## III. ULTRASTRUCTURAL ABNORMALITIES

The types of ultrastructural abnormality include loss of both dynein arms, loss of the inner or outer arms, radial spoke defect, and translocation defect of peripheral doublets into a central position with missing central microtubules [5].

With virtually complete dynein arm deficiency, ciliary motion appears chaotic. These cilia showed two fairly distinct types of abnormality; they either rotated about their long axis or showed a rigid vibratory motion. These changes are much less marked if only the inner arms are missing. Dynein arm deficiency is by far the most common abnormality and accounts for about 90% of PCD.

Radial spokes appear to stabilize the cilium longitudinally and transmit the sliding and bending motion. Rather than being absent as was thought in the past, the radial spokes appear to be collapsed and the resulting motion is rigid and corkscrew like.

The translocation deformity results in a rigid grabbing motion with slight bending of the distal third of the cilia, a poor recovery stroke and an incomplete effector stroke.

It should be stressed that with ciliary malorientation as the only abnormality, the TEM may be reported as normal unless the reporting pathologist is aware of this defect and is prepared to undertake the evaluation of the angles between adjacent cilia on the same cell. A minimum of 50 cilia should be evaluated. The average normal variation in ciliary orientation is 10° and the maximum that has been recorded in normals is 28°. Values above that must be considered abnormal and in keeping with failure or gross impairment of MCT [6]. Studies have shown that ciliary malorientation may occur as a secondary phenomenon in the presence of severe infection with improvement to normal orientation some weeks after the infection has been successfully treated [16].

## IV. SUMMARY AND CONCLUSIONS

The diagnosis of PCD depends in part upon exclusion of more common clinical look-alikes using appropriate studies (e.g., sweat electrolytes, gamma globulin, and immunoglobulin A deficiency, etc). Asthma, $\alpha_1$ antitrypsin deficiency, and gastroesophageal reflux disease should also be considered in the differential diagnosis since these may resemble PCD or co-exist with PCD and require specific therapy. Above all, the diagnosis of PCD should be considered, whether situs inversus is present or not, if recurring upper and lower respiratory tract infections leading to bronchitis and/or bronchiectasis, sinusitis, and otitis media start in infancy or early childhood and persist or progress inexorably. The diagnosis would be strongly supported if the TTT is greater than 45 minutes in the absence of grossly purulent sinusitis or rhinitis at the time the study is carried out. As yet, the TTT has not been adequately standardized nor have sensitivity or specificity been established but an international committee has been formed recently to try to achieve consensus on methodology. [17]

Nasal brushings for CBF and wave form evaluation can be carried out using phase-contrast motion analysis video microscopy while clinical and radiographic assessment are being undertaken. Furthermore, ciliated cells placed in Earl's balanced salt solution and in insulated containers cooled with ice packs can readily be couriered to centers where facilities for motion analysis and ciliary beat frequency exist. Ciliary ultrastructural changes or malorientation on TEM will almost always provide histologic confirmation of the diagnosis and should be rapidly available from relatively painless and rapid nasal brushings. In males suspected of PCD, sperm may be assessed for lack of normal flagellar motility and progression, but absence of these features does not exclude the diagnosis of PCD in all cases since malorientation has no effect on sperm flagellae, and in these patients, despite clinical features of respiratory tract PCD, fertility is not impaired. There is rarely a need to do isotope studies of nasal or tracheobronchial clearance to study mucus transport in PCD (except for research purposes) but such studies would confirm absence of MCT and would allow evaluation of cough clearance of mucus in PCD patients.

## ACKNOWLEDGMENTS

I would like to thank Carol Rossman and Carole Chambers who carried out many of the studies on which this review is based. I also much appreciate the efforts of Janice Rutgers whose assistance in the preparation of this manuscript was invaluable.

## REFERENCES

1. Camner P, Mossberg B, Afzelius B. Evidence of congenital non-functioning cilia in the immotile–cilia syndrome. Am Rev Respir Dis 1975; 112:807–809.
2. Kartagener M. Zur Pathogenese der Bronchiektasien: Bronchiektasien bei Situs viscerum inversus. Beitr Klin Tuberk. 1933; 83:489–501.
3. Newhouse MT. Primary ciliary dyskinesia: What has it taught us about pulmonary disease? Eur J Respir Dis 1983; 64(Suppl 127):151–156.
4. Boucher RC. Human airway ion transport. Am J Respir Crit Care Med 1994; 150:271–281.
5. Rossman CM, Newhouse MT. Primary ciliary dyskinesia: evaluation and management. Pediatr Pulmonol 1988; 5:36–50.
6. Rutland J, de Iongh RU. Random ciliary orientation. A cause of respiratory tract disease. N Engl J Med 1990; 323:1681–1684.
7. Rutman A, Cullinan P, Woodhead M, Cole PJ, Wilson R. Ciliary disorientation: a possible variant of primary ciliary dyskinesia. Thorax 1993; 45: 770–771.
8. Sturgess JM, Turner JAP. Ultrastructural pathology of cilia in the immotile cilia syndrome. Perspect Pediatr Pathol 1984; 8:133–161.
9. Sontag SJ, O'Connell S, Khandelwal S, Miller T, Nemchausky B, Schnell TG, Serlovsky R. Most asthmatics have gastroesophageal reflux with or without bronchodilator therapy. Gastroenterology 1990; 99:613–620.
10. Canciani M, Barlocco EG, Mastella G, de Santi MM, Gardi C, Lungarella G. The saccharin method for testing mucociliary function in patients suspected of having primary ciliary dyskinesia. Pediatr Pulmonol 1988; 5: 210–214.
11. Rossman CM, Forrest JB, Ruffin RE, Newhouse MT. Immotile cilia syndrome in individuals with and without Kartagener's syndrome. Am Rev Respir Dis 1980; 121:1011–1016.

12. Wanner A, Matthias S, O'Riordan TG. Mucociliary clearance in the airways. Am J Respir Crit Care Med 1996; 154:1868–1902.
13. Rossman CM, Forrest JB, Lee RMKW, Newhouse AF, Newhouse MT. The dyskinetic cilia syndrome. Abnormal ciliary motility in association with abnormal ciliary ultrastructure. Chest 1981; 80S:860–864.
14. Rossman CM, Lee RMKW, Forrest JB, Newhouse MT. Nasal ciliary ultrastructure and function in patients with primary ciliary dyskinesia compared with that in normal subjects and in subjects with various respiratory diseases. Am Rev Respir Dis 1984; 129:161–167.
15. Rossman CM, Lee RMKW, Forrest JB, Newhouse MT. Nasal cilia in normal man, primary ciliary dyskinesia and other respiratory diseases: analysis of motility and ultrastructure. Eur J Respir Dis 1983; 64(Suppl 127):5–10.
16. Armengot M, Juan G, Carda C, Basterra J, Cano B. The prevalence of primary dyskinetic ciliary syndromes in patients with sinusitis and bronchiectasis. An Otorrinolaringol Ibero Am 1995; 22:85–92.
17. Working Group on standardization of taste transit time established at the International Congress on Cilia, Mucus and Mucociliary Interactions, Jerusalem, Israel, Feb 23–28, 1997.

# 44
# Diagnosis of Primary Ciliary Dyskinesia

**Jonathan Rutland and Lucy Morgan**
*Concord Hospital, Concord, and The University of Sydney,*

**Karen A. Waters and Peter van Asperen**
*Royal Alexandra Hospital for Children, Sydney, New South Wales, Australia*

**Robbert U. de Iongh**
*Concord Hospital, Concord, and University of Sydney, Sydney, New South Wales, Australia*

## I. INTRODUCTION

Primary ciliary dyskinesia (PCD), which includes Kartagener's syndrome [1] and the immotile-cilia syndrome [2] is a genetic congenital defect associated with heterogeneous abnormalities of ciliary axonemes and defective ciliary function that result in chronic respiratory disease. The clinical features range widely in severity and extent but usually include chronic or recurrent upper and/or lower respiratory tract infections (sinusitis, otitis media, bronchitis often leading to suppuration and bronchiectasis), male infertility (due to similar structural abnormalities of sperm axonemes causing reduced or absent motility), and organ malrotation (dextrocardia or situs inversus) in approximately 50% of cases [3]. Kartagener's original description [1] was of a triad consisting of chronic sinusitis, bronchiectasis, and situs inversus totalis (now commonly interpreted as the presence of any organ malrotation–dextrocardia and/or situs inversus viscerum).

The clinical features of PCD were first linked to defective cilia [4,5] in 1976 and a range of ciliary ultrastructural abnormalities and motility patterns (from normal to immotile) have been identified [4–13]. The diagnosis of PCD requires demonstration of abnormal ciliary function and/or ultrastructure. Ciliary function can be assessed by measuring ciliary beat frequency (CBF) in vitro [14]. Measurements of mucociliary clearance (MCC) are done in vivo but give less direct evidence of impaired ciliary function. In adults MCC can be assessed quickly by the nasal saccharin method [15–17] but this method is not useful in young children. More sophisticated methods of measuring MCC and CBF are not always readily available and many patients need to be referred to specialist units for investigation. Ciliary ultrastructure is assessed by electron microscopy and many case reports have described various structural ciliary defects that are thought to be the cause of abnormal ciliary motility in patients with PCD [4–13]. These defects include absent or reduced numbers of inner and or outer dynein arms [4–7], absent radial spokes [8], translocation of microtubule doublets [9], abnormal length cilia [11], abnormal basal apparatus [12], random ciliary orientation (not necessarily linked to axonemal defects) [18–20], and even ciliary aplasia [13]. There are also reports of patients with clinical signs of PCD in whom ciliary ultrastructure is normal [21,22]. The interpretation of electron microscopic findings of respiratory tract cilia from patients with chronic respiratory disease has often been limited by the absence of quantitative analysis of variations in ciliary ultrastructure in normal subjects. Determination of the normal range and the incidence of structural defects in human respiratory tract cilia is essential for reliable diagnosis of PCD. Because analysis of ciliary ultrastructure is time consuming and expensive it would be beneficial to ascertain whether there are clinical features of recurrent respiratory disease (RRTD) that are more closely associated with a positive diagnosis of PCD and that might assist in the identification of children who warrant evaluation of ciliary function and ultrastructure.

In this paper we report how the ranges of ciliary ultrastructural defects (dynein arm numbers, variation in ciliary orientation, and microtubular defects), quantified in normal subjects, can be used for the laboratory diagnosis of primary ciliary dyskinesia and also report the clinical features that we found most closely associated with a diagnosis of PCD in patients referred for study because of recurrent respiratory tract infections.

## II. MATERIALS AND METHODS

### A. Laboratory Studies—Ciliary Function and Ultrastructure

Ciliary beat frequency (CBF) and ciliary ultrastructure were studied in 62 normal subjects aged <1–74 years (mean ± SD; 40 ± 20 years) without any history of significant respiratory disease and in 55 patients with PCD aged <1–71 years (15 ± 16 years). The healthy subjects were grouped according to tobacco consumption (nonsmokers, smokers, and exsmokers) and, since there were no significant differences between the groups, the results were pooled as reported previously [23]. Patients with recurrent respiratory tract disease (such as bronchiectasis and sinusitis) were referred for study at Concord Hospital because of a clinical suspicion of PCD.

Ciliary beat frequency was measured photometrically in samples of ciliated epithelium obtained from the surface of the inferior turbinate by nasal mucosal brushing [14,24]. Further samples of ciliated epithelium were processed for electron microscopy [25,26] and the incidences of the following ultrastructural parameters were quantified from electron micrographs as described previously [23,25]: compound cilia (CC), central (CT), and peripheral (PT) microtubule defects, inner (DAi) and outer (DAo) dynein arm numbers seen per cilium (usually less than the theoretical maximum of nine) and ciliary disorientation. Ciliary disorientation was measured by a method described previously [25]. On micrographs enlarged ×65,000, lines were drawn through the two central tubules to define an axis perpendicular to the ciliary beat axis (see Figure 1). The angles (0–180°) made between these lines and an arbitrary reference line (usually the horizontal axis for convenience), were digitized by computer-aided image analysis to derive a mean ciliary angle. The variation (or standard deviation) of the ciliary beat axis from the mean ciliary angle was calculated [25].

Nasal mucociliary clearance time was measured using the saccharin technique [15–17] in those patients old enough to cooperate (usually >7 years). In all patients with PCD clearance times were greater than 45 minutes (normal range <30 minutes [16]).

## B. Clinical Features

Clinical information was available for 53 of the PCD patients (pediatric <16 years, $n = 39$; adult ≥16, $n = 14$), whose diagnosis was made on the basis of abnormal ciliary function and/or ultrastructure. This was compared to 43 other patients (mean age 17 ± 19 years), referred because of a similar history of recurrent respiratory tract infections (RRTD) and who all had normal CBF, nasal mucociliary clearance times, and/or ciliary ultrastructure.

Thirty-five clinical features were recorded at presentation or obtained from case notes. Data collected included (a) age at presentation including neonatal events (such as neonatal respiratory distress); (b) upper respiratory tract symptoms (including ear, nose, and sinus), radiological

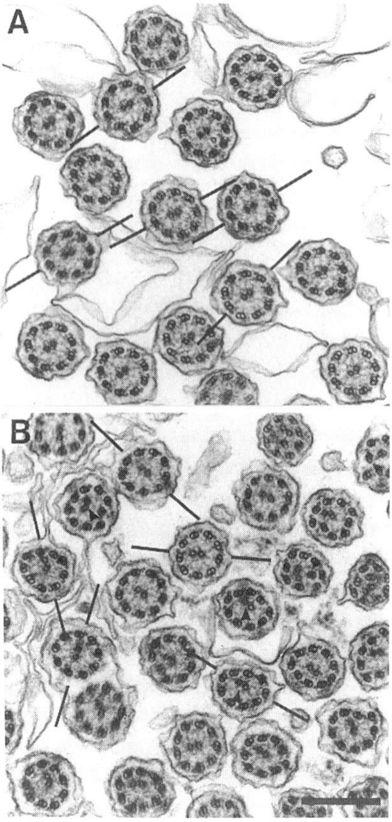


**Figure 1** Cilia from a normal subject (A) showing aligned orientation of ciliary axes and from a PCD patient (B) showing random orientation of the ciliary axes and also eccentrically located central tubules (arrowheads). The ciliary axes are shown for five cilia in each field. Scale: 0.3 μm. From de Iongh and Rutland [23].

evidence of sinusitis, auditory acuity, surgery (antrostomy, grommets, polypectomy); (c) lower respiratory tract symptoms (such as cough, wheeze), radiology, surgery; (d) associated clinical features such as clubbing, dextrocardia, situs inversus, infertility, deformities, and growth parameters; (e) family history; (f) detailed past medical history (encompassing any relevant previous medical event such as lobar collapse, pneumonia, the need for ventilation, gastroesophageal reflux); (g) investigations carried out, either to diagnose or assess the patient (lung function tests, microbiology, bronchoscopy) or to exclude other diseases (sweat test, immunology, $\alpha_1$-antitrypsin, cardiac investigations).

None of the patients had identifiable conditions such as cystic fibrosis, defective immune function, $\alpha_1$-antitrypsin deficiency, or severe allergic disease. Results for each feature were expressed as a percentage of the incidence of all patients for whom the information was available.

### C. Statistical Analyses

Mann-Whitney U tests or Student's *t*-tests were used to test the significance of differences in ciliary beat frequency and ultrastructural parameters between the PCD and normal subjects according to the distribution of data. Contingency tables and Chi-squared analysis were used to compare each of the clinical features between the two groups. The significance level was set at $p < 0.05$.

## III. RESULTS

### A. Laboratory Studies—Ciliary Function and Ultrastructure

Cilia in patients with PCD showed a range of ultrastructural defects such as reduced numbers of dynein arms including inner (Figure 2B), outer (Figure 2C) and both inner and outer dynein arms (Figure 2D) as well as eccentrically located central microtubules, extra single microtubules, and displaced peripheral doublet microtubules (Figure 2E) and compound cilia (Figure 2F). Other microtubule defects seen commonly included absence of one or more peripheral doublets, supernumerary doublets, and absence of the 'B' tubule in peripheral doublets. Abnormal orientation of cilia was also commonly seen in PCD patients (Figure 1).

The results of CBF measurements and ciliary ultrastructure for the 62 normal subjects compared to results from PCD patients are shown in Table 1. For the PCD patients, CBF was significantly slower than for the healthy subjects. The incidence of compound cilia did not differ between the two groups but PCD patients had significantly more microtubular defects (central and peripheral), significantly fewer dynein arms (both inner and outer), and significantly greater ciliary disorientation than the normal subjects.

The frequency distributions for CBF and ciliary ultrastructure are shown in Figure 3. The distributions for CBF, dynein arm numbers, and ciliary orientation were distinctly different in patients with PCD compared to healthy subjects and these measurements were more useful in differentiating between the groups than were compound cilia numbers or incidence of microtubular defects.

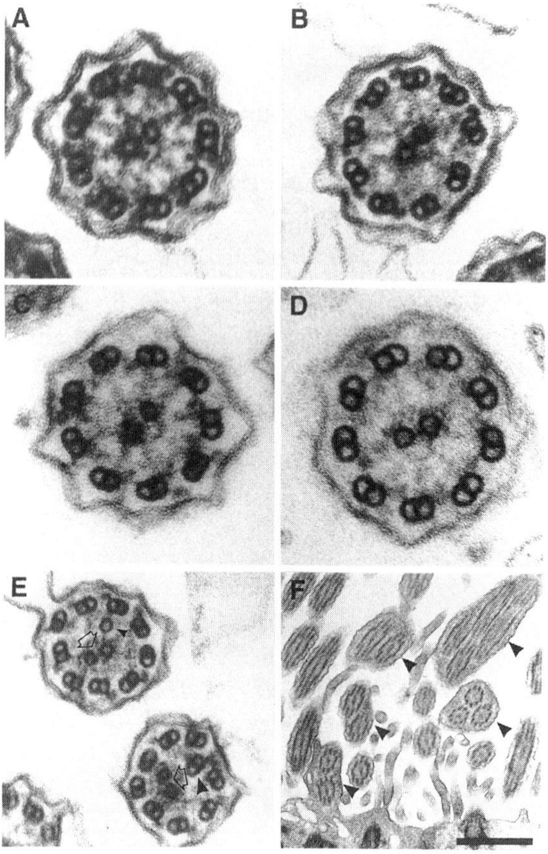


**Figure 2** Electron micrographs of cilia from a normal subject (A), four PCD patients (B, C, D, E) and an RRTD patient (F). (A) Cilium from a normal subject with the normal complement and arrangement of peripheral and central microtubules, inner and outer dynein arms, radial spokes and nexin links. (B) Cilium with a deficiency of inner dynein arms. (C) Cilium with reduced numbers of outer dynein arms. (D) Cilium showing reduction of both inner and outer dynein arms, other components of the axoneme appear normal. (E) Two cilia with absent inner dynein arms, eccentrically located central microtubules (open arrows), displaced peripheral doublet (larger arrowhead), and an extra single microtubule (smaller arrowhead) inside the peripheral ring of doublets. (F) Compound cilia (arrowheads) from a patient with RRTD, not due to PCD. Scale bar: A–D, 0.1 μm; E, 0.13 μm; F, 0.6 μm. From de Iongh and Rutland [23].

**Table 1** Quantitation of Ciliary Ultrastructure in Normal Subjects and Patients with Primary Ciliary Dyskinesia

| Subjects/ patients | Age (years) | CBF (Hz) | Axonemal defects (%) | | | Dynein arm numbers | | Disorientation (degrees) |
|---|---|---|---|---|---|---|---|---|
| | | | CC | PT | CT | Inner | Outer | |
| Normal subjects | | | | | | | | |
| Mean | 40 | 11.6 | 0.4 | 2.2 | 1.2 | 5.4 | 8.5 | 14.3 |
| ±SD | ±20 | 1.5 | ±0.6 | ±1.7 | ±1.4 | ±1.0 | ±0.3 | ±3.7 |
| (Range) | (1–74) | (9.1–15.3) | (0–3.7) | (0–8.9) | (0–7.8) | (3–7.1) | (7.4–9.0) | (7.9–28.5) |
| PCD | | | | | | | | |
| Mean | 14.5 | 3.6 | 0.2 | 7.5 | 5.6 | 3.0 | 3.1 | 30.9 |
| ±SD | ±15.6 | 3.7 | ±0.6 | ±9.5 | ±11.0 | ±2.5 | ±2.6 | ±9.9 |
| (Range) | (0.3–71) | (0–12.9) | (0–2.7) | (0–36.5) | (0–46.5) | (0–8.3) | (0–8.9) | (14.4–51.9) |

CC—Compound cilia; CT—Central tubule defects; PT—Peripheral tubule defects; SD—Standard deviation, and PCD—Primary ciliary dyskinesia.

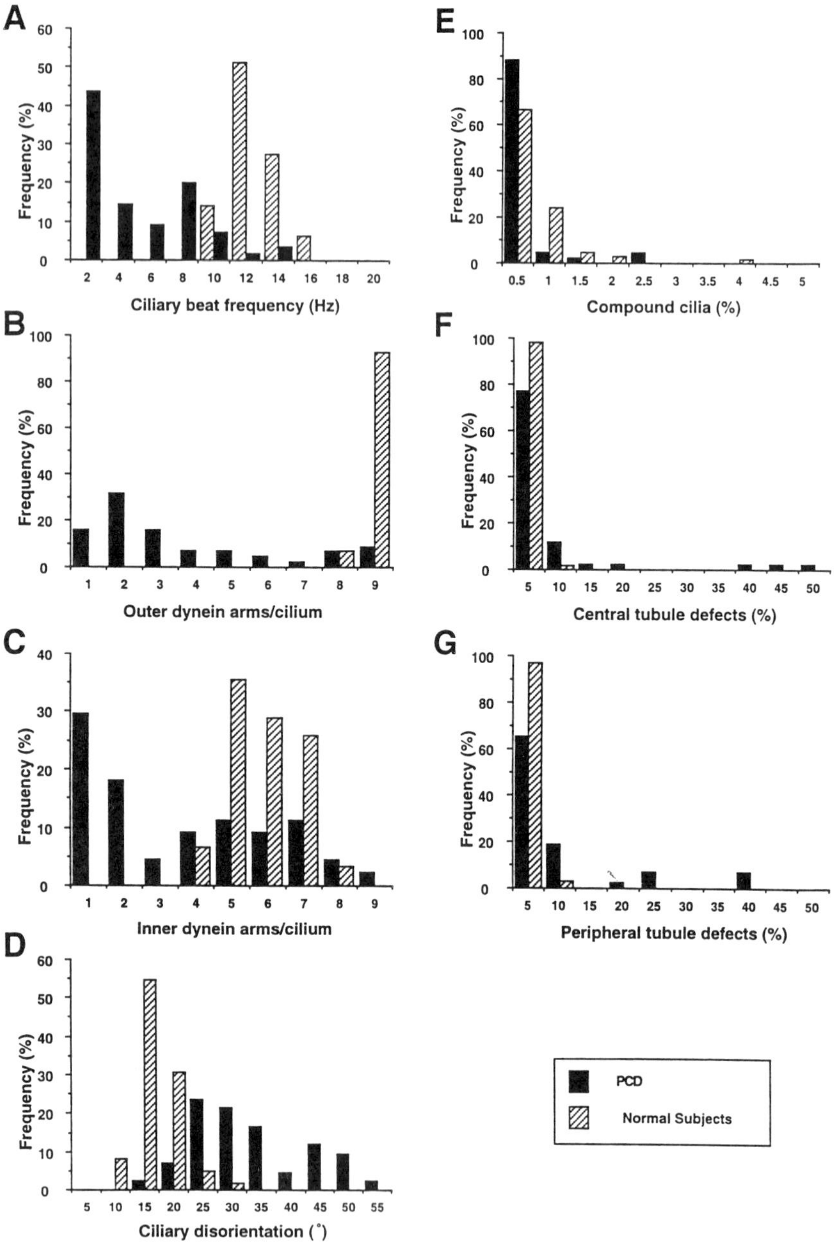


**Figure 3** Frequency distributions of the ciliary parameters measured from normal subjects and PCD patients. These histograms demonstrate the usefulness of

The relationships between CBF and ultrastructural features as well as between different ultrastructural features were examined. For the PCD patients, CBF was correlated with outer but not with inner dynein arm numbers (Figure 4). Central tubule defect numbers were positively correlated with peripheral tubule defects and there was a weak correlation between central tubule defects and ciliary orientation (Figure 4).

## B. Clinical Features

The clinical features that were significantly different between the PCD patients and those with RRTD not due to PCD are shown in Figures 5 and 6 for the pediatric and adult groups, respectively. Some other features often considered as a basis for identifying patients who might have PCD are also shown.

For children, a positive sibling history, a history of neonatal respiratory distress, otitis media (as well as glue ear and deafness), purulent rhinitis, and the presence of organ malrotation (dextrocardia and/or situs inversus) occurred significantly more often in the PCD patients. Clinical features that were not significantly different between the two groups included the incidences of lobar collapse, sinusitis, bronchiectasis, digital clubbing, and cough. Fertility is not relevant for this group. Of the classical features of Kartagener's syndrome only organ malrotation occurred significantly more often in the PCD group.

In the adult group, only otitis media and wheeze occurred significantly more often in the PCD patients compared to the RRTD patients. There was not a significantly higher incidence of infertility in the adult PCD patients but neonatal respiratory distress, clubbing, and sinusitis were of borderline significance. Bronchiectasis occurred more often in the non-PCD patients. None of the classical features of Kartagener's syndrome occurred significantly more often in the PCD group.

By Chi-square analysis Kartagener's triad was found to be less useful in predicting a diagnosis of primary ciliary dyskinesia than were the three

---

each parameter in differentiating between patients with PCD and normal subjects. There was some overlap for all parameters, but CBF (A), outer dynein arm numbers (B), inner dynein arm numbers (C), and ciliary disorientation (D) are more useful discriminators for PCD than compound cilia (E) or microtubular defects (F, G). Adapted from de Iongh and Rutland [23].

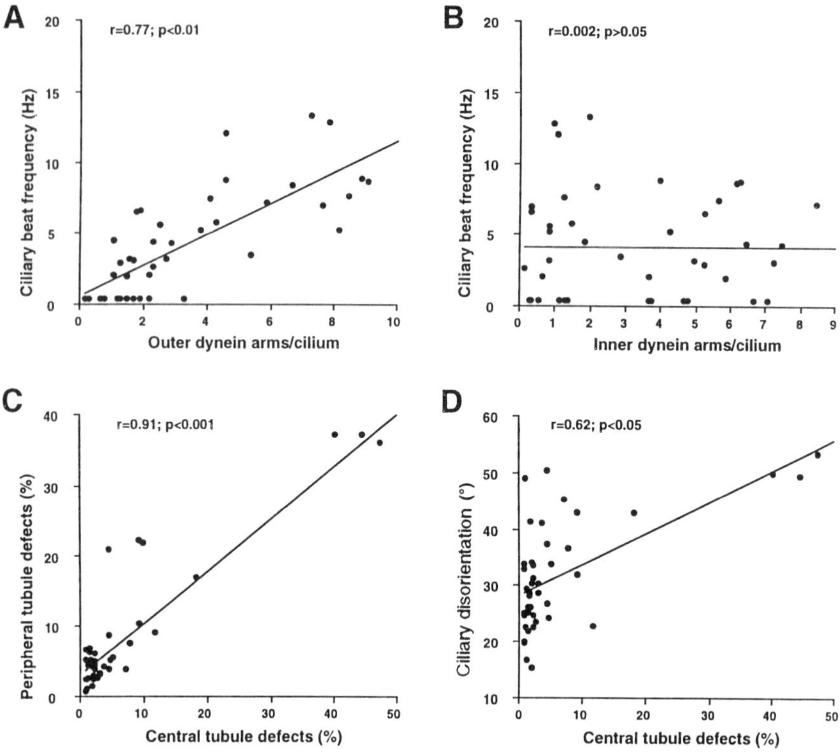


**Figure 4** Scattergrams showing correlations of ciliary beat frequency with inner (A) and with outer (B) dynein arms and correlations of central microtubule defects with peripheral microtubule defects (C) and with ciliary disorientation (D) in patients with PCD. Adapted from de Iongh and Rutland [23].

most significant clinical features (otitis media, purulent rhinitis, sinusitis) in the PCD group or a combination of the three features (neonatal respiratory distress, otitis media, grommets) that individually were most significant between the two groups (Table 2).

## IV. DISCUSSION

The diagnostic tests of primary ciliary dyskinesia has evolved over the last 65 years since Kartagener's initial description [1] in 1933, as has the

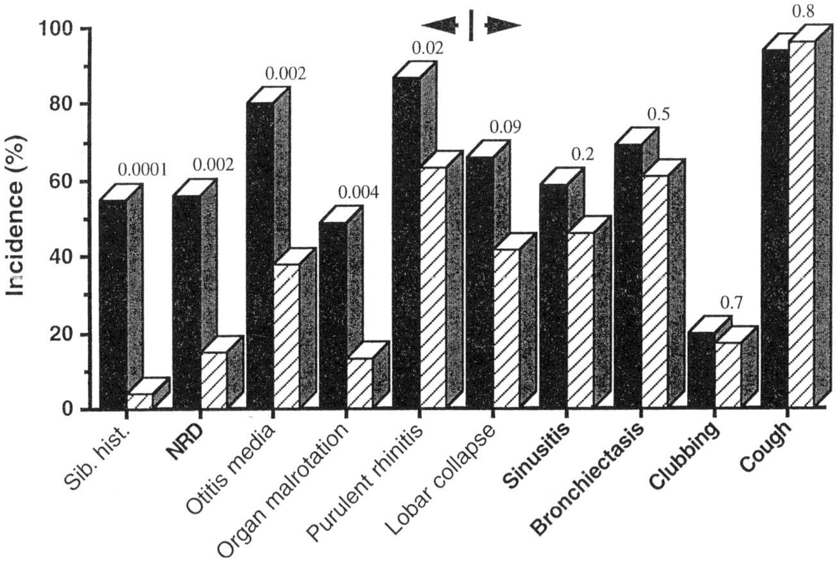


**Figure 5** Histogram showing the incidences of ten clinical features for the pediatric (aged ≤ 16 years) PCD (■) and RRTD (▨) patients. The p values for each of the comparisons are shown above the bars and the level of significance ($p = 0.05$) is indicated by a vertical line and arrows. Comparisons to the left were significantly different.

nomenclature [27]. Kartagener, following Siewert [28] in earlier years, reported a triad of symptoms (bronchiectasis, sinusitis and situs inversus totalis) and these provided a clinical basis for making a diagnosis for almost half a century. In 1976, two separate groups (Afzelius and co-workers [4], immotile human cilia; Pedersen and Mygind [5[, absence of dynein arms in Kartagener's syndrome) reported defective ciliary function and ultrastructure in what came to be called the "immotile-cilia" syndrome [2]. In 1981 primary ciliary dyskinesia was suggested as the most appropriate terminology following the demonstration of variations in ciliary motility ranging from complete absence to normal beat frequency [27]. Paralleling this, a range of ultrastructural defects has been reported [4–13,18–20]. Initially, judgements of ciliary beating were made on a subjective basis until accurate measurements of beat frequency became available in clinical practice. Similarly, for many years, ciliary ultrastructure was evalu-

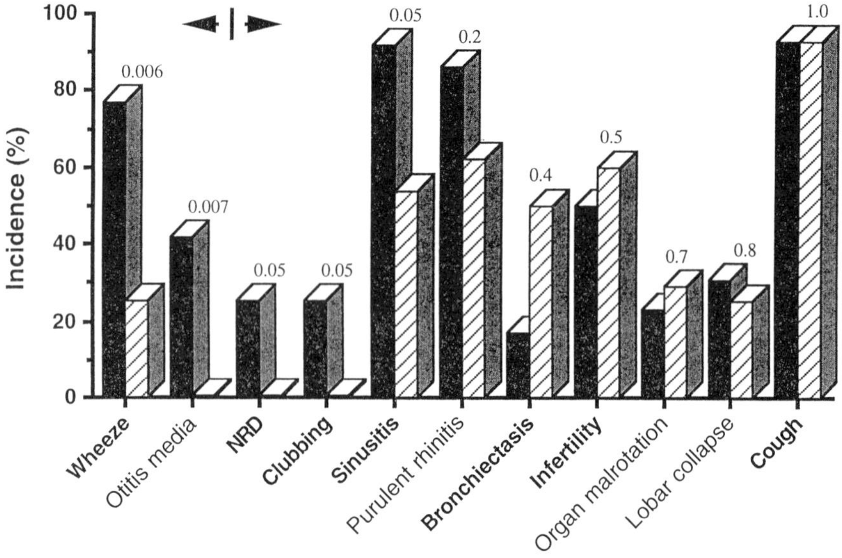


**Figure 6** Histogram showing the incidences of eleven clinical features for the adult (aged > 16 years) PCD (■) and RRTD (▨) patients. The p values for each of the comparisons are shown above the bars and the level of significance (p = 0.05) is indicated by a vertical line and arrows. Comparisons to the left were significantly different.

**Table 2** Comparison of Clinical Features in PCD and RRTD Patients

| Clinical features | % PCD | % RRTD | p value |
|---|---|---|---|
| Kartagener's triad (Bronchiectasis, sinusitis, organ malrotation) | 29 | 8 | 0.14 |
| Neonatal respiratory distress | 48 | 1 | <0.001 |
| Otitis media | 70 | 26 | <0.001 |
| Most prevalent (Otitis media, purulent rhinitis, sinusitis) | 70 | 7 | 0.002 |
| Most significant (Neonatal respiratory distress, otitis media, grommets) | 34 | 3 | 0.002 |

ated subjectively. In order to appreciate the pathological significance of ultrastructural defects, careful quantification of ciliary defects in PCD patients and normal subjects is required [29–33].

## A. Laboratory Studies

Some abnormal cilia are present in all normal subjects [29–33]. To permit accurate and reliable identification of significant structural defects and objective diagnosis of PCD patients, we quantified ciliary ultrastructure in a group of normal subjects to provide normal ranges for the various axonemal components [23].

As shown in Table 1, all ultrastructural defects quantified, except compound cilia, were significantly different in the PCD patients compared to the normal subjects. Although a wide range of ultrastructural defects has been reported in PCD, their relevance to ciliary function and their clinical significance has not yet been determined fully. Some information such as the relationship between dynein arms and CBF is beginning to emerge [7,23].

Most clinical features in PCD result from impaired mucociliary clearance, which may be due to many factors including slowed ciliary beat frequency (the one aspect of ciliary function that can be measured easily in clinical practice). Effective ciliary beating is dependent on an intact axoneme, in particular dynein arms, and on ciliary orientation that is required for mucus to be cleared in a consistent direction.

Diagnostic criteria for PCD, which include clinical features, abnormalities of ciliary function, and/or ciliary ultrastructure, are still evolving. It is possible that further defects of ultrastructure and function will be identified in the future. For the purposes of this study, we have taken the position that PCD is present in any patient with a history of recurrent respiratory tract disease who has defects of any identifiable axonemal component outside the normal range. Most such patients have defective ciliary beating but in a few, CBF lies within the normal range. In this situation electron microscopy is mandatory to make the diagnosis. When electron microscopy is not available, PCD may be strongly suspected in the presence of grossly defective ciliary beating. It is recognised that this is not ideal and may need to be modified as further data becomes available from ongoing research. However, in our present state of knowledge there are no other diagnostic criteria—no gold standard—on which to base a definitive diagnosis. For all the ultrastructural features quantified, even when the mean values were significantly different, there was an overlap between the ranges for normal subjects and PCD patients indicating the importance of examining more than one feature. Doing this with both CBF and ultrastructural features allowed a discrimination to be made between PCD patients and normal subjects.

The usefulness of the various laboratory measurements were examined for the PCD patients. A diagnosis could be made on the basis of abnormal CBF alone in 93%. Using ultrastructural features, outer dynein arms were abnormal in 90% of patients, inner dynein arms in 55%, and ciliary orientation in 50%. Peripheral tubules were abnormal in 19% of patients, central tubules in 17%, and compound cilia in none. A diagnosis of PCD based on the presence of abnormal inner and/or outer dyein arm numbers and/or abnormal ciliary orientation detected all of the PCD patients. Dynein arm numbers, CBF, and ciliary orientation thus proved to be much more useful in diagnosing PCD than did the incidence of microtubular defects or compound cilia. This is further highlighted by the frequency distributions of CBF and ciliary ultrastructural parameters for PCD patients and normal subjects (Figure 3). The frequency distributions for parameters such as CBF, inner and outer dynein arm numbers, and ciliary deviation were distinctly different in patients with PCD when compared with healthy subjects. In contrast the distributions for compound cilia and central and peripheral tubule defects in PCD patients were similar to those found in normal subjects.

The limitations of calculating a sensitivity to examine the usefulness of various diagnostic criteria when these criteria themselves (in the absence of a gold standard) are used to make that diagnosis are recognized. There is no alternative at present given the absence of another reliable means of making the diagnosis. It is possible that further research will reveal molecular or biochemical defects, possibly in the absence of ultrastructural defects. Molecular biology may well provide the next advances in PCD diagnosis. Several gene linkage studies are in progress in Great Britain, Poland, and Switzerland that may help elucidate the genetic basis of this clinically homogeneous yet structurally heterogeneous [34] disease.

It is well recognized that ultrastructural defects, particularly microtubular defects, may be secondary to respiratory tract infections and other insults to the ciliated epithelium [35–40]. Verra and colleagues [41] have suggested that when there is a high incidence of ultrastructural defects (especially with dynein arm abnormalities), studies of nasal cilia are usually sufficient to make a diagnosis. If there is low incidence of structural defects or predominantly microtubular defects are present, studies should be conducted on epithelium obtained from different anatomical sites (such as upper and lower respiratory tract) to reliably diagnose PCD. Alternatively, a repeat study, particularly when no acute respiratory tract infection is present, may be undertaken. This sometimes poses difficulties in practice. In postpubertal males, sperm ultrastructure may be useful but is little used in clinical practice. Caution also needs to be exercised in extrapolating findings from spermatozoa to cilia and vice versa as there are reports suggesting that the genetic control of axonemal structure in different parts of the body may be regulated by different genes [10,42].

A relationship between ciliary beating and dynein arms, particularly the outer dynein arms, has been reported in animal studies [43]. In humans a relationship between outer, but not inner, dynein arm numbers and ciliary beat frequency has been demonstrated previously [7]. The same relationship between outer dynein arm numbers and CBF was found in the present study. This suggests that outer dynein arms may be more important than inner dynein arms in generating ciliary motility. Although inner dynein arm numbers were not correlated with CBF this can be the sole ultrastructural defect in patients with PCD, suggesting that inner dynein arms are related to another component of ciliary beating [44].

The limited value of microtubule abnormalities in making a diagnosis of PCD is consistent with previous reports [29,32,35–39]. The incidence of central and peripheral microtubule defects were significantly correlated, possibly reflecting a common mechanism in the assembly of central and peripheral microtubules. The significant correlation between central tubule defects and ciliary disorientation (Figure 4D) may have become apparent since central tubules were used to measure orientation but may also suggest that ciliary disorientation is related to abnormal tubule assembly. This could be explored further by examining the relationship of ciliary disorientation measured from basal bodies to central tubule defects. Hard [45] has suggested that disruption of microtubules interconnecting the basal feet of cilia may result in loss of ciliary coordination and orientation of the basal feet.

What value is there in making a definitive diagnosis of PCD? As far as the individual patient is concerned, the demonstration of an underlying cause for symptoms, sometimes of many years' standing, may limit the need for further investigation. Since the natural history of PCD is less progressive than some other clinical conditions that can manifest in a similar way (such as cystic fibrosis), a definite diagnosis of PCD, suggesting a relatively good prognosis may limit the anxiety of all concerned. In the individual patient, the use of physical means of overcoming impaired mucociliary clearance (such as postural drainage) and early vigorous treatment of acute respiratory infections may slow deterioration. A diagnosis of PCD constitutes a relative contraindication to lung resection in bronchiectasis of limited extent, since the underlying defect is present throughout the tracheobronchial tree and chronic bronchial suppuration is likely to recur. For the individual patient, anticipation of possible infertility, particularly in males, and information about genetic counseling may be relevant.

Finally, identification of patients with PCD allows clinical studies of new therapeutic modalities.

## B. Clinical Features

For decades a diagnosis of PCD was based on clinical features such as those described by Kartagener [1]. These classical clinical indicators have been widely used when referring patients with recurrent respiratory tract disease for investigation. With the development of less invasive means of measuring CBF and ultrastructure, more patients are undergoing laboratory studies in special centers. These can be time consuming and expensive, particularly electron microscopy, and patients may need to travel considerable distances. The usefulness of Kartagener's triad in predicting the presence of PCD—and whether other clinical features should be considered as a basis for instigating investigation—was examined.

Few of the clinical features studied differentiated between our group of patients with PCD and the patients with RRTD not due to PCD. In children (Figure 5), a sibling history of recurrent respiratory tract infections, but not of any other family history, occurred significantly more often in the PCD patients. Neonatal respiratory distress (occurring in a term infant with no other apparent cause) occurred significantly more often in PCD. This was first reported by Whitelaw and colleagues in 1983—five of their six PCD patients presented in this way [46]. Otitis media, as well as its sequelae of glue ear and deafness, occurred significantly more often as did purulent rhinitis, both secondary to defective upper respiratory tract cilia. Organ malrotation, (dextrocardia and/or situs inversus) occurred significantly more often in PCD. The incidence of all other clinical features studied, including lobar collapse, sinusitis, bronchiectasis, digital clubbing, and cough did not differ between the two groups. Fertility is not relevant in this age group. Of the classical features of Kartagener's syndrome, organ malrotation alone occurred significantly more often in the PCD group.

In the adult group (Figure 6), only otitis media and wheeze occurred significantly more often in the PCD patients. Wheeze may have occurred more often with the adults because of the development of bronchial hyperreactivity with time. Neonatal respiratory distress ($p = 0.05$) may have occurred in more of these patients but they may have forgotten or not been told of it. There was a higher incidence of digital clubbing, sinusitis (clinical and radiological), and purulent rhinitis in PCD but these did not reach statistical significance. It is possible that these features might reach statistical significance if larger numbers of adult patients are studied. Bronchiectasis occurred less often in PCD, again a non significant difference. In the PCD patients there was not a higher incidence of infertility, organ malrotation, lobar collapse, cough, or any other feature studied. None of the classical features of Kartagener's syndrome occurred significantly more often in PCD.

The usefulness of Kartagener's triad in predicting the presence of PCD was examined in the whole group (Table 2). Bronchiectasis, sinusitis, and organ malrotation were present in 29% of patients with PCD compared to 8% of patients with RRTD. This difference was not statistically significant. This applied to adults and children when analysed separately, indicating that using the presence of Kartagener's syndrome alone did not discriminate adequately between the two groups. For those patients who manifested Kartagener's syndrome, 86% had PCD but for the PCD patients only 29% had Kartagener's syndrome. In this study the use of Kartagener's triad as a basis for diagnosing PCD resulted in a low sensitivity (0.29) but a high specificity (0.92). The use of a history of neonatal respiratory distress alone was more useful in discriminating between PCD and RRTD groups as was the use of otitis media alone (Table 2). The presence of Kartagener's syndrome is a relatively good predictor of PCD; however, its absence by no means excludes it.

The patients studied were selected in one way—almost all were referred by respiratory physicians and had a history of recurrent respiratory tract disease. Although few patients were referred because of an upper respiratory tract problem alone, those with PCD had a significantly higher incidence of upper respiratory tract disease than did the RRTD group. This was emphasized when the three most prevalent features (otitis media, purulent rhinitis, and sinusitis) were compared between PCD (present in 70%) and RRTD (7%), a statistically significant difference (Table 2). When the conjunction of the three most significantly different features (neonatal respiratory distress, otitis media, and a history of grommet insertion) were compared between PCD (34%) and RRTD (3%), the difference was again statistically significant. This emphasizes the importance of upper respiratory tract disease as well as the presence of neonatal respiratory distress (NRD) in discriminating PCD from RRTD. Apart from the incidence of NRD (or the presence of wheeze) no other indicator of lower respiratory tract disease differentiated between the two groups.

There are a number of reports in the literature of PCD patients in which the incidence of various clinical features is documented [47–50] but these have not been compared to a control group of patients with similar clinical presentations (Table 3). Barlocco and colleagues [51] in a paper dealing mainly with ultrastructure, compared four clinical features in PCD patients with controls. We compared thirty-five clinical features with a control group and found comparable results to previous studies (Table 3).

**Table 3** Studies of the Incidence of Clinical Features in PCD

| Clinical feature | Turner et al., (1981) [47] $n = 21$ | Mygind, Pedersen, Stafanger (1983) [48,49] $n = 27$ | Greenstone et al., (1988) [50] $n = 30$ | Barlocco et al., (1991) [51] $n = 28$ | Rutland et al., (1997) $n = 53$ |
|---|---|---|---|---|---|
| NRD | — | — | 43 | — | 48[a] |
| Chronic rhinitis | — | 100 | — | 89[a] | 86[a] |
| Otitis media | 100 | >89 | 50 | 86[a] | 61[a] |
| Sinusitis (X-ray) | 100 | >74 | 92 | — | 91[a] |
| Nasal polyps | 19 | — | — | — | 10 |
| Bronchiectasis | 29 | 57 | 57 | 86[a] | 65 |
| Situs inversus | 48 | 53 | 53 | 39 | 38 |

[a] $p < 0.05$

The numbers of PCD patients in these studies have been limited. Studies of larger numbers of patients might well demonstrate further significant differences between PCD and other patients. This could be explored further by combining information from multicenter studies. To permit such a study, uniform diagnostic criteria for PCD will be required. Previous reports of PCD have based the diagnosis on various criteria including the presence of consistent clinical features, impaired mucociliary clearance, and various observations and/or measurements of ciliary function and ciliary ultrastructure. A consensus of criteria to diagnose PCD has yet to be reached.

We believe that primary ciliary dyskinesia is underrecognized. These patients may present not only to their family practitioners, but also to pediatricians, respiratory physicians, ENT specialists, infertility clinics, and a variety of medical services. Education of these groups and a wider awareness of PCD as a cause of recurrent upper and lower respiratory tract disease as well as for organ malrotation and infertility will improve diagnosis of patients with PCD. The presence of Kartagener's triad alone, while still an indicator for referral and study, is an inadequate basis for detecting PCD patients. This study has demonstrated the importance of recognizing the significance of upper respiratory tract disease and neonatal respiratory distress when considering a diagnosis of primary ciliary dyskinesia.

## REFERENCES

1. Kartagener M. Zur pathogenese der bronchiectasien. Beitr Klin Tuberk 1933; 83:489–501.
2. Eliasson R, Mossberg B, Camner P, Afzelius BA. The "immotile–cilia" syndrome. A congenital abnormality as an etiologic factor in chronic airway infection and male sterility. N Engl J Med 1977; 297:1–6.
3. Schidlow DV. Primary ciliary dyskinesia (the immotile cilia syndrome). Ann Allergy 1994; 73:457–470.
4. Afzelius BA. A human syndrome caused by 'immotile–cilia'. Science 1976; 193:317–319.
5. Pedersen M, Mygind N. Absence of axonemal arms in nasal mucosa cilia in Kartagener's syndrome. Nature 1976; 121:494–495.
6. Rossman CM, Lee RMKW, Forrest JB, Newhouse MT. Nasal cilia in normal man, primary ciliary dyskinesia and other respiratory diseases. Analysis of motility and ultrastructure. Eur J Resp Dis 1983; 64(Suppl 127):74–70.
7. Rutland J, Cox T, Dewar A, Cole PJ. Screening for ciliary dyskinesia—a spectrum of defects of motility and structure. Eur J Resp Dis 1983; 64(Suppl 127):71–77.
8. Sturgess JM, Chao J, Wong J, Aspin N, Turner JAP. Cilia with defective radial spokes. A cause of human respiratory disease. N Engl J Med 1980; 300:53–57.
9. Sturgess JM, Chao J, Turner JAP. Transposition of ciliary microtubules. Another cause of impaired ciliary motility. N Engl J Med 1980; 303:318–322.
10. Lungarella G, Fonzi L, Burrini AG. Ultrastructural abnormalities in respiratory cilia and sperm tails in a patient with Kartagener's syndrome. Ultrastruct Path 1982; 3:319–323.
11. Afzelius BA, Gargani G, Romano C. Abnormal length of cilia as a possible cause of defective mucociliary clearance. Eur J Resp Dis 1985; 166:173–180.
12. Lungarella G, de Santi MM, Palatresi R, Tosi P. Ultrastructural observations on basal apparatus of respiratory cilia in immotile cilia syndrome. Eur J Resp Dis 1985; 66:165–172.
13. Fonzi L, Lungarella G, Palatresi R. Lack of kinocilia in the 'immotile–cilia' syndrome. Eur J Resp Dis 1982; 63:558–563.
14. Rutland J, Cole PJ. Non-invasive sampling of nasal cilia for measurement of beat frequency and study of ultrastructure. Lancet 1980; 2:564–565.
15. Anderson IB, Camner P, Jensen PL, et al. Nasal clearance in monozygotic twins. Am Rev Respir Dis 1974; 110:301–351.

16. Rutland J, Cole P. Nasal mucociliary clearance and ciliary beat frequency in cystic fibrosis compared with sinusitis and bronchiectasis. Thorax 1981; 36:654–658.
17. Canciani M, Barlocco EG, Mastella G, de Santi MM, Gardi C, Lungarella G. The saccharin method for testing mucociliary function in patients suspected of having primary ciliary dyskinesia. Pediatr Pulmonol 1988; 5: 210–214.
18. Rutland J, de Iongh R. Random ciliary orientation. A cause of respiratory tract disease. N Engl J Med 1990; 323:1681–1684.
19. Rutman A, Cullinan P, Woodhead M, Cole PJ, Wilson R. Ciliary disorientation: a possible variant of primary ciliary dyskinesia. Thorax 1993; 36: 654–658.
20. Rayner CFJ, Rutman A, Dewar A, Greenstone M, Cole PJ, Wilson R. Ciliary disorientation alone as a cause of primary ciliary dyskinesia syndrome. Am J Respir Crit Care Med 1996; 153:1123–1129.
21. Escudier E, Escalier D, Homassen JP, Pinchon MC, Bernaudin JF. Unexpectedly normal cilia and spermatozoa in an infertile man with Kartagener's syndrome. Eur J Respir Dis 1983; 70:180–186.
22. Greenstone MA, Dewar A, Cole PJ. Ciliary dyskinesia with normal ultrastructure. Thorax 1983; 38:875–876.
23. de Iongh R, Rutland J. Ciliary defects in healthy subjects, bronchiectasis and primary ciliary dyskinesia. Am J Respir Crit Care Med 1995; 151:1559–1567.
24. Rutland J, Griffin WM, Cole PJ. Nasal brushing and measurement of ciliary beat frequency. An 'in vitro' method for evaluating pharmacological effects on human cilia. Chest 1981; 80:865–867.
25. de Iongh R, Rutland J. Orientation of respiratory tract cilia in patients with primary ciliary dyskinesia, bronchiectasis, and in normal subjects. J Clin Pathol 1989; 42:613–619.
26. Rutland J, Dewar A, Cox T, Cole PJ. Nasal brushing for the study of ciliary ultrastructure. J Clin Pathol 1982; 35:356–359.
27. Sleigh MA et al. Primary ciliary dyskinesia (Letter). Lancet 1981; 2:476.
28. Siewert AK. Ueber einen fall von bronchiektasie bei einem patienten mit situs inversus viscerum. Berliner Klin Wochenschr 1903; 41:139–141.
29. Fox B, Bull TB, Arden GB. Variations in the ultrastructure of human nasal cilia including abnormalities found in retinitis pigmentosa. J Clin Pathol 1980; 33:327–335.
30. Nielsen MH, Pedersen M, Christensen B, Mygind N. Blind quantitative electron microscopy of cilia from patients with primary ciliary dyskinesia and from healthy subjects. Eur J Respir Dis 1983; 64(Suppl 127):19–30.
31. Eavey RD, Nadol JB, Holmes LB, Laird NM, Lapey A, Joseph MP, Strome M. Kartagener's syndrome. A blinded controlled study of cilia ultrastructure. Arch Otolaryngol Head Neck Surg 1986; 112:646–650.
32. Smallman LA, Gregory J. Ultrastructural abnormalities of cilia in the human respiratory tract. Hum Pathol 1986; 17:848–855.
33. Van der Baan S, Veerman AJP, Bezemer PD, Feenstra L. Primary ciliary

dyskinesia: quantitative investigation of the ciliary ultrastructure with statistical analysis. Ann Otol Rhinol Laryngol 1987; 96:264–272.

34. Mossberg B. The immotile cilia syndrome: ultrastructurally heterogeneous and clinically homogeneous. [Editorial]. Eur Respir Dis 1985; 66:161–163.
35. Rutland J, Cox T, Dewar A, Cole PJ, Warner JO. Transitory ultrastructural abnormalities of cilia. Br J Dis Chest 1982; 76:185–188.
36. Afzelius BA, Camner P, Mossberg B. Acquired defects compared to those seen in the immotile–cilia syndrome. Eur J Resp Dis 1983; 64(Suppl 127): 5–10.
37. Carson JL, Collier AM, Hu SS. Acquired ciliary defects in nasal epithelium of children with acute viral upper respiratory infections. N Engl J Med 1985; 312:463–468.
38. Lurie M, Rennert G, Goldenberg S, Rivlin J, Greenberg E, Katz I. Ciliary Ultrastructure in primary ciliary dyskinesia and other chronic respiratory conditions: the relevance of microtubular abnormalities. Ultrastruct Pathol 1992; 16:547–553.
39. Carson JL, Collier Am, Fernald GW, Hu SS. Microtubular discontinuities as acquired defects in airway epithelium of patients with chronic respiratory diseases. Ultrastruct Pathol 1994; 18:327–332.
40. Rayner CF, Rutman A, Dewar A, Cole PJ, Wilson R. Ciliary disorientation in patients with chronic upper respiratory tract inflammation. Am J Respir Crit Care Med 1995; 151:800–804.
41. Verra F, Fleury-Feith J, Boucherat M, Pinchon MC, Bignon J, Escudier E. Do nasal ciliary changes reflect bronchial changes? An ultrastructural study. Am Rev Respir Dis 1993; 147:908–913.
42. Jonsson MS, McCormick JR, Gillies CG, Gondos B. Kartagener's syndrome with motile spermatozoa. N Engl J Med 1982; 307:1131–1133.
43. Satir P. Modelling outer dynein arm activity and its relation to the ciliary beat cycle. In: Baum J, ed. Cilia, Mucus, and Mucociliary Interaction. New York: Marcel Dekker, 1998.
44. Wilton LJ, Teichtahl H, Temple-Smith PD, de Kretser DM. Structural homogeneity of the axonemes of respiratory cilia and sperm flagella in normal men. J Clin Invest 1985; 75:825–831.
45. Hard R. Regulation of respiratory ciliary activity. International Congress on Cilia, Mucus and Mucociliary Interactions, Jerusalem, Israel, Feb 23–28, 1997.
46. Whitelaw A, Evans A, Corrin B. PCD a new cause of neonatal respiratory distress. Arch Dis Child 1981; 56:432–435.
47. Turner JAP, Corkey CWB, Lee JYC, Levison H, Sturgess J. Clinical expressions of 'immotile–cilia' syndrome. Pediatrics 1981; 67:805–810.
48. Mygind N, Pedersen M. Nose-, sinus-, and ear—symptoms in 27 patients with primary ciliary dyskinesia. Eur J Respir Dis 1983; 64(Suppl 127):96–101.
49. Pedersen M, Stafanger G. Bronchopulmonary symptoms in primary ciliary dyskinesis. A clinical study of 27 patients. Eur J Respir Dis 1983; 64(Suppl 127):118–128.

50. Greenstone M, Rutman A, Dewar A, Mackay I, Cole PJ. Primary Ciliary Dyskinesia: cytological and clinical features. Quart J Med 1988; 67:405–430.
51. Barlocco EG, Valletta EA, Canciani M, Lungarella G, Gardi C, De Santi MM, Mastella G. Ultrastructural ciliary defects in children with recurrent infections of the lower respiratory tract. Pediatr Pulmonol 1991; 10:11–17.

# 45
# Mucociliary Clearance in Health and Disease: Qualitative and Quantitative Analysis

**Toyoharu Isawa**
*Yokohama Higashi National Hospital, Yokohama, Japan*

## I. AIRWAY MUCUS AND CILIA

Mucociliary clearance is the first line of defense in the respiratory system. The mucus overlying the epithelial surface of the respiratory tract is propelled upward toward the larynx by coordinated beating of the cilia of the airway epithelium. The cilia line the epithelium of the respiratory mucosa from the nasal passages to the terminal bronchioles. In and distal to the terminal bronchiole the cilia are either poorly developed or absent [1].

The bronchial surface (the ciliary surface) is covered with a mucus layer. Actually the cilia are immersed in the periciliary fluid above which there is a mucus layer. By a coordinated ciliary motion at the rate of about 1000 beats/min the mucus layer is normally transported toward the oropharynx. Mucociliary clearance or transport is orchestrated by interaction between the mucus layer and the beating cilia [2,3].

## II. THE PRINCIPLE OF MUCOCILIARY IMAGING

For the study of mucociliary clearance mechanisms, bronchoscopic or radiographic methods have been used but they are mostly invasive requiring anesthesia of the airways and insertion of a bronchoscope [4–7].

In order to study mucociliary clearance by external measurement, radioactive materials have been placed on the mucosal surface of the airways and sequential imaging carried out [8–10]. Documenting the transport of the radioactive substance placed on the mucus layer provides an objective measure of mucociliary clearance.

Acute toxicity of cigarette smoke to mucociliary clearance has been easily evaluated in dogs by this method as described later. A large field view gamma camera is used for imaging.

Radioactivity is introduced onto the mucosal surface either as a radioactive droplet, as a marker tagging an inert carrier or as inhaled radioaerosols. The former 2 methods are more or less invasive because they require anesthesia of the airways and bronchoscopic procedures, whereas the latter, radioaerosol inhalation, is noninvasive and readily applicable in clinical practice.

In addition to the lung imaging immediately following radioaerosol inhalation, sequential or follow-up lung images are taken at 30 min, 1, 2, and 3 hr to evaluate how mucociliary transport is taking place in the lung, especially in the large airways like the trachea. A qualitative assessment of mucociliary clearance is possible with the delayed imaging method.

An inevitable disadvantage of this delayed imaging, however, is the break in continuity from image to image. There is no proof that a "hot spot" seen on one image is the same one seen on a previous or subsequent image. Thus conclusions regarding clearance rate based on movement of such "hot spots" may be in error.

In this sense the delayed imaging method is not an adequate means to study mucociliary clearance mechanisms in vivo in humans.

Transport direction or velocity can be determined by this method only when an adequate hot spot is discretely recognized. In order to avoid this drawback, radioactivity of the entire thorax including the trachea is continuously measured from immediately following radioaerosol inhalation for 90 to 120 min in a sequential 10 sec frame mode in 64 × 64 matrix. Each sequential 10-sec data are recorded and stored in a computer.

The raw data are either compiled for cinematographic display that we call "Radioaerosol inhalation lung cinescintigraphy" or utilized for quantitative analysis [11–13].

Radioaerosol inhalation lung cine-scintigraphy is displayed in cine mode on a CRT screen at the rate of 18 frames per sec and the cinematographic display is recorded by a movie camera or on videotape.

This cinematographic display has enabled visualization of the features of actual mucous transport in the airways in vivo [11].

## III. MECHANISMS OF AEROSOL DEPOSITION

Inhaled aerosol deposits on the airways by impaction, sedimentation, and diffusion. When turbulence occurs in the airways at the sites of stenosis or narrowing caused by mucous deposition, cancerous protrusion of the airway mucosa, excessive narrowing of the airways in obstructive airways disease, etc., excessive deposition of inhaled aerosol or "hot spots" appear regionally corresponding to the stenotic lesions [14].

## IV. AGENTS USED FOR STUDYING MUCOCILIARY CLEARANCE

### A. Radioactive Droplet

Any agent, $^{99m}TcO_4^-$ (technetium pertechnetate), $^{99m}$Tc-HSA (human serum albumin), $^{99m}$Tc-millimicrosphere, $^{99m}$Tc-albumin microsphere, or $^{99m}$Tc-MAA (macroaggregated albumin), that stays on the airway mucous layer following placement through bronchoscopy can be used [15]. A droplet size on the order of 0.025–0.05 mL of $^{99m}$Tc-MAA placed through a tube inserted into the bronchoscope is practical and convenient.

Anesthesia of the airway mucosa as well as the oropharynx is required for placing a radioactive droplet on the tracheal or bronchial mucosa. This is tedious and cumbersome and gives so much discomfort to patients that the actual clinical application of this method is limited.

### B. Radioaerosol

Radioaerosol—generated by either a jet nebulizer, an ultrasonic nebulizer, a spinning-disc atomizer under special circumstances, or other technology—is inhaled through a mouthpiece with the nose clipped. An agent that is not absorbed or dispersed through the airway mucosa is ideal. Either $^{99m}$Tc-sulfur colloid, $^{99m}$Tc-phytate, or $^{99m}$Tc-HSA (human serum albumin) can be used for the study of mucociliary clearance mechanisms. We have usually used $^{99m}$Tc-HSA aerosol generated with a jet nebulizer or an ultrasonic nebulizer.

Aerosols with geometric standard deviation (GSD or $\sigma g$) of 1.1 or less are termed monodisperse aerosols, while those with GSD of greater than 1.1 are called polydisperse aerosols.

The size of inhaled radioaerosol, hygroscopy, and breathing pattern determine the sites of deposition in the lungs [14,16]. The size of various aerosols is shown in Table 1. We use normal tidal breathing for inhaling the radioaerosol.

For studying mucociliary clearance the aerosol size (active median aerodynamic diameter, AMAD) of less than 5–6 micrometer ($\mu$m) but greater than 1 $\mu$m is most suitable as shown by Morrow and Yu [17]. The authors personally use $^{99m}$Tc-HSA aerosol of AMAD of 1.93 $\mu$m with

**Table 1** Active Median Aerodynamic Diameter (AMAD) and Geometric Standard Deviation ($\sigma g$) Under Standard Conditions

| Nebulizer | AMAD (μm) | $\sigma g$ |
|---|---|---|
| Jet nebulizers | | |
| OEM-1 (USA) | 1.96 | 1.65 |
| OEM-2 (USA) | 1.19 | 1.86 |
| Ultravent (USA) | 1.04 | 1.71 |
| Penicillin nebulizer glass (Japan) | 1.76 | 1.70 |
| BARC with reservoir (India) | 0.84 | 1.73 |
| BARC without reservoir | 1.57 | 1.80 |
| Ultrasonic nebulizers | | |
| Mistogen EN-142 (USA) | 1.93 | 1.52 |
| Omuron-NE-U11 (Japan) | 1.62 | 1.50 |
| Devilbiss (USA) | 1.78 | 1.60 |

(Temperature 37°C, relative humidity 100%, and test agent: $^{99m}$Tc-HSA) measured by cascade impactor (Andersen sampler).

GSD of 1.52 inhaled by normal tidal breathing for studying mucociliary clearance in the lungs.

## V. EXPERIMENTAL STUDIES IN DOGS

When a single radioactive bolus is placed on the carina or the main bronchus of a normal dog through a vinyl tube under bronchoscopic guidance, the radioactive bolus is transported upward toward the vocal cords at a nearly constant velocity.

The transport can be well appreciated by sequential or follow-up imaging as shown in Figure 1. The transport velocity is calculated by dividing the migrating distance of the radioactive bolus by the time required for transport. The transport velocity of the particular dog in Figure 1 was 11.3 mm/min.

When dogs are forced to smoke cigarettes, the transport velocity tends to slow down as the number of cigarettes smoked increases (Figure

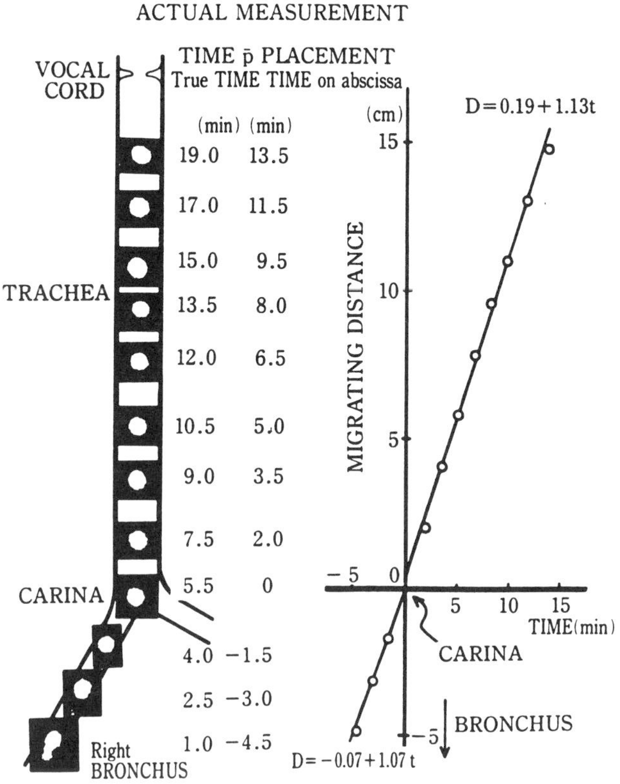


**Figure 1** Radioactive droplet ($^{99m}$Tc-MAA) placed on the large airway of a dog is transported upward with time. In this dog, the transport velocity along the trachea was 11.3 mm/min.

2). This indicates that the toxicity of the cigarette smoke acutely hinders mucociliary clearance in normal dogs [10]. Cauterization of the airway mucosa by applying silver nitrate in the normal dog can also damage mucociliary clearance [15]. It takes about 2 weeks for the recovery of normal mucociliary clearance from either cigarette smoke or cauterization [10,15]. Filtering cigarettes does not prevent the mucociliary clearance function from being damaged once the number of cigarettes smoked exceeds a certain level [10].

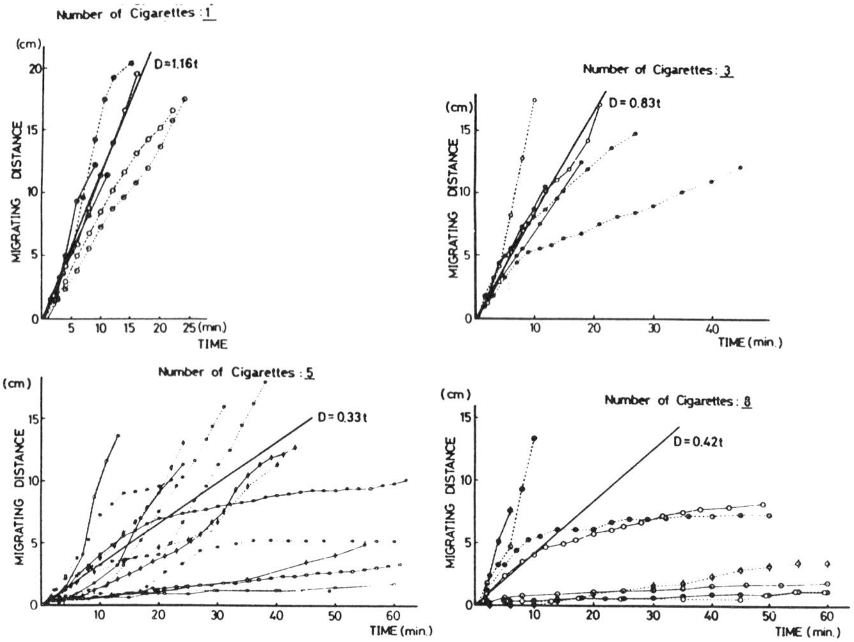


**Figure 2** Migrating or mucociliary transport distance versus time when the dog smokes 1, 3, 5, and 8 nonfiltered cigarettes. The shaded band indicates normal range. The transport velocity generally becomes slower as the number of cigarettes smoked increases.

## VI. HUMAN STUDIES

### A. Delayed Imaging

Because anesthesia of the oropharynx and the upper airways is absolutely necessary for bronchoscopic insertion, the placement of a radioactive droplet in the large airways under bronchoscopic guidance itself is not practical in most clinical situations. Instead inhalation of radioaerosol that is not absorbable from the mucosal surface becomes the method of choice in daily clinical practice.

When delayed imaging is repeated following radioaerosol inhalation in patients with bronchogenic carcinoma, some show a hot spot in later images either unchanged or larger than that seen on the image produced immediately after inhalation (Figure 3). In others a hot spot observed on

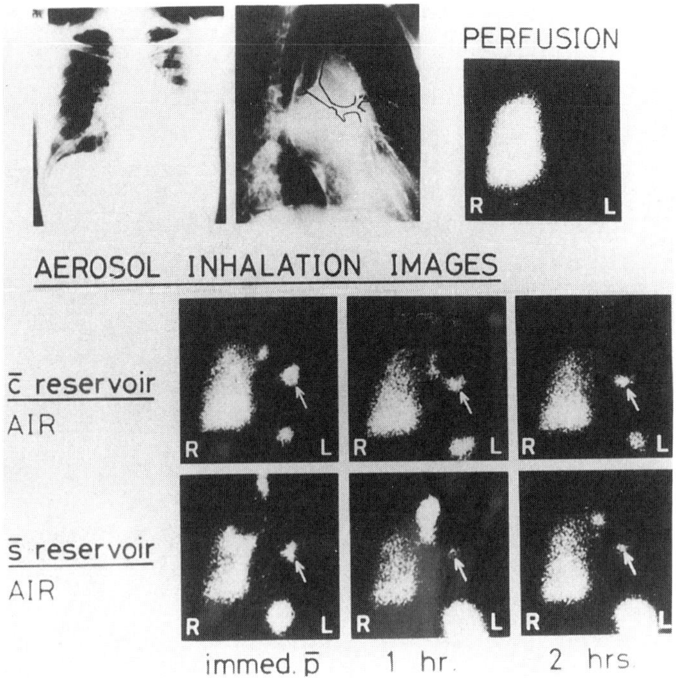


**Figure 3** A hot spot at the site of cancer invasion on the left main bronchus persists for 2 hrs or longer indicating disturbed mucociliary clearance. 55-year-old male with large cell carcinoma.

the immediate image disappears on delayed images (Figure 4). The former indicates that the airway mucosa at the site of the hot spot is invaded by cancer cells, while in the latter it is not invaded and the mucociliary clearance remains intact.

While studying mucociliary clearance function by repeated delayed imaging following radioaerosol inhalation, we found the case shown in Figure 5 puzzling with regard to interpretation of the behavior of a mucous glob. Immediately after aerosol inhalation there was no radioactivity in the left main bronchus. One hour later delayed imaging revealed a radioactive glob on the left main bronchus that disappeared 2 hours following inhalation. If mucus is always transported in cephalad direction toward the vocal cord, the presence of radioactivity in the left main bronchus is a mystery difficult to interpret. This finding suggests that there must be

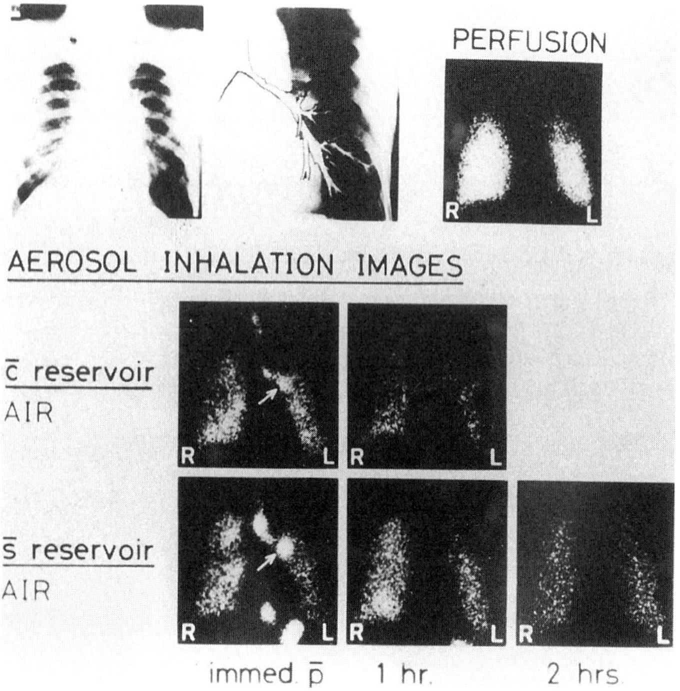


**Figure 4** A hot spot recognized immediately following radioaerosol inhalation at the site of cancer protrusion disappears after 1 hr, indicating intact mucociliary clearance. 46-year-old male with small cell carcinoma.

some inherent defect in this methodology for studying mucociliary clearance. Continuous measurement of radioactivity following radioaerosol inhalation instead of repeated delayed imaging is necessary to know exactly what happens to the inhaled aerosols in terms of mucociliary clearance.

## B. Radioaerosol Inhalation Lung Cine-Scintigraphy

Following inhalation of ultrasonically generated $^{99m}$Tc-HSA aerosol by tidal breathing through a mouthpiece with the nose clipped, a subject is placed either under or above a gamma camera in the supine position and radioactivity is measured over the entire thorax including the trachea for 120 min in sequential 10-sec frame mode with $64 \times 64$ matrices. The data is recorded and stored in a computer. The data is then displayed in cine-

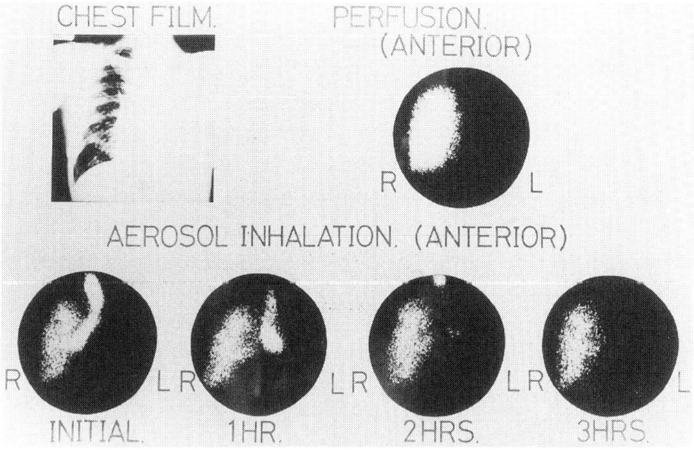


**Figure 5** Immediately after radioaerosol inhalation, no radioactivity deposits in the left main bronchus indicating no ventilation in the left lung. However, in 1 hr a hot spot has appeared in the left main bronchus. By 3 hrs the hot spot has completely disappeared. 50-year-old male with adenocarcinoma. The left lung shows a blackout on chest x-ray and absent perfusion on perfusion lung image.

mode on a cathode ray tube (CRT) screen at the rate of 18 frames per sec and the cinematographic display recorded with a movie camera. By cinematographic display, dynamic mucous transport patterns in the large airways like the trachea or the main bronchus can be observed and evaluated. This provides the first opportunity to observe the actual dynamic mucous transport in vivo in human lungs [11,12,13].

## VII. QUANTITATIVE EVALUATION

### A. Extrapulmonary Airways

Under an experimental situation as described above (experimental studies in the dogs), transport velocity of a radioactive droplet can be assessed by measuring peak radioactivity of the droplet at two different measuring points and the time required for transport between the two spots. Then the distance transported divided by the time required for the transport is equivalent to the velocity of the mucociliary clearance. This is only true when the transport is straight and upward in direction and constant in transport velocity.

Under clinical conditions, however, the insertion of bronchoscopy itself requires anesthesia, and placing a radioactive droplet on the sensitive trachea gives great discomfort to patients and is thus not practical and nor feasible. Furthermore transport patterns are not always straight and upward in direction. That is why radioaerosol inhalation lung cine-scintigraphy has been developed. Various patterns of mucous transport have been observed and appreciated by radioaerosol inhalation lung cine-scintigraphy [11,12,13].

Quantitative analysis was carried out by the following methods [18].

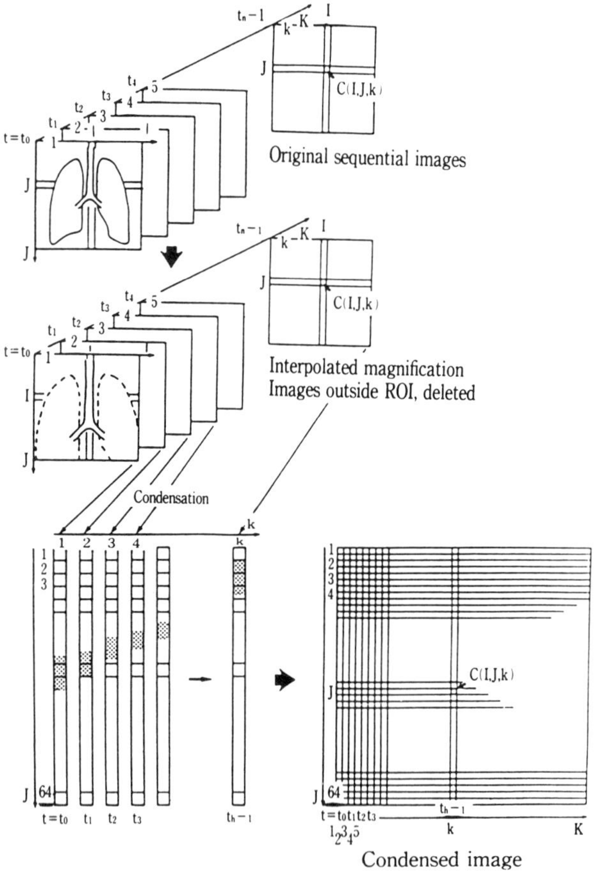


**Figure 6** Diagram of condensed image mode. Only the tracheal portion is arranged sequentially and the hot spot is imaged. (From Reference 18.)

## Condensed Image Mode

The tracheal portion of each frame data is selected and arranged sequentially on a computer as shown in Figure 6 and displayed on the CRT with its x axis as time and the y axis as the distance. When mucous transport is cephalad and straight in direction and constant in velocity, the transport path becomes a diagonal line (Figure 7). When mucous transport is greatly disturbed and the velocity becomes nearly zero, a horizontal line is depicted (Figure 8). With radioaerosol inhalation, wide bands of trajectories are depicted (Figure 9) [18].

## Trajectory Mode

In order to analyze the transport of a mucous glob more accurately and quantitatively, the following method was devised and named "trajectory mode" [18].

At first a point of interest (POI) is set at a "hot spot" on a frame datum that was magnified from the original 64 × 64 matrix data by interpo-

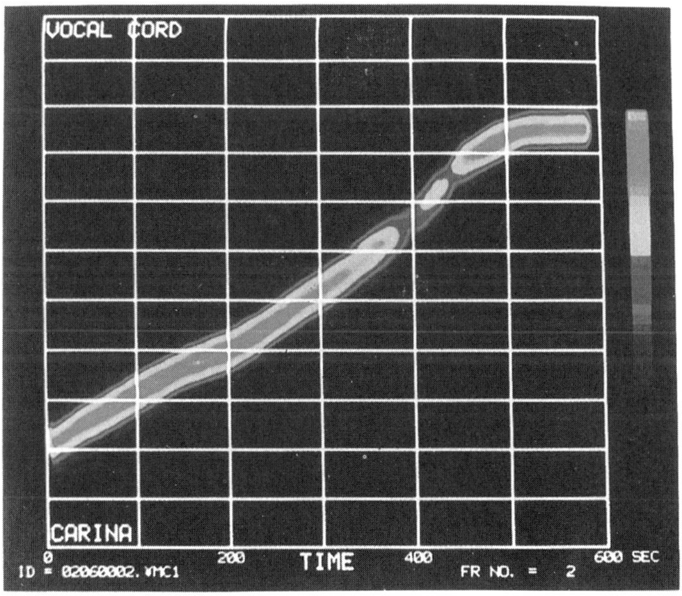


**Figure 7** Transport of a radioactive droplet with mucus of a normal dog from the carina to the vocal cord. (From Reference 18.)

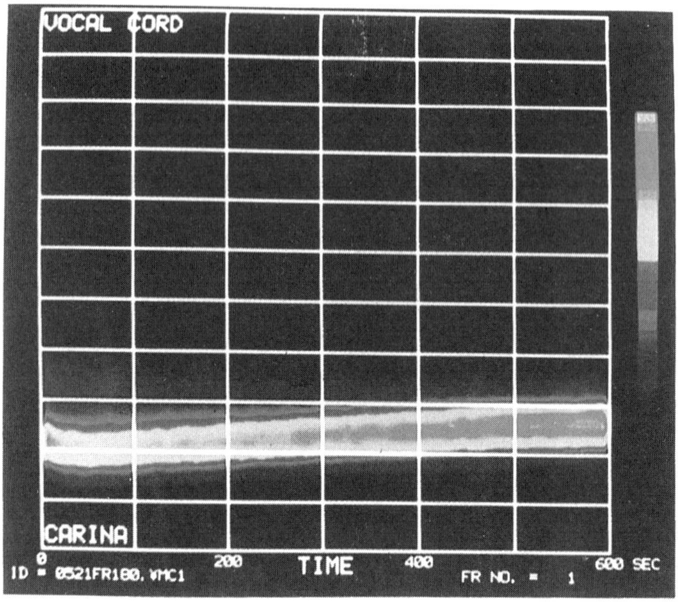


**Figure 8** Transport of a radioactive droplet with mucus of a normal dog who is forced to smoke 5 cigarettes. Mucus transport is greatly disturbed. (From Reference 18.)

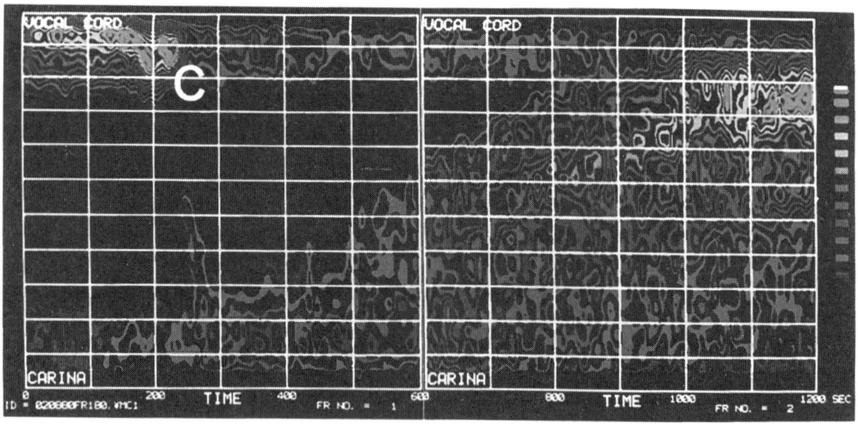


**Figure 9** Condensed image of a 29-year-old normal nonsmoker following radioaerosol inhalation. Wide bands towards the upper right are seen. C indicates that a mucus glob is expectorated by cough from the vocal cord. (From Reference 18.)

lation, and its location mathematically determined and recorded in a computer. On the next frame, also similarly magnified by interpolation, the mathematically changed location of the same hot spot is measured and recorded. This procedure is repeated frame by frame and each mathematical location is connected using a B spline function, which makes a trajectory of the transport of the hot spot (Figures 10 and 11). This trajectory enables a detailed analysis of mucociliary transport in the trachea [18].

Using this trajectory, vector analysis is possible and various parameters can be calculated using various values indicated in Figure 12. For example, effective velocity (Veff) and apparent velocity (Vapp) (as defined in Table 2) of tracheal mucous transport in normal nonsmokers, smokers, and patients with chronic obstructive pulmonary diseases (COPD) are shown in Figure 13. Fractions of forward transport, backward transport, and stasis in mucous transport can easily be calculated. Stasis ratios between healthy smokers, nonsmokers, and patients with COPD are shown in Figure 14.

### Iso-Count Display

Iso-count display of a hot spot in sequential frames is also possible by delineating iso-count lines in each frame. The direction of the transport can be assessed, but this method, too, is only feasible when a radioactive droplet is placed in the trachea. Thus this method is not practically applicable to aerosol inhalation studies [18].

## B. Ciliated Airways in Lung Parenchyma

When radioaerosol is inhaled, radioactivity deposits in the extrapulmonary ciliated airways (A), intrapulmonary ciliated airways (B), in the nonciliated small airways including the alveolar space (C), and in the esophagus and the stomach or the gastrointestinal (GI) tract (D) as illustrated in Figure 15 [12,13].

Disregarding the radioactivity in the stomach and/or the GI tract (D), the radioactivity at time zero or immediately after aerosol inhalation is finished can be written as follows;

$$A_0 + B_0 + C_0 = T_0 \tag{1}$$

where, A, B, C represents the compartments in Figure 15 and T, the total radioactivity in all three compartments. At time $t$, radioactivity at each compartment would be

$$A_t + B_t + C_t = T_t \tag{2}$$

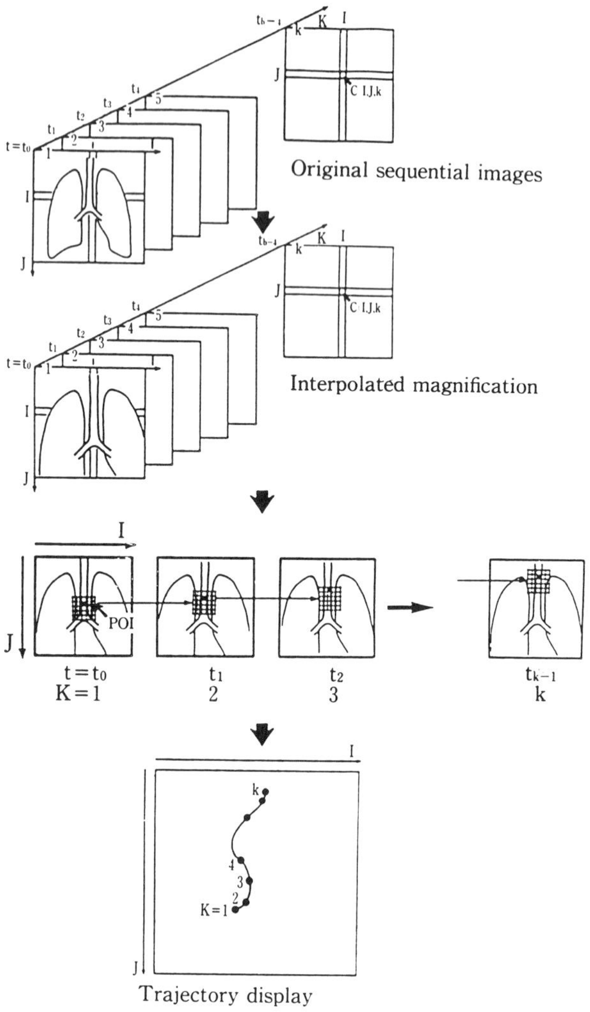


**Figure 10** Diagram showing the principle of the trajectory mode. The location of the point of interest (POI) is sequentially determined frame by frame and each mathematical location is connected using a B spline function. (From Reference 18.)

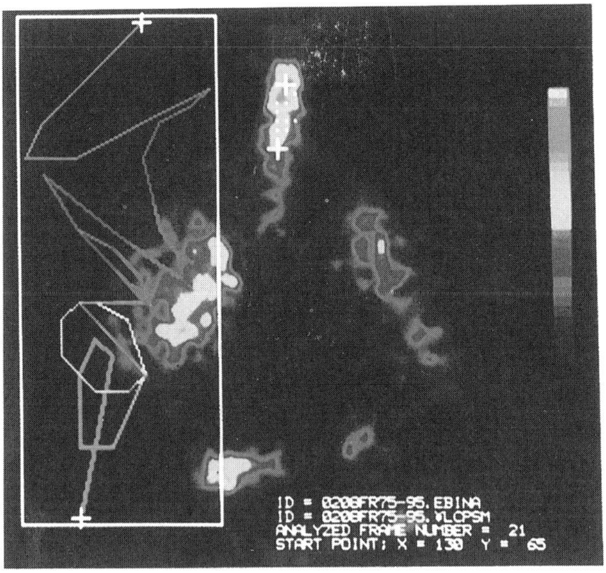


**Figure 11** An example of the trajectory in a 29-year-old normal nonsmoker beginning at the lower cross (red line) to the upper cross (blue line). (From Reference 18.)

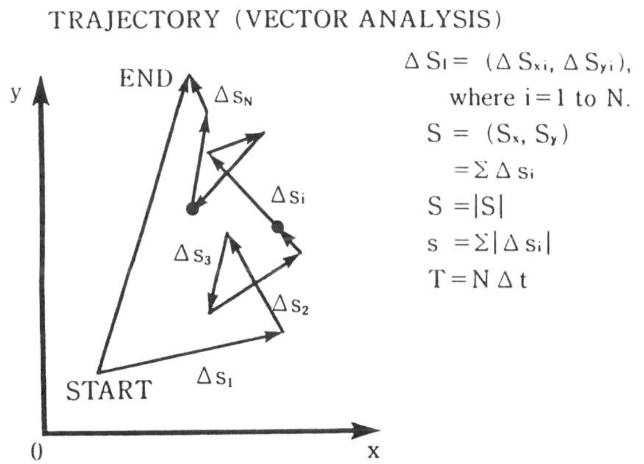


**Figure 12** Vector analysis of a trajectory. From START to END the actual pathway is comprised of $s_1$, $s_2$, .... $s_N$, although macroscopically it looks like a straight line between START and END. Each $s_1$, $s_2$, .... $s_N$ is a vector and mathematical calculation is possible. (From Reference 18.)

**Table 2** Definition of Indexes

| | |
|---|---|
| Effective transport velocity, $V_{eff}$ = S/T | |
| Apparent transport velocity, $V_{app}$ = s/T | |
| Traveling pathway index, $\alpha$ = s/S | |
| Stasis ratio, SR = $\Sigma$(number of time ($\Delta s_i$ = 0)/N) | |

If radioactivity is corrected for physical decay, Eq. (1) becomes

$$A_{tc} + B_{tc} + C_{tc} = T_{tc} \quad (3)$$

If we define the radioactivity remaining in the lung at 24 hrs is the amount of radioactivity deposited in the nonciliated space of the lung, $C_0$, corrected for physical decay should be the same as $C_{tc}$; $C_0 = C_{tc}$.

Practically speaking, it is extremely difficult to measure $A_t$ without contamination by radioactivity in the esophagus behind the trachea that has been swallowed. In evaluating the disappearance of radioactivity with time from the lung parenchyma, only radioactivity in the extrapulmonary

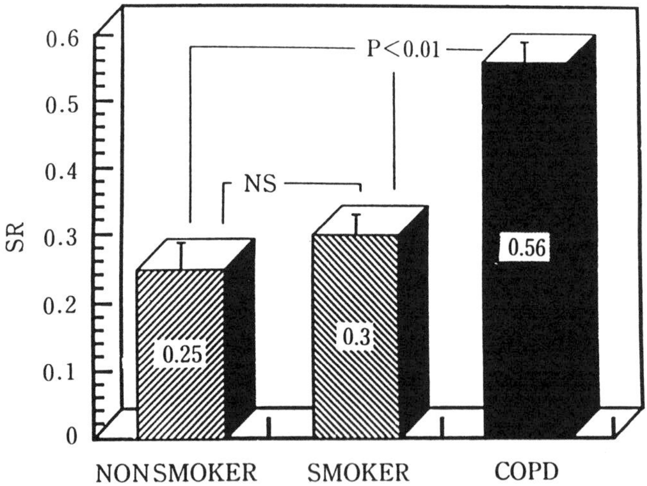


**Figure 13** Although there is no statistically significant difference in effective velocity ($V_{eff}$) between nonsmokers, smokers, and patients with chronic obstructive pulmonary disease (COPD), apparent velocity ($V_{app}$) significantly differs between normal subjects (smokers and nonsmokers) and patients with COPD. (From Reference 18.)

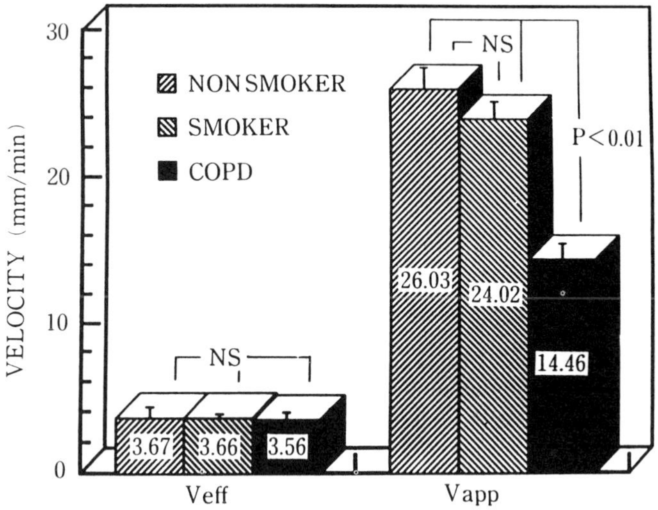


**Figure 14** Stasis ratio (SR) significantly differs between normal (nonsmokers and smokers) and patients with COPD. (From Reference 18.)

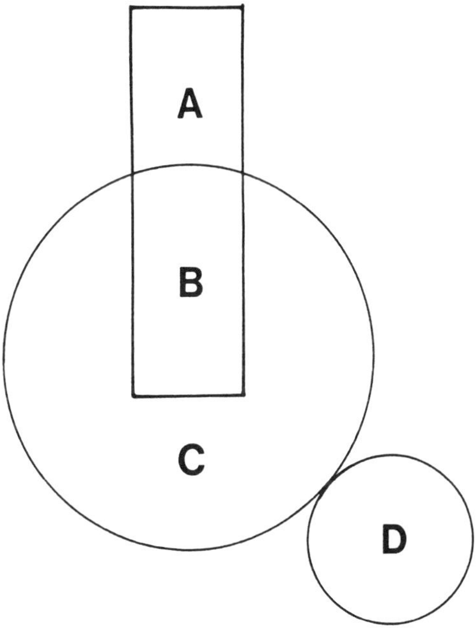


**Figure 15** Diagram of deposition sites of inhaled aerosol such as A, the trachea and the proximal portions of the major bronchi; B, the intrapulmonary ciliated airways; C, the nonciliated small airways including the alveolar space; and D, the digestive canal like the mouth, the esophagus, the stomach, etc.

airways should be taken into consideration. Radioactivity in compartment A should be excluded from consideration. Thus the above formulae should be rewritten as follows:

$$B_0 + C_0 = T_0 \tag{1'}$$

$$B_t + C_t = T_t \tag{2'}$$

$$B_{tc} + C_{tc} = T_{tc} \tag{3'}$$

Sequential measurements of radioactivity in the lung parenchyma, or in the thorax excluding the extrapulmonary mediastinal region, and the radioactivity in the same lung parenchyma at 24 hrs are the only values required to calculate the following indices.

### Lung Retention Ratio (LRR)

$$\text{LRR (\%)} = \frac{T_{tc}}{T_0} \times 100$$

This ratio expresses the amount of radioactivity remaining in the lungs at time $t$ relative to the total radioactivity initially deposited.

### Airway Deposition Ratio (ADR)

$$\text{ADR (\%)} = \frac{B_{tc}}{T_0} \times 100 = \frac{T_{tc} - C_0}{T_0} \times 100$$

The ADR indicates the amount of radioactivity throughout the ciliated airways relative to the total radioactivity initially deposited in the lungs.

### Airway Retention Ratio (ARR)

$$\text{ARR (\%)} = \frac{B_{tc}}{B_0} \times 100 = \frac{T_0 - T_{tc}}{T_0 - C_0} \times 100$$

The ARR indicates what percentage of radioactivity initially deposited on the ciliated airways still remains there at time $t$.

### Airway Clearance Efficiency (ACE)

$$\text{ACE (\%)} = \frac{B_0 - B_{tc}}{B_0} \times 100 = \frac{T_0 - T_{tc}}{T_0 - C_0} \times 100$$

The ACE indicates what percentage of the radioactivity initially deposited on the ciliated airways has already been cleared by time $t$.

**Table 3** Lung Retention Ratio (LRR) in the Right and Left Lungs in 13 Normal Nonsmokers, 15 Normal Smokers, and 28 Total Normal Subjects

| Time (min) | Nonsmoker ($n$ = 13) | | Smoker ($n$ = 15) | | Normal ($n$ = 28) | |
|---|---|---|---|---|---|---|
| | Left (%) | Right (%) | Left (%) | Right (%) | Left (%) | Right (%) |
| 10 | 100 | 100 | 100 | 100 | 100 | 100 |
| 20 | 93.9 ± 4.4 | 95.1 ± 4.3 | 89.9 ± 3.4 | 92.1 ± 3.5 | 91.8 ± 4.3 | 93.5 ± 4.1 |
| 30 | 90.0 ± 6.9 | 93.4 ± 5.3 | 84.1 ± 5.1 | 87.5 ± 5.9 | 86.8 ± 6.6 | 90.2 ± 6.3 |
| 40 | 87.3 ± 8.9 | 92.2 ± 6.3 | 79.7 ± 5.7 | 83.2 ± 6.1 | 83.2 ± 8.2 | 87.4 ± 7.6 |
| 50 | 85.2 ± 10.4 | 90.8 ± 6.7 | 77.4 ± 6.9 | 80.5 ± 7.0 | 80.9 ± 9.3 | 85.3 ± 8.5 |
| 60 | 83.7 ± 10.6 | 88.8 ± 7.0 | 75.3 ± 7.0 | 78.4 ± 7.1 | 79.4 ± 9.8 | 83.3 ± 8.7 |
| 70 | 82.6 ± 9.9 | 87.5 ± 7.1 | 73.9 ± 7.8 | 76.7 ± 7.8 | 78.0 ± 9.7 | 81.7 ± 9. |
| 80 | 82.0 ± 8.7 | 86.5 ± 7.5 | 71.9 ± 7.9 | 74.6 ± 7.8 | 76.8 ± 9.6 | 80.3 ± 9. |
| 90 | 80.8 ± 8.2 | 86.7 ± 7.7 | 70.4 ± 8.2 | 73.9 ± 9.7 | 75.4 ± 9.6 | 79.7 ± 10. |

## Alveolar Deposition Ratio (ALDR)

$$\text{ALDR (\%)} = \frac{C_0}{T_0} \times 100$$

The ALDR indicates the percentage of the total initial radioactivity remaining in the lung parenchyma at 24 hrs or the percentage of radioactivity deposited in the nonciliated space of the lungs including the alveolar space

**Table 4** Airway Deposition Ratio (ADR) in the Right and Left Lungs in 13 Normal Nonsmokers, 15 Normal Smokers, and 28 Total Normal Subjects

| Time (min) | Nonsmoker ($n$ = 13) | | Smoker ($n$ = 15) | | Normal ($n$ = 28) | |
|---|---|---|---|---|---|---|
| | Left (%) | Right (%) | Left (%) | Right (%) | Left (%) | Right (% |
| 10 | 56.2 ± 4.0 | 56.2 ± 3.9 | 64.3 ± 10.6 | 64.7 ± 10.8 | 60.5 ± 9.1 | 60.7 ± 9. |
| 20 | 50.1 ± 5.4 | 51.2 ± 5.2 | 54.2 ± 10.8 | 56.8 ± 9.9 | 52.3 ± 8.8 | 54.2 ± 8. |
| 30 | 46.2 ± 6.7 | 49.5 ± 5.3 | 49.1 ± 13.2 | 52.1 ± 10.4 | 47.7 ± 10.6 | 50.9 ± 8 |
| 40 | 43.5 ± 8.1 | 48.3 ± 6.1 | 43.9 ± 12.6 | 47.9 ± 11.0 | 43.7 ± 10.5 | 48.1 ± 8 |
| 50 | 41.3 ± 9.3 | 46.9 ± 6.5 | 41.5 ± 12.9 | 45.2 ± 11.1 | 41.4 ± 11.1 | 46.0 ± 9 |
| 60 | 39.8 ± 9.6 | 45.0 ± 6.5 | 39.5 ± 12.9 | 43.0 ± 11.6 | 39.7 ± 11.3 | 43.9 ± 9 |
| 70 | 38.8 ± 8.9 | 43.7 ± 6.6 | 38.1 ± 12.7 | 41.3 ± 11.5 | 38.4 ± 10.9 | 42.4 ± 9 |
| 80 | 38.2 ± 7.6 | 42.6 ± 6.8 | 36.1 ± 13.3 | 39.1 ± 11.5 | 37.1 ± 10.8 | 41.1 ± 8 |
| 90 | 36.9 ± 7.0 | 42.8 ± 7.0 | 34.5 ± 13.9 | 37.6 ± 12.0 | 35.7 ± 11.0 | 40.1 ± 10 |

**Table 5** Airway Retention Ratio (ARR) in the Right and Left Lungs in 13 Normal Nonsmokers, 15 Normal Smokers, and 28 Total Normal Subjects

| Time (min) | Nonsmoker ($n$ = 13) | | Smoker ($n$ = 15) | | Normal ($n$ = 28) | |
|---|---|---|---|---|---|---|
| | Left (%) | Right (%) | Left (%) | Right (%) | Left (%) | Right (%) |
| 10 | 100 | 100 | 100 | 100 | 100 | 100 |
| 20 | 89.1 ± 7.9 | 91.2 ± 7.6 | 83.9 ± 5.5 | 88.0 ± 5.4 | 86.3 ± 7.1 | 89.5 ± 6.6 |
| 30 | 82.5 ± 11.7 | 88.3 ± 9.0 | 74.8 ± 7.9 | 80.7 ± 9.1 | 78.3 ± 10.6 | 84.2 ± 9.7 |
| 40 | 77.8 ± 14.8 | 87.9 ± 8.9 | 67.3 ± 10.1 | 73.6 ± 9.8 | 72.2 ± 13.4 | 80.8 ± 11.7 |
| 50 | 74.0 ± 17.2 | 83.9 ± 11.4 | 63.5 ± 11.6 | 69.7 ± 10.7 | 68.4 ± 15.1 | 76.3 ± 13.0 |
| 60 | 71.2 ± 17.6 | 80.3 ± 11.8 | 60.6 ± 12.1 | 66.1 ± 11.6 | 65.5 ± 15.6 | 72.7 ± 13.6 |
| 70 | 69.5 ± 16.1 | 77.8 ± 11.6 | 58.5 ± 12.4 | 63.8 ± 12.5 | 63.6 ± 15.0 | 70.3 ± 13.8 |
| 80 | 68.3 ± 14.0 | 75.7 ± 12.6 | 55.1 ± 13.2 | 60.1 ± 12.5 | 61.5 ± 14.9 | 67.6 ± 14.6 |
| 90 | 66.4 ± 13.1 | 76.3 ± 12.9 | 52.9 ± 13.9 | 57.9 ± 13.6 | 59.4 ± 14.9 | 66.7 ± 16.0 |

at the completion of aerosol inhalation. Compartment C lacks mucociliary clearance.

Using human serum albumin aerosol, the AMAD of which is 1.9 micrometer (μm) with geometric standard deviation of 1.7, normal values are shown in Tables 3 to 6 [19]. The ALDR values can be different when different sized aerosols are inhaled. The smaller the aerosol size, the greater the ALDR [16]. It is ideal to establish the normal ranges of each parameter at each laboratory according to the aerosols used.

**Table 6** Airway Clearance Efficiency (ACE) in the Right and Left Lungs in 13 Normal Nonsmokers, 15 Normal Smokers and 28 Total Normal Subjects

| Time (min) | Nonsmoker ($n$ = 13) | | Smoker ($n$ = 15) | | Normal ($n$ = 28) | |
|---|---|---|---|---|---|---|
| | Left (%) | Right (%) | Left (%) | Right (%) | Left (%) | Right (%) |
| 10 | 0 | 0 | 0 | 0 | 0 | 0 |
| 20 | 10.9 ± 7.9 | 7.3 ± 6.8 | 16.1 ± 5.5 | 12.0 ± 5.4 | 13.7 ± 7.1 | 9.8 ± 6.4 |
| 30 | 17.5 ± 11.7 | 10.9 ± 8.7 | 25.4 ± 8.4 | 19.3 ± 9.1 | 21.7 ± 10.6 | 15.4 ± 9.8 |
| 40 | 22.2 ± 14.8 | 12.9 ± 10.3 | 32.7 ± 10.1 | 26.4 ± 9.8 | 27.8 ± 13.4 | 20.1 ± 12.0 |
| 50 | 26.0 ± 17.2 | 15.2 ± 11.0 | 36.5 ± 11.6 | 30.3 ± 10.7 | 31.6 ± 15.1 | 23.3 ± 13.1 |
| 60 | 28.8 ± 17.6 | 18.8 ± 11.3 | 39.4 ± 12.1 | 33.9 ± 11.6 | 34.5 ± 15.6 | 26.9 ± 13.6 |
| 70 | 30.5 ± 16.1 | 21.3 ± 11.2 | 41.5 ± 12.4 | 36.2 ± 12.5 | 36.4 ± 15.0 | 29.3 ± 13.9 |
| 80 | 31.7 ± 14.0 | 23.6 ± 12.4 | 44.9 ± 13.2 | 40.1 ± 12.6 | 38.5 ± 14.9 | 32.1 ± 14.8 |
| 90 | 32.8 ± 14.3 | 23.0 ± 12.6 | 47.1 ± 13.9 | 42.1 ± 13.6 | 40.3 ± 15.6 | 32.9 ± 16.1 |

# VIII. MUCOCILIARY CLEARANCE IN HEALTH AND DISEASE

## A. Large Airways: Trachea and Major Bronchi

### Normal Subjects

In nonsmoking normal subjects the transport of inhaled radioaerosol deposited in the airways is always axial and cephalad in direction and steady and constant in its transport velocity, showing no stagnation of radioactivity in the trachea or the bronchi. However, in smokers and some exsmokers, although radioactive transport is still cephalad in direction and transport velocity nearly constant, temporary accumulation of radioactivity is seen over the bronchi near or over the carina [12]. Such stasis, however, never persists very long. There is no visible retrograde transport or retreat, stasis or stagnation of mucous globs, frequent up-and-down motion of radioactive globs in the trachea or bronchi, or migration into the other regions of the same lung or into the bronchus of the opposite lung—findings that are often observed in patients with obstructive airway disease [13].

However, the trajectory in normal subjects is tortuous and not a simple trajectory as shown in Figure 11, which indicates that there is microscopically mixed forward and retrograde transport mixed with stasis but that the overall transport direction is toward the oropharyx [18].

### Obstructive Airway Disease

Transport of radioactivity over the trachea and the major bronchi in COPD patients is extremely protean in its direction and patterns. Of 21 patients with obstructive airway disease studied by radioaerosol inhalation lung cine-scintigraphy, 14 showed temporary but frequent stopping and starting of movement of radioactivity in the airways in the course of lung clearance.

Even after radioactivity begins to migrate up the trachea, it tends to stop on the way. Thus, stopping and renewed migration were repeated many times in the course of mucous transport. Migration was often accelerated by coughing and/or clearing the throat, with the radioactive mucus finally swallowed into the stomach or expectorated. Sometimes radioactivity remains at the same spot without migration until cleared by coughing. In this sense cough appears to be the only means of upward propulsion of the mucus. In 10 cases there was reversal of mucus flow. In another 5 cases, there was migration or straying of radioactivity from one bronchus to that of the opposite lung, bypassing the trachea, followed by shuttling between the right and left main bronchus, and finally mucus was coughed up. In 4 cases there was spiral or zigzag transport of radioactivity as shown in Figure 16.

In another series of patients with obstructive airway disease not only was there shuttling of mucus between the right and left main bronchus, but there was also migration of mucus from one region of the lung into different regions of the same lung frequently observed by radioaerosol inhalation lung cine-scintigraphy [13].

Using trajectory mode analysis, the trajectory in patients with COPD is seen to be simpler in shape than that in normal subjects because the stasis ratio becomes higher as shown in Figure 14 [18].

## Bronchiectasis

In bronchiectatic lung regions, deposition of inhaled radioaerosols is diminished and inhomogeneous. Radioaerosol inhalation lung cine-scintigraphy has revealed that transport of inhaled radioactivity from the bronchiectatic regions is greatly deranged. Regional stasis was observed in 12 of the 20 patients studied (12/20), regurgitation or reversed transport in 14/20, straying in 8/20, spiral or zigzag motion in 1/20. The transport patterns were more or less the combinations of these 4 basic abnormal patterns. When coughs occur, regurgitation and straying become more marked in the bronchiectatic regions. Only coughs can squeeze out radioactive mucus from inside the bronchiectatic regions to more proximal areas.

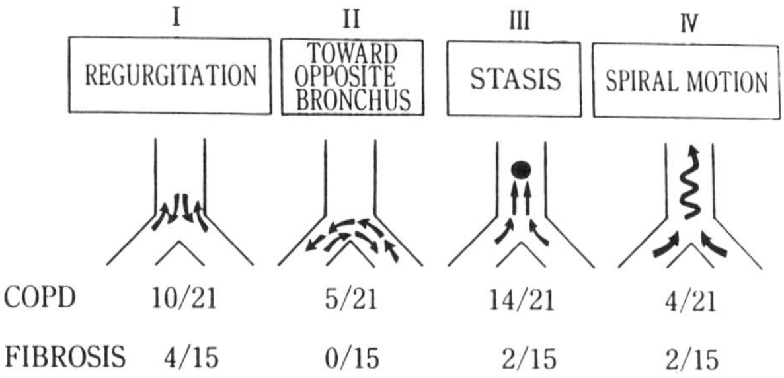


**Figure 16** Four abnormal mucus transport patterns in the trachea. Although steady transport cephalad in direction is seen in normal nonsmokers, the above 4 abnormal mucus transport patterns are observed in combination in pathologic states. In straying, only shuttling transport between the opposite bronchi is shown here. Straying transport from one region to a different portion inside the same lung is also seen. The numbers below indicate frequency observed in 20 patients with COPD and 15 patients with idiopathic interstitial fibrosis.

Mucociliary clearance from the bronchiectatic regions is very inefficient without the help of coughing. These regional abnormalities in mucociliary transport seem to be responsible for the development of infections and hemoptysis in the bronchietatic regions [20].

### Bronchogenic Carcinoma

In bronchogenic carcinoma, abnormal mucociliary transport patterns such as regurgitation, straying, stasis, and spiral or zigzag motions are seen especially when obstructive airway disease complicates bronchogenic carcinoma. Abnormal mucous transport patterns have nothing to do with histological diagnosis of bronchogenic carcinoma but rather with the degrees of functional and anatomical airway obstruction.

Bronchial invasion or protrusion of cancer in the large airways is often recognized as *a* "hot spot" that persists or disappears over time according to the degree mucosal damage. When a tumor is covered with intact ciliary mucosa, the hot spot disappears with time, but it persists when the bronchial mucosa is denuded by the tumor [21].

### Idiopathic Interstitial Fibrosis and Pulmonary Vascular Disease

Mucus transport in the large airways for patients with these conditions is not different from normal subjects unless complicated with obstructive airway disease. If complicated with COPD, similar transport patterns to those encountered in patients with COPD are observed [22,23].

## B. Ciliated Airways in Lung Parenchyma

### Normal Subjects

As shown in Tables 3–6, the lung retention ratio (LRR) at 30, 60, and 90 min is in the range of about 85–90, 80–85, and 75–80%, respectively. The alveolar deposition ratio (ALDR) is equivalent to the LRR at 24 hrs and amounts to about 40%. The airway deposition ratio (ADR) immediately after inhalation is about 60% and decreases with time to about 50, 45, and 40% at 30, 60, and 90 min, respectively. The airway retention ratio (ARR) is in the range of about 80–85, 65–70, 60–65% at 30, 60, and 90 min, respectively. The ACE is in the range of 15–20, 30–35, 35–40% at 30, 60, and 90 min, respectively. Normal smokers show a slightly faster clearance than nonsmokers [19]. The ALDR is significantly larger in normal nonsmokers than in normal smokers as shown in Table 7.

**Table 7** Case, Sex, Smoker or Not, Age, Pack-Year (PY), Alveolar Deposition Ratio (ALDR), and $FEV_{1.0}\%$

| Case | Sex | Smoking Habit | Age | PY | ALDR | $FEV_{1.0\%}$ |
|---|---|---|---|---|---|---|
| 1 | M | E | 43 | 7 | 47 | 85.5 |
| 2 | M | E | 33 | 6 | 30 | 72.9 |
| 3 | M | E | 32 | 3 | 47 | 88.8 |
| 4 | M | N | 29 | 0 | 44 | 77.5 |
| 5 | M | N | 30 | 0 | 47 | 81.3 |
| 6 | M | N | 76 | 0 | 40 | 76.6 |
| 7 | M | N | 76 | 0 | 41 | 76.6 |
| 8 | M | N | 29 | 0 | 43 | 86.4 |
| 9 | M | N | 31 | 0 | 48 | 91.1 |
| 10 | M | N | 25 | 0 | 50 | 75.9 |
| 11 | M | N | 28 | 0 | 43 | 75.9 |
| 12 | M | N | 33 | 0 | 46 | 90.6 |
| 13 | M | N | 41 | 0 | 44 | 78 |
| 14 | M | S | 36 | 24 | 37 | 75.3 |
| 15 | M | S | 32 | 18 | 38 | 81.2 |
| 16 | M | S | 37 | 26 | 44 | 81.1 |
| 17 | M | S | 51 | 50 | 31 | 82.4 |
| 18 | M | S | 38 | 18 | 37 | 76.3 |
| 19 | M | S | 47 | 41 | 34 | 78 |
| 20 | M | S | 28 | 8 | 55 | 86.9 |
| 21 | M | S | 38 | 27 | 34 | 86 |
| 22 | M | S | 58 | 65 | 15 | 72.7 |
| 23 | M | S | 38 | 27 | 35 | 86 |
| 24 | M | S | 58 | 65 | 22 | 73.6 |
| 25 | M | S | 32 | 13 | 35 | 83.6 |
| 26 | M | S | 28 | 10 | 39 | 93 |
| 27 | M | S | 34 | 21 | 32 | 83.2 |
| 28 | M | S | 29 | 9 | 39 | 88.5 |
| Nonsmoker and exsmoker |  | Mean | 38.9 | 0 | 43.8 | 81.3 |
| ($n$ = 13) |  | S.D. | 17.2 | 0 | 5 | 6.3 |
| Smoker |  | Mean | 38.9 | 28.1 | 35.1 | 81.9 |
| ($n$ = 15) |  | S.D. | 10 | 18.8 | 9 | 5.8 |
| Normal |  | Mean | 38.9 | 0 | 39.2 | 81.6 |
| ($n$ = 28) |  | S.D. | 13.5 | 0 | 8.5 | 6 |

M = male; E = exsmoker; N = nonsmoker; S = smoker.

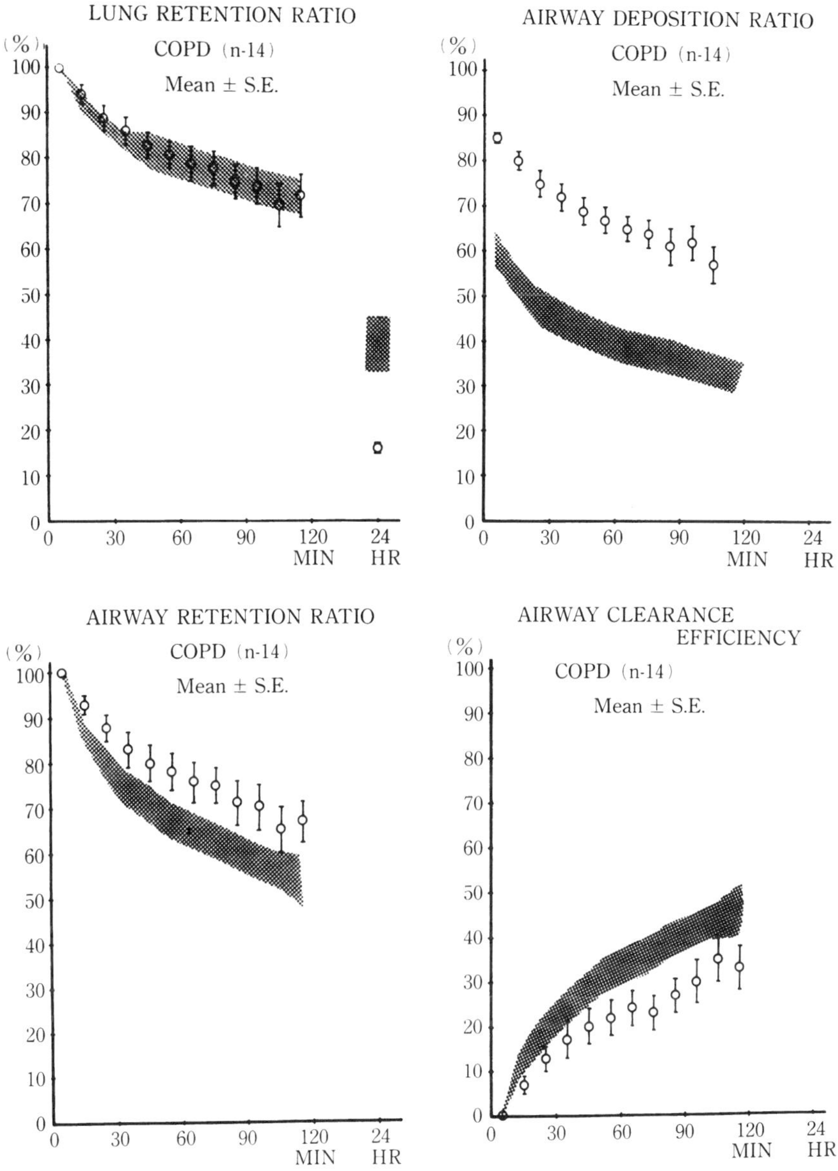


**Figure 17** Four parameters in patients with COPD. ALDR equivalent to LRR at 24 hr and ACE are significantly decreased; ADR and ARR, increased. LRR remains within normal limits. Each band represents mean value and 95% confidence interval.

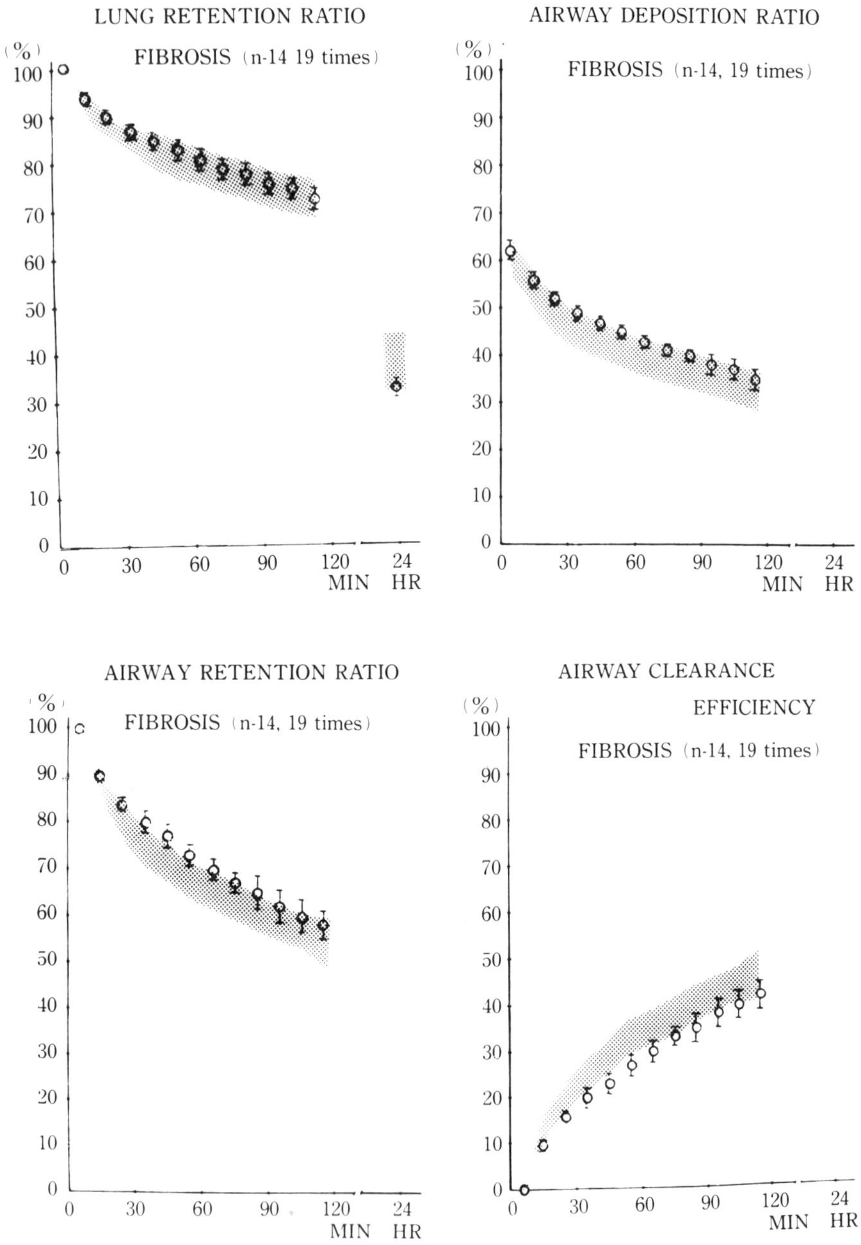


**Figure 18** Four parameters in patients with interstitial fibrosis. All 4 parameters remain within normal range in patients with interstitial fibrosis. Each band represents mean value and 95% confidence interval.

### Obstructive Airway Disease

As shown in Figure 17, the LRR itself is not distinguishable from the normal range, but the ALDR is significantly less than normal range. Both the ADR and the ARR are higher and the ACE is less than the normal range, indicating that a larger fraction of inhaled aerosol deposits in the ciliated airways and that mucociliary clearance is less efficient [13].

### Idiopathic Interstitial Fibrosis and Pulmonary Vascular Disease

As shown in Figure 18, all the parameters remain in the normal range in idiopathic interstitial fibrosis [22] and are not deranged from the normal range in pulmonary vascular diseases such as pulmonary embolism or pulmonary vascular anomalies [23].

## IX. PHARMACOLOGICAL EFFECTS

Clinical effects of mucolytic agents and bronchodilators have been evaluated rather subjectively on the basis of patients' sense of clinical improvement. Quantitative evaluation seems possible applying the present methodology of radioaerosol inhalation and subsequent sequential imaging and measurement of radioactivity of the lungs. Reports regarding drug effects are conflicting [24–27].

### A. Bromhexine

Bromhexine is claimed to liquefy mucus by breaking the mucopolysaccharide chains in the mucus, facilitating its removal. Airway clearance efficiency (ACE) was evaluated by radioaerosol inhalation lung scintigraphy in 10 patients with various chest diseases before and after 7 days of administration of 8 mg of bromhexine orally three times a day (t.i.d.). As shown in Figure 19, ACE was statistically barely significantly improved by chi-square test ($p=0.05$). Pulmonary function testing revealed little change before and after the administration except for maximum midexpiratory flow rate (MMF) becoming slightly larger and RV/TLC ratio (residual volume)/(total lung capacity) becoming slightly smaller [28].

### B. Beta 2-Stimulators

It is known that ciliary beat frequency in vitro becomes increased by administration of salbutamol [29]. Administration of oral salbutamol, 8

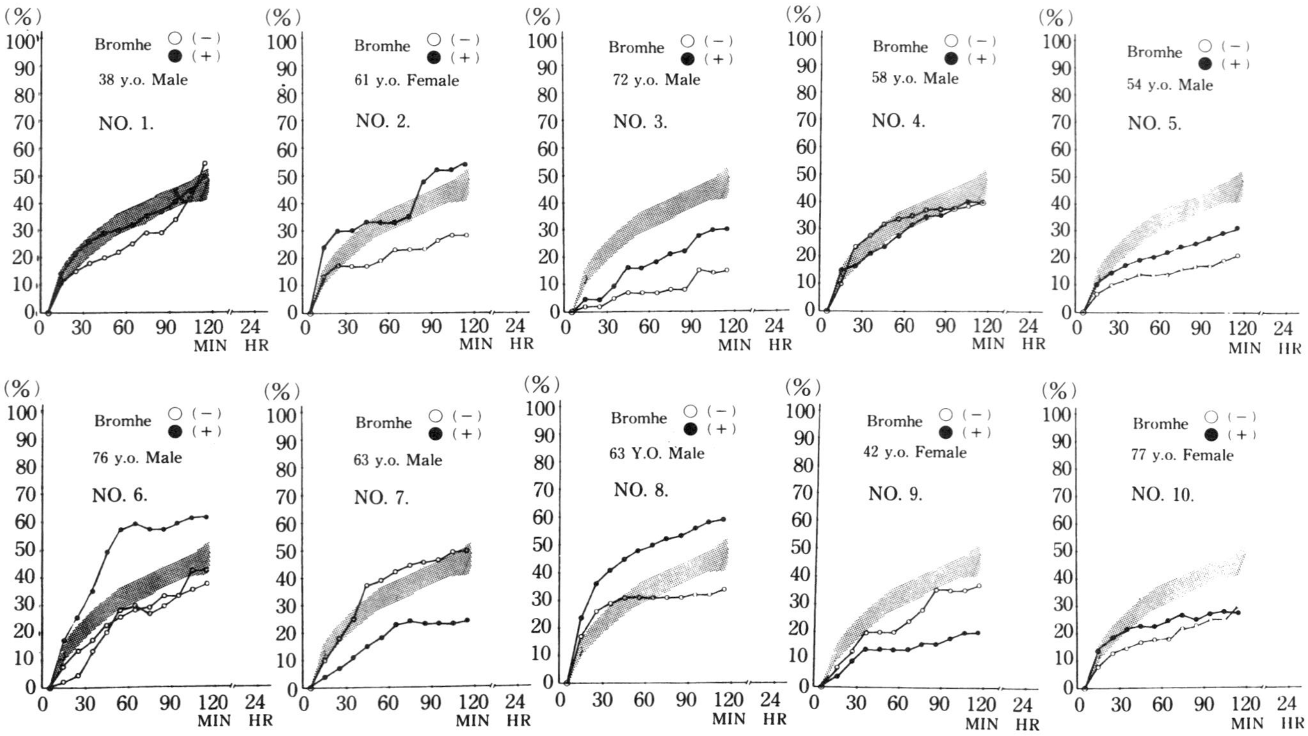


**Figure 19** ACE in 10 patients before and after treatment with oral bromhexine 8 mg t.i.d. for 7 days. ACE barely significantly improves (p = 0.05) after treatment. Each band represents mean value and 95% confidence interval.

mg t.i.d. for 7 days, did not change either radioaerosol inhalation lung images or quantitative parameters as compared with before administration, although significant bronchodilation occurred by pulmonary function test (PFT) after the 7 days of administration [30].

In the midst of radioactive measurement following radioaerosol inhalation a beta 2-stimulator, procaterol, was inhaled to see if there was any change in shape of the time–activity curve following drug inhalation. If the time–activity curve over the lungs becomes steeper, we could surmize that the mucociliary clearance was accelerated by the medication.

There was neither significant acceleration in mucous transport by radioaerosol inhalation lung cine-scintigraphy nor significant changes in the slope of time–activity curves. No changes were observed in any quantitative parameters after drug inhalation in patients with bronchial asthma in remission, although spirometry indicated significant bronchodilation [31].

### C. Aminophylline and Beta 2-Stimulators

Radioaerosol inhalation lung cine-scintigraphy and PFT were performed on 10 patients with bronchial asthma in remission before and after 250 mg of aminophylline infusion followed by inhalation of salbutamol.

The bronchodilating effect of the combined treatment was significant: inhaled aerosol deposited more homogeneously and less centrally in the lungs. The penetration index increased from 31 ± 3% to 49 ± 7% and the alveolar deposition ratio from 29 ± 2% to 39 ± 1%. The ADR decreased from 72 ± 2% to 61 ± 1% immediately after the treatment. Spirometry indicated significant bronchodilation. However, there was little qualitative or quantitative improvement in mucociliary clearance after the treatment [32].

## REFERENCES


1. Sleigh MA. The nature and action of respiratory tract cilia. In: Brain JD, Proctor DF, Reid LM, eds. Respiratory Defense Mechanisms. Part 1. New York: Marcel Dekker, 1977: 247–288.
2. Lucas AM, Douglas LC. Principles of underlying ciliary activity in the respiratory tract. II. A comparison of nasal clearance in man, monkey and other mammals. Arch Otolaryngol 1934; 20:518–541.
3. Wanner A. Clinical aspects of mucociliary transport. Am Rev Respir Dis 1977; 116:73–125.
4. Hilding AC. Ciliary streaming in the bronchial tree and the time element in carcinogenesis. N Engl J Med 1957; 256:634–640.

5. Berke HL, Roslinski LM. The roentgenographic determination of tracheal mucociliary transport rate in the rat. Am Ind Hyg Assoc J 1971; 32:174.
6. Sackner MA, Rosen MJ, Wanner A. Estimation of tracheal mucous velocity by bronchofiberscopy. J Appl Physiol 1973; 34:495–499.
7. Friedman M, Scott FD, Poole DO, Dougherty R, Chapman GA, Watson H, Sackner MA. A new roentgenographic method for estimating mucous velocity in airways. Am Rev Respir Dis 1977; 115:67–72.
8. Sakakura Y, Proctor DF. The effect of various conditions on tracheal mucociliary transport in dogs. Proc Soc Exp Biol Med 1972; 140:870–879.
9. Baetjer AM. Effect of ambient temperature and vapor pressure on cilia–mucus clearance rate. J Appl Physiol 1967; 23:498–504.
10. Isawa T, Hirano T, Teshima T, Konno K. Effect of non-filtered and filtered cigarette smoke on mucociliary clearance mechanism. Tohoku J Exp Med 1980; 130:189–197.
11. Isawa T, Teshima T, Hirano T, Ebina A, Konno K. Radioaerosol inhalation lung cine–scintigraphy: a preliminary report. Tohoku J Exp Med 1981; 134: 245–255.
12. Isawa T, Teshima T, Hirano T, Ebina A, Konno K. Mucociliary clearance mechanisms in smoking and nonsmoking normal subjects. J Nucl Med 1984; 25:352–359.
13. Isawa T, Teshima T, Hirano T, Ebina A, Motomiya M, Konno K. Lung clearance mechanisms in obstructive airways disease. J Nucl Med 1984; 25: 447–454.
14. Isawa T, Wasserman K, Taplin GV. Lung scintigraphy and pulmonary function studies in obstructive airways disease. Am Rev Respir Dis 1970; 102: 161–172.
15. Hirano T. Mucociliary clearance mechanisms. I. Canine tracheal mucous velocity and different particles sizes (Jpn). Kohkenshi 1988; 40:107–113.
16. Miki M, Isawa T, Teshima T, Anazawa Y, Motomiya M. Difference in inhaled aerosol deposition patterns in the lungs due to three different sized aerosols. Nucl Med Commun 1992; 13:553–562.
17. Morrow PE, Yu CP. Models of aerosol behavior in airways and alveoli. In: Moren F, Dolovich MB, Newhouse MT, Newman SP, eds. Aerosols in Medicine. Principles, Diagnosis and Therapy. Amsterdam: Elsevier Science Publishers BV, 1993:157–193.
18. Teshima T, Isawa T, Hirano T, Anazawa Y, Miki M, Motomiya M. Image processing for mucociliary transport system (Jpn). Respiration 1989; 8: 828–835.
19. Isawa T, Teshima T, Hirano T, Anazawa Y, Miki M, Konno K, Motomiya M. Normal values for quantitative parameters for evaluation of mucociliary clearance in lungs. Tohoku J Exp Med 1989; 158:119–131.
20. Isawa T, Teshima T, Hirano T, Anazawa Y, Miki M, Konno K, Motomiya M. Mucociliary clearance and transport in bronchiectasis: global and regional assessment. J Nucl Med 1990; 31:543–548.
21. Isawa T, Teshima T, Hirano T, Ebina A, Konno K. Mucociliary clearance mechanisms in lung cancer (Jpn). Kohkenshi 1986; 38:147–157.

22. Isawa T, Teshima T, Hirano T, Ebina A, Konno K. Mucociliary clearance mechanism in interstitial lung disease. Tohoku J Exp Med 1986; 148:169–178.
23. Isawa T, Teshima T, Hirano T, Ebina A, Anazawa Y, Konno K. Mucociliary clearance in pulmonary vascular disease. Ann Nucl Med 1988; 2:41–47.
24. Sackner MA. Effects of respiratory drugs on mucociliary clearance. Chest 1978; 73(suppl):958–964.
25. Matthys H, Koehler D. Effect of theophylline on mucociliary clearance in man. Eur J Respir Dis 1980; 109(Suppl):98–102.
26. Wanner A. Alteration of tracheal mucociliary transport in airway disease. Effect of pharmacologic agents. Chest 1981; 80(Suppl):967–870.
27. Pavia D, Sutton PP, Lopez-Vidriero, MT, Agnew JE, Clarke SW. Drug effects on mucociliary function. Eur J Respir Dis 1983; 64(Suppl 128):304–317.
28. Isawa T, Teshima T, Hirano T, Ebina A, Konno K. Evaluation of mucociliary clearance mechanisms by radioaerosol inhalation lung scintigraphy—Effect of oral bromhexine (Jpn). J Jpn Soc Chest Dis 1984; 22:899–909.
29. Van As. The role of selective beta 2-adrenoreceptor stimulants in the control of ciliary activity. Respiration 1974; 31:146–151.
30. Isawa T, Teshima T, Hirano T, Ebina A, Konno K. Effect of oral salbutamol on mucociliary clearance mechanisms in the lungs. Tohoku J Exp Med 1986; 150:51–61.
31. Isawa T, Teshima T, Hirano T, Anazawa Y, Miki M, Konno K, Motomiya M. Does a beta 2-stimulator really facilitate mucociliary transport in the human lungs in vivo? Am Rev Respir Dis 1990; 141:715–720.
32. Isawa T, Teshima T, Hirano T, Ebina A, Anazawa Y, Konno K. Effect of bronchodilation on the deposition and clearance of radioaerosol in bronchial asthma in remission. J Nucl Med 1987; 28:1901–1906.

# 46

# Human Ciliated Epithelium In Vitro—Mechanisms of Injury and Protection

**Charles Feldman**
*Johannesburg Hospital and University of the Witwatersrand, Johannesburg, South Africa*

**Ronald Anderson**
*University of Pretoria, Pretoria, South Africa*

**Andrew Rutman, Peter J. Cole, and Robert Wilson**
*National Heart and Lung Institute, and Royal Brompton Hospital, London, United Kingdom*

## I. INTRODUCTION

The primary function of the lungs is to breathe, and with each breath microorganisms and other potentially toxic agents are moved in and out of the airways [1]. The bronchial epithelium is the primary barrier between the environment and the conducting airways and the mucociliary escalator serves as the first-line defense mechanism, clearing the lower respiratory tract of potentially harmful particles and keeping it sterile [1]. There are a number of factors, both bacterial and host derived, that have been shown in vitro to cause injury to the epithelium and its mucociliary mechanism, and various agents have been identified that protect against this injury (Table 1). This overview discusses these mechanisms of injury and protection focusing on in vitro studies undertaken by the authors. In these studies epithelium was obtained by brushings taken from the inferior nasal turbinate of healthy volunteers [2] and organ cultures were prepared using

**Table 1** Factors Documented To Cause Injury To Respiratory Epithelium and its Mucociliary Mechanism and Agents that May Protect Against This Injury In Vitro

| Injury |
|---|
| Bacterial factors |
| *Pseudomonas aeruginosa* |
| *Streptococcus pneumoniae* |
| *Haemophilus influenzae* |
| *Mycobacterium tuberculosis* |
| Others |
| Reactive oxidants |
| Hydrogen peroxide |
| Hypochlorous acid |
| Bioactive phospholipids |
| Platelet-activating factor (PAF) |
| Lyso-PAF |
| Lysophosphatidylcholine |
| Protease enzymes |
| Neutrophil elastase |
| **Protection** |
| Salmeterol |
| Antibiotics |
| Azalide/macrolide |
| Others |
| Additional agents |
| Theophylline |
| Phosphodiesterase inhibitors |
| Vitamin E |
| Others |

resected nasal turbinate or adenoid tissue [3] as described previously. Effects on ciliary beating were measured by a photometric [4] or phototransistor technique [5] and injury to the epithelium quantified using a visual scoring index under light microscopy and by transmission and scanning electron microscopy.

## II. MECHANISMS OF EPITHELIAL INJURY

### A. Bacterial Factors

#### Bacterial Interaction with Cilia

Several bacteria associated with either acute or chronic respiratory infections in humans, including *Pseudomonas aeruginosa, Streptococcus pneumoniae, Haemophilus influenzae*, and others [1] have been documented to produce factors that perturb the mucociliary mechanism and cause injury to ciliated epithelium in vitro.

Wilson and co-workers noted that the phenazine pigments of *P. aeruginosa*, pyocyanin and 1-hydroxyphenazine, slowed cilia in a dose-dependent manner [6]. Pyocyanin caused a gradual onset of progressive ciliary slowing, ultimately associated with widespread ciliostasis and epithelial damage, while 1-hydroxyphenazine caused a rapid onset of ciliary slowing with subsequent recovery [6]. Marked ciliary dyskinesia was seen with the latter. Using radiolabeled erythrocytes in anesthetised guinea pigs, they also documented that these toxins slowed tracheal mucus velocity [7]. The phenazine pigments have been noted to have diverse proinflammatory interactions with human phagocytes, which may intensify both bacterial- and neutrophil-mediated epithelial injury [8]. *P. aeruginosa* proteases [9] and rhamnolipid [10] also inhibit ciliary beating and cause epithelial disruption.

Studies with *S. pneumoniae* have documented that its protein toxin, pneumolysin, is responsible for the effects on ciliated epithelium (Figure 1) [11,12]. Pneumolysin has distinct cytotoxic and complement-activating moeties resident in separate domains on the molecule, and we have demonstrated that the effects on ciliated epithelium are associated with the cytotoxic region (Figure 2) [12]. Pneumolysin has additional proinflammatory activities. In an experimental model of lobar pneumonia, pneumolysin alone induced the salient histological features of pneumococcal infection [13], and prior immunization with Freund's adjuvant and pneumolysin reduced the severity of the pneumonia [14].

We unfortunately do not have the ability to test these in vitro observations accurately in humans in vivo. However, *P. aeruginosa* phenazine pigments have been found in the sputum sol phase of patients with cystic fibrosis and bronchiectasis, who are colonized with this organism, in sufficient concentrations to inhibit ciliary beating in vitro [15]. Factors perturbing mucociliary clearance may enhance bacterial colonization of the airway, which may be an important precursor of both acute and chronic infections of the lower respiratory tract.

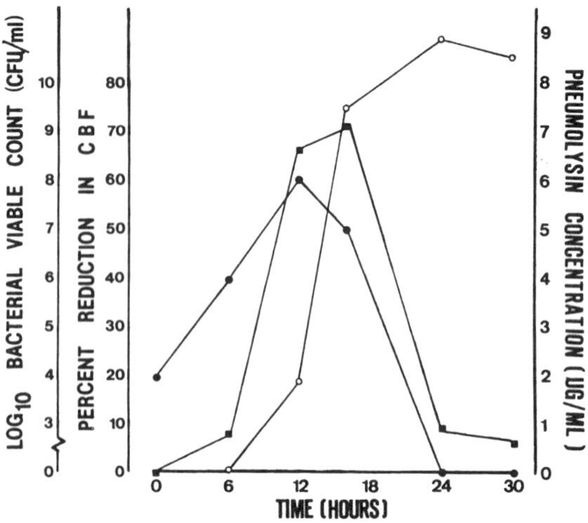


**Figure 1** The properties of aliquots removed from a 30-hour broth culture of a laboratory type-3 strain of *Streptococcus pneumoniae*. The figure demonstrates the organism viable count (●; cfu/mL), percent reduction in ciliary beat frequency (■; CBF) caused by culture filtrates, and the pneumolysin concentration in the supernatant (○; μg/mL) at six time intervals. (Reprinted from Feldman et al. The effect of *Streptococcus pneumoniae* pneumolysin on human respiratory epithelium in vitro. Microbial Pathogenesis 1990;9:275–284, with permission from Publishers Academic Press Limited, London, England.)

The cellular mechanisms by which these toxins may cause ciliary dysfunction and epithelial disruption have been studied further. In a study using sheep trachea it was suggested that the mechanism of epithelial injury with pyocyanin may be related to enhanced release of reactive oxidants from polymorphonuclear leucocytes (PMNL) within the epithelial surface [16]. In contrast, using human nasal ciliated epithelium, our studies have suggested that the mechanism of ciliary slowing is due to depletion of intracellular cyclic AMP and ATP and the effects do not appear to be mediated through neutrophils [17]. Ciliary slowing was prevented by agents that increased intracellular cAMP and by the cAMP analogue, dibutyryl cAMP [17]. These agents also delayed the onset of epithelial disruption [17].

## Bacterial Adherence to Ciliated Epithelium

An important initial event in the pathogenesis of most infections is the attachment of bacteria to mucosal surfaces. We studied these interactions

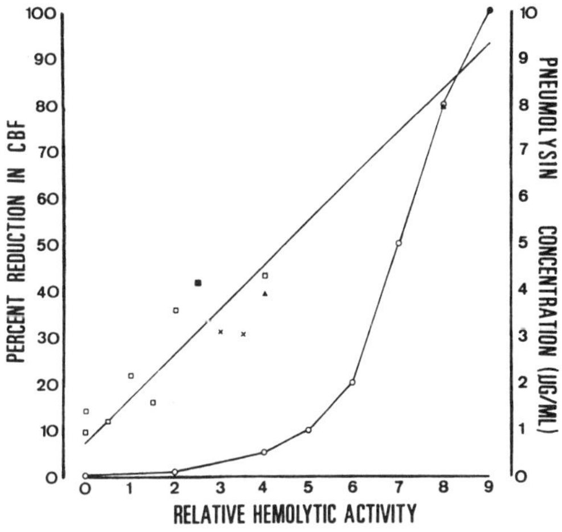


**Figure 2** The straight line represents the percent reduction in ciliary beat frequency (CBF) and relative hemolytic activity of ten culture filtrates of *Streptococcus pneumoniae* (□); seven filtrates from different bacteria and three separate filtrates of a laboratory type-3 strain (■ and x) and three different pneumolysin concentrations in broth (▲). The curve represents the relative hemolytic activity of eight concentrations of recombinant pneumolysin (μg/mL) in medium 199 (●). (Reprinted from Feldman et al. The effect of *Streptococcus pneumoniae* pneumolysin on human respiratory epithelium in vitro. Microbial Pathogenesis 1990;9: 275–284, with permission from Publishers Academic Press Limited, London, England.)

in organ culture in vitro. *S. pneumoniae* adhered initially to the thick gelatinous mucus layer (Figure 3a, 3b) and caused progressive ciliary slowing, presumably due to the effects of pneumolysin [18]. Adherence to epithelial cells or cilia was not seen, and only minor and patchy epithelial damage noted. Later, epithelial disruption was noted with separation of tight junctions through which these organisms appeared to invade [19].

*H. influenzae* was also initially associated with the mucus layer of the epithelial surface and later with epithelial damage that appears to be a prerequisite for the association of this organism with respiratory epithelium [3]. In the organ culture, *P. aeruginosa* adhered preferentially to mucus, damaged epithelium, and exposed basement membrane [20].

More recently we have documented the in vitro interaction of *Mycobacterium tuberculosis* and human ciliated mucosa [21]. By 18 hours, a 50% loss of ciliation was seen together with marked epithelial damage and internalization of these organisms in the epithelial cells [22].

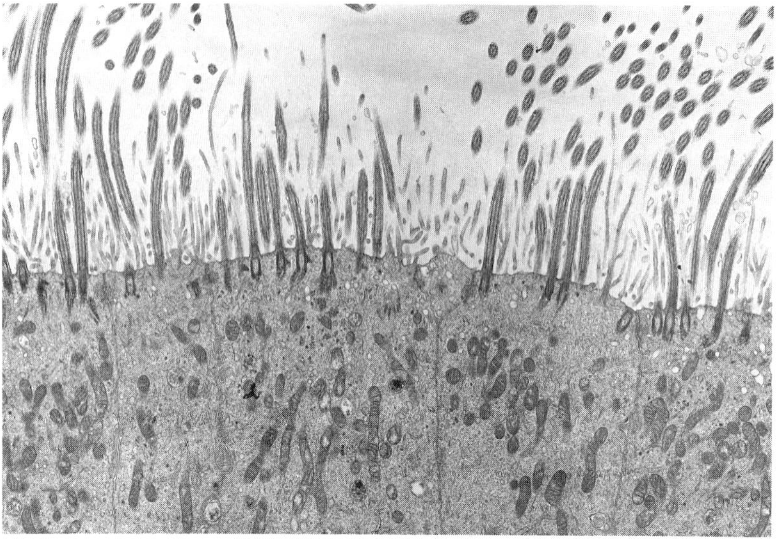

(a)

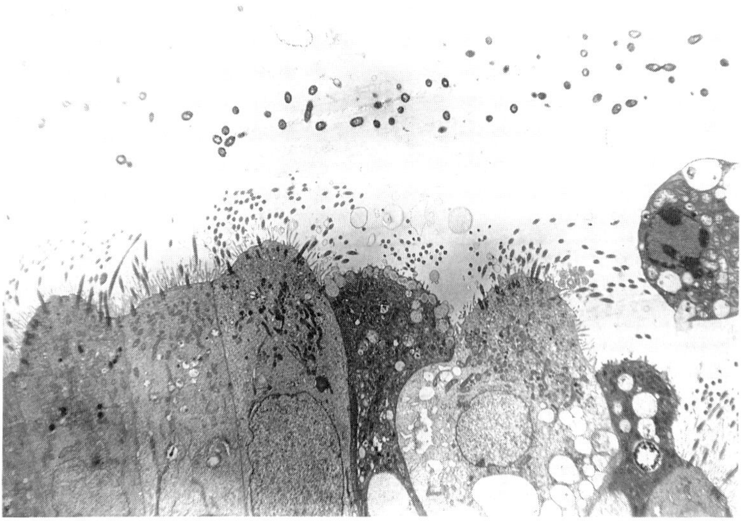

(b)

**Figure 3** (a) Transmission electron micrograph of normal human nasal turbinate organ culture at 24 hours (1 cm = 1.00 μm). (b) Transmission electron micrograph of human nasal turbinate organ culture incubated with *Streptococcus pneumoniae* for 24 hours. Many *S. pneumoniae* are seen well above the tips of the cilia. The epithelium shows only mild cell extrusion and some mitochondrial abnormalities (1 cm = 3.33 μm). (Reprinted with permission from Reference 18).

## B. Host Factors

A number of host-derived factors have been noted to cause ciliary dysfunction and epithelial injury.

### Reactive Oxidants

Using enzymatically generated oxidant and reagent preparations we noted that hydrogen peroxide ($H_2O_2$) and hypochlorous acid caused ciliary dyskinesia and slowing at concentrations $\geq 100$ μm with complete ciliostasis occurring at higher concentrations [22]. While phagocyte-derived oxidants may be an important source of these molecules, it has been recognized that the pneumococcus itself produces $H_2O_2$ in significant amounts, and this may be an important virulence determinant of the organism [23].

### Bioactive Phospholipids

Endogenously produced bioactive phospholipids, platelet-activating factor (PAF), lyso-PAF, and lysophosphatidylcholine (LPC) may play a role in noninfective inflammatory diseases of the airway such as asthma. Bioactive phospholipids in vitro cause significant and dose-dependent slowing of ciliary beating and epithelial injury at concentrations $\geq 1$ μg/mL [24,25]. These effects are enhanced in the presence of PMNL, probably due to enhanced reactive oxidant production by these cells [24]. In an experimental model, PAF-activated guinea pig eosinophils were shown to disrupt guinea pig tracheal epithelium by a mechanism related, at least partly, to reactive oxidant production [26].

### Protease Enzymes

Elastase is another neutrophil product that damages epithelium. High concentrations (500 μg/mL) progressively reduced ciliary beat frequency and caused marked epithelial disruption, while lower concentrations (20 μg/mL) also caused epithelial disruption but without slowing ciliary beat [9].

# III. MECHANISMS OF PROTECTION

Several agents have been identified that may protect against or reverse these harmful effects on ciliated epithelium in vitro.

## A. Salmeterol

Salmeterol reduced the pyocyanin-induced slowing of human ciliary beating in vitro by 53% and lessened the fall in intracellular cAMP and ATP

[27]. These effects appeared to be mediated through stimulation of the $B_2$-adrenoceptors [27]. In organ culture, salmeterol and rolipram had no effect on the distribution of *P. aeruginosa* organisms on the mucosa, but did reduce the amount of epithelial damage and loss of ciliated cells [28]. We have recently demonstrated that salmeterol antagonizes the proinflammatory, pro-oxidative activity of the bioactive phospholipids by a membrane-stabilizing mechanism [29].

### B. Antibiotics

Low, sub-minimum inhibitory concentration (MIC) concentrations of amoxycillin, loracarbef, and ciprofloxacin have been noted to attenuate the effects of *H. influenzae* on human ciliated epithelium in vitro [30], while erythromycin was shown to suppress the production by *P. aeruginosa* of toxins that either damage epithelium or stimulate PMNL cytoxicity [31]. The macrolide/azalide antibiotics (azithromycin, clarithromycin, and roxithromycin) attenuated the effects of bioactive phospholipids on human ciliated epithelium predominantly by inhibition of reactive oxidant production [25]. These agents are effective inhibitors of reactive oxidant production by activated PMNL due to a membrane-stabilizing activity [32].

### C. Other Agents

We have identified and are currently investigating a number of other agents that through their effects on cAMP, reactive oxidant production, or membrane stabilization protect human ciliated epithelium in vitro [33].

## IV. CONCLUSION

Mucociliary dysfunction is present in a number of respiratory diseases, especially those that involve the airways, and is a consequence of the disease process [34]. While it may itself cause considerable morbidity, currently available treatment modalities for this component of the disorder are not very effective, and are usually ignored [34]. A better understanding of factors implicated in mucociliary dysfunction and modalities by which these may be successfully reversed, as investigated in studies such as those described above, may help identify novel treatment strategies for this component of various respiratory disorders.

## REFERENCES


1. Wilson R, Dowling RB, Jackson AD. The biology of bacterial colonization and invasion of the respiratory mucosa. Eur Respir J 1996; 9:1523–1530.
2. Rutland J, Cole PJ. Non invasive sampling of nasal cilia for measurement of beat frequency and study of ultrastructure. Lancet 1980; ii:564–565.
3. Read R, Wilson R, Rutman A, Lund V, Todd HC, Brain APR, Jeffery PK, Cole PJ. Interaction of nontypable *Haemophilus influenzae* with human respiratory mucosa *in vitro*. J Infect Dis 1991; 163:549–558.
4. Greenstone M, Logan-Sinclair R, Cole PJ. An automated method of reading ciliary beat frequency. IRCS Med Sci 1984; 12:715–716.
5. Feldman C, Voorvelt A. A phototransistor technique for the measurement of the ciliary beat frequency of human ciliated epithelium *in vitro*. S Afr J Sci 1994; 90:555–556.
6. Wilson R, Pitt T, Taylor G, Watson D, MacDermot J, Sykes D, Roberts D, Cole P. Pyocyanin and 1-hydroxyphenazine produced by *Pseudomonas aeruginosa* inhibit the beating of human respiratory cilia *in vitro*. J Clin Invest 1987; 79:221–229.
7. Munro NC, Barker A, Rutman A, Taylor G, Watson D, McDonald-Gibson WJ, Towart R, Taylor WA, Wilson R, Cole PJ. Effect of pyocyanin and 1-hydroxyphenazine on *in vivo* tracheal mucus velocity. J Appl Physiol 1989; 67:316–323.
8. Ras GJ, Anderson R, Taylor GW, Savage JE, Van Niekerk E, Wilson R, Cole PJ. Proinflammatory interactions of pyocyanin and 1-hydroxyphenazine with human neutrophils *in vitro*. J Infect Dis 1990; 162:178–185.
9. Amitani R, Wilson R, Rutman A, Read R, Ward C, Burnett D, Stockley RA, Cole PJ. Effects of human neutrophil elastase and *Pseudomonas aeruginosa* proteinases on human respiratory epithelium. Am J Respir Cell Mol Biol 1991; 4:26–32.
10. Read RC, Roberts P, Munro N, Rutman A, Hastie A, Shryock T, Hall R, McDonald-Gibson W, Lund V, Taylor G, Cole PJ, Wilson R. Effect of *Pseudomonas aeruginosa* rhamnolipids on mucociliary transport and ciliary beating. J Appl Physiol 1992; 72:2271–2277.
11. Steinfort C, Wilson R, Mitchell T, Feldman C, Rutman A, Todd H, Sykes D, Walker J, Saunders K, Andrew PW, Boulnois GJ, Cole PJ. Effect of *Streptococcus pneumoniae* on human respiratory epithelium *in vitro*. Infect Immun 1989; 57:2006–2013.
12. Feldman C, Mitchell TJ, Andrew PW, Boulnois GJ, Read RC, Todd HC, Cole PJ, Wilson R. The effect of *Streptococcus pneumoniae* pneumolysin on human respiratory epithelium *in vitro*. Microbial Pathogenesis 1990; 9: 275–284.
13. Feldman C, Munro NC, Jeffery PK, Mitchell TJ, Andrew PW, Boulnois GJ, Guerreiro D, Rohde JAL, Todd HC, Cole PJ, Wilson R. Pneumolysin induces the salient histologic features of pneumococcal infection in the rat lung *in vivo*. Am J Respir Cell Mol Biol 1991; 5:416–423.

14. Roberts P, Jeffery PK, Mitchell TJ, Andrew PW, Boulnois GJ, Feldman C, Cole PJ, Wilson R. Effect of immunization with Freund's adjuvant and pneumolysin on histologic features of pneumococcal infection in the rat lung *in vivo*. Infect Immun 1992; 60:4969–4972.
15. Wilson R, Sykes DA, Watson D, Rutman A, Taylor GW, Cole PJ. Measurement of *Pseudomonas aeruginosa* phenazine pigments in sputum and assessment of their contribution to sputum sol toxicity for respiratory epithelium. Infect Immun 1988; 56:2515–2517.
16. Jackowski JT, Szepfalusi ZS, Wanner DA, Seybold ZS, Sielczak MW, Lauredo IT, Adams T, Abraham WM, Wanner A. Effects of *P. aeruginosa*-derived bacterial products on tracheal ciliary function: role of $O_2$ radicals. Am J Physiol 1991; 260:L61–L67.
17. Kanthakumar K, Taylor G, Tsang KWT, Cundell DR, Rutman A, Smith S, Jeffery PK, Cole PJ, Wilson R. Mechanisms of action of *Pseudomonas aeruginosa* pyocyanin on human ciliary beat *in vitro*. Infect Immun 1993; 61:2848–2853.
18. Feldman C, Read R, Rutman A, Jeffery PK, Brain A, Lund V, Mitchell TJ, Andrew PW, Boulnois GJ, Todd HC, Cole PJ, Wilson R. The interaction of *Streptococus pneumoniae* with intact human respiratory mucosa *in vitro*. Eur Respir J 1992; 5:576–583.
19. Rayner CFJ, Jackson AD, Rutman A, Dewar A, Mitchell TJ, Andrew PW, Cole PJ, Wilson R. Interaction of pneumolysin-sufficient and -deficient isogenic variants of *Streptococcus pneumoniae* with human respiratory mucosa. Infect Immun 1995; 63:442–447.
20. Tsang KWT, Rutman A, Tanaka E, Lund V, Dewar A, Cole PJ, Wilson R. Interaction of *Pseudomonas aeruginosa* with human respiratory mucosa *in vitro*. Eur Respir J 1994; 7:1746–1753.
21. Feldman C, Blumberg L, de Kok U, de Bruin E, Steel H, Cantrell J, Phillips JI, Anderson R, Wadee A, Koornhof HJ. The interaction of *Mycobacterium tuberculosis* (MTB) with human respiratory mucosa *in vitro*. Am J Respir Crit Care Med 1996; 153:A807.
22. Feldman C, Anderson R, Kanthakumar K, Vargas A, Cole PJ, Wilson R. Oxidant-mediated ciliary dysfunction in human respiratory epithelium. Free Radical Biology and Medicine 1994; 17:1–10.
23. Duane PG, Rubins JB, Weisel HR, Janoff EN. Identification of hydrogen peroxide as a *Streptococcus pneumoniae* toxin for rat alveolar epithelial cells. Infect Immun 1993; 61:4392–4397.
24. Feldman C, Anderson R, Theron A, Cole PJ, Wilson R. The effects of platelet activating factor (PAF) and lysophosphatidycholine (LPC) on human ciliated epithelium *in vitro*. Am J Respir Crit Care Med 1995; 151:A65.
25. Feldman C, Theron AJ, Ramafi G, Anderson R. Macrolides and azalides attenuate the effects of bioactive phospholipids on human ciliated epithelium *in vitro* (abstr). The third International Conference on macrolides, azalide and streptogramins, Lisbon, Portugal, January 24–27, 1996.
26. Yukawa T, Read RC, Kroegel C, Rutman A, Chung KF, Wilson R, Cole

PJ, Barnes PJ. The effects of activated eosinophils and neutrophils on guinea pig airway epithelium *in vitro*. Am J Respir Cell Mol Biol 1990; 2:341–353.

27. Kanthakumar K, Cundell DR, Johnson M, Wills PJ, Taylor GW, Cole PJ, Wilson R. Effect of salmeterol on human nasal epithelial cell ciliary beating: inhibition of the ciliotoxin, pyocyanin. Br J Pharmacol 1994; 112:493–498.
28. Dowling RB, Johnson M, Cole PJ, Wilson R. Effect of salmeterol and rolipram on *Pseudomonas aeruginosa* (PA) infection of the respiratory mucosa. Am J Respir Crit Care Med 1996; 153:A258.
29. Anderson R, Feldman C, Theron AJ, Ramafi G, Cole PJ, Wilson R. Anti-inflammatory, membrane-stabilizing interactions of salmeterol with human neutrophils *in vitro*. Br J Pharmacol 1996; 117:1387–1394.
30. Tsang KW, Rutman A, Kanthakumar K, Belcher J, Lund V, Roberts DE, Read RC, Cole PJ, Wilson R. *Haemophilus influenzae* infection of human respiratory mucosa in low concentrations of antibiotics. Am Rev Respir Dis 1993; 148:201–207.
31. Tanaka E, Kanthakumar K, Cundell DR, Tsang KWT, Taylor GW, Kuze F, Cole PJ, Wilson R. The effect of erythromycin on *Pseudomonas aeruginosa* and neutrophil mediated epithelial damage. J Antimicrob Chemother 1994; 33:765–775.
32. Anderson R, Theron AJ, Feldman C. Membrane-stabilising, anti-inflammatory interactions of macrolides with human neutrophils. Inflammation 1996; 20:693–705.
33. Goolam Mahomed A, Anderson R, Theron AJ, Feldman C. Anti-oxidant interactions of theophylline and aminophylline with human neutrophils *in vitro* are related to cyclic AMP elevation. Eur Respir J 1996; 9:401S.
34. Salathe M, O'Riordan TG, Wanner A. Treatment of mucociliary dysfunction. Chest 1996; 110:1048–1057.

# 47

# Regulation of Mucociliary Activity in the Upper Airways by Nitric Oxide—Clinical Implications

**Sven Lindberg, Birgitta Carlén, Anders Cervin, Catarina Cramnert, and Thomas Runer**
*University Hospital, Lund, Sweden*

## I. INTRODUCTION

Endogenous nitric oxide (NO) is a messenger molecule generated from the amino acid L-arginine by NO synthase (NOS), which transforms L-arginine to citrulline. There is increasing evidence that NO regulates epithelial function in the airways measured as the ciliary beat frequency (CBF) in vitro and mucociliary (MC) activity in vivo. Thus, pretreatment with competitive inhibitors of NOS blocked the increase of CBF in cultured tracheal cells produced by various ciliostimulators such as the $\beta_2$-adrenoceptor agonists isoproterenol and salbutamol, or substance P and bradykinin [1,2]. Moreover, the effects of nonspecific stimulator of CBF, ethanol, was shown to be NO dependent in cultured bovine tracheal epithelium [3]. The substrate for NOS, L-arginine, was shown to increase CBF in specimens from the rabbit maxillary sinus, and an NO donor, sodium nitroprusside, stimulated MC activity in the same sinus and on the human nasal septum [4,5].

Three isoforms of NOS have been described so far: constitutive brain or neuronal NOS (bNOS) [6], constitutive endothelial NOS (eNOS) [7,8], and inducible NOS (iNOS) [9]. The latter, which is independent of calcium/calmodulin in contrast to the constitutive isoforms [10], is also called macrophage NOS because a high content of this type of NOS is found in macrophages. Endothelial NOS is predominantly membrane bound, whereas bNOS and iNOS are mainly cytosolic fraction enzymes [10].

There is some controversy regarding the type of NOS that dominates in the ciliated epithelium of the airways. Using antibodies against iNOS and an in situ hybridization technique, which labeled the epithelial cells with mRNA NOS probes, Lundberg and coworkers found that iNOS is the predominant NOS in the human maxillary sinus and the nose [11]. On the other hand, a recent study in the rat lung, which included transmission electron microscopy, showed that eNOS immunoreactivity was found in ciliated cells associated with the basal microtubule membrane of the cilia, but nonciliated cells were not stained [12].

High concentrations of luminal NO has been reported from the upper airways [11,13,14]. In the lower airways, the concentration is lower, but shows marked flow dependence [15]. Luminal NO is probably produced by the covering epithelium itself, as NO released to tissues is rapidly inactivated by binding to metalloproteins, mainly hemoglobin [16]. Because the concentrations of NO is especially high in the maxillary sinus (5–10 fold the nasal NO concentration), a role for the paranasal sinuses as a NO generator in the airways has been proposed [11]. It has been shown that the nasal NO concentration is low in patients with chronic sinusitis [17].

Nasal NO is conveniently measured in all ages using a chemiluminescence technique [17]. The nasal concentration of NO is very low in patients suffering from primary ciliary dyskinesia (PCD) and acquired conditions [13,17,18]. The close relation between luminal nasal NO concentration and the MC function in the nose was confirmed when it was found in a recent study that a low nasal NO concentration strongly correlated with an impairment of MC function, both in terms of CBF and the saccharine test [18].

The aim of the morphological part of this study was therefore to investigate the type of NOS in the ciliated epithelium in the paranasal sinuses and its cellular localization, in particular its relation to the cilia. Although the previous studies strongly suggest NO to be a regulator of mucociliary function in the upper airways, the clinical importance of this finding is yet to be substantiated. Thus, it is not known whether a high level of NO in nasal air represents a well-functioning mucociliary system, or whether a low level signifies malfunction of the system. Accordingly, a further aim of the present investigation was to ascertain whether a relationship exists between the nasal NO concentration in patients with respiratory tract disease and two measurements of the mucociliary function in clinical use: ciliary beat frequency (CBF) and saccharine transit time.

## II. METHODS

### A. Morphology

Using a Blakesley cup forceps, specimens from the sphenoidal sinus were obtained from patients undergoing surgery for pituitary tumours. The investigation was approved by the local Ethics Committee. The specimens were immediately placed in physiological saline and transported to the laboratory. The specimens were then frozen in 99% isopenthane on carbonic ice and 5 μm thin sections were cut in a cryostat. The sections were dried during 30 minutes and placed in cold 4% formaldehyde in PBS (phosphate-buffered saline) at pH 7.4. They were then rinsed several times in TBS (tris-buffered saline) at pH 7.6, then washed once in 0.2% Triton X-100 and kept for 40 minutes.

The primary antibodies against eNOS were either a monoclonal mouse antibody against a peptide fragment of the amino acids 1030-1209 of human eNOS (1:50 in 0.2% Triton X-100, N30020 Transduction Labs., Lexington, KY, USA) or a polyclonal rabbit antibody against the same protein fragment of human eNOS (1:100 in 0.2% Triton X-100, N30030 Transduction Labs., Lexington, KY, USA). The primary antibody against iNOS was a polyclonal rabbit antibody against a peptide fragment of the amino acids 961–1144 of mouse inducible NOS (iNOS) (1:100 in 0.2% Triton X-100, N32030 Transduction Labs., Lexington, KY, USA). Specimens were incubated with the primary antibodies overnight at 4°C and then washed 10 minutes in TBS at pH 7.6. The secondary antibodies were swine antirabbit (1:300, No. E353, DAKO, Copenhagen, Denmark) against iNOS or polyclonal eNOS and rabbit antimouse (1:200, No. E354, DAKO, Copenhagen, Denmark) against monoclonal eNOS. Sections were washed in TBS at pH 7.6 and then incubated with the antiperoxidase kit (ABComplex HRP, No. K0355, DAKO, Copenhagen, Denmark) for 30 minutes. After renewed washing with TBS, immunostaining of all specimens was accomplished by DAB (3,3-diaminobenzidine tetrahydrochloride). Finally, the sections were counterstained with Mayer's hematoxyline and examined in a light microscope. Specimens incubated with TBS instead of a primary antibody but otherwise processed identically with the same secondary antibodies were used as controls.

### B. Nasal NO Concentration

The nasal NO concentration was determined in both nostrils using an NO chemiluminescence analyzer (CLD 700 AL med, Eco Physics AG, Dürnten, Switzerland). In brief, this analyzer works by mixing NO with ozone, producing an activated $NO_x$ molecule which in turn produces a light quantum. The light intensity is proportional to the concentration of NO in the analyzed gas mixture. According to the manufacturer, the minimum detectable concentration of NO is 1 ppb (part per billion). The NO analyzer, which has a sampling flow rate of 0.66 L/min, was calibrated with known concentrations of NO (700 ppb and 20 ppm). All measurements were made at room temperature (i.e., 21–24°C). The tube from the NO analyzer, with a nasal olive-shaped adapter at the end, was held gently in contact with one of the nostrils, but avoiding contact with the mucosa. The subjects were asked to breathe through the mouth with the contralateral nostril open during measurements, and to try to avoid closing the soft palate. When the reading of the NO concentration has stabilized, usually within one minute, this plateau level was considered to be the nasal NO concentration. The NO measuring device was then reapplied to the contralateral nostril after a pause to allow the readings to fall to background levels. The mean nasal NO concentration from both nostrils was used in the calculations.

## C. Ciliary Beat Frequency (CBF)

Brush samples were taken from the septal and conchal mucosa at the level of the middle turbinate without the use of local anesthesia. The brush was immediately placed in HEPES buffer containing NaCl 135, KCl 4.6, $MgSO_4$ 1.2, $CaCl_2$ 1.5, glucose 11, and HEPES (*n*-2-hydroxyethylpiperazine, *n*-2-ethane, sulfonic acid) 10 m*M*. Imprints were made on a slide containing a shallow well filled with HEPES buffer and kept at 33°C in an incubator. After an interval of one hour, the slide was placed on a phase contrast microscope (Nikon Diaphot Phase Contrast-2 LWD 0.52, Tokyo, Japan) at 33°C to visualize the beating cilia at 400× magnification. The beating cilia produced light pertuberations that were sensed by a photodiode connected to the microscope and converted into electrical analog signals that were displayed on an oscilloscope. The signal obtained was further analyzed by a computer (Hewlett-Packard Vectra 486DX2 66MHz, Grenoble, France) equipped with an analog to digital converter (DAS 1401, Keithley Metrabyte, Taunton, MA, USA) using fast Fourier transformations (FFT). The sampling frequency for the FFT analysis was 100 Hz and the sample window was open during 10 seconds. The software used for the FFT analysis was Viewdac™(Keithley Data Acquisition Division, Taunton, MA, USA). The signal amplitude and frequency spectrum displayed by the computer program enabled quality assessment of the recordings. Three to five microexplants were studied from each brush sample. Five sites were investigated on each microexplant. Six consecutive recordings of 10 seconds each were obtained from each site, and the median CBF of these recordings was calculated. The five median values were averaged for each microexplant, the mean microexplant value being used in the calculations. As the mean CBF for normal controls at our laboratory is $11.1 \pm 1.1$ Hz, in the present study a CBF value below 9.0 Hz [mean − (1.96 × standard deviations)] was considered to be pathologic. In some cases, with either a history indicating a severe disease in the mucociliary system or pathologic results of measurements of CBF and the saccharine test, an extra brush sample was taken for examination of ciliary ultrastructure by transmission electron microscopy.

## D. Saccharine Transit Time

The mucus transport rate was studied by introducing a 0.5 mm (diameter) particle of saccharine dyed with indigo blue and measuring the time until the patient reported sensing the taste of saccharine. The investigator then verified the appearance of a blue color in the nasopharynx. As with investigations using radiolabeled particles, the saccharine test is flawed by the great variation of the results and the poor reproducibility. One possible reason for this is that the particle may be placed in the nose under nonstandardized conditions. To minimize this problem, the saccharine particle was placed in a plastic pipette. When the pipette was compressed, the particle was ejected without the investigator needing to touch the mucosa. The particle was placed on the septal mucosa in the middle meatus at the level of the anterior end of the middle turbinate while monitoring with an endoscope (Panoview Plus 4 mm, Richard Wolf GmbH, Knittlingen, Germany). The reason for choosing the septum instead of the turbinates is that the septal mucosa has less vasoactive tissue than the turbinates. The patient was instructed to swallow once per minute. If no sweet taste was noticed within one hour, the test was interrupted and a particle of saccharine placed on the tip of the tongue to ascertain whether the subject was able to sense the taste of saccharine. If so, the patient was told to return for a repeat saccharine test after at least one month. It has been reported that an episode of the common cold may affect the mucociliary system up to one month after the beginning of symptoms [19] and the presence of a common cold might explain the observed impairment of transport. The reference value for normal controls at our laboratory is $22 \pm 7$ min (mean ± SD), suggesting that saccharine transit times in excess of 40 min may represent pathology in the mucociliary system.

## E. Patients

All subjects in this study were patients referred to us for investigation of the mucociliary system. The main indication for this investigation was either chronic sinusitis (including failure of endoscopic surgery) or recurrent airway infections in the lower airways (usually recurrent pneumonia in nonsmokers) where normal results had been obtained at investigations for immunodeficiency or cystic fibrosis. The pneumonia patients were investigated in a disease-free interval. Two patients with previously diagnosed primary ciliary dyskinesia (PCD) were included in the series (both had been treated earlier for chronic sinusitis). The investigations commenced with measurements of nasal NO, followed by the saccharine test, and finally one or two nasal brush samples were taken. Sixteen women and nine men were investigated. Their age was 43.8 ± 19.0 years (mean ± SD, range 8–71 years).

## F. Statistics

The results are expressed as means and standard deviations (SD). The Mann-Whitney U-test for unpaired data was used for comparisons and correlation coefficients were calculated according to Spearman. Differences between groups were calculated with Fisher's exact test. Neither nonparametric methods (as the numbers of subjects were small) nor raw data demonstrated Gaussian distribution. P values $<0.05$ were considered significant.

# III. RESULTS

## A. Morphology

Both eNOS and iNOS immunoreactivity were found within the ciliated epithelium in the sphenoidal sinuses. However, the distributions of the two types of NOS were different. The monoclonal antibody against eNOS stained the apical part of some ciliated cells whereas the polyclonal antibody against eNOS evenly stained the ciliated cells (Figure 1 and Figure 2). The eNOS-like immunoreactivity was patchy in contrast to iNOS-like immunoreactivity which stained all ciliated cells. The iNOS-like immunoreactivity appeared predominantly in the apical part of these cells, i.e., close to the basal spurs of cilia (Figure 3). The negative control sections immunostained with omission of the primary antibody showed no staining in the ciliated cells.

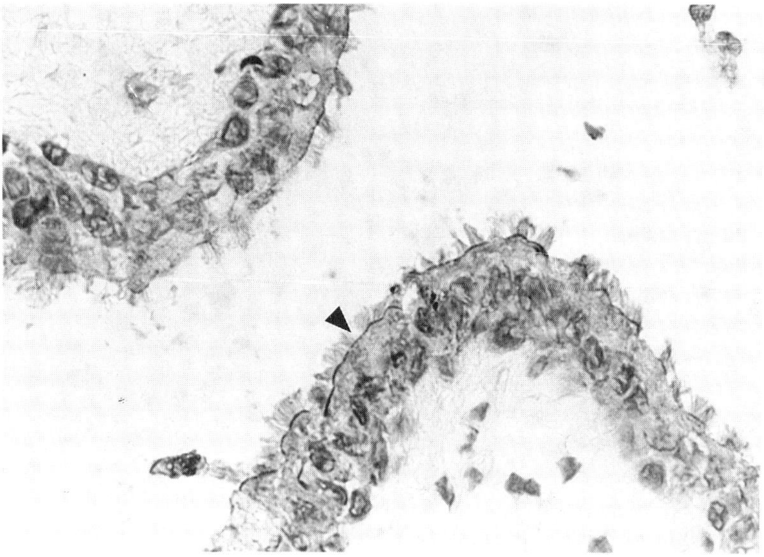

**Figure 1** Section from ciliated epithelium in the human sphenoidal sinus. A patchy eNOS immunoreactivity (monoclonal mouse IgG1) is detected on the tops of ciliated cells facing the airway lumen (arrowhead) close to the attachment of the cilia. The lamina propria is not stained. Original magnification × 640.

## B. Patient Study

The underlying assumption for this investigation was that measurement of nasal NO yields information regarding the status of the mucociliary system in the upper airways. Thus, in the analysis of data, the concentration of nasal NO was the independent variable and the CBF measurement or saccharine transit time the dependent variable. As a previous study in 66 healthy subjects had indicated 100 ppb to be a cut-off level between a normal nasal NO concentration and a possibly pathologic one, the present series was subdivided accordingly [7].

### Normal NO (≥100 ppb)—15 Patients

Median nasal NO concentration in this group was 185 ppb (range 120–373 ppb), which is not significantly lower than the median NO in age-matched normal controls, 237 ppb (range 110–412 ppb), $p = 0.10$ [17]. The patients

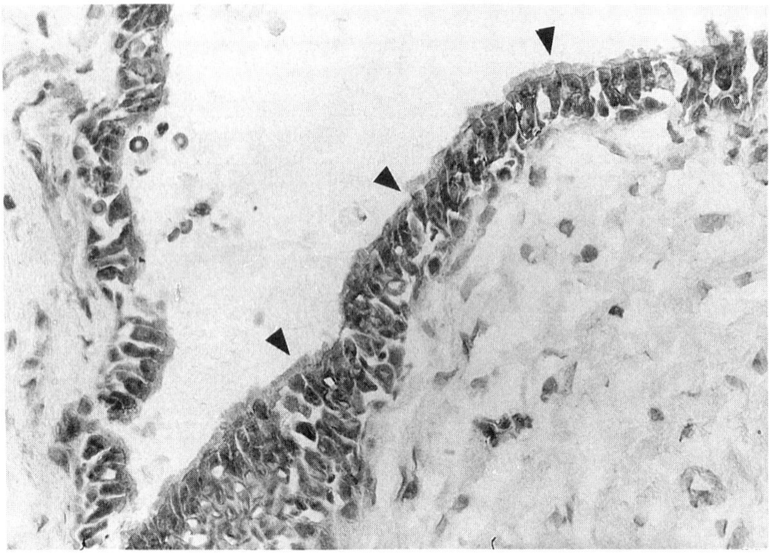

**Figure 2** Section from ciliated epithelium in the human sphenoidal sinus. eNOS immunoreactivity (polyclonal purified antibody) is evenly distributed in the ciliated cells (arrowheads). There is no concentration of eNOS-like immunoreactivity close to the cilia. The lamina propria is not stained. Original magnification × 400.

in this group had a median CBF of 10.4 Hz, and median saccharine transit time in this group was 23 min. In four patients with normal NO values the occurrence of mucus transport could not be verified with the saccharine test.

## Pathologic NO (<100 ppb)—10 Patients

The median nasal NO concentration in this group was 40.3 ppb (range 10–89 ppb). Their median CBF was 8.4 Hz ($p<0.001$), i.e. significantly lower then that of the group with normal nasal NO concentrations. Their median saccharine transit time was more than 60 min ($p<0.01$). Nine patients had a saccharine time longer than 40 min, and in seven of them no mucus transport could be verified with the saccharine test. Nine patients in this group were examined with transmission electron microscopy (TEM) and six of them had significant changes in their cilia (Table 1).

Analysis of the data with Fisher's exact test showed the group with subnormal nasal NO levels to differ significantly from the group with

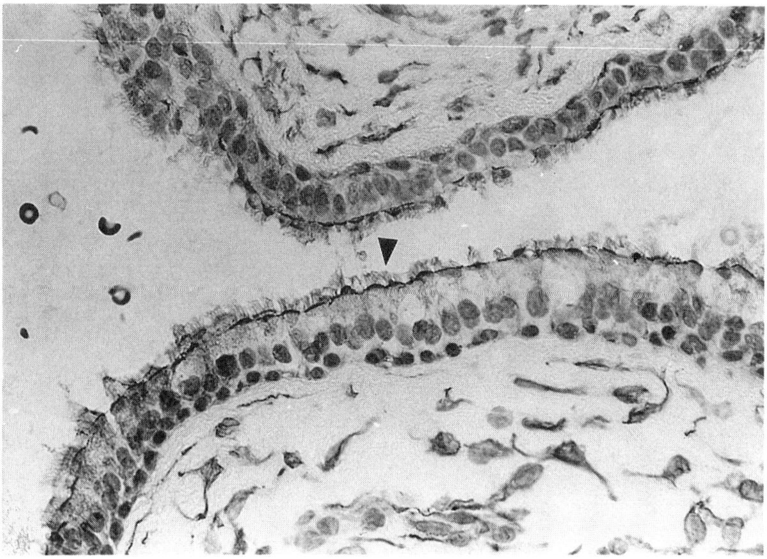

**Figure 3** Section from ciliated epithelium in the human sphenoidal sinus. Strong iNOS immunoreactivity (polyclonal purified antibody) is detected in apical parts of all cells (arrowhead) facing the airway lumen and close to the attachment of the cilia. The lamina propria is not stained. Original magnification × 400.

normal NO levels both in CBF ($p<0.01$) and in saccharine transit time ($p<0.01$). The difference in nasal NO concentration between patients suffering from the two main indications for the investigations, chronic sinusitis (median 78 ppb, $N=13$) or recurrent pneumonia (median 147 ppb, $N=7$), was significant, $p<0.05$.

## Correlation Between NO and CBF

Two patients with no beating cilia were excluded from this analysis, as including them might have yielded false correlation. It cannot be ruled out that lack of beating cilia might have been due to the investigator's inability to detect them and not to the true status of the mucociliary system. The correlation coefficient for the relationship between NO and CBF was 0.74 (Spearman's *rho*), the equation for best fit being CBF (in Hz) $= 7.22 + [0.22 \times \text{Nasal concentration of NO (in ppb)}^{0.5}]$.

**Table 1** Individual Data Including Indications for Investigation of the Mucociliary System

| No. | Age | Sex | Smoker | Saccharine test | Mean CBF | NO | Indications |
|---|---|---|---|---|---|---|---|
| 1 | 45 | F | N | 16 min | 9.7 Hz | 153 ppb | Chron sin |
| 2 | 59 | F | Y | 61 min | 8.0 Hz | 87 ppb | Chron sin, IgG subclass deficiency, TEM: Mild ACD |
| 3 | 30 | F | N | No transport | No beating cilia | 16 ppb | Chron sin, severe bronchiectasies, TEM: ACD |
| 4 | 59 | F | N | 15 min | 10.8 Hz | 206 ppb | Chron sin |
| 5 | 54 | F | Y | 33 min | 8.7 Hz | 66.5 ppb | Chron sin, chron bronch, TEM: normal cilia |
| 6 | 15 | F | N | No transport | 9.4 Hz | 10 ppb | Chron sin, atelectasis, TEM: ACD |
| 7 | 64 | M | N | 45 min | No beating cilia | 78 ppb | Chron sin, refused to return for TEM |
| 8 | 12 | M | N | 13 min | 10.6 Hz | 137 ppb | Chron sin |
| 9 | 60 | F | N | No transport | 6.0 Hz | 48.5 ppb | Chron sin, chron otitis media, TEM: PCD |
| 10 | 23 | F | N | No transport | 10.4 Hz | 192.5 ppb | Recurrent upper airway infection, TEM: normal cilia |
| 11 | 32 | M | N | No transport | 8.7 Hz | 21 ppb | Chron sin, chron otitis media, TEM: normal cilia |
| 12 | 25 | F | N | No transport | Not measured | 30.5 ppb | Recurrent otitis media, chron bronch, TEM: PCD |
| 13 | 66 | F | N | 24 min | 13.1 Hz | 185 ppb | Recurrent pneumonia |
| 14 | 66 | F | N | 20 min | 9.5 Hz | 147 ppb | Recurrent pneumonia |
| 15 | 71 | F | N | 29 min | 10.4 Hz | 122.5 ppb | Recurrent pneumonia |
| 16 | 54 | F | Y | 28 min | 12.5 Hz | 269.5 ppb | Recurrent pneumonia |
| 17 | 34 | M | N | No transport | 10.6 Hz | 89 ppb | Recurrent pneumonia, sinusitis, TEM: normal cilia |
| 18 | 60 | M | N | No transport | 11.2 Hz | 349.5 ppb | Secretory rhinitis |
| 19 | 64 | F | N | No transport | 9.4 Hz | 131.5 ppb | Recurrent pneumonia |
| 20 | 22 | M | N | 7 min | 10.8 Hz | 241 ppb | Acute sinusitis |
| 21 | 38 | F | N | 23 min | 11.7 Hz | 337.5 ppb | Surgery for pituitary gland adenoma |
| 22 | 47 | F | Y | No transport | 7.2 Hz | 157 ppb | Chron sin, TEM: normal |
| 23 | 38 | M | N | 19 min | 9.8 Hz | 120 ppb | Chron sin |
| 24 | 8 | M | N | No transport | Not measured | 32 ppb | Chron sin, TEM: suspect PCD |
| 25 | 48 | M | N | 10 min | 10.1 Hz | 372.5 ppb | Recurrent pneumonia |

Abbreviations: F = female, M = male, N = no, Y = yes, chron sin = chronic sinusitis, TEM = transmission electron microscopy, ACD = acquired ciliary dysfunction, PCD = primary ciliary dyskinesia, chron bronch = chronic bronchitis.

Transmission electron microscopy (TEM) was performed only for those patients who either had a history indicating a severe disease in the mucociliary system, or the results of measurements of CBF and the saccharine test indicated that this could be the case. Investigation for cystic fibrosis (CF) in patient no. 3 showed this patient to be heterozygot for the CF gene delta F508, minor CF genes were negative. Patient no. 6 had a completely normal CF work-up.

## Correlation Between NO and Saccharine Transit Time

Eleven patients with no demonstrable mucus transport were excluded from this analysis, as including them with an improvised value (for example 60 or 90 min) might have yielded false correlation. It cannot be ruled out that lack of demonstrable mucus transport could have been due to investigator error and not to the true status of the mucociliary system. The correlation coefficient was −0.60 (Spearman's *rho*), the equation for best fit being saccharine time (in min) = 5.83 + [2597.5/Nasal concentration of NO (in ppb)].

The strong correlation between nasal NO concentrations and corresponding values for mucociliary function are shown in Figures 4 and 5 where CBF and saccharine transit time, respectively, are plotted as functions of the nasal NO concentration. Details of the medical histories and results of the individual measurements, including the results of transmission electron microscopy, are given in Table 1.

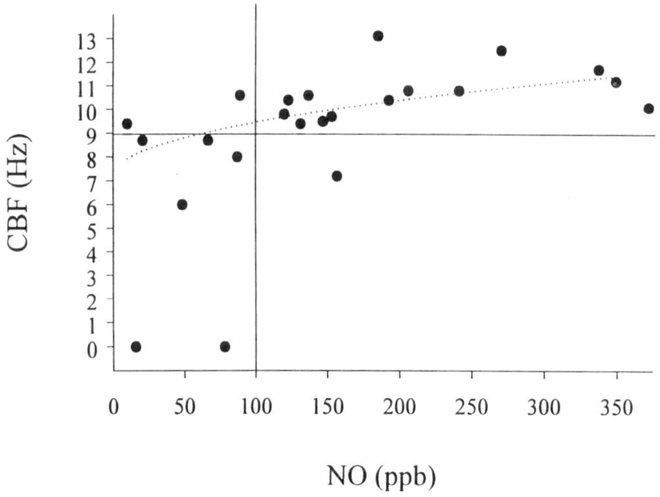


**Figure 4** CBF measurements. Mean CBF is plotted as a function of the nasal NO concentration. The straight lines represent the approximate 95% lower confidence intervals for normal controls (9.0 Hz and 100 ppb, respectively). The dotted line represents the equation for best fit (omitting the two patients with no beating cilia).

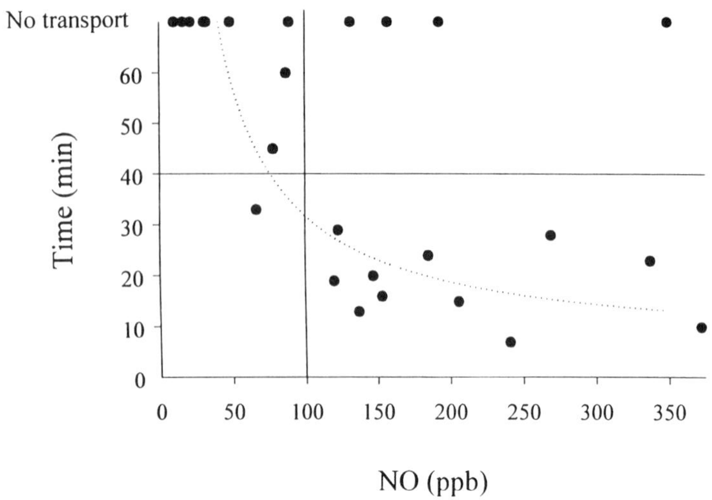


**Figure 5** The results of the saccharine test. Time in minutes until the patient sensed the sweet taste of a saccharine particle placed on the septum at the level of the anterior end of the middle turbinate is plotted as a function of the nasal NO concentration. The straight lines represent the approximate 95% lower confidence intervals for normal controls (40 min and 100 ppb, respectively). The dotted line represents the equation for best fit (omitting the patients with no transport).

## IV. DISCUSSION

The results of the morphological part of this study confirm the previous view put forward by Lundberg and coworkers that the NOS type found in the vicinity of cilia in the upper airways is mainly iNOS-like. These investigators found a distinct immunohistochemical staining for iNOS in the apical part of the upper airway epithelium [11], i.e. a similar finding as in the present study. In contrast, Xue and coworkers found intense immunostaining of eNOS in rat bronchial epithelium, also close to the cilia [12]. They used the same monoclonal antibody against eNOS as we used. However, in the sphenoidal sinus the staining against eNOS was more pathcy than for iNOS, i.e. not all ciliated cells were stained for eNOS. The two types of NOS had different distributions: a polyclonal antibody against iNOS stained the apical part of the ciliated cell whereas polyclonal antibody eNOS staining was found more evenly distributed throughout the ciliated cells. A patchy staining against eNOS was found on top of some ciliated cells using the monoclonal eNOS antibody. Hopefully investigations with transmission electron microscopy will clarify the exact subcellular localization of both iNOS and eNOS. The present results support the view that NOS in the upper airways is both iNOS- and eNOS-like, with iNOS being predominant in the apical part of the ciliated cell and subsequently closest to the cilia.

The functional investigation showed strong correlation to exist between the nasal NO concentration and mucociliary function in terms of CBF and saccharine transit time. Patients with subnormal nasal NO concentrations were found to significantly differ in mucociliary function from those with normal nasal NO concentrations. Nitric oxide has previously been shown to act as a local mediator between the cilia and various ciliostimulators such as substance P, bradykinin, isoproterenol, and salbutamol [2,20]. We have previously shown that an NO donor, nitroprusside, stimulates mucociliary activity in vivo in the rabbit maxillary sinus and in the human nose. Moreover, the NOS substrate L-arginine stimulated ciliary beat frequency in explants from the rabbit maxillary sinus in vitro [4,5]. As these observations suggested a link to exist between NO and the mucociliary system in the upper airways, the clinical study reported here was undertaken to ascertain whether this is the case.

It may seem confusing that the measurement of a gas can yield important information regarding the function of the ciliated epithelium in the upper airways. Nonetheless, the measurement of nasal NO makes sense if the ciliated epithelium of nasal and paranasal sinuses is the main source of NO in nasal air. It has been shown that NO is spontaneously released from airway epithelium [2]. Thus, a damaged ciliated epithelium will have cilia that beat slowly (measured as a decrease in CBF), produce a less effective mucus transport as compared to a healthy epithelium (seen as a prolonged saccharine test time), and release little NO to the lumen.

A major question is the existence of a causal relationship between mucociliary function and nasal NO levels, i.e. if low nasal NO levels cause impaired mucociliary function or if nasal NO levels are low because of impaired mucociliary function. One may argue that an impaired mucociliary function causes blockage of the ostium with retention of mucus, subsequently causing lower nasal NO levels because the paranasal sinuses represent the major source of the NO found in the nasal cavities [11]. Blockage of the ostium as the sole explanation to the present results seems less likely as all patients suffering from chronic sinusitis had been operated on previously and were known to have patent neo-ostia. However, if the area of airway epithelium in contact with the air stream is reduced, one would expect NO levels to drop. The extreme situation is a total blockage of the ostia to the paranasal sinuses, which probably results in reduced nasal NO levels as the nasal cavities are flooded by NO derived from the paranasal sinuses. Indeed, measurements of the nasal NO concentration in patients suffering from severe nasal polyposis accompanied by intolerance towards acetylsalicylic acid, suggest that almost 50% of these patients have a nasal NO concentration <100 ppb. Similarly, nasal NO levels are lowered after nasal provocation with histamine or allergen challenge, if these challenges result in profound swelling of the nasal mucosa (unpublished observations). These findings indicate that nasal NO measurements may have a high sensitivity but rather low specificity for disease affecting the mucociliary system. Thus, the evaluation of a given low nasal NO concentration requires the inclusion of nasal endoscopy in order to rule out severe nasal blockage as the cause of the observed low nasal NO concentration.

Rather than reflecting loss of NO producing epithelium, a low nasal NO level probably reflects impaired epithelial function. Thus, the CBF in nasal brush samples correlated with nasal NO concentrations, despite the fact that the recordings of CBF were performed in vitro in a fluid medium more than one hour after harvest. It is likely that the influence on CBF of luminal NO had vanished at this point in time. The strong correlation between the nasal NO level and the CBF therefore supports the view that the nasal NO concentration primarily reflects epithelial function rather than low levels of nasal NO causing a decrease of the CBF. However, the results of previous pharmacological studies suggest that NO derived from the paranasal sinuses may also act as an airborne messenger in the airway epithelium, causing an increase in the CBF and probably the mucociliary transport rate as well.

The cut-off level for normal CBF of $\geq$9.0 Hz was calculated as the normal mean minus 1.9 $\times$ SD of the mean (representing the 95% confidence interval). At our laboratory, the CBF in nasal brush samples from healthy individuals is 11.1 $\pm$ 1.1 Hz, a value consistent with that of 11.3 $\pm$ 1.0 Hz for CBF in nasal brush samples reported from the Mayo Clinic [21]. Reference values for the saccharine transit time are more heterogeneous than those for CBF. This is probably due to a multiplicity of factors, such as the size of the particle used, the site and the method of application. In order to standardize the saccharine test and minimize random influences on the result, we use a rigid nasal endoscope and a specially designed pipette for the application. By using gentle pressure on the pipette, the particle is expelled onto the mucosa of the nasal septum, thus avoiding contact between the pipette and the mucosal surface. However, despite these precautions, there was obviously a greater variation in the saccharine test results than in the measurements of CBF, with some patients having normal nasal NO concentrations and CBF, but without demonstrable mucus transport (Table 1). Findings of no mucus transport in normal subjects have been reported previously. Thus, using both the saccharine test and a more sophisticated isotope technique for the measurement of mucosal transport, Danish investigators noted approximately 20% of healthy individuals to be characterized by an absence or very slow rate of transport [22]. At present, there is no satisfactory explanation for the absence of transport in otherwise healthy individuals. This finding obviously complicates the interpretation of a suspected pathologic result of the saccharine test and probably explains why some of our patients in the group with normal nasal NO values lacked normal mucosal transport as measured with the saccharine test. A brush sample from one of them (subject number 10) was investigated by transmission electron microscopy. The cilia examined were found to be normal in ultrastructure, which—together with the results of measurements of NO and CBF—indicated the function of her mucociliary system to be normal.

An important issue is whether measurements of the nasal NO concentration might be used for screening purposes when suspected disease affecting the mucociliary system in the upper airways is investigated. As the present series is small, calculations of sensitivity or specificity are meaningless. However, our data strongly suggest the feasibility of using measurements of nasal NO for screening purposes. The method has several advantages in the clinical situation, because it is fast (an investigation takes less than five minutes to perform) and noninvasive, and therefore well tolerated by patients including children. Measurements of NO can also be used to monitor the status of the epithelium in the nose and its sinuses on an individual basis with the patients acting as their own controls, for example after surgical and medical treatment of disease.

It is not known whether measurements of the nasal NO concentration are representative or even useful for the evaluation of disease in the lower airways. Patients suffering from recurrent pneumonia were found to have normal nasal NO concentrations, except in one case. Although the number of subjects studied was small, these results suggest that dysfunction of the mucociliary system in the upper airways is not the cause of recurrent pneumonia. It might be argued that all pneumonia-prone patients were investigated in a disease-free interval and that investigation during a pneumonia episode would yield different results. In this context, an interesting observation in the present series was that two of the patients with acquired ciliary dysfunction, as verified by functional and ultrastructural investigation, had an even lower nasal concentration than the patients with primary ciliary dyskinesia. The two patients with acquired mucociliary dysfunction were characterized by involvement of the lower airways, including atelectasis and bronchiectasis, in contrast to our primary ciliary dyskinesia patients who have not developed comparable severe pulmonary problems. It is open to speculation whether a very low nasal NO concentration is an ominous sign of the possible presence of chronic mucociliary dysfunction in the lower airways too. Entirely a matter of speculation is the question of whether a severe viral infection of the airways may result in lack of NOS in the epithelium, in turn creating a vicious circle of events with increased risk of new viral and bacterial infections, culminating in acquired ciliary dysfunction with severe complications as in two of our patients (numbers 3 and 6). It has been reported that, in addition to its functional importance to the mucociliary system in the upper airways, NO also has antibacterial [23] and antiviral effects [24].

## V. CONCLUSION

In summary the results of the present investigation strongly support the view, originally deduced from the results of immunohistochemical investigations and pharmacological experiments, that NO is a regulator of mucociliary function. Measurement of the nasal NO concentration should be included in investigations of the mucociliary system in the upper airways because a normal level of nasal NO indicates that the ciliated epithelium is functioning properly.

A report containing part of the investigation in patients is to be published by Acta Otolaryngologica (Stockholm) 1997.

## ACKNOWLEDGMENTS

This investigation was supported by grants from the Swedish Medical Research Council (project nos. 7940, 12032), the Swedish Society of Medicine (reference nos. 152.0 and 155.0), the Torsten Söderberg and Ragnar Söderberg Trusts, the Crafoord Foundation, and the Medical Faculty of the University of Lund.

## REFERENCES

1. Jain B, Rubinstein I, Robbins RA, Sisson JH. TNF-α and IL-1β upregulate nitric oxide-dependent ciliary motility in bovine airway epithelium. Am J Physiol 1995; 268:L911–917.
2. Tamaoki J, Chiyotani A, Kondo M, Konno K. Role of NO generation in β-adrenoceptor-mediated stimulation of rabbit airway ciliary motility. Am J Physiol 1995; 268:C1342–1347.

3. Sisson JH. Ethanol stimulates apparent nitric oxide-dependent ciliary beat frequency in bovine airway epithlial cells. Am J Physiol 1995; 268:L596–600.
4. Runer T, Lindberg S. Effect of nitric oxide on blood flow and mucociliary activity in the human nose. Ann Otol Laryngol, accepted 1997.
5. Runer T, Cervin A, Lindberg S, Uddman R. Nitric oxide is a regulator of mucociliary activity in the upper respiratory tract. Otolaryngol Head Neck Surg, accepted 1997.
6. Bredt DS, Hwang PM, Glatt CE, Lowenstein C, Reed RR, Snyder SH. Cloned and expressed nitric oxide synthase structurally resembles cytochrome P-450 reductase. Nature 1991; 351:714–718.
7. Janssens SP, Shimouchi A, Quertermous T, Bloch DB, Bloch KD. Cloning and expression of cDNA encoding human endothelium-derived relaxing factor/nitric oxide synthase. J Biol Chem 1992; 267:14519–14522.
8. Lamas S, Marsden PA, Li GK, Tempst P, Michel T. Endothelial nitric oxide synthase: molecular cloning and characterization of a distinct constitutive enzyme isoform. Proc Natl Acad Sci USA 1992; 89:6348–6352.
9. Xie QW, Cho HJ, Calaycay J, Mumford RA, Swiderek KM, Lee TD, Ding A, Troso T, Nathan C. Cloning and characterization of inducible nitric oxide synthase from mouse macrophages. Science 1992; 256:225–228.
10. Nathan C. Nitric oxide as a secretory product of mammalian cells. FASEB J 1992; 6:3051–3064.
11. Lundberg JON, Farkas-Szallasi T, Weitzberg E, Rinder J, Lidholm J, Ängaård A, Hökfelt T, Lundberg JM, Alving K. High nitric oxide production in human paranasal sinuses. Nature Med 1995; 1:370–373.
12. Xue C, Botkin SJ, Johns RA. Localization of endothelial NOS at the basal microtubule membrane in ciliated epithelium of rat lung. J Histochem Cytochem 1996; 44:463–471.
13. Lundberg JON, Weitzberg E, Nordvall SL, Kuylenstierna R, Lundberg JM, Alving K. Primarily nasal origin of exhaled nitric oxide and absence in Kartagener's syndrome. Eur Respir J 1994; 7:1501–1504.
14. Schedin U, Frostell C, Persson MG, Jakobsson J, Andersson G, Gustafsson LE. Contribution from upper and lower airways to exhaled endogenous nitric oxide in humans. Acta Anaesthesiol Scand 1995; 39:327–332.
15. Silkoff PE, McClean PA, Slutsky AS, Furlott HG, Hoffstein E, Wakita S, Chapman KR, Szalai JP, Zamel N. Exhaled nitric oxide measured without a nasal component shows marked flow dependence. Am J Respir Crit Care Med. In press 1997.
16. Deliconstantinos G, Villiotou V, Stavrides JC. Scavening effects of hemoglobin and related heme containing compounds on nitric oxide, reactive oxidants and carcinogenic volatile nitrosocompounds of cigarette smoke. A new method for protection against the dangerous cigarette constituents. Anticancer Res 1994; 14:2717–2726.
17. Lindberg S, Cervin A, Runer T. Nitric oxide (NO) production from the upper airways is decreased in chronic sinusitis. Acta Otolaryngol (Stockh) 1997; 117:113–117.
18. Lindberg S, Cervin A, Runer T. Low levels of nitric oxide correlate to an

impaired mucociliary function in the upper airways. Acta Otol (Stockh). In press 1997.
19. Pedersen M, Sakakura Y, Winther B, Brofeldt S, Mygind N. Nasal mucociliary transport, number of ciliated cells, and beating pattern in naturally acquired common colds. Eur J Respir Dis 1983; 64(Suppl 128):355–365.
20. Jain B, Rubinstein I, Robbins RA, Leise KL, Sisson JH. Modulation of airway epithelial cell ciliary beat frequency by nitric oxide. Biochem Biophys Res Commun 1993; 191:83–88.
21. Phillips PP, McCaffrey TV, Kern EB. Measurement of human nasal ciliary motility using computerized microphotometry. Otolaryngol Head Neck Surg 1990; 103:420–426.
22. Andersen I, Lundqvist G, Proctor DF. Human nasal mucosal function in a controlled climate. Arch Environ Health 1971; 23:408–420.
23. Mancinelli RL, McKay CP. Effects of nitric oxide and nitrogen dioxide on bacterial growth. Appl Environ Microbiol 1983; 46:198–202.
24. Croen KD. Evidence for an antiviral effect of nitric oxide. Inhibition of herpes simplex virus type 1 replication. J Clin Invest 1993; 91:2446–2452.

# 48

# Ciliary Stimulation by Extracellular ATP in Frog Esophagus and Palate Epithelium

**Alex Braiman and Zvi Priel**
*Ben-Gurion University of the Negev, Beer-Sheva, Israel*

## I. INTRODUCTION

Extracellular ATP produces a wide range of physiological responses in cardiac and smooth muscle, neurons, exocrine and endocrine glands, mast cells, platelets, ciliary cells, and other tissues [1,2]. According to the classification introduced by Burnstock, extracellular ATP acts through purinergic receptors of $P_2$ type [3]. These receptors are activated by ATP and ADP and exhibit a potency order of ATP>ADP>AMP>adenosine. The existence of a very diverse set of physiological responses suggested involvement of multiple receptor subtypes [4]. The $P_2$ purinergic receptors have been further subdivided into 6 major families. The discrimination between the families is based on the relative ability of the ATP structural analogs to activate them. The $P_2$ receptor families differ structurally and functionally. Two of them, $P_{2X}$ and $P_{2Z}$, are ion channels, whereas the others, $P_{2Y}$, $P_{2U}$, $P_{2T}$, $P_{2D}$, are coupled through G-proteins to phospholipase C, adenylyl cyclase, possibly other cellular enzymes [1,2].

The pronounced effect of extracellular ATP on ciliary transport rate and beat frequency has been known for a long time [5]. It has been found that extracellular ATP is one of the most potent ciliary activators, stimulating at micromolar concentrations a variety of ciliary cells, including those isolated from human nasal polyps [6–8], rabbit trachea [9–11], rabbit oviduct [12], frog palate, and esophagus [13]. It has been shown in dog trachea that extracellular ATP in vivo also produces its effect at micromolar concentrations [14].

The molecular events underlying the purinergic stimulation of ciliary activity is a subject for intensive investigation on a worldwide scale. This article summarizes our contribution to the understanding of the mechanism of ciliary stimulation and the physiological responses induced by extracellular ATP in mucociliary epithelium of frog esophagus and palate.

## II. METHODS

The methods used in our experiments are presented here briefly since they are described extensively in the original articles.

### A. Measurements of Ciliary Beat Frequency

These experiments were carried out on tissue culture from frog esophagus or palate [15]. The method is based on simultaneous measurement of scattered light from small ciliated areas under a microscope [16]. The light is collected by optical fibers (50 μm in diameter) and detected by a photomultiplier. The signal is amplified and low-pass filtered. The ciliary beat frequency, the wavelength, the wave direction, the wave velocity, and the degree of correlation between small ciliated areas at a given distance apart can be obtained using two or three optical fibers simultaneously [16–20].

### B. Simultaneous Measurement of Intracellular Calcium Concentration and Ciliary Beat Frequency

These experiments were carried out on tissue culture from frog esophagus or palate. The cells were loaded with the fluorescent calcium indicator Fura-2 [10,21–23]. The loaded cells were illuminated by excitation light from a 150W Xenon lamp filtered alternatively by 340 nm and 380 nm interference filters. The emitted light filtered by a 510 nm interference filter was collected by a photon-counting head. The intracellular calcium concentration was calculated from the 340/380 fluorescent ratio using a calibration curve created by measurement of the Fura-2 fluorescence in solutions with known calcium concentration [10,24]. To measure the ciliary beat frequency, the same ciliated cells were transilluminated simultaneously with light at 600 nm. The scattered light was collected and analyzed as described above.

This method allows measurement of the cause (intracellular calcium) and the physiological response (ciliary beating) simultaneously from the same cells.

### C. Photoaffinity Labeling of Intact Cells by [$^{32}$P]BzATP

Intact pieces of frog esophagus or cells in suspension were incubated in the dark with [$\alpha$-$_{32}$P]BzATP. Then the samples were irradiated using a 15-W UV lamp for 3 min and the reaction was stopped. The samples were analyzed by SDS-polyacrylamide gel electrophoresis and autoradiography of the dried gel [25].

### D. Measurements of Membrane Fluidization

The freshly excised frog palate was labeled with the fluorescent probe TMA-DPH or DPH [26]. Front-face fluorescent measurement was carried out with an MPF-44 Perkin Elmer spectrofluorimeter [27]. The sample was illuminated with excitation light at 360 nm and the emission was measured at 430 nm. The fluorescence polarization obtained in this measurement reflects the probe rotational mobility, which depends on the fluidity of the membrane. The membrane fluidization was calculated by comparing the values of fluorescent polarization obtained before and after the application of a particular agent [26].

### E. Measurements of Membrane Potential

These experiments were performed on tissue culture from frog esophagus or palate using a 25–50-MΩ microelectrode, filled with 4 M potassium acetate, connected to a high impedance electrometer via Ag–AgCl wire. The voltage was monitored using an oscilloscope [22].

## III. RESULTS

### A. Characterization of a Receptor for Extracellular ATP

To determine a subtype of the receptor involved in purinergic stimulation in frogs, we used different structural analogs of ATP. The experiments revealed that the most potent ciliary activators, along with ATP itself, are ADPβS, 2MeSATP, UTP, and GTPγS [25]. The former two analogs show high specificity for the $P_{2Y}$ subtype, while the latter two activate preferentially the $P_{2U}$ subtype [1,2].

Another set of experiments involved photolabeling of the cells with [$^{32}$P]BzATP—a radioactive analog that bonded covalently to the receptor under UV illumination. Two polypeptides were found to be labeled by the analog, p46 and p96. The former was proposed to be a purinergic receptor [25]. Preincubation with ATP almost completely prevented the labeling of p46, while the other four active analogs reduced the labeling only by 60–70% each [25]. These results suggested that p46 represents two distinct purinergic receptors that mediate the ciliary stimulation by extracellular ATP—one of them is $P_{2Y}$ and the other is $P_{2U}$.

## B. Involvement of $Ca^{2+}$ in Purinergic Ciliary Stimulation

The role of $Ca^{2+}$ as a ciliary-beat stimulator is well established. The simultaneous measurement of intracellular calcium concentration and ciliary beat frequency [10] has shown that application of extracellular ATP causes immediate strong elevation of the intracellular calcium concentration $[Ca^{2+}]_i$ [22,23,26]. The effect of ATP is blocked by intracellular calcium pool depletion produced by thapsigargin [22]. Reduction of the extracellular calcium concentration to quite low values (10 $\mu M$) has a moderate influence on the initial elevation of the ciliary beat frequency produced by ATP. Further decrease in the external calcium depresses the response to ATP, but it can be restored by replacing the external calcium by magnesium. Calcium channel blockers $Cd^{2+}$, $Ni^{2+}$, $La^{3+}$, and nitrendipine fail to inhibit the maximal response to ATP [13]. All these results suggest that immediate elevation of $[Ca^{2+}]_i$ and the beat frequency is produced by the discharge of intracellular calcium stores. This conclusion is consistent with the fact that $P_{2Y}$ and $P_{2U}$ receptors are known to mobilize calcium from internal stores via phospholipase C activation [1,2].

Similar results have been obtained in rabbit trachea [11] and in human nasal polyps [7,8] suggesting that such a mechanism of ciliary response initiation to the purinergic stimulation is probably universal in mucociliary systems.

The presence of a divalent cation, $Ca^{2+}$ or $Mg^{2+}$, in the external medium is probably important for ATP binding to the receptor, since the labeling of the cells with [$^{32}$P]BzATP also requires either $Ca^{2+}$ or $Mg^{2+}$ in the medium [25]. There is no evidence for involvement of $Ca^{2+}$ channels in the initial stage of the ciliary response to ATP.

Following the initial rise, the intracellular calcium concentration decays within approximately 4 minutes to a slightly elevated plateau [22], while the ciliary beat frequency remains highly activated for a very long time (up to 30 minutes). Hence, the release of $Ca^{2+}$ from the internal stores can not explain the duration of the purinergic ciliary stimulation.

## C. Protein Kinase C Induces Calcium Influx and Prolongs Purinergic Ciliary Stimulation

The activation of phospholipase C by purinergic receptors inevitably leads to diacylglycerol (DAG) production and protein kinase C (PKC) activation

[28]. We have investigated the possible role of PKC in the purinergic ciliary stimulation [23]. Direct activation of PKC by phorbol ester TPA or by dioctanoyl glycerol, a DAG cell-permeable analog, causes calcium influx induction and strong ciliary beat frequency elevation. The PKC-induced calcium influx is inhibited by low extracellular calcium concentration ($<10^{-5}$ *M*) and by $La^{3+}$, but it is not affected by a voltage dependent calcium channel blocker verapamil and by $Ni^{2+}$, which suggests involvement of nonvoltage-dependent calcium channels.

In the cells that have been treated with PKC inhibitors (chelerythrine, calphostin C, GF109203X) or have been subjected to PKC down-regulation, extracellular ATP induces a normal immediate response. However, the slightly elevated $[Ca^{2+}]_i$ plateau normally produced by ATP disappears and the beat frequency rapidly decays to the basal level [23]. These results suggest that extracellular ATP activates PKC, which induces calcium influx through nonvoltage-dependent calcium channels. This influx maintains the prolonged ciliary stimulation produced by ATP.

## D. Extracellular ATP Induces Membrane Hyperpolarization

The membrane potential measurements using a microelectrode technique have revealed that extracellular ATP induces significant hyperpolarization (from −28 to −35 mV) of the ciliary cell membrane [22]. The hyperpolarization is inhibited by a calcium-activated potassium channel blocker quinidine, by $Cs^{+}$, and by high extracellular $K^{+}$ concentration. The intracellular calcium stores depletion produced by thapsigargin inhibits the ATP-induced hyperpolarization, but the elevation of intracellular calcium concentration caused by thapsigargin during the depletion results in transient membrane hyperpolarization. These results indicate that the hyperpolarization is produced by calcium-activated potassium channel activity.

The $K^{+}$-flux inhibition, preventing hyperpolarization, shortens the purinergic ciliary beat frequency enhancement and causes rapid decay of the frequency to the initial level [13]. The phorbol ester induced frequency elevation is also inhibited by the $K^{+}$-flux blockage, albeit the calcium influx is only slightly diminished [23]. In addition, amiloride-induced hyperpolarization, by itself, produces significant elevation of the beat frequency [22,29]. All these facts imply that the ATP-induced hyperpolarization plays an important role in the mechanism of ciliary beat frequency enhancement. It should be mentioned that membrane depolarization achieved by replacing $Cl^{-}$ ions by $SO_4^{2-}$ or gluconate prior to ATP application does not affect the ciliary response to purinergic stimulation [22]. Hence, the extracellular ATP effect does not depend on the initial membrane potential, but rather requires a shift of the potential to a more negative value than the initial value.

The ATP-produced membrane hyperpolarization provides indirect evidence that voltage-activated calcium channels can not be involved in purinergic mucociliary stimulation since they require depolarization for opening.

### E. Extracellular ATP Induces Membrane Fluidization in Ciliary Cells

Addition of extracellular ATP to ciliary epithelia from the frog palate produces a profound increase of membrane fluidity [26]. This effect is blocked by intracellular calcium stores depletion. The ATP-induced membrane fluidization is also inhibited by quinidine, a calcium-activated potassium channel blocker, and by a high extracellular $K^+$ concentration. Moreover, amiloride-induced membrane hyperpolarization by itself causes significant fluidization of the membrane. Therefore, the $[Ca^{2+}]_i$ elevation and the membrane hyperpolarization produced by ATP play an important role in the membrane fluidity elevation [26].

Ionomycin, a calcium ionophore that elevates $[Ca^{2+}]_i$ to the same extent as ATP, also induces significant membrane fluidization. However, ionomycin-induced fluidization is only half as strong as the effect produced by ATP. Hence, the extracellular ATP activates additional effectors that contribute to the increase in the membrane fluidity [26].

## IV. CONCLUSION

Based on our findings, the following mechanism of purinergic ciliary stimulation can be suggested (Figure 1): extracellular ATP interacts, in the presence of $Ca^{2+}$ or $Mg^{2+}$, with purinergic receptors that probably belong to the $P_{2Y}$ and $P_{2U}$ families. The receptors activate phospholipase C that produces two second messengers, $IP_3$ and DAG. The former releases calcium from internal stores, whereas the latter activates protein kinase C that induces $Ca^{2+}$ influx through nonvoltage-dependent calcium channels. The calcium released from the internal stores is responsible for immediate but short-term frequency elevation, while the calcium influx maintains the prolonged ciliary stimulation produced by ATP. Elevation of $[Ca^{2+}]_i$ opens calcium-activated potassium channels and hyperpolarizes the membrane. The intracellular calcium elevation and the membrane hyperpolarization seem to be necessary for the prolonged ciliary stimulation induced by ATP. Concurrently, high $[Ca^{2+}]_i$, the membrane hyperpolarization,

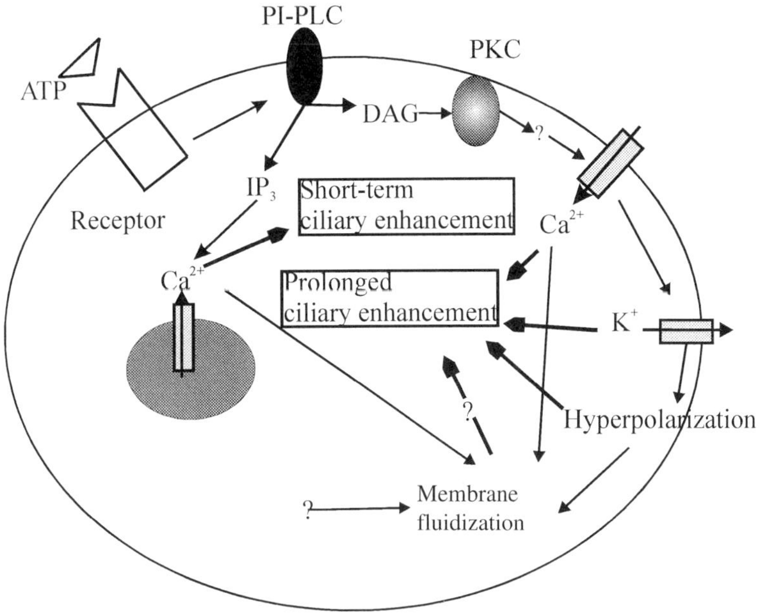


**Figure 1** The scheme of the suggested model.

and presumably some additional factors cause significant membrane fluidization. The role of this effect in purinergic ciliary stimulation is still unclear and requires further investigation.

All these molecular events ultimately lead to the observed physiological response—tremendous increase in ciliary activity, continuing for 20–30 minutes. The extracellular ATP increases ciliary beat frequency two to three fold [13]. The wave component parallel to the effective stroke direction and the velocity of the wave propagation in the direction parallel to the effective stroke increase markedly. The angle between the metachronal wave and the effective stroke direction decreases, approaching 90°. The number of adjacent cilia joining the propulsive effort during the effective stroke increases. All these changes lead to a dramatic rise in ciliary transport efficiency [20].

It seems that enhanced ciliary activity constantly requires a high concentration of intracellular calcium. The rapid massive calcium release from the intracellular stores triggered by ATP causes rapid elevation of the calcium concentration throughout the cell that enables an immediate, strong ciliary response. This stage probably involves rapid mobilization of cellular resources, translocation of essential enzymes to the membrane, and other events, preparing the cell for sustained activation. However prolonged elevation of calcium concentration all over the cell may be extremely hazardous, triggering different molecular events that eventually lead to cell injury or death. Therefore following the initial elevation, the calcium concentration drops to its initial level in the major volume of the cell, but, due to the calcium influx activation, the calcium concentration near the membrane remains high and maintains the high level of ciliary activity.

## ACKNOWLEDGMENT

This work was supported in part by a grant from the Israel Science Foundation.

## REFERENCES

1. Harden TK, Boyer JL, Nicholas RA. $P_2$-purinergic receptors: Subtype-associated signaling responses and structure. Ann Rev Pharmacol Toxicol 1995; 35:541–579.
2. Dubyak GR, El-Moatassim C. Signal transduction via $P_2$-purinergic receptor for extracellular ATP and other nucleotides. Am J Physiol 1993; 265: C577–C606.
3. Burnstock G. A basis for distinguishing two types of purinergic receptors. In: Straub RW, Bolis L eds. Cell Membrane Receptors for Drugs and Hormones: A Multidisciplinary Approach. 1978; 107–118.
4. Burnstock G, Kennedy C. Is there a basis for distinguishing two types of $P_2$-purinoceptors? Gen Pharmacol 1985; 16:433–440.
5. Varhaus EF, Deyrup IJ. The effect of ATP on the cilia of the pharyngeal mucus of the frog. Science 1953; 118:553–554.
6. Mason SJ, Paradiso AM, Boucher RC. Regulation of transepithelial ion transport and intracellular calcium by extracellular ATP in human normal and cystic fibrosis airway epithelium. Br J Pharmacol 1991; 103:1649–1656.
7. Korngreen A, Priel Z. Signal transduction in ciliary systems: The effect of exogenous ATP on intracellular calcium in human nasal epithelium. Biophys J 1993; 64:A76.
8. Korngreen A, Braiman A, Priel Z. Induction of $Ca^{2+}$ release and $Ca^{2+}$ influx in ciliated human polyps by exogenous ATP. Biorheol 1994; 32:210–211.
9. Lansley AB, Sanderson MJ, Dirksen ER. Control of the beat cycle of respiratory tract cilia by $Ca^{2+}$ and cAMP. Am J Physiol 1992; 263:L232–L242.
10. Korngreen A, Priel Z. Simultaneous measurement of ciliary beating and intracellular calcium. Biophys J 1994; 67:377–380.
11. Korngreen A, Priel Z. Purinergic stimulation of rabbit ciliated airway epithelia: control by multiple calcium sources. J Physiol 1996; 497(1):53–66.
12. Villalon M, Hinds TR, Verdugo P. Stimulus-response coupling in mammalian

ciliated cells. Demonstration of two mechanisms of control for cytosolic [$Ca^{2+}$]. Biophys J 1989; 56:1255–1258.
13. Weiss T, Gheber L, Shoshan-Barmatz V, Priel Z. Possible mechanism of ciliary stimulation by extracellular ATP: involvement of calcium-dependent potassium channels and exogenous $Ca^{2+}$. J Membrane Biol 1992; 127: 185–193.
14. Wong LB, Yeates DB. Luminal purinergic regulatory mechanism of tracheal ciliary beat frequency. Am J Respir Cell Mol Biol 1992; 7:447–454.
15. Eshel D, Grossman Y, Priel Z. Spectral characterization of ciliary beating: variations of frequency with time. Am J Physiol 1985; 249:C160–C165.
16. Eshel D, Priel Z. Characterization of metachronal beating of cilia on frogs palate epithelium in tissue culture. J Physiol 1987; 388:1–8.
17. Priel Z. Direct measurement of velocity of metachronal wave in beating cilia. Biorheol 1987; 24:599–603.
18. Gheber L, Priel Z. Synchronization between beating cilia. Biophys J 1989; 55:183–191.
19. Gheber L, Priel Z. Ciliary activity under normal conditions and under viscous load. Biorheol 1990; 27:547–557.
20. Gheber L, Priel Z. Metachronal activity of cultured mucociliary epithelium under normal and stimulated conditions. Cell Motil Cytoskeleton 1994; 28: 333–345.
21. Tsien RY, Rink TJ, Poenie M. Measurement of cytosolic free $Ca^{2+}$ in individual small cells using fluorescence microscopy with dual excitation wavelength. Cell Calcium 1985; 6:145–157.
22. Tarasiuk A, Bar-Shimon M, Gheber L, Korngreen A, Grossman Y, Priel Z. Extracellular ATP induced hyperpolarization and motility stimulation of ciliary cells. Biophys J 1995; 68:1163–1169.
23. Levin R, Braiman A, Priel Z. Protein kinase C induced calcium influx and sustained enhancement of ciliary beating by extracellular ATP. Cell Calcium 1997; 21(2):103–113.
24. Grinkiewicz G, Poenie M, Tsien RY. A new generation of $Ca^{2+}$ indicators with improved fluorescence properties. J Biol Chem 1985; 260:3440–3450.
25. Gheber L, Priel Z, Aflalo C, Shoshan-Barmatz V. Extracellular ATP binding proteins as potential receptors in mucociliary epithelium: characterization using [$^{32}P$]2′-0-(4-Benzoyl) benzoyl ATP, a photoaffinity label. J Membrane Biol 1995; 147:83–93.
26. Alfahel E, Korngreen A, Parola AH, Priel Z. Purinergically induced membrane fluidization in ciliary cells: characterization and control by calcium and membrane potential. Biophys J 1996; 70:1045–1053.
27. Nathan I, Fleischer G, Livne A, Dvilansky A, Parola AH. Membrane microenvironmental changes during activation of human blood platelets by thrombin; a study with a fluorescent probe. J Biol Chem 1979: 254:9822–9828.
28. Nishizuka Y. Intracellular signaling by hydrolysis of phospholipids and activation of protein kinase C. Science 1992; 258:607–614.
29. Bar-Shimon M, Tarasiuk A, Grossman Y, Priel Z. On membrane potential and ciliary stimulation. Bioelectrochemistry and Bioenergetics. 1997; 42: 133–140.

# 49
# The Messianic Relationship Between Inspiration and Expectoration

**Irwin Ziment**
*Olive View-UCLA Medical Center, Sylmar, and University of California, Los Angeles, School of Medicine, Los Angeles, California*

> *"Take thou also unto thee principal spices . . . Pure myrrh . . . sweet cinnamon . . . sweet calamus . . . and of cassia . . . and of olive oil . . . it shall be an holy anointing oil."*
> *—Exodus 30:23–25*

> *"Take sweet spices, stacte, and onycha, and galbanum pure frankincense . . . And make an incense . . . seasoned with salt . . ."*
> *—Exodus 30:34–35*

## I. INTRODUCTION

There is no doubt that among the primitive human's earliest pharmacologic experiments, the inhaling of strong smelling odors and scents from plants would have played an important role [1]. Undoubtedly, the pungent odor of smoke from burning aromatic wood could induce sneezing and coughing and would facilitate the expulsion of abnormal mucoid secretions. The Hebrews learned about incenses and perfumes from the Egyptians, Assyrians, and Babylonians and developed a high regard for the incense mixtures, or ointments, that were used in religious practices [2,3].

## II. BIBLICAL DRUGS

About 110 plants are named in the Bible [4], but biblical names do not always enable experts to identify specific plants. Only about 75 of the named agents were utilized for possible medical purposes [5]. The ingredients for the sacred, pure, holy incense were the sweet or principal spices or herbs: stacte, onycha, galbanum, and pure frankincense; these were blended, and the mixture was seasoned with salt [6,7]. Of greater importance was the holy anointing oil that contained choice, finest, sweet, or principal spices—pure myrrh, sweet or fragrant cinnamon, sweet calamus (also called aromatic cane), and cassia—blended as a perfume in olive oil.

The Lord gave instructions to anoint ("māshah" in Hebrew) Aaron and his sons, thereby sanctifying and consecrating them to serve as priests [6,8]. The word "anointed" is "māssiah" in Hebrew and "messiah" in Greek. Jesus, the Messiah, was never consecrated by anointment as a priest, but was anointed in Bethany with ointment containing spikenard by a woman who poured it on his head (Mark 14:3). After Jesus's death, Mary Magdalene, Mary the mother of James, and Salome brought sweet spices to anoint his body, but Jesus had arisen from the sepulcher and the frightened women fled (Mark 16:1–8) [7].

## III. CINNAMON AND CASSIA

The cinnamon of today is mainly derived from the bark of the tree *Cinnamomum zeylanicum*, which is native to Ceylon (Sri Lanka), but it also grows in India, Indonesia, the Seychelles, and China [9,10]. Cassia, also from China, is typically coarser in appearance and taste; it is obtained from the bark of *Cinnamomum cassia* [11]. The importance of the trade from China is revealed by the names given to cinnamon in countries such as India and Persia—darchini and dalchini, meaning the wood from China [10,11]. Cinnamon and cassia were imported from India or Ceylon only since the 13th century A.D. [12]. It is unlikely that Chinese cinnamon reached ancient Egypt or the Israelites in the time of Exodus by way of a sea journey of more than 4000 miles [11,13]. Reliable overland routes from China were not thought to exist until the end of the second century A.D. [14,15]. Nevertheless, there is a fascinating possibility that there was an active trade reaching biblical Egypt and the Holy Land from sources in China and Indonesia.

The name cinnamon may have come from the Malayan "kaya manis", meaning "sweet wood"; the Hebrew is "qinnamōn." The Hebrew for cassia are the words "ketziah" and "kiddah." It is possible that the qinnamōn and kiddah of the Bible were not related to the spices we know, but were obtained from trees growing in Arabia or near Punt in Africa [11,12,16].

## IV. MYRRH, STACTE, AND FRANKINCENSE

True myrrh is *Commiphora* (formerly known as *Balsamodendron*) *myrrha*. Other names include didin, mukul (or bdellium), Balm of Gilead, myrobalanum, and Balm (or Balsam) of Mecca. The word myrrh comes from the Arabic "murr" or the Hebrew "môr" meaning bitter [13]. A related tree, *C. kataf,* produces the resin opopanax ("sweet myrrh" or "false myrrh") [13]. Storax and styrax have been confused with myrrh [16,17]. Stacte is a gummy exudate from myrrh trees.

Frankincense or olibanum is the true ("frank") incense; it is derived from trees of the *Boswellia* family. It was also known in Arabic as "lubān nad," and in Hebrew as "lebonah" (l'vo-nah). These words are derived from the white (la-van) color of the fresh gum-resin; Lebanon also derives its name from this source word. It is of interest that the Latin word for frankincense is "tus" [13]. The Indian guggul may be the same as frankincense, but it has also been confused with bdellium—which, in turn, has been confused with myrrh!

## V. CALAMUS AND SPIKENARD

Calamus (*Acorus calamus*) is probably native to India [11]. It has been identified as sweet flag, sweet rush, sweet cane, sweet myrtle, sweet sedge, cinnamon sedge, and lemon-grass [11,16]. Spikenard (nard) also was imported from India by the Arabs and was confused with lemon-grass and ginger-grass [15].

## VI. GALBANUM AND ONYCHA

The Talmud suggests that an atoning congregation contains sinners—thus justifying these malodorous incense ingredients [18]. Moreover, the Talmud suggests that the holy incense may have contained a total of eleven to thirteen ingredients, including nard, costus, balsam, fuller's herb, saffron, and wine from Cyprus [16,18].

Galbanum or bubonion is the product of the oozing juice of the plant *Ferula galbaniflua*; it is related to the fetid plant asafetida [19]. Onycha is a particularly intriguing choice as a holy ingredient. The Hebrew word is "scheheleth," which is the origin of the word shell [14]. Onycha may have been the operculum of the foot of a mollusc, *Strombus lentiginosus* [20,21]. Needham suggests that a related mollusc was discovered and used in China as an incense [22]. It is possible that onycha could have been ladana (also known as ladanum, ledanon, labdanum, and lot), a gum-resin derived from a species of rockrose [16]. It is impossible to be sure what the actual concept of onycha was in the Old Testament.

## VII. USES OF BIBLICAL SPICES IN RESPIRATORY THERAPEUTICS

Most of the ingredients of the sacred tabernacle perfumes—myrrh, frankincense, calamus, cinnamon, cassia, galbanum, and salt—had a long-persisting reputation of effectiveness as expectorants. Other biblical herbs, such as garlic, spikenard, labadum, hyssop, balms, storax, aloe, and mandrake are still used in folk remedies [4,19].

The Ebers Papyrus mentions the use of frankincense in a prescription for the treatment of asthma [23]. Frankincense and myrrh were used as fumigants and for treating coughs, phlegm, and other chest ailments including tuberculosis [13]. Dioscorides mentions cinnamon, sweet flag, and resins from conifers for coughs and opobalsamum for difficulty in breathing [24]. He describe the use of hyssop for coughs, catarrh, and asthma; sweet flag, turpentine, and coltsfoot [25] were burned and their smoke inhaled. Avicenna recommended cassia bark as a treatment for stasis in the lungs [25]. Maimonides flavored his chicken soup with spices that included spikenard [25]. In 1678, Salmon recommended myrrh and stacte for stinking breath, hoarseness, coughs, and asthma and olibanum and opoponax for thick, tough phlegm and for coughs, asthma, and difficulty in breathing [26]. In India, galbanum has been used for coughs and asthma; labdanum with bdellium is recommended for coughs, spitting of blood, and fluxes from the lungs [28].

Over 100 years ago, James [28] wrote of stimulant expectorants with antiseptic properties including essential oils, balsams, and turpenes; among this group are cinnamon, frankincense, myrrh, styrax, storax, asafetida, cloves, and balsams. Frankincense was used as an expectorant and myrrh in bronchitis.

In Chinese medicine, cinnamon is said to decrease expectoration and to improve pulmonary function [29]. In Indian folk medicine, cinnamon is

**Table 1** Drugs that Improve Mucociliary Clearance

| Effect | Agents/Actions | Examples |
|---|---|---|
| Regulate water and electrolytes | Topical diluents | Hydrating agents, e.g., water, salt solutions, bronchorrheics |
| | Stimulators of respiratory mucosa | |
| | Irritants | Incense smoke, aromatic vapors, ultrasonic particles |
| | Hyperosmolar solutions | Hypertonic drug or saline solutions |
| | Stimulators of mucus secretion | Bronchomucotropics, e.g., iodide |
| | Stimulators of the gastropulmonary vagal reflex | Expectorants |
| | | Nauseants e.g., ipecac, guaifenesin |
| | | Bitters e.g., myrrh, galbanum |
| Regulate glycoconjugate secretion or structure | Breakdown (lyse) protein or DNA | True mucolytics |
| | Thiols (split disulfide bonds) | Acetylcysteine |
| | Decomplexing agents (break mucoprotein-DNA complexes) | Hypertonic salt solutions |
| | Proteolytic enzymes | α-Dornase, streptokinase |
| | Activators of natural proteases | Iodides, electrolyte solutions |
| | Normalize biochemical production of mucus | |
| | Mucoregulators (alter mucoprotein synthesis) | Acetylcysteine, bromhexine, terpenes, sobrerol |
| | Antimicrobials (decrease DNA production) | Antibiotics, fumigants |
| | Nonspecific | Corticosteroids |
| Improve airway patency | Bronchodilators | Sympathomimetics, theophyllines |
| | Anti-inflammatory agents | Corticosteroids, cromolyn, nedocromil |
| | Mucosal constrictors | Phenylephrine, epinephrine |
| Increase ciliary effectiveness | Ciliary stimulants | Sympathomimetics, theophyllines |
| Increase cough | Irritants | Aromatic smoke or vapor (?) |
| | | Hypertonic aerosols |
| | | Acrid smokes, incenses |
| | | Bitter or pungent drinks |

used to treat common or influenzal colds and asthma; smoke from the burning root of cassia encourages a copious discharge in the common cold [30,31]. The bark oil of cinnamon is fungitoxic against respiratory tract mycoses [32]. The oil has also been given by inhalation to treat tuberculosis [31].

Frankincense or olibanum is used in some countries for treating asthma and as a respiratory tract antiseptic. In India, it was given internally and by inhalation to treat chronic pulmonary affections such as bronchorrhea and for chronic laryngitis [27]. In current Arabian practice, myrrh is used as an expectorant, and its smoke is inhaled to treat chest colds [31,32]. In India, it has been used to treat coughs, asthma, and bronchitis [27,33]. Calamus is used in India as an expectorant in treating asthma, for whooping cough, and for the common cold [30,34].

Potential expectorants account for over half of the 23 drugs, spices, incenses, and perfumes listed by Zohary in his "Plants of the Bible" and perhaps 20 of the 80 products discussed in de Waal's "Medicinal Herbs in the Bible" [5]. Oral expectorants generally act by stimulating afferent receptors in the throat or stomach to bring about a vagal reflex that increases bronchial gland secretion (Table 1) [36,37]. This reflex is similar to that involved in vomiting, and it appears that subemetic doses of emetics cause expectoration. Thus, unpleasant tasting drugs, such as myrrh and galbanum, would evoke the gastropulmonary mucokinetic reflex. Some oral agents will have a further mucokinetic effect when absorbed into the blood if they act on the bronchial glands directly; terpin, guaiacol, and other phenolic drugs appear to act in this way, and presumably the biblical balms and storax would also.

Inhaled agents, such as saline, and incense smoke from frankincense and myrrh can loosen up airway secretions by a mucolytic effect or by stimulating bronchorrhea to improve the effectiveness of coughing. The effect of inhaled smokes on ciliary activity is variable [35]. Some drugs, including tobacco products in small amounts, may improve ciliary beating, while larger doses or prolonged exposures inhibit or damage the cilia. It could be postulated that appropriate anointing with incense perfumes and their smoke have additional benefits beyond inspiration.

## VIII. CONCLUSION

It is surprising that the holy incense, and the anointing oil as described in the Old Testament that supposedly served to consecrate the Messiah, contained agents that were subsequently widely used as expectorants. The potential benefit, or harm, to the lungs of repeated exposure to consecrational incenses has not been explored. The long-standing popularity of these agents for treating respiratory diseases suggests that they may have a therapeutic effect on mucokinesis and mucociliary clearance.

## REFERENCES


1. Morris ET. Fragrance: The Story of Perfume from Cleopatra to Chanel. New York: Charles Scribner's Sons, 1984.
2. Majno G. The Healing Hand. Man and Wound in the Ancient World. Cambridge, Mass: Harvard University Press, 1975.
3. Parry JW. Spices. The Story of Spices. The Spices Described. Vol. 1. New York: Chemical Publishing Company, Inc., 1969.
4. Zohary M. Plants of the Bible. Cambridge: Cambridge University Press, 1982.
5. de Waal M. Medicinal Herbs in the Bible. York Beach, Maine: Samuel Weiser, Inc., 1980.
6. Hertz JH. The Pentateuch and Haftorahs. 2nd ed. London: Soncino Press, 1987.
7. The Holy Bible. Authorized King James Version. Chicago: John A. Dickson Publishing Company, 1946.
8. Myers AC, ed. The Eerdman's Bible Dictionary. Grand Rapids, Michigan: William B. Eerdmans Publishing Co, 1987.
9. Rosengarten F Jr. The Book of Spices. New York: Pyramid Books, 1969.
10. Pruthi JS. Spices and Condiments. New Delhi: National Book Trust, 1976.
11. Miller JI. The Spice Trade of the Roman Empire. 29 B.C. to A.D. 641. Oxford: Clarendon Press, 1969.
12. Flückiger FA, Hanbury D. Pharmacographia. A History of the Principal Drugs of Vegetable Origin Met With in Great Britain and British India. London: Macmillan and Co., 1879.
13. Groom N. Frankincense and Myrrh. A Study of the Arabian Incense Trade. London: Longman Group Limited, 1981.
14. Needham J. Science and Civilization in China. Introductory Orientations. Vol. 1. Cambridge: Cambridge University Press, 1954.
15. Warmington EH. The Commerce Between the Roman Empire and India. London: Curzon Press, 1974.
16. Moldenke HN, Moldenke, AL. Plants of the Bible. New York: Dover Publications, Inc., 1952.
17. Groom N St. J. The Perfume Handbook. London: Chapman and Hall, 1992.
18. Jacob B. The Second Book of the Bible: Exodus. Hoboken: Ktav Publishing House, 1992.
19. Grieve M. A Modern Herbal (Leyel CF, Ed.) Harmondsworth, England: Penguin Books Ltd., 1980.
20. Abrahams HJ. Onycha. Ingredient of the ancient Jewish incense. Econ Botany 1979;33:233–230.

21. Hastings J, ed. Dictionary of the Bible. New York: Charles Scribner and Sons, 1963.
22. Needham J, Gwei-Djen L. Science and Civilization in China. Chemistry and Chemical Technology, Part II. Spagyrical Discovery and Invention. Magisteries of Gold and Immortality. Vol 5. Cambridge: Cambridge University Press, 1974.
23. Mann RD. Modern Drug Use. An Enquiry on Historical Principles. Lancaster, England: MTP Press Limited, 1984.
24. Gunther RT. The Greek Herbal of Discorides. New York: Hafner Publishing Co., 1959.
25. Ziment I. Historic overview of mucoactive drugs. In: Braga PC, Allegra L, eds. Drugs in Bronchial Mucology. New York: Raven Press, 1989:1–33.
26. Salmon W. Pharmacopoeia Londinensis. Or, The New London Dispensatory. London: Thomas Dawks, 1678.
27. Said HM, ed. Pharmacographia Indica. A History of the Principal Drugs of Vegetable Origin Met With in British India. Hamdard Vol 15, No. 1–12, Jan–March 1972.
28. James P. The Therapeutics of the Respiratory Passages. New York: William Wood and Company, 1884.
29. Hsu H-Y, Hsu C-S. Commonly Used Chinese Herb Formulas with Illustrations. Long Beach, California: Oriental Healing Arts Institute, 1980.
30. Bakhru HK. Herbs that Heal. Delhi, India: Orient Paperbacks, 1992.
31. Bellamy D, Pfister A. World Medicine. Plants, Patients and People. Oxford: Blackwell, 1992.
32. Singh HB, Srivastava M, Singh AB, Srivastava AK. Cinnamon bark oil, a potent fungitoxicant against fungi causing respiratory tract mycoses. Allergy 1995; 50:995–999.
33. Ghazanfar SA. Handbook of Arabian Medicinal Plants. Boca Raton, Florida: CRC Press, 1994.
34. Jain SK. Dictionary of Indian Folk Medicine and Ethnobotany. New Delhi: DEEP Publications, 1991.
35. Ziment I. Respiratory Pharmacology and Therapeutics. Philadelphia: WB Saunders Company, 1978.
36. Ziment I. Pharmacologic control of mucus secretion. In: Junod A, Olivieri D, Pozzi E, eds. Endothelial and Mucus Secretory Cells. Milan: Masson, 1981:269–283.

# 50

# An Approach to Imaging Interpretation in Paranasal Sinus Disease

**Arnold M. Noyek, Edward Earl Kassel, Ian James Witterick, and George Wortzman**
*Mount Sinai Hospital, and University of Toronto, Toronto, Ontario, Canada*

**Dan M. Fliss**
*Soroka Medical Center, and Ben-Gurion University of the Negev, Beer-Sheva, Israel*

**Yehudah Roth**
*The Edith Wolfson Medical Center, Holon, Israel*

## I. INTRODUCTION

The paranasal sinuses are inaccessible for direct evaluation, other than through endoscopic sinus surgery. Medical imaging therefore is an important component in the qualitative and quantitative assessment of paranasal sinus disease. Medical imaging, whether through conventional x-rays or high technology studies such as computed tomography (CT), requires close communication between the rhinologist and the medical imager.

The correlative significance of the diagnostic information obtained is only as good as the level of communication between imager and surgeon and their ability to focus accurately on the key diagnostic problem. Clinician and imager cooperation is essential for effective treatment decisions and follow-up evaluations, especially in endoscopic techniques in which the osteomeatal complex must be thoroughly studied by CT.

Conventional plain film examinations are the most common diagnostic imaging studies in otolaryngology and head and neck surgery. The target structures and views for the paranasal sinuses are summarized in the literature [1,2,3]. A complete series of sinus x-ray films usually includes Waters, Caldwell, lateral, and submentovertex views; right and left oblique orbital views are obtained if the orbital views are obtained if the orbital apex or optic canal approximates the area of concern. Additional views may be required to study the nasal bones or zygomatic arches. Panoramic tomography may be especially useful.

Before the advent of CT, complex motion tomography provided critical anatomical information, especially in cases in which bone destruction or bone displacement were in question. Contiguous sections may be performed as close as 1 mm apart in coronal or lateral projections to detail regions of interest after a preliminary screening examination has been carried out employing 0.5 cm contiguous sections, as in assessing a blow-out fracture of the orbital floor. This technique is still used if CT scanning is unavailable. Although allowing clear morphologic visualization of bony abnormalities, this conventional technique has a limited gray scale.

## II. COMPUTED TOMOGRAPHY

Computed tomography has been advocated for improved diagnosis and is now considered the best method for evaluating the paranasal sinuses. Inflammatory conditions, trauma, mucocele, and tumor are critical conditions in which the use of CT is mandatory. Understanding the orbital-sinus and cranial-sinus relationships is essential in assessing the pretreatment stage of carcinoma of the nasal cavities or paranasal sinuses, response to radiation therapy, or postoperative tumor recurrence.

Coronal examinations are optimal for determining superoinferior extension into the orbit and brain. Axial examinations are essential for observing breaching of the posterior wall of the maxillary sinus. Although these CT scans can be performed without contrast enhancement, enhancement may differentiate between obstructive secretions and a mass. Tumor extensions beyond the sinuses into the orbit, brain, or retromaxillary region are optimally seen with contrast enhancement.

Contrast enhancement techniques are important in evaluating vascular lesions. Angiofibroma of the nasopharynx requires this technique to identify the epicenter and extension of angiofibroma from the region of the sphenopalatine foramen, medially into the nasopharynx or laterally into the infratemporal fossa and anteriorly into the maxillary sinus or superiorly into the middle cranial fossa.

Cerebrospinal fluid leak and meningocele are two neurogenic disorders that may require intrathecal CT enhancement. Computer-generated three-dimensional CT images are increasingly employed in evaluating complex facial fractures and major craniofacial anomalies. These images may be significant in tumor management.

## III. MAGNETIC RESONANCE

The advent of magnetic resonance (MR) has influenced the diagnosis, treatment, and followup of tumors of the nasal cavities, paranasal sinuses, and nasopharynx. The MR technique has demonstrated significant advantages over CT in the assessment or characterization of soft tissue mass lesions; for example, MR depicts vascular structures without the use of intravenous contrast agents. Inflammatory lesions produce bright signals on T2-weighted images, and tumor masses tend to have low or intermediate T2-weighted signal intensities. If the MR scan indicates a noninflammatory mass, an immediate biopsy should be considered.

## IV. OTHER DIAGNOSTIC MODALITIES

Other diagnostic modalities currently employed in studying the paranasal sinuses have a lesser but selective significance. Ultrasound of the maxillary and frontal sinuses has limited usefulness but can identify fluid when conventional radiographs show opacification.

Radionuclide bone scanning using Tc-methylene diphosphonate demonstrates osteoblastic activity in several situations, particularly in the identification of focal manifestations of systemic disease, such as Paget's Disease. Detection of these nonspecific findings is highly sensitive, and the bone-scan technique shows physiologic changes before morphologic studies employing conventional x-ray films and CT scans record them. Gallium citrate scans, if used in combination with the bone scan, allows diagnosis of osteomyelitis and follow-up assessment of treatment effectiveness. Successful treatment is indicated by the gallium scan returning to normal, because this modality specifically images the infective focus, unlike the bone scan that images the osteoblastic response about that focus.

Radiologic imaging signs attempt to demonstrate the status of the air content of the paranasal sinuses and changes involving the mucous membrane and the bony walls. Adjacent structures (orbit, skullbase, brain, dentition) are also critically assessed. Imaging studies provide both structural and functional information and imaging signs reflect anatomic, physiologic, and combined information.

Conventional x-ray films, complex motion tomography, and CT scans share common radiographic signs (see listing below). The radiologic signs listed below reflect primarily structural disease. The reader is referred to a number of references in this regard [1–5]. Eleven signs are listed as follows: (1) decreased aeration and luminal opacification, (2) mucosal thickening, (3) cyst formation, (4) soft tissue mass, (5) air-fluid level, (6) emphysema, (7) calcification, (8) ossification, (9) one or more embryonic or adult teeth, (10) foreign body(ies), and (11) alterations in bony walls.

The radiologic signs associated with alterations in bony walls can be further subdivided into (1) decalcification; (2) osteolysis/destruction; (3) dehiscence; (4) fracture(s); (5) osteoblastosis or hyperostosis; (6) expansion and displacement of bony walls; (7) decreased luminal volume: (a) bony hypoplasia, (b) decreased pneumatization, and (c) hyperostotic encroachment of thickened bony walls; and (8) sequestrum(a).

## V. SUMMARY

In summary, the reader should recall that every radiologic examination is a consultation and its value is directly proportional to the clarity of communication between the clinician and imager. Generally, high technology imaging (i.e., CT, MR, angiography, ultrasound, radionuclide scans) supplement rather than replace conventional radiographs. However, there is an increasing tendency towards the use of coronal CT as both the best modality for evaluating the paranasal sinuses and frequently the only modality used in evaluation. This varies from geographic region to geographic region.

This brief manuscript lists the radiologic signs encountered in paranasal sinus disease, stresses the role of good communication, and asks the reader to consider pitfalls in radiologic interpretation that can mislead the clinician and imager.

## ACKNOWLEDGMENT

The above work was a Canada-International Scientific Exchange Program (CISEPO) Project.

## REFERENCES

1. Noyek AM, Wortzman G, Kassel EE, Diagnostic imaging in rhinology. In: Goldman JL, ed. The Principles and Practice of Rhinology. New York: John Wiley and Sons, 1987.
2. Yanigisawa E, Smith HW. Normal radiographic anatomy of the paranasal sinuses. Otolaryngol Clin North Am 1973; 6:429.
3. Zinreich S, Abidin M, Kennedy D. Cross-sectional imaging of the nasal cavity and the paranasal sinuses. Operative Techniques in Otolaryngol Head Neck Surg, 1990; 1(2):93–99.
4. Som PM, Curtin HD. Head and Neck Imaging. Vol. 1. 3rd Ed. Mosby–Yearbook Inc., 1996.
5. Noyek AM, Witterick I, Fliss DM, Kassel EE. Diagnostic imaging. In: Bailey BJ, ed. Vol. 1. 2nd Ed. Head and Neck Surgery—Otolaryngology. J.B. Lippincott Company, 1997.

# 51

# Chronic Sinusitis in Relation to the Lower Airway

**Saul Frenkiel and Chantal Giguère**
*Sir Mortimer B. Davis–Jewish General Hospital, and McGill University, Montreal, Quebec, Canada*

## I. INTRODUCTION

There exists more than a casual relationship between the sinonasal passages and the lower airway. This relationship is evidenced by common or interrelated disease states and treatments that may similarly benefit each area. The basis for this commonality may stem from anatomy as a passage in continuity, an identical mucociliary cover, similar mucosal or submucosal molecular events, and a neurophysiological interplay (Table 1).

A number of diseases show an affinity for all parts of the respiratory tract. Examples are the mucociliary deficiencies and abnormalities of mucus production that may cause chronic sinusitis and lower airway abnormalities. Other diseases such as allergic rhinitis, nasal polyps, acetylsalicylic acid (ASA) sensitivity, and asthma may coexist in the same patient. Whether the relationship is cause and effect or similar disease process occurring in both areas remains to be determined.

Historically, authors have talked about the physiological and pathophysiological relationship between the upper and lower airways. In 1880, Voltalini [1] reported a patient whose asthma improved after a nasal polypectomy. Other authors [2–4] in the 1920s and 1930s further developed the concept of sinusitis as a provoking factor in asthma production. Presently, there is considerable research in progress concerning the inflammatory process that may trigger both chronic sinusitis and bronchial asthma. The

**Table 1** Mechanisms that May Trigger Asthma in a Chronic Sinusitis Patient

| |
|---|
| Physiological factors |
| Nasopulmonary reflexes |
| Pharyngopulmonary reflexes |
| Mediators of chronic infection |
| Postnasal discharge |
| Bacterial induced β-blockade |
| Immunological mediators |
| Activated eosinophils |
| IgE mediated hypersensitivity |
| Th-2 cytokines |
| Other chemical mediators |

following sections will discuss chronic sinusitis and the lower airway in terms of underlying mechanisms and clinical data that interrelate at both anatomical sites.

## II. CAUSE-AND-EFFECT OR COMMON DISEASE

A cause-and-effect relationship between chronic sinusitis and asthma may exist. As early as 1925, Gottlieb [5] described the classical nasobronchial reflex whereby sinusitis was able to trigger lower airway bronchospasm. Later work by Bucca et al. [6] gave evidence of bronchoconstrictive reflexes originating in the extra thoracic airways. They concluded that upper respiratory tract diseases such as postnasal drip, pharyngitis, laryngitis, and sinusitis were able to produce asthma-like symptoms. Evidence emerged that increased viscoelasticity of tracheobronchial mucus secretions in patients with chronic sinusitis may be the important factor in sinobronchial syndrome.

Indirect evidence for the causal relationship stems from the clinical observation that medical and surgical treatment of sinusitis frequently improves bronchial asthma. Numerous studies have shown that traditional surgeries such as Caldwell–Luc procedures and the more recent techniques of endoscopic sinus surgery have not only improved chronic sinusitis but have also had a positive effect on the status of the lower airways [7].

The similar disease theory may also be valid. The benefit of steroids in the treatment of both chronic sinusitis and asthma gives credence to

**Table 2** Common Diseases Affecting the Upper and Lower Airways

| |
|---|
| ASA triad |
| Cystic fibrosis |
| Primary ciliary dyskinesia (Kartagener's, Young's) |
| Vasculitides (Churg–Strauss, Wegener's) |
| Immune deficiency states |

the possibility of an identical histopathological process. There is commonality of many diseases that afflict the airways as indicated in Table 2. The ASA triad links nasal polyposis, bronchial asthma, and ASA sensitivity. Cystic fibrosis occurs in the pediatric age group and represents an abnormality of exocrine glands. Because of their inspissated secretions, these patients tend to develop chronic sinusitis and bronchial infections. Other entities that affect the entire respiratory tract are primary ciliary dyskinesias such as Kartagener's and Young's syndromes and vasculitides such as Churg–Strauss syndrome or Wegener's granlomatosis. Various immunological defects such as panhypo-gamma-globulinemia, or deficiencies in IgA or IgG subclasses will contribute to chronic infections in the sinuses and bronchial passages.

## III. ANATOMY AND HISTOLOGY

Despite an arbitrary division of the human airway into an upper or extrathoracic portion and a lower or intrathoracic part, the entire tract is a continuous passage with an identical epithelium. The sinuses form a network of cavities and labyrinths that conditions the incoming airstream for entry into the lungs. The nasal passages are lined by a ciliated pseudostratified columnar epithelium containing goblet cells and supported by connective tissue where numerous serous and mucinous glands are found. The vasomotor response is primarily governed by nervous innervation from the parasympathetic system through the vidian nerve.

The lower airway begins with similar surface histology. More inferiorly, the columnar cells begin to flatten, goblet cells and glandular elements diminish and increasing amounts of surrounding smooth muscle are found as airway diameters decrease. Cartilagineous rings provide a supportive framework preventing airway collapse during ventilation. The cartilage eventually disappears in the tiny bronchioles before they blossom into alveoli where gas exchange occurs. The intrathoracic airways and lungs are under sympathetic control via the vagus nerve.

Histopathological changes that occur with chronic sinusitis and asthma are noted to be strikingly similar. Harlin et al. [8] described a number of common features such as eosinophilic infiltration, basement membrane thickening, glandular hyperplasia, and stromal edema. These inflammatory hallmarks are evident in both disease states and possibly indicate a common pathogenesis.

## IV. THE PHYSIOLOGICAL RELATIONSHIP

Besides acting as a simple conduit for inspiratory and expiratory airflow, the nasal chambers perform a number of other important functions (Table 3). They are involved in temperature regulation bringing inspired air to within a degree of body temperature before it leaves the nasopharynx. They humidify the incoming airstream, which serves an important purpose in dryer climates. The nose and sinuses provide a protective function through their secretions, filtration mechanism, and initiation of reflexes. A well-defined mucociliary transport mechanism is an important part of this air-conditioning process.

In turn, the lower airways and lungs also have important functions (Table 4). Proper upper airway conditioning is vital for adequate gas exchange at the alveolar level. Approximately 50% of total airway resistance is developed within the nasal cavity and mostly at the level of the nasal valve. It is necessary to have such adequate resistance for airflow in order to properly expand the lungs. Reflex bronchial constriction may develop subsequent to an upper airway event. The resulting clinical manifestation is usually a cough. The lower ciliary escalator serves to evacuate larger

**Table 3** Nasal and Sinus Physiology

| |
|---|
| Temperature regulation |
| Cooling and warming |
| Humidification of inspired air |
| Airflow |
| Provides 50% of total airway resistance |
| Protection |
| Filtration, secretions, reflexes, immunologic activity |
| Mucociliary transport |

**Table 4** Physiology of Lower Airway

| |
|---|
| Airflow and resistance |
| Alveolar gas exchange |
| Protection |
| Cough, reflex bronchial constriction, alveolar marcophages, immunological processes |
| Ciliary escalator |

particles as part of its surface action. In the submucosa, pulmonary alveolar macrophages are also an important part of the protective mechanism.

The nasopulmonary physiological link has been described by numerous authors. Ogura and Harvey [9] in 1971 performed both human and animal studies that revealed that changes in upper airway resistance resulted in comparable effects on the lower airway. In some cases, they discovered that proper pulmonary function was returned after the surgical correction of a nasal septal deviation. Other descriptions of nasopulmonary and nasosystemic reflexes were described by Haight and Cole [10a] in human experiments using head-out body plethysmography as a measure of nasal resistance.

Such evidence clearly indicates a physiological interplay between the nasal chambers and the intrathoracic airway. It is likely that mucosal erectile tissue is regulated by autonomic nervous stimulation and possibly by biochemical mediators in the submucosa. Mechanical stimuli at any point of origin in the airways will further influence airflow and airflow resistance.

## V. IMMUNOLOGICAL CONCEPTS

Considerable evidence exists suggesting that an immunological process is responsible for development of diseases of the upper and lower airways. Studies by Frenkiel and co-workers [10b–12] described the importance of IgE in the pathogenesis of nasal polyps. These investigators showed that elevated IgE occurred both locally within nasal tissue and systemically in chronic sinusitis. In addition, chronic sinusitis in association with asthma showed significantly elevated IgE in comparison to the nonasthmatic sinusitis group.

The eosinophil has been described as the important inflammatory cell in some types of bronchial asthma and in forms of chronic sinusitis associated with nasal polyposis [8,10b]. It is likely that the eosinophil acts as the effector cell for the inflammatory process in both upper and lower respiratory tract mucosa. Mast-cell degranulation as initiated through antigen stimulation and IgE production leads to the release of various mediators that cause the typical allergic reaction. The common clinical manifestations of this IgE-mediated response are rhinitis and asthma.

More recently, research has concentrated on the role of TH-2 type cytokines in antibody production and IgE mediated atopic inflammation [13–15]. Cytokines such as IL-4, IL-5, IL-13, and GM-CSF have been shown to be important regulators of eosinophil growth and survival. Specifically, IL-5 may be responsible for eosinophil activation and GM-CSF may be the promotor of macrophage activity at peripheral tissue sites. Chemotactic type cytokines may play a role in the mediation of the late-phase inflammatory response that occurs both in the nasal passages and the lower airways [16].

## VI. THE ROLE OF INFECTION

Another postulated link between chronic sinusitis and asthma is the potential role played by infection. Two mechanisms have been proposed that could account for the increased bronchial responsiveness. One theory involves the mucopurulent discharge from the sinuses that drips into the larynx and trachea. It results in inflammation that could then trigger an acute asthmatic reaction. Using a rabbit model, Irvin [17] was able to demonstrate lower airways hyperresponsiveness in the presence of maxillary sinusitis after histamine was inhaled. He explained this phenomenon of postnasal drip as the elaboration of inflammatory mediators in the upper, which are then transports to the lower, respiratory tract. However, in a recent study using radionuclide scanning in humans, pulmonary aspiration of nasal mucopurulent secretions could not be demonstrated [18].

A second possible mechanism relates to the concept of enhanced β-adrenergic blockade [19]. Because a partial blockade may be present with asthma and because infection can contribute to the blockade, it is postulated that chronic sinusitis may result in an increased hyperreactivity of the bronchial tree. This represents an unproven theoretical possibility.

## VII. CLINICAL DATA

Inflammatory diseases of the nasal chambers and bronchial asthma are each commonly occurring entities in the community. Yet, their coexist-

**Table 5** Clinical Evidence Relating Chronic Sinusitis to Lower Airway Disease

| |
|---|
| Presence of common diseases in respiratory type mucosa |
| High associated incidence of nasal polyposis and asthma |
| Improvement of asthma after successful sinusitis therapy |

ence in the same patient is striking (Table 5). As much as 80% of asthmatics may suffer from some type of rhinitis and possibly up to 15% of people with perennial rhinitis may have asthma [20]. Other data indicate that over 50% of patients with nasal polyps have lower airway obstructive disease as documented by pulmonary function testing [10b].

Data has indicated that there occurs improvement in lower airway resistance after treatment of sinusitis. Rachelefsky, Katz, and Siegel [21] studied 48 children with sinusitis and lower airway disease and demonstrated improvement in the asthmatic state after medical therapy of the sinuses. Friedman, Ackerman, and Wald [22] described 8 patients with asthma exacerbation after maxillary sinusitis. Seven cases improved their asthma after antibiotic treatment and 5 showed objective improvement according to pulmonary function tests. Similar results are evident after surgical treatment of chronic sinusitis as reported by Slavin [20]. Manning, Wasserman, Silver, and Phillips [23] described a reduction in hospitalization and school days missed in asthmatics who had undergone functional endoscopic sinus surgery. They found improved asthma symptom scores and reduced requirement of steroid use. Yet, their study group showed no significant improvement in pulmonary function testing 12 months postoperatively. Similarly, Nishioka, Cook, Davis, and McKinsey [24] reported that the value of endoscopic sinus surgery for chronic sinusitis is an important adjunct in the treatment of coexistent asthma.

## VIII. CONCLUSION

The relationship between upper and lower airways is evidenced by common disease entities and similar treatments that improve both anatomical areas. Whether this represents a causal relationship or a common process is presently under debate. It is likely that both of these theoretical concepts are valid in selected cases.

## REFERENCES


1. Voltalini R. Ueber nasenpolypen und deren ooperation. Alg Med Cetr Ztg. 1880; 126.
2. Gottlieb MS. Relation of intranasal sinus disease in the production of asthma. JAMA 1925; 85:105–109.
3. Bullen SS. Incidence of asthma in 400 cases of chronic sinusitis. J Allergy 1932; 4:402–408.
4. Weille FL. Studies in asthma XIX: the nose and throat in 500 cases of asthma. N Engl J Med 1936; 215:235–238.
5. Gottlieb MJ. Relation of intranasal disease in the production of bronchial asthma. JAMA 1925; 85:105.
6. Bucca C, Rolla G, Brussino L, De Rose V, Bugiani M. Are asthma-like symptoms due to bronchial or extrathoracic airway dysfunction? Lancet 1995; 23:346 (8978):791–795.
7. Forgsgren K, Fukami M, Penttila M, Kumlien J, Stierna P. Endoscopic and Caldwell–Luc approaches in chronic maxillary sinusitis: a comparative histopathologic study on preoperative and postoperative mucosal morphology. Annals of Otology 1995; 104(5):350–357.
8. Harlin SL, Ansel DG, Lane SR, Myers J, Kephart GM, Gleich GJ. A clinical and pathologic study of chronic sinusitis: the role of the eosinophil. J Allergy Clin Immunol 1988; 81:867–875.
9. Ogura JH, Harvey JZ. Nasopulmonary mechanisms, experimental evidence on the influence of the upper airway upon the lower. Acta Otolaryngol 1971; 71:123–132.

10a. Haight J, Cole P. Unilateral nasal resistance and asymmetrical body pressure. J Otolaryngol 1986; (15suppl16):1–31.

10b. Frenkiel S, Small P, Rochon L, Cohen C, Darragh D, Black M. Nasal polyposis—a multidisciplinary study. J Otolaryngol 1982; 11:275–278.

11. Frenkiel S, Chagnon F, Small P, Rochon L, Cohen C, Black M. The immunological basis of nasal polyp formation. J Otolaryngol 1985; 14:89–91.
12. Jones E, Frenkiel S, Small P, Rochon L. Immunopathological characteristics of nasal polyps. J Otolaryngol 1987; 16:19–22.
13. Durham SR, Ying S, Varney VA, Jacobson MR, Sudderick RM, Mackay IS, Kay AB, Hamid Q. Cytokine mRNA expression for IL-3, IL-4, IL-5 and GM-CSF in the nasal mucosa after local allergen provocation: relationship to tissue eosinophilia. J Immunol 1992; 148:2390–2394.
14. Al-Ghamdi K, Ghaffar O, Small P, Frenkiel S, Hamid Q. IL-4 and IL-13 expression in chronic sinusitis. Relationship with cellular infiltrate and the effect of topical corticosteroid treatment. J Otolaryngol 1997; 26:236–240.
15. Wright ED, Frenkiel S, Small P, Al-Ghamdi K, Ghaffar O, Hamid Q. Il-4, IL-5 and GM-CSF receptor expression in chronic sinusitis and response to topical steroids. Otolaryngol—Head and Neck Surg 1997. In press.
16. Wright ED, Frenkiel S, Ghaffar O, Al-Ghamdi K, Small P, Hamid Q. Monocyte chemotactic protein expression in allergy and non-allergy–associated chronic sinusitis. J Otolaryngol 1997. Submitted.

17. Irvin CG. Sinusitis and asthma: an animal model. J Allergy Clin Immunol 1992; 90:521–533.
18. Borden PG, Van Hearden BB, Joubert JR. Absence of pulmonary aspiration of sinus contents in patients with asthma and sinusitis. J Allergy Clin Immunol 1990; 86:82–88.
19. Szentivanvi A. The beta adrenergic theory of the atopic abnormality in bronchial asthma. J Allergy 1968; 42:203.
20. Slavin RG. Asthma and sinusitis. J Allergy Clin Immunol 1992; 90:534–537.
21. Rachelefsky GS, Katz RM, Siegel SC. Chronic sinus disease with associated reactive airway disease in children. Pediatrics 1984; 73:526–529.
22. Friedman R, Ackerman M, Wald E. Asthma and bacterial sinusitis in children. J Allergy Clin Immunol 1984; 74:185–189.
23. Manning SC, Wasserman RL, Silver R, Phillips DL. Results of endoscopic sinus surgery in pediatric patients with chronic sinusitis and asthma. Arch Otolaryngol—Head and Neck Surg 1994; 120:1142–1145.
24. Nishioka GJ, Cook PR, Davis WE, McKinsey JP. Functional endoscopic sinus surgery in patients with chronic sinusitis and asthma. Otolaryngol—Head and Neck Surg 1994; 110:494–500.

# 52

# Immune Responses in the Chronic Inflammatory Process Within the Paranasal Sinus Mucosa

**Erin Daniel Wright and Qutayba Hamid**
*Meakins–Christie Laboratories, and McGill University, Montreal, Quebec, Canada*

**Saul Frenkiel**
*Sir Mortimer B. Davis–Jewish General Hospital, and McGill University, Montreal, Quebec, Canada*

## I. INTRODUCTION

In the last few years our understanding of the pathophysiology underlying upper respiratory inflammation and chronic sinusitis has improved considerably. Significant roles for $T_h$-2 cytokines as well as newer classes of intercellular mediators have been suggested. What follows is a presentation of the relevant literature concerning the immunopathology of chronic sinusitis as well as the current laboratory methodology. Some recent data will be summarized and the implications discussed.

## II. CHRONIC SINUSITIS

Chronic sinusitis (CS), whether allergy-associated or not, is a common clinical condition that has a considerable impact on the healthcare system in developed countries [1]. This disease is characterized by variable degrees of mucosal thickening, alteration of goblet cells, thickening of the basement membrane with or without polyp formation, and persistent inflammation [2]. Recent studies have revealed that the composition of the inflammatory infiltrate in both allergy-associated (ACS) and non-allergy–associated (NCS) chronic sinusitis is similar to that which has been documented in chronic allergic inflammation of the nasal mucosa [3] and asthmatic airways [4]. Lymphocytes, plasma cells, macrophages, and most notably activated eosinophils are prominent in the sinus mucosa of subjects with chronic sinusitis [5]. The accumulation of activated eosinophils may provoke the chronic changes observed in ACS and NCS through the release of cytotoxic mediators such as Major Basic Protein (MBP) and Eosinophil Cationic Protein (ECP). It is thought that CD4+ T-lymphocytes regulate the inflammatory event by producing particular cytokines.

## III. $T_H$-2 TYPE CYTOKINES

Two general subsets of CD4+ T-lymphocytes have been described, each with different cytokine and functional profiles (Figure 1). The $T_h$-1 subset is primarily involved in cell-mediated immune responses while the $T_h$-2 subset is primarily involved in antibody production and IgE-mediated atopic inflammation [6]. It should also be pointed out that there are certain cytokines that are produced by members of both subsets, such as GM-CSF, IL-13, and TNF-α.

Numerous in vitro studies support the hypothesis that $T_h$-2 cytokines play a significant role in the pathophysiology of chronic sinusitis. We have recently shown that in allergic inflammation there is increased local expression of IL-4 [7], which is essential for isotype switching in favor of IgE production [8]. The IL-4 cytokine has also been demonstrated to upregulate vascular cell adhesion molecule (VCAM-1) on endothelial cells [9], which is the counterligand for very late activation antigen-4 (VLA-4). The VLA-4 antigen is expressed on eosinophils and thus IL-4 promotes eosinophil transendothelial migration [10]. The hemotopoietic growth factors IL-5 and GM-CSF have been shown to be important regulators of eosinophil growth and survival [11–13]. In addition, IL-5 is primarily responsible for the terminal differentiation and activation of eosinophils [14,15], whereas GM-CSF may also be important in the activation of macrophages at peripheral tissue sites [16,17].

## IV. $T_h$-2 TYPE CYTOKINE RECEPTORS

The effects of $T_h$-2 cytokines are mediated through cell membrane associated receptors that are all members of the cytokine receptor superfamily

| $T_h$-1 | $T_h$-2 |
|---|---|
| IL-2 | IL-4 |
| IFN-γ | IL-5 |
| IL-12 | IL-6 |
| IL-10 | IL-10 |
| IL-13 | IL-13 |
| TNF-α,β | TNF-α |
| IL-3 | IL-3 |
| GM-CSF | GM-CSF |
| CMI/DTH | Ab/IgE |

**Figure 1** T-helper lymphocyte cytokine profiles. Note that $T_h$-1 subsets are primarily involved in cell-mediated immunity (CMI) or delayed-type hypersensitivity (DTH). $T_h$-2 subsets are primarily involved in antibody (Ab) production and IgE-mediated atopic inflammation.

[18]. These receptors are composed of both shared and specific subunits. The IL-4 receptor is a heterodimer that has a specific α- subunit as well as a γ- subunit that it shares with IL-2 and IL-7 [19]. The IL-5R and GM-CSFR are also heterodimers with a specific α subunit and a shared β subunit [20–22] and have been shown to be expressed by eosinophils [23–25].

Studies examining cytokine receptor expression in vivo have demonstrated upregulation of IL-5R and GM-CSFR in the airway mucosa of allergy-associated asthma [26,27]. It would thus seem reasonable to expect an increase in the local production of αIL-4R, αIL-5R, and αGM-CSFR in the mucosa of subjects with chronic sinusitis.

## V. EFFECTS OF TOPICAL CORTICOSTEROID TREATMENT

Allergic inflammation has been demonstrated to be amenable to treatment with topical steroids [28]. The evidence suggests that this medication acts to inhibit both the early and late responses of type-I hypersensitivity [28]. The mechanism for this response appears to be at least partially due to the inhibition of mRNA coding for IL-4 [29]. An issue of clinical interest was to know the effects of topical steroids on the chronic inflammation in chronic sinusitis and whether there is a differential response to steroids between ACS and NCS subjects.

## VI. CHEMOKINES

Chemotactic cytokines or chemokines are a novel group of inflammatory mediators that appear to possess significant specificity for different inflammatory cells. They have the ability to recruit particular leukocyte populations to a site of inflammation (Figure 2) [30,31]. Inflammatory cytokines such as IL-4 and IL-5 have been shown to elaborate chemokine release, and neutralization of IL-4 has been demonstrated to reduce production of certain chemokines (MIP-1α, RANTES) [32]. This information supports the theory that chemokines play a role in the mediation of the late-phase inflammatory response [33] via a mechanism that appears to be a function of promoting adhesion to endothelial cells and the promotion of extravasation of these inflammatory cells [34].

Monocyte chemotactic proteins (MCP) are a subfamily of chemokines that include MCP-1, MCP-2, MCP-3, and MCP-4. They show significant structural similarity, sharing greater than 60% amino acid homology [35,36]. The primary target for the MCP family are mononuclear cells such as basophils, eosinophils, monocytes, and lymphocytes [37]. The primary cell sources of MCPs are mononuclear cells, fibroblasts, endothelial cells, and epithelial cells [38–40]. From the perspective of biological activity, all of the MCP family members attract monocytes, T cells, and NK cells [41,42]. Their different biological activities relate to their action on basophils, eosinophils, and dendritic-type cells. The MCP-1 protein has been demonstrated to be a potent activator of basophils but does not attract eosinophils [43]. The MCP-2 protein has a slightly broader range of activities, with action on basophils as well as eosinophils [44,45], and MCP-3

| | MCP-1 | MCP-2 | MCP-3 | MCP-4 |
|---|---|---|---|---|
| Basophils | + | + | + | + |
| *Eosinophils* | | + | + | + |
| Dendritic cells | | | + | |
| Neutrophils | | | + | |
| Monocytes | | | | + |

**Figure 2** MCP family chemotactic target cells. Table illustrates activity of the various members of the MCP family with respect to their ability to attract inflammatory cells. (+) indicates documented ability to attract cells.

seems to be the chemokine with the broadest spectrum of action: it is active on eosinophils, basophils, and dendritic cells as well as neutrophils [46]. Finally, MCP-4 was found to be a potent chemoattractant for monocytes and eosinophils and stimulated histamine release from basophils [36].

To date there have been limited descriptions of MCP-1 in sinonasal conditions [47] despite ample association with other inflammatory diseases such as asthma, arthritis, and inflammatory skin diseases [32,48,49]. Further, little is known about the role for MCP 3 and MCP-4 in pathology except that they appear to act primarily at epithelial surfaces and in allergic inflammation [37]. An exciting new hypothesis casts MCPs as significant chemoattractants for eosinophils in chronic sinusitis.

## VII. CURRENT LABORATORY METHODOLOGY

### A. Patient Populations

The upper respiratory tract, by virtue of its ease of examination and procurement of tissue samples, affords a unique opportunity for study of mucosal inflammation. Specifically, endoscopic visualization of the nose and paranasal sinuses during surgery allows directed biopsies for study using various staining and molecular biology techniques. In several recent studies we have used these techniques to study cytokines and their receptors as well as chemokines.

Our patients were recruited from the nasal clinic at a tertiary care teaching hospital. The diagnosis of chronic sinusitis was made on the basis of ongoing symptomatology of at least six-months duration including nasal congestion, rhinorrhea, anosmia or hyposmia, and mid-facial pains. In all cases, the sinus CAT scan indicated mucosal abnormalities involving at least the ethmoid sinus with additional features of ostiomeatal complex obstruction. The imaging frequently revealed at least partial or complete obstruction of the other sinus chambers. Allergic subjects were defined by positive intradermal tests to a panel of common seasonal and perennial allergens. Those atopic patients treated with topical corticosteroids received fluticasone proprionate, 200 μg once daily in each nasal chamber, for a period of at least 3 months prior to surgery. Subjects in the control group did not have sinusitis nor were they atopic. Patients were excluded if they had received previous immunotherapy or were taking regular oral anti-inflammatory medication.

Biopsy specimens were obtained at the time of surgery. Subjects in the atopic and nonatopic study groups were undergoing endoscopic sinus surgery for treatment of their chronic sinusitis. Control subjects were those undergoing other endonasal procedures such as septoplasty. In all cases, mucosal biopsy specimens measuring at least 3 mm in diameter were obtained from the anterior ethmoidal region. In each case, written informed consent was obtained and the study was approved by the hospital Ethics Committee.

## B. Immunocytochemistry

Tissue preparation for immunocytochemistry (ICC) involved immersion in 15% phosphate buffered saline (PBS) on ice for 15–20 minutes. The tissue was then blocked and frozen in liquid-nitrogen-cooled isopentane. Cryostat sections of 5 μm thickness were obtained from the specimens and allowed to dry for 1–2 hours. After fixation in a mixture of 60% acetone and 40% methanol, the slides were dried and stored at −80°C until further processing.

We employed an alkaline phosphatase antialkaline phosphatase technique (APAAP) [50]. The slides were retrieved from the −80°C freezer and allowed to warm to room temperature before incubation with the primary antibodies. Staining was performed with either commercially available antibodies or those obtained from generous colleagues. Examples of antibodies used include anti-CD3 as a marker for T-lymphocytes, anti-CD4 as a marker for T-helper lymphocytes, anti-MBP as a marker for eosinophils, anti-MCP-1, anti-MCP-3, and anti-MCP-4.

## C. In Situ Hybridization

In general complementary DNA (cDNA) was used as a template for synthesizing RNA probes coding for the desired cytokine or cytokine receptor [21,51]. The riboprobes synthesized contained the sequence complementary to the exons coding for the desired protein. The cDNA template was inserted into a pGEM vector and linearized with appropriate enzymes. Digoxigenin-labeled antisense (or sense) transcripts were synthesized as previously described [52]. For in situ hybridization the slides were first treated with a proteinase K solution to allow the probes to enter the cells. Negative controls were slides treated with RNase to digest all tissue mRNA.

Digoxigenin-labeled riboprobes, both antisense and sense controls, were added to each section. The section were covered and incubated overnight at 40°C in a humid chamber. Slides were then washed to remove any unhybridized RNA. After a brief wash in TBS, sections were incubated with 3% bovine serum albumin/TBS to reduce nonspecific background. The tissue was then incubated with 1:500 to 1:5000 dilutions of ⟨Dig⟩-AP conjugate (sheep polyclonal anti-Digoxigenin antiserum conjugates with alkaline phosphatase) overnight at room temperature. Color development was achieved by adding a freshly prepared substrate solution consisting of X-phosphate-5-bromo-4-chloro-3-indoly phosphate (BCIP) and nitroblue tetrazoleum (NBT) salt to the slides. Slides were counterstained with hematoxylin and mounted with glycerol gelatin.

### D. Quantification and Statistical Analysis

Two forms of quantification were used. The first is termed number of positively staining subepithelial cells and refers to the mean count of positive cells per high-power field under the light microscope. Generally, at least two sections and a median of four fields were counted using a graticule oriented along the epithelial basement membrane. This measurement gave an impression of inflammatory cell infiltrate in the mucosa. The second measurement possible is an epithelial score, which is based on the percentage of positive cells in the epithelium. This second measurement gives insight into events occurring at the epithelial surface.

Comparisons between cell counts in the various groups (e.g., ACS, NCS, steroid treated, and controls) were performed using ANOVA analysis and Student's *t*-test. Correlations were performed using linear regression and reported as $R^2$ and p values. The p values $<0.05$ were considered statistically significant.

## VIII. RECENTLY ACCUMULATED DATA

### A. $T_h$-2 Cytokines in Chronic Sinusitis

Recent experiments by our group have demonstrated differential expression of cytokines in ACS and NCS subjects. We have shown the augmented expression of $T_h$2-type cytokines such as IL-4 and IL-5 in the sinus mucosa of ACS but not NCS subjects [53]. Conversely, IL-13, which shares biological properties with IL-4, was expressed in elevated levels in both ACS and NCS subjects [54]. In this same study it was also found that treatment with topical steroids significantly reduced the expression of both IL-4 and IL-13 in the sinus mucosa of ACS subjects. The GM-CSF has also been demonstrated to have elevated expression in both ACS and NCS subjects.

### B. $T_h$-2 Cytokine Receptors in Chronic Sinusitis [55]

The IL-4R mRNA (Figure 3a) has been demonstrated to be expressed in both the epithelium and lamina propria of sinus mucosal biopsies. The level of expression in ACS subjects was significantly higher than in NCS

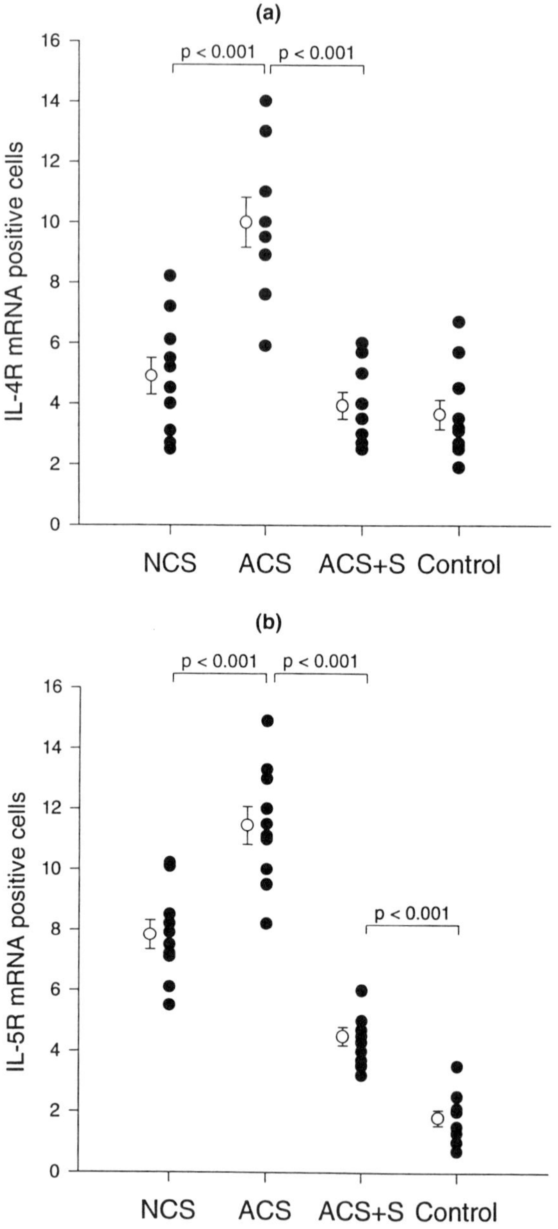


**Figure 3** Expression of $T_h$-2 type cytokine receptors in chronic sinusitis and in response to topical steroids. Reported as number of positively staining subepithelial cells per high power field. (a) IL-4 receptor expression is increased in ACS

and control subjects. Interestingly, in patients treated with topical corticosteroids (ACS + S), there was significantly lower expression of IL-4R mRNA compared to ACS subjects.

The IL-5R mRNA (Figure 3b) has been shown to be expressed in the subepithelium. Both ACS and NCS subjects had significantly greater levels of expression as compared with controls. ACS subjects had a significantly greater level of expression of IL-5R than NCS patients.

The GM-CSFR mRNA (Figure 3c) was expressed in greater levels in ACS and NCS subjects as compared with controls. In contrast to IL-4R and IL-5R, the expression of GM-CSFR was significantly greater in NCS patients as compared to ACS patients. There was no difference between ACS patients and those treated with topical steroids.

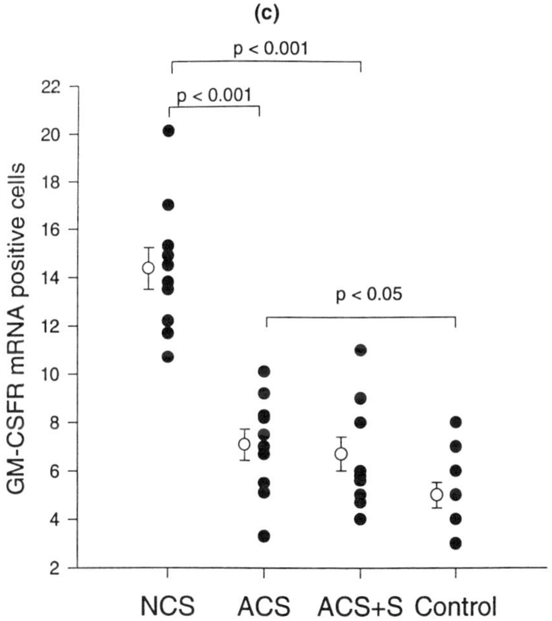


subjects. Expression is decreased in response to treatment with topical steroids. (b) IL-5 receptor expression is increased primarily in ACS subjects and also in NCS subjects. Expression is decreased in response to treatment with topical steroids. (c) GM-CSF receptor expression is increased primarily in NCS subjects.

## C. Chemokine Expression in Chronic Sinusitis [56]

For MCP-1 expression in sinus mucosal biopsies, there were no statistical differences between NCS, ACS, and control groups with respect to the number of positive subepithelial cells (Figure 4). A similar trend was observed for the epithelial score (Figure 5).

As measured by both positive subepithelial cells and epithelial score, there was a significant difference between the three groups for the expression of MCP-3. The NCS subjects had significantly greater expression of MCP-3 than controls. An identical trend was observed between ACS subjects and controls. There was no significant difference for MCP-3 expression between NCS and ACS subjects.

As measured by both positive subepithelial cells and epithelial score, there was a significant difference between the three groups for the expression of MCP-4. The NCS subjects had a significantly greater expression of MCP-4 as compared with controls. The ACS subjects also had a significantly greater expression of MCP-4 as compared with controls. There was no significant difference for MCP-4 expression between NCS and ACS subjects.

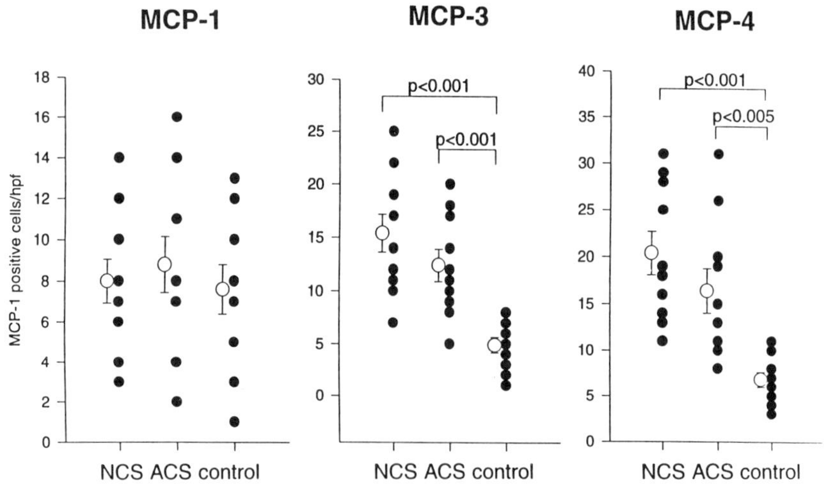


**Figure 4** MCP expression in sinus epithelium as detected by immunocytochemistry. Reported as number of positively staining cells (per high power field) in the subepithelium. There is increased expression of MCP-3 and MCP-4, but not MCP-1, in subjects with CS (either ACS or NCS).

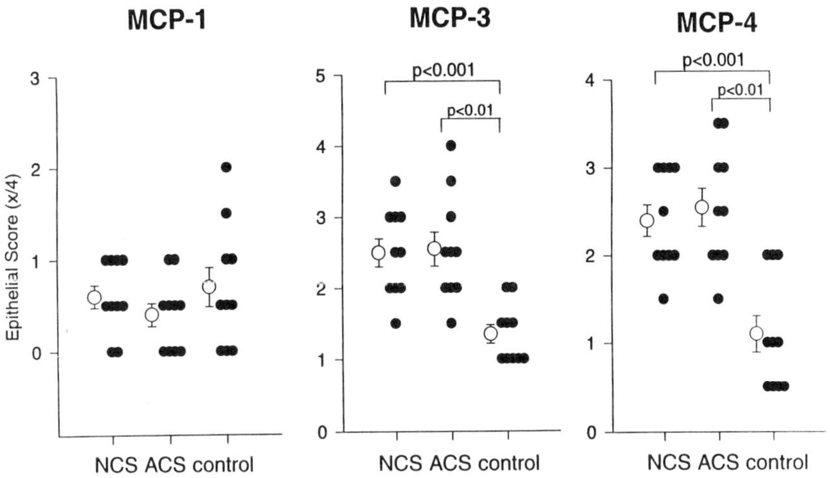


**Figure 5** MCP expression in sinus epithelium as detected by immunocytochemistry. Epithelial score represents percentage of positively staining cells in the epithelium (score out of 4). There is increased expression of MCP-3 and MCP-4, but not MCP-1, in subjects with CS (either ACS or NCS).

An important addition to the study of chemokine expression was the correlation with inflammatory cell infiltrate (Figure 6). Eosinophilic infiltrate, as measured by MBP-positive cells, was significantly increased in both NCS and ACS subjects as compared with controls. Further, both MCP-3 and MCP-4 expression correlated with eosinophil infiltrate. Addi-

| | MCP-1 | MCP-3 | MCP-4 |
|---|---|---|---|
| MBP (Eosinophils) | $R^2 = 0.08$<br>$p = 0.14$ | *$R^2 = 0.30$*<br>*$p < 0.005$* | *$R^2 = 0.39$*<br>*$p < 0.001$* |
| CD3+ T-cells | $R^2 = 0.07$<br>$p = 0.16$ | *$R^2 = 0.37$*<br>*$p < 0.001$* | *$R^2 = 0.41$*<br>*$p < 0.001$* |
| CD4+ T-cells | $R^2 = 0.039$<br>$p = 0.30$ | *$R^2 = 0.38$*<br>*$p < 0.001$* | *$R^2 = 0.42$*<br>*$p < 0.001$* |
| CD8+ T-cells | $R^2 = 0.016$<br>$p = 0.50$ | $R^2 = 0.06$<br>$p = 0.19$ | $R^2 = 0.002$<br>$p = 0.84$ |

**Figure 6** Correlations between MCP expression and inflammatory cell infiltrate. There were positive correlations between MCP expression and eosinophils (MBP+), CD3+, and CD4+ T lymphocytes but not CD8+ T lymphocytes.

tionally, there was a statistically significant correlation between CD3-positive cells and both MCP-3 and MCP-4-positive cells. Of significant interest is the fact that both MCP-3 and MCP-4 correlated with CD4+ (T-helper lymphocyte) infiltrate while there was no correlation with CD8+ cells and MCP-3/MCP-4.

## IX. DISCUSSION

Based on our recent data, eosinophil infiltration may be an important common factor in the pathogenic mechanisms of chronic sinusitis in allergic and nonallergic individuals. Further, our data suggests that the prominence of T-cells is an integral feature in the immunopathology of chronic sinusitis. This is not inconsistent with data from allergic rhinitis and asthmatic inflammation. T-cells may function to initiate and sustain the inflammation associated with chronic sinusitis through the production of cytokines such as IL-4 and IL-5.

There appear to be similar but distinct pathophysiologic mechanisms driving the inflammation in ACS and NCS subjects. In ACS patients, significant roles have been demonstrated for the cytokines IL-4, IL-5, and IL-13. The receptors for these cytokines display similar patterns of expression, as would be expected. The cytokines and receptors thus drive the IgE response and the eosinophilia thought to play a significant role in allergic inflammation. The NCS patients have been shown to have different signals for a similar pathologic result. The inflammation and eosinophilia in NCS appears driven primarily by IL-13, GM-CSF, and their receptors. Thus, a common final pathological profile is achieved by different mechanisms in ACS and NCS subjects.

It appears that chemokines such as MCP-3 and MCP-4 may play a significant role in accounting for the common cellular infiltrate seen in ACS and NCS. They may be the final common mediators, although signals for their production are likely different depending upon the etiology of the inflammation. This will likely be an area of intense research interest in the next few years.

A commonly used class of medications in upper and lower respiratory tract inflammation is topical steroid medications. From an immunomodulatory perspective, steroids appear to work primarily by decreasing expression of $T_h$-2 type cytokines and receptors in ACS subjects. Little effect has been demonstrated in NCS subjects. This observation suggests that new therapeutic interventions should address the common processes in ACS and NCS subjects.

## X. CONCLUSION

In recent years, our understanding of the pathophysiology of upper respiratory tract inflammation and allergy has improved significantly. Our appreciation of the role cytokines and their receptors play may soon allow us to engineer new, more specific treatment modalities for diseases such as allergic rhinitis and chronic sinusitis. Exciting discoveries of new mediators in chronic sinusitis such as the chemokines hold even more potential for the future.

## REFERENCES

1. Gliklich RE, Metson R. The health impact of chronic sinusitis in patients seeking otolaryngologic care. Otolaryngology—Head and Neck Surgery 1995; 113(1):104–109.
2. Van Nostrand AWP, Goodman WS. Pathologic aspects of mucosal lesions of the maxillary sinus. Otolaryngol Clin North Am 1976; 9:21–27.
3. Varney V, Jacobson M, Sudderick R et al. Immunohistology of the nasal mucosa following allergen-induced rhinitis. Identification of activated T lymphocytes, eosinophils, and neutrophils. Am Rev Respir Dis 1992; 145: 170–175.
4. Wardlaw AJ, Dunnettes S, Gleich GJ et al. Eosinophils and mast cells in bronchoalveolar lavage in mild asthma. Relationship to bronchial hyper-reactivity. Am Rev Respir Dis 1988; 137:62–69.
5. Harlin SL, Lane SR, Myers J et al. A clinical and pathologic study of chronic sinusitis: the role of the eosinophil. J Allergy Clin Immunol 1988; 81:867–75.
6. Mosmann TR, Coffman RL. TH1 and TH2 cells: different patterns of lymphokine secretion lead to different functional properties. Annual Review of Immunology 1989; 7:145–173.
7. Durham SR, Sun Ying, Varney VA, Jacobson MR, Sudderick RM, Mackay IS, Kay AB, Hamid Q. Cytokine mRNA expression for IL-3, IL-4, IL-5 and GM-CSF in the nasal mucosa after local allergen provocation: relationship to tissue eosinophilia. J Immunol. 1992; 148:2390–2394.
8. Del Prete G, Maggi E, Parronichi P, Chretien I, Tiri A, Macchia D, Ricci M, Banchereau J, De Vries J, Romagnani S. IL-4 is an essential factor for the IgE synthesis induced in vitro by human T cell clones and their supernatants. J Immunol 1988; 140:4193–198.
9. Moser R, Fehr J, Bruijnzeel PLB. IL-4 controls the selective endothelial-driven transmigration of eosinophils from allergic individuals. J Immunol 1992; 149:1432–1438.
10. Dobrina A, Menegazzi R, Carlos TM et al. Mechanisms of eosinophil adherence to cultured vascular endothelial cells. Eosinophils bind to the cytokine-

induced endothelial ligand vascular cell adhesion molecule-1 via the very late activation antigen-4 integrin receptor. J Clin Invest 1991; 88:20–26.
11. Yamaguchi Y, Hayashi Y, Sugama Y et al. Highly purified murine interleukin-5 (IL-5) stimulates eosinophil function and prolongs in vitro survival. J Exp Med 1988; 167:1737–1742.
12. Metcalf D. The granulocyte-macrophage colony stimulating factor. Science 1985; 229:16–22.
13. Metcalf D. The molecular biology and functions of the granulocyte–macrophage colony-stimulating factors. Blood 1986; 67:257–267.
14. Clutterbuck EJ, Hirst EMA, Sanderson CJ. Human interleukin-5 (IL-5) regulates the production of eosinophils in human bone marrow cultures: comparison and interaction with IL-1, IL-3, IL-6, and GM-CSF. Blood 1988; 73: 1504–1513.
15. Lopez AF, Sanderson CJ, Gamble JR, Campbell HR, Young IG, Davas MA. Recombinant human interleukin-5 is a selective activator of human eosinophil function. J Exp Med 1988; 167:219–224.
16. Bentley AM, Meng Q, Robinson DS, Hamid Q, Kay AB, Durham SR. Increases in activated T lymphocytes, eosinophils, and cytokine mRNA expression for Interleukin-5 and Granulocyte/Macrophage Colony-stimulating factor in Bronchial Biopsies after allergen challenge in Atopic Asthmatics. Am J Resp Cell Mol Biol 1993; 8:35–42.
17. Corrigan CJ, Hamid Q, North J, Barkans J, Moqbel R, Durham SR, Gemou-Engesaeth V, Kay AB. Peripheral blood CD4 but not CD8 T-lymphocyte in patients with exacerbation of asthma transcribe and translate messenger RNA encoding cytokines which prolong eosinophil survival in the context of a Th-2 type pattern: effect of glucocorticoid therapy. Am J Respir Cell Mol Biol 1995; 12:567–578.
18. Bazan JF. Structural design and molecular evolution of a cytokine receptor superfamily. Proc Natl Acad Sci USA 1990; 87:6934–6938.
19. Izerda R, March C, Mosley B et al. Human interleukin 4 receptor confers biological responsiveness and defines a novel receptor superfamily. J Exp Med 1990; 171:861–873.
20. Lopez AF, Vadas MA, Woodcock JM et al. Interleukin-5, interleukin-3, and granulocyte–macrophage colony stimulating factor cross-compete for binding to cell surface receptors on human eosinophils. J Biol Chem 1991; 266:24741–24747.
21. Tavernier J, Devos R, Cornelis S et al. A human high affinity interleukin-5 receptor (IL-5) is composed of an IL-5-specific $\alpha$ chain and a $\beta$ chain shared with the receptor for GM-CSF. Cell 1991; 66:1175–1184.
22. Murata Y, Takaki S, Migita S, Kikuchi Y, Tominaga A, Takatsu K. Molecular cloning and expression of the human interleukin-5 receptor. J Exp Med 1992; 175:341–351.
23. Lopez AF, Williamson DJ, Gamble JR et al. Recombinant human granulocyte–macrophage colony stimulating factor stimulates in vitro mature human eosinophil and neutrophil function, surface receptor expression and survival. J Clin Invest 1986; 78:1220–1228.
24. Miyajima A, Mui AL-F, Ogorochi T, Sakamaki K. Receptors for granulo-

cyte–macrophage colony-stimulating factor, interleukin-3, and interleukin-5. Blood 1993; 82:1960–1974.
25. Wognum AW, Westerman Y, Visser TP, Wagemaker G. Distribution of receptors for granulocyte–macrophage colony-stimulating factor on immature CD34+ bone marrow cells, differentiating monomyeloid progenitors, and mature blood cell subsets. Blood 1994; 84:764–774.
26. Yasruel Z, Humbert M, Kotsimbos T et al. Expression of membrane-bound and soluble interleukin-5 alpha receptor mRNA in the bronchial mucosa of atopic and non-atopic asthmatics. Amer J Resp Crit Care Med 1996. In press.
27. Kotsimbos ATC, Humbert M, Minshall E et al. Upregulation of αGM-CSF–receptor in non-atopic but not in atopic asthma. J Allerg Clin Immunol 1996. In press.
28. Pipkorn U, Proud D, Lichtenstein LM, Kagey-Sobotka A, Norman PS, Naclerio RM. Inhibition of mediator release in allergic rhinitis by pretreatment with topical glucocorticosteroids. N Engl J Med 1987; 316:1506–1510.
29. Masuyama K, Jacobson MR, Rak S, Meng Q, Sudderick RM, Kay AB, Lowhagen O, Hamid Q, Durham SR. Topical glucocorticosteroid (fluticasone propionate) inhibits cells expressing cytokine mRNA for interleukin-4 in the nasal mucosa in allergen-induced rhinitis. Immunology 1994; 82: 192–199.
30. Oppenheim JJ, Zachariae COC, Mukaida N, Matsushima K. Properties of the novel proinflammatory supergene "intercrine" cytokine family. Annu Rev Immunol 1991; 9:617–648.
31. Miller MD, Krangel MS. Biology and biochemistry of the chemokines: a family of chemotactic and inflammatory cytokines. Crit Rev Immunol 1992; 12:17–46.
32. Lukacs NW, Strieter RM, Chensue SW, Kunkel SL. Activation and regulation of chemokines in allergic airway inflammation. J Leuk Biol 1996; 59: 13–17.
33. Kaplan AP, Kuna P, Reddigari SR. Chemokines in the allergic response. Exp Dermatol 1995; 4:260–265.
34. Bacon KB, Schall TJ. Chemokines as mediators of allergic inflammation. Int Arch Allergy Immunol 1996; 109:97–109.
35. Van Damme J, Proost P, Lenaerts JP, Opdenakker G. Structural and functional identification of two human, tumor-derived monocyte chemotactic proteins (MCP-2 and MCP-3) belonging to the chemokine family. J Exp Med 1992; 176:59–65.
36. Garcia-Zepeda EA, Combadiere C, Rothenberg ME, Srafi M, Lavigne F, Hamid Q, Murphy PM, Luster AD. Human monocyte chemoattractant protein (MCP)-4 is a novel CC chemokine with activities on monocytes, eosinophils and basophils induced in allergic and non-allergic inflammation that signals through the CC chemokine receptors (CCR)-2 and -3. J Immunology 1996; 157:5613–5626.
37. Proost P, Wuyts A, Van Damme J. Human monocyte chemotactic proteins-2 and -3: structural and functional comparison with MCP-1. J Leuk Biol 1996; 59:67–74.
38. Van Damme J, Proosr P, Put W et al. Induction of monocyte chemotactic

proteins MCP-1 and MCP-2 in human fibroblastsand leukocytes by cytokines and cytokine inducers. J Immunol 1994; 152:5495–5502.
39. Shyy YJ, HsiehHJ, UsamiS, Chien S. Fluis shear stress induces a biphasic response of human monocyte chemotactic protein 1 gene expression in vascular endothelium. Proc Natl Acad Sci USA 1994; 91:4678–4682.
40. Schmouder RL, Strieter RM, Kunkel SL. Interferon-γ regulation of human renal cortical epithelial cell-derived monocyte chemotactic peptide-1. Kidney Int 1993; 44:43–49.
41. Taub DD, Proost P, Murphy WJ et al. Monocyte chemotactic protein-1 (MCP-1), -2 and -3 are chemotactic for human T lymphocytes. J Clin Invest 1995; 95:1370–1376.
42. Allavens P, Bianchi G, Zhou D et al. Induction of natural killer cell migration by monocyte chemotactic protein-1, -2, and -3. Eur J Immunol 1994; 24: 3233–3236.
43. Leonard EJ, Yoshimura T. Human monocyte chemoattractant protein-1 (MCP-1). Immunol Today 1990; 11:97–101.
44. Noso N, Proost P, Van Damme J, Schroder JM. Human monocyte chemotactic proteins-2 and -3 (MCP-2 and MCP-3) attract human eosinophils and desensitize the chemotactic responses towards RANTES. Biochem Biophys Res Commun 1994; 200:1470–1476.
45. Alam R, Forsythe P, Stafford S et al. Monocyte chemotactic protein-2, monocyte chemotactic protein-3 and fibroblast-induced cytokine, three new chemokines, induce chemotaxis and activation of basophils. J Immunol 1994; 153:3155–3159.
46. Sozzani S, Sallusto F, Luini et al. Migration of dendritic cells in response to formyl peptides, C5a and a distict set of chemokines. J Immunol 1995; 155:3292–3295.
47. Kuna P, Lazarovich M, Kaplan AP. Chemokines in seasonal allergic rhinitis. J Allergy Clin Immunol 1996; 97:104–112.
48. Kunkel SL, Lukacs N, Kasama T, Strieter RM. The role of chemokines in inflammatory joint disease. J Leuk Biol 1996; 59:6–12.
49. Schroder JM, Noso N, Sticherling M, Christophers E. Role of eosinophil–chemotactic C-C chemokines in cutaneous inflammation. J Leuk Biol 1996; 59:1–5.
50. Frew AJ, Kay AB. The relationship between infiltrating CD4 lymphocytes, activated eosinophils and the magnitude of the allergen-induced late phase cutaneous reaction in man. J Immunol 1988; 141:4158–4164.
51. Nakagawa Y, Kosugi H, Miyajima A, Arai K, Yokota T. Structure of the gene encoding the alpha subunit of the human granulocyte–macrophage colony-stimulating factor receptor. J Biol Chem 1994; 269:10905–10912.
52. Ying S, Durham SR, Barkans J et al. T cells are the principal source of interleukin-5 mRNA in allergen-induced rhinitis. Am J Respir Cell Mol Biol 1993; 9:356–360.
53. Hamilos D, Leung D, Wood R et al. Evidence for distinct cytokine expression in allergic vs. nonallergic chronic sinusitis. J Allergy Clin Immunol 1995; 96:537–544.

54. Al-Ghamdi K, Ghaffar O, Small P, Frenkiel S, Hamid Q. IL-4 and IL-13 expression in chronic sinusitis. Relationship with cellular infiltrate and the effect of topical corticosteroid treatment. J Otolaryngology 1997;26:236–240.
55. Wright ED, Frenkiel S, Small P, Al-Ghamdi K, Ghaffar O, Hamid Q. IL-4, IL-5 and GM-CSF receptor expression in chronic sinusitis and response to topical steroids. Otolaryngology—Head and Neck Surgery 1997. In press.
56. Wright ED, Frenkiel S, Ghaffar O, Al-Ghamdi K, Small P, Hamid Q. Monocyte chemotactic protein expression in allergy and non-allergy–associated chronic sinusitis. J Otolaryngology 1997. Submitted.

# 53

# Character of Bronchopulmonary Changes in Patients with Primary Ciliary Dyskinesia

**Jacek Pawlik, Jerzy Żebrak, and Andrzej Pogorzelski**
*Institute of Tuberculosis and Lung Diseases, Rabka, Poland*

**Michał Witt**
*Polish Academy of Sciences, Poznan, Poland*

## I. INTRODUCTION

The association of bronchiectasis, sinusitis, and situs inversus was first described by Siewert in 1903. The triad of symptoms is now called Kartagener's syndrome after Manes Kartagener who described (in 1933) four patients with these clinical features [1]. In the classical description Kartagener considered bronchiectasis was due to congenital deficiency of bronchial elastic tissue, which allowed infection to destroy pulmonary tissue [1].

In 1976 Afzelius described the ultrastructural defect of cilia visible through the electron microscope in patients with situs viscerum inversus [2,3]. Now Kartagener's syndrome is described as a part of the wider syndrome: primary ciliary dyskinesia—PCD (previously named immotile cilia syndrome and dyskinetic cilia syndrome) [3–7,9]. Abnormal ciliary function observed in this disease, depending on ultrastructural changes of cilia of acquired origin, leads to impaired mucociliary clearance, which is a cause of recurrent bronchopulmonary infections and causes a chronic obstructive pulmonary disease in patients with PCD syndrome [4,5,9,10]. Frequency of PCD was established as 1:16000 (1:32000). Kartagener's syndrome is twice as rare as the PCD syndrome [3,4,9,10]. Clinical and radiological manifestations of PCD have been well described [9–12]. We have not found, however, papers describing the evolution of bronchial changes in patients with PCD.

## II. AIM OF THE STUDY

The aim of the study was to estimate the usefulness of bronchological procedures (bronchoscopy and bronchography) in evaluating and localizing bronchial damage in patients with primary ciliary dyskinesia (PCD) syndrome and to relate the findings to the patients' age.

## III. MATERIAL

Ninety-five patients—41 men and 54 women, aged from 1 to 41 years—were included into the study. All the patients were examined and treated in the period 1993–1996 for primary ciliary dyskinesia/immotile cilia syndrome in the Institute of Tuberculosis and Lung Diseases, Clinic of Bronchology and Cystic Fibrosis in Rabka. In 69 out of 95 patients situs viscerum inversus totalis (Kartagener's syndrome) was diagnosed. In 26 out of 95 patients the probable diagnosis of PCD was established based on clinical history, ENT examination, physical and bronchological examination. We paid particular attention to the following: (a) Kartagener's syndrome previously diagnosed; (b) Kartagener's syndrome in PCD patients' families; (c) dextrocardia/dextroposition of the heart—family analysis; and (d) children with recurrent chronic bronchitis, recurrent chronic sinusitis, recurrent chronic otitis media, in whom light microscopy did not reveal motile cilia—family analysis.

## IV. METHODS

In all cases a good history was obtained. All patients were diagnosed using our standard patients' examination card. When indicated, chest x-rays and/or x-rays of paranasal sinuses were performed.

We performed bronchological examinations in all patients using standard indications. Bronchological examinations of younger children were performed using general anaesthesia: premedication with Atropine (0.01 mg/kg IV or SC), specific anesthesia with Halothane; sometimes Midazolam or Metohexital. In younger children we often used a rigid bronchoscope (Storz). In older children bronchoscopy was performed using flexible instruments (OLYMPUS or PENTAX) with premedication with Atropine (0.01 mg/kg). Sometimes before bronchofiberoscopy Dolargan (1.0–1.5 mg/kg IV) was given to calm the patient. Topical anesthesia either during rigid or flexible procedures was performed with 2% Xylocaine. Bronchography, when indicated, was performed using Hytrast (Byk Gulden).

Frequency of appearance and localization of bronchographic changes were evaluated. We took into account the following changes: bronchiectasis, bronchial deformations, signs of retention of secretions, and signs of fibrosis of the lung tissue. We compared bronchial changes with the normal anatomical picture of bronchi. We routinely examined and took biopsies from the right middle lobe no matter which part of the lung was involved.

In all the patients we performed examination of ciliary motility using the light microscope, defining only whether there was motility or not. Material (mucus membrane) was taken from one of the bifurcations of segmental bronchi (mostly from the bifurcation of the sixth segment of right or left lung) during bronchoscopy, using biopsy forceps. If the material was taken during ENT examination the place of biopsy was an edge of the lower nasal turbinate. If there was at least one place with motile cilia we considered the result as positive. We did not analyze the amplitude of motion or the length of cilia—due to lack of possibilities. When one patient had more than one examination, we examined ciliary motility every time. During bronchoscopy we collected samples of bronchial mucus membrane to analyze with electron microscopy (examinations were performed concurrently). The protocol of the study was approved by the Ethics Committee of our institute.

## V. GROUP CHARACTERISTICS

Physical examination and history revealed the following symptoms in our group of 95 PCD patients: recurrent/chronic bronchitis, 100% of patients; recurrent/chronic sinusitis, 100% of patients; recurrent/chronic otitis media, 100% of patients; situs viscerum inversus totalis, 72.6% (69 patients); bronchiectasis, 58.9% (56 patients); bronchial deformations only, 15.8% (15 patients); and nasal polyps, 4.2% (4 patients).

## VI. CILIARY MOTILITY EXAMINATION

Examination of ciliary motility was performed in 89 cases. We found motile cilia in 16 out of 89 examined patients (18.6%).

**Table 1** Localization of Bronchographic Changes in Patients with PCD in the Age Group Below 10 Years ($n = 26$)

| Localization | Bronchi-ectasis | Bronchial deformations | Retention of secretions and fibrosis of the lung tissue | Without changes |
|---|---|---|---|---|
| Right upper lobe | 1 (3.8%) | 1 (3.8%) | 5 (19.2%) | 19 (73.1%) |
| Right middle lobe | 6 (23.1%) | 3 (11.5%) | 7 (26.9%) | 10 (38.5%) |
| Right lower lobe | 2 (7.7%) | 2 (7.7%) | 9 (34.6%) | 13 (50%) |
| Left upper lobe | — | 1 (3.8%) | 4 (15.4%) | 21 (80.8%) |
| Left lower lobe | 1 (3.8%) | 1 (3.8%) | 5 (19.2%) | 19 (73.1%) |
| **Character of changes** | | | **Right lung** | **Left lung** |
| Bronchiectasis | | | 6 (23.1%) | 1 (3.8%) |
| Bronchial deformations | | | 3 (11.5%) | 2 (7.7%) |
| Retention of secretions and fibrosis of the lung tissue | | | 7 (26.9%) | 5 (19.2%) |
| Bronchography without changes | | | 10 (38.5%) | 18 (69.2%) |

**Table 2** Localization of Bronchographic Changes in Patients with PCD in the Age Group Between 10–18 Years ($n = 36$)

| Localization | Bronchi-ectasis | Bronchial deformations | Retention of secretions and fibrosis of the lung tissue | Without changes |
|---|---|---|---|---|
| Right upper lobe | 1 (2.8%) | 2 (5.6%) | 6 (16.7%) | 27 (75%) |
| Right middle lobe | 14 (38.9%) | 7 (19.4%) | 7 (19.4%) | 8 (22.2%) |
| Right lower lobe | 2 (5.6%) | 6 (16.7%) | 19 (52.8%) | 9 (25%) |
| Left upper lobe | 2 (5.6%) | 4 (11.1%) | 11 (30.6%) | 19 (52.8%) |
| Left lower lobe | 7 (19.4%) | 2 (5.6%) | 18 (50%) | 9 (25%) |
| **Character of changes** | | | **Right lung** | **Left lung** |
| Bronchiectasis | | | 15 (41.7%) | 9 (25%) |
| Bronchial deformations | | | 9 (25%) | 6 (16.7%) |
| Retention of secretions and fibrosis of the lung tissue | | | 10 (27.8%) | 14 (38.9%) |
| Bronchography without changes | | | 2 (5.6%) | 7 (19.4%) |

**Table 3** Localization of Bronchographic Changes in Patients with PCD in the Age Group Above 18 Years ($n$ = 33)

| Localization | Bronchi-ectasis | Bronchial deformations | Retention of secretions and fibrosis of the lung tissues | Without changes |
|---|---|---|---|---|
| Right upper lobe | 1 (3%) | 1 (3%) | 8 (24.2%) | 23 (69.7%) |
| Right middle lobe | 21 (63.6%) | 5 (15.1%) | 4 (12.1%) | 3 (9.1%) |
| Right lower lobe | 10 (30.3%) | 7 (21.2%) | 8 (24.2%) | 8 (24.2%) |
| Left upper lobe | 11 (33.3%) | 4 (12.1%) | 8 (24.2%) | 10 (30.3%) |
| Left lower lobe | 9 (27.3%) | 1 (3%) | 17 (51.5%) | 6 (18.2%) |
| Character of changes | | | Right lung | Left lung |
| Bronchiectasis | | | 23 (69.7%) | 13 (39.4%) |
| Bronchial deformations | | | 4 (12.1%) | 3 (9.1%) |
| Retention of secretions and fibrosis of the lung tissue | | | 5 (15.1%) | 12 (36.4%) |
| Bronchography without changes | | | 1 (3%) | 5 (15.1%) |

## VII. RESULTS

Taking into account the dynamics of development of bronchial changes in PCD patients, results were analyzed by dividing the study group into three age groups (below 10 y; from 10–18 y; and above 18 y) - see Tables 1–3.

## VIII. DISCUSSION

In the literature we found that irreversible injuries of bronchi occurred in 77% of PCD patients in the retrospective observation of Reyes de la Rocha et al. [10]. Sturgess and Turner [7] found that bronchiectasis was present in approximately 50% of children aged 10 to 18 years but in their previous article from 1981 the percentage was lower—only 29% of 21 patients with PCD had bronchiectases. In 1991 Barlocco et al. [8] found that bronchiectases were present in 85.7% of examined PCD patients. Our results are in agreement with these findings. In our material the frequency of irreversible bronchial changes was 58.9%. The results of our work indicate that the frequency of irreversible changes of bronchi increases with the patients' age. The incidence of bronchiectases in the age group below 10 y is 23%, in the age group from 10 to 18, 55.6%, and in adulthood, 87.9%. So the observed difference in frequency of incidence of the changes may depend on the patients' age.

We observed a low percentage of nasal polyps in our patients. We found only 4 patients (4%) with nasal polyps, while in other studies we found that the frequency was 19% [4] and 26.5% [7]. In contrast to these studies Zalewska found nasal polyps in only 2 children (4.8%) in the series of 42 Kartagener's syndrome patients [11].

Frequency of situs viscerum inversus (SVI) in the examined group is different in different articles but it is approximately 50% of cases [3–7,9]. Barlocco [8] found 39.3% of SVI in a series of 28 PCD patients. This confirms that the SVI appears in PCD patients as a facultative feature resulting from the lack of motility of cilia during embriogenesis. In our material the frequency of SVI was 72% mainly due to the fact that we diagnosed PCD without SVI only when patient's complaints were evident. In cases with abnormal mucociliary clearance coexisting with situs

**Table 4** Abnormalities Found in PCD Patients in Relation to Situs Inversus

| | Kartagener's syndrome | PCD situs solitus |
|---|---|---|
| Number of patients | 69 (72.6%) | 26 (27.6%) |
| Recurrent/chronic bronchitis | 69 (100%) | 26 (100%) |
| Recurrent/chronic sinusitis | 69 (100%) | 26 (100%) |
| Recurrent/chronic otitis media | 69 (100%) | 26 (100%) |
| Motile cilia | 14 (20.3%) | 2 (7.7%) |
| Bronchiectasis | 42 (60.8%) | 14 (53.8%) |
| Only deformations of bronchi | 10 (14.5%) | 5 (19.2%) |
| Retention of secretions and fibrosis of the lung tissue | 12 (17.4%) | 5 (19.2%) |
| Bronchography without changes | 5 (7.2%) | 2 (7.7%) |

viscerum inversus we can be sure that the diagnosis of PCD is correct. (Table 4).

A lower percentage of Kartagener's syndrome patients in younger age groups—below 10 years and from 10–18 years—is evidence of better diagnostic procedures in PCD in cases with chronic bronchitis without situs viscerum inversus (Table 5). We suppose that our 26 patients without situs viscerum inversus have, in reality, PCD according to the presence of appropriate symptoms (in spite of the lack of results of examination of ultrastructure of cilia) (Table 4).

A lower percentage of irreversible deformative bronchial changes in younger age groups confirms the theory that bronchial changes in PCD patients are of acquired origin and gives evidence of the progressive character of the disease (of 26 children with PCD below 10 years of age only 23% had confirmed bronchiectasis) (Table 5). From previous examinations of patients with Kartagener's syndrome made in our institute in 1980 it was concluded that irreversible bronchial changes occurred in patients under long-term observation as a result of recurrent bronchopulmonary infections [11].

In the current group of 95 patients, bronchographic changes were mainly located in the right middle lobe, which fully confirms the role of the right middle lobe bronchus as a locus minoris resistentiae in diseases with impaired mucociliary clearance.

**Table 5** Abnormalities Found in PCD in Relation to Patients Age

| Patients age | Below 10 years $n = 26$ | From 10–18 y $n = 36$ | Above 18 y $n = 33$ | Total $n = 95$ |
|---|---|---|---|---|
| Kartagener's syndrome | 61.5%(16) | 66.7%(24) | 87.9%(29) | 69 (72.6%) |
| PCD without situs viscerum inversus | 38.5%(10) | 33.3%(12) | 12.1%(4) | 26 (27.4%) |
| Motile cilia | 26.9%(7) | 15.2%(5) | 12.9%(4) | 16 (18%) |
| Bronchiectasis | 23.1%(6) | 55.6%(20) | 87.9%(29) | 55 (57.9%) |
| Only deformations of bronchi | 15.4%(4) | 30.5%(11) | 3%(1) | 16 (16.8%) |
| Retention of secretions and fibrosis of the lung tissue | 34.6%(9) | 13.9%(5) | 9.1(3) | 17 (17.9%) |
| Bronchography without changes | 26.9%(7) | none (0) | none (0) | 7 (7.4%) |

## IX. ACKNOWLEDGMENT

The authors wish to thank Dr. Waldemar Tomalak for his support during the translation of this paper.

## REFERENCES

1. Kartagener M. Zur Pathogenese der bronchiectasien. Beitr Klin Tuberk 1933; 83:489–501.
2. Afzelius BA. A human syndrome caused by immotile cilia. Science 1976; 193:317–319.
3. Afzelius BA. Genetical and ultrastructural aspects of the immotile cilia syndrome. Am J Hum Gen 1981; 33:852–864.
4. Sturgess J, Turner P. The immotile cilia syndrome. In: Fraser P, ed. Diseases of the chest. 1991; 675–682.
5. Greenstone M, Rutman A et al. Primary ciliary dyskinesia: clinical and cytological features. Q J Med 1988; 67:405–430.
6. Rossmann CM, Lee RMKW, Forrest JB, Newhouse MT. Nasal ciliary ultrastructure and function in patients with primary ciliary dyskinesia compared with that in normal subjects and in subjects with various respiratory diseases. Am Rev Resp Dis 1984; 129:161–167.
7. Sturgess JM, Turner JAP. Ultrastructural pathology of cilia in the immotile cilia syndrome. Perspect Pediatr Path 1984; 8:133–161.
8. Barlocco E, Valetta E, Canciani M, Lungarella G et al. Ultrastructural ciliary defects in children with recurrent infections of the lower respiratory tract. Ped Pulm 1991; 10:11–17.
9. Rossman CM, Newhouse MT. Primary ciliary dyskinesia: evaluation and management. Ped Pulm 1980; 5:36–50.
10. Reyes de la Rocha S, Pysher T, Leonard J, Dyskinetic cilia syndrome: clinical, radiographic and scintigraphic findings. Ped Radiol 1987; 17:97–103.
11. Majewska, Zalewska H. Zespót Kartagenera u dzieci. Ped Pol 1980; LV:7.
12. Nadel HR, Stringer DA, Levison H, Turner JAP, Sturgess JM. The immotile cilia syndrome. Radiological manifestations. Radiology 1985; 154:651–655.

# 54

# Genetic Counseling of Adults with Primary Ciliary Dyskinesia or Cystic Fibrosis

**Hans-Dieter Rott**
*University of Erlangen-Nuremberg, Erlangen, Germany*

## I. INTRODUCTION

Primary ciliary dyskinesia (PCD) and cystic fibrosis (CF) are the two most common genetic diseases exhibiting abnormalities of mucociliary clearance. In both diseases the pattern of inheritance is autosomal recessive with a recurrence risk in sibs of 25% [1]. Recently, as a consequence of improving therapy for CF as well as of male sterility in PCD and CF there has been an increasing demand for genetic counseling by affected adults who wish to have their own children. This paper deals with genetic problems arising from these patients.

## II. RECURRENCE RISK ASSESSMENT

The occurrence risk for PCD and CF in children of an affected parent is 50% or zero, depending on whether the spouse is heterozygotic for the appropriate gene or not. The frequency ($f_h$) of heterozygotic individuals in a population can be estimated from the incidence [I] of the disease. According to Hardy–Weinbergs equilibrium [2] for low incidences is $f_h \cong 2\sqrt{I}$. Assuming an incidence of PCD of about 1:15 000 [3] the frequency of heterozygotes is about 1:61 and the occurrence risk for PCD in a child is about 0.8%. Assuming genetic heterogeneity of PCD this risk is somewhat

**Table 1** CF Incidence and Occurrence Risks for Children from CF Affected Patients[a] in Different Jewish Ethnic Groups in Israel

| Group | CF incidence | Calculated probability for | |
|---|---|---|---|
| | | Heterozygosity | CF affected child |
| Ashkenazi Jews | 1:3300 | 1:29 | 1.7% |
| Nonashkenazi Jews from | | | |
| Lybia | 1:2700 | 1:26 | 1.9% |
| Georgia | 1:2700 | 1:26 | 1.9% |
| Greece and Bulgaria | 1:2400 | 1:24 | 2.0% |
| Yemen | 1:8800 | 1:47 | 1.1% |
| Morocco | 1:15,000 | 1:61 | 0.8% |
| Iraq | 1:32,000 | 1:89 | 0.6% |
| Iran | 1:39,000 | 1:99 | 0.5% |

[a] Parental consanguinity excluded.
Source: Reference 6.

overestimated. In Germany the risk figures for CF are as follows: the incidence is about 1:2 000, the prevalence of heterozygotes 1:22, and the occurrence risk in children about 2%. The situation for PCD is probably the same in the United States and Israel as in Germany. The estimation of CF occurrence risks must take into consideration the different prevalences of heterozygotic individuals in various subpopulations. The ethnic differences are mostly due to the different prevalence of the Δf508 mutation, which originates from northern Europe. In the United States the prevalence of this mutation varies considerably from 75% and 35% of all mutations in the cystic fibrosis transmembrane regulator (CFTR) gene [4]. For Jewish ethnic groups the situation is similar (Table 1). Moreover, for both diseases the occurrence risks suppose no consanguinous mating and absence of affected family members in the family of the spouse. For these special situations individual risk estimations are necessary that consider the respective pedigree constellation.

## III. HETEROZYGOSITY TESTING

The above mentioned risks can be specified if the genetic status of the spouse has been examined. This is especially important in the case of consanguineous mating. For PCD heterozygosity testing is not feasible at present, because the gene loci are yet to be identified. This situation might, however, change in the near future [5]. For CF, more than 400 different mutations have been identified in the CFTR gene [6]. In Germany, a series of 15 mutations are routinely tested for that identifies about 85% of all CF mutations. A complete mutation screening for all mutations is not practicable at present for routine testing. If a spouse is a proven carrier, the CF risk for a child increases to 50%. In the majority of spouses fortunately no mutation can be found. In this case the remaining risk of heterozygosity is about 1% and the risk for an affected child decreases to about 0.2%.

## IV. MALE STERILITY

Male sterility due to immotility of sperm occurs in the majority of males with PCD. Affected women may conceive spontaneously, although overall fertility seems to be reduced. In CF males sperm are usually normal, but a substantial portion of CF males are sterile due to aplasia of the vas deferens. In both diseases in vitro fertilization by intracytoplasmic sperm injection (ICSI) is the only promising treatment. In CF, a testicular biopsy must be performed to obtain the necessary sperm. The ICSI technique has an efficiency of about 30%, provided that three fertilized eggs are implanted. In patients with aplasia of the vas deferens, ICSI is currently a routine clinical treatment. For PCD males at least two successful fertilizations have been observed, one resulting the birth of a healthy child [7]. However, ICSI implies an increased risk for twin pregnancies. Moreover, there is possibly an increased fetal risk for chromosomal aberrations by undermining the natural selection against sperm with additional chromosomes. Prenatal chromosome analyses of ICSI pregnancies have indeed shown slightly increased numbers of sex chromosome aberrations. In contrast, malformation syndromes and genetic diseases are apparently not increased in frequency. Until the genetic risks of ICSI are fully evaluated, it appears wise to discuss and offer the full spectrum of prenatal diagnostic procedures [8,9].

## V. PREGNANCY RISKS IN FEMALES

Before planning a family, the potential effect of pregnancy on maternal disease and vice versa have to be considered. For PCD as well as for CF, good pulmonary function must be proven with a sufficient capacity for the last trimester of pregnancy. In women receiving long-term medication teratogenic effects have to be excluded. On the positive side pregnancy does not influence either mucus properties or ciliary function.

## VI. PRENATAL DIAGNOSIS

For PCD there is at present no prenatal diagnosis available. For CF prenatal diagnosis depends on whether the parental mutations have been identified. The following situations have to be considered: (a) One or both mutations in the affected parent are identified, heterozygosity testing of the healthy spouse gave normal findings. In this case the occurrence risk is $< 1\%$; prenatal diagnosis of CF is not possible. (b) Both mutations in the affected parent are identified, the healthy spouse is proven to be heterozygotic. In this situation the occurrence risk is 50%. The pedigree information is informative, prenatal diagnosis can be offered. (c) Only one mutation of the affected parent is identified, the healthy spouse is proven heterozygotic for a different mutation. For prenatal diagnosis this constellation is comparable to (b) and prenatal diagnosis can be offered. (d) Only

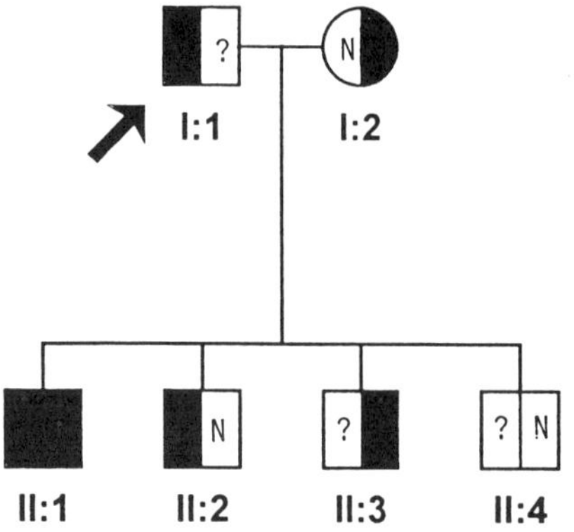


**Figure 1** Hypothetical semi-informative pedigree constellation in prenatal diagnosis of CF. I:1 (arrow): CF Patient with one identified mutation (black) and one unidentified mutation (?). I:2: Heterozygotic spouse with the same mutation as her husband. The second gene must be normal, because she is not affected. In the fetus alternatives II:1 and II:4 permit definitive diagnoses: II:1 is proven homozygotic and affected; II:4 is heterocygotic and not affected. II:2 and II:3 cannot be differentiated: II:2 would be heterozygotic and not affected, II:3 would be homozygotic, although with different mutations, and affected. Therefore in case of fetal heterozygosity for the identified mutation the CF risk is 50%.

one mutation of the affected parent is identified, the spouse is proven to be heterozygotic for the same mutation. This is a semi-informative pedigree that gives a definitive result in only 50% of the cases (for details see Figure 1).

In addition to molecular genetic studies, sonographic examination may give some additional prenatal information. In PCD fetuses, half the cases have situs inversus, which can be seen by ultrasound. The CF fetuses may have sonographic signs of meconium ileus. Of course, exclusion of PCD or CF by sonography is not possible.

## ACKNOWLEDGMENT

The author is greatly indebted to Dr. Maggie Meeks, London, for critical review of the manuscript.

## REFERENCES

1. McKusick VA, Francomano CA, Antonarakis SE. Mendalian Inheritance in Man. Catalogs of autosomal dominant, autosomal recessive and X-linked phenotypes. 10th. ed. Baltimore, London: The John Hopkins University Press, 1992.
2. Hardy GH. Mendelian proportions in a mixed population. Science 1908; 28: 49–50.
3. Rott HD. Kartagener syndrome and the syndrome of immotile cilia. Hum Genet 1979; 46:249–261.
4. The Cystic Fibrosis Genetic Analysis Consortium. World wide survey of the Δf508 mutation—Report from the The Cystic Fibrosis Analysis Consortium. Am J Hum Genet 1990; 47:354–359.
5. Meeks MG, Rott HD, Gardiner RM, Chung E. Mapping a gene for primary ciliary dyskinesia (PCD) using the technique of homozygosity mapping in a large consanguineous family. International Congress on Cilia, Mucus and Mucociliary Interactions, Jerusalem, February, 23–28, 1997.
6. Kerem E, Kalman YM, Yahav Y, Shoshani T, Abeliowich D, Szeinberg A, Rivlin J, Blau H, Tal A, Ben-Tur L, Springer C, Augarten A, Godfrey S, Lerer I, Branski D, Friedman M, Kerem B. Highly variable incidence of cystic fibrosis and different mutation distribution among different Jewish ethnic groups in Israel. Hum Genet 1995; 96:193–197.
7. Rott HD. Personal observation, 1997.
8. Engel W, Schmid M. Gibt es genetische Risiken der mikroassistieren Reproduktion? Fertilität 1995; 11:214–228.
9. Meschede D, De Geyter C, Nieschlag E, Horst J. Genetic risk in micromanipulative assisted reproduction. Human Reproduction 1995; 11:2880–2886.

# 55

# Preliminary Exclusion of Chromosome 7 as a Potential Location of a Gene for Primary Ciliary Dyskinesia

**Michał Witt and Ewa Rutkiewicz**
*Polish Academy of Sciences, Poznan, Poland*

**Elizabeth Gillanders, Shengbiao Wang, Diana Freas-Lutz, Cui-e Sun, and Scott R. Diehl**
*National Institutes of Health, Bethesda, Maryland*

**Jerzy Żebrak and Jacek Pawlik**
*Institute of Tuberculosis and Lung Diseases, Rabka, Poland*

## I. INTRODUCTION

Primary ciliary dyskinesia (PCD), formerly known as immotile cilia syndrome (ICS), is a genetically determined group of disorders characterized by dysmotility or immotility of the cilia in airway epithelial cells, spermatozoa, and other ciliated cells of the body [1–3]. Kartagener's syndrome is a subgroup of PCD characterized by a classic triad of symptoms: situs inversus, bronchiectasis, and chronic sinusitis. The clinical consequences of PCD include chronic cough and expectoration, bronchiectasis and/or obstructive lung disease, chronic rhinitis, nasal polyposis, chronic or recurrent sinusitis, often agenesis of the frontal sinuses, otosalpingitis, and otitis media. Sterility usually occurs in males and infertility also frequently occurs in females. Ciliary immotility is caused by various ultrastructural defects of cilia, predominantly by a total absence of dynein arms (70–80% of cases). Other less common ultrastructural abnormalities include partial absence or fragmentation of dynein arms, radial spoke defects (total absence, absent spoke head and inner dynein arm), absence of nexin links and inner dynein arms with axonemal disorganization, and microtubular transposition. Thus far, at least 11 subgroups of defects of ciliary ultrastructure have been described for PCD [1].

The incidence of PCD has been calculated to range from 1/16,000 to 1/60,000 live births [1], and PCD is likely to be a genetically heterogeneous disease. Inheritance in most cases is autosomal recessive, although some examples of dominant or X-linked modes of inheritance have been reported [4]. In some PCD families, about half of the affected sibs also display situs inversus. These families are categorized as Kartagener's Syndrome (KS) and the incidence of this subform of PCD is estimated to range from 1/32,000 to 1/120,000. The incidence of situs inversus without any respiratory or other abnormality is 1/8,000 to 1/11,000. It appears that homozygotes of the recessively transmitted KS gene have lost the usual developmental controls that result in the normal direction of asymmetry, so that a random process determines whether they exhibit situs inversus or situs solitus. Nearly 200 different polypeptides have been identified within the ciliary axoneme of lower organisms; at least the same number of proteins can be expected in axonemes of humans [5]. It is likely that mutations within many of the 2 genes coding for various ciliary proteins might cause the same or similar pathologic consequences of ciliary dysfunction. But if this were true, one might expect that the incidence of PCD would be much higher than actually occurs. This might suggest involvement of a more limited number of genes in PCD pathology. Several animal models of primary ciliary dyskinesia have been found: mice, [6], rats [7], dogs [8], and pigs [9]. Previous attempts to localize the assumed gene for PCD have been unsuccessful. However, none of these previous studies used very large samples of subjects or applied the currently available modern large-scale mapping technologies.

Currently, several chromosomal regions and genes can be considered as candidates for PCD. These include (a) the HLA region (chromosome 6p) containing the β-tubulin gene (TUBB) [10] (however, the motylin gene (MLN), which resides in the same region, did not appear to be involved in PCD etiology in another set of families [11]); (b) the chromosome 7q33-q34 region, which is syntenic to a fragment of murine chromosome containing the hop mutation [12] (previously named hpy [6]) (mice homozygous for this mutation have a dynein defect in cilia and flagella [dynein arms missing from A-tubules of the outer doublets] similar to that seen in some cases of PCD); and (c) the chromosome 14q32 containing Usher syndrome type 1A gene [13] and 14qter region containing dynein heavy chain gene [14].

Many common diseases such as diabetes, ischemic heart disease, asthma, and schizophrenia are likely to be etiologically and genetically heterogeneous, where the same clinical disorder is caused by different susceptibility genes and environmental factors. Data from several authors suggest that PCD may be similarly complex in etiology [2]. Successful dissection of complex traits by genome-wide linkage mapping currently can be applied to any organism chosen for study. The characterization of microsatellite marker loci by the polymerase chain reaction (PCR) [15], construction of high-resolution human genetic linkage maps [16], and the application of fluorescence-based, automated DNA fragment-sizing technology [17,18] have now greatly facilitated human genome-wide mapping [19].

## II. PATIENTS, MATERIALS, AND METHODS

The study group consisted of 32 families of Polish origin in which at least one child was diagnosed with PCD. These included 20 families that were informative for linkage analysis (i.e., with diagnostic evaluation and DNA available for a sufficient number of family members to allow inferences about linkage). Our sample consisted primarily of nuclear families but some included more distant relatives. Sixteen of our informative families contained one or more members displaying situs inversus thus characterizing them as Kartagener's syndrome, while 4 informative families exhibited other PCD signs and symptoms. Microsatellite marker genotypes were analyzed with the use of ABI PRISM Linkage Mapping Set on an automatic DNA fragment analyzer equipped with GeneScan software [17,18]. Alleles were scored using the GENOTYPER program correlating fluorescent intensity of labeled DNA fragments (PCR products) with marker loci PCR product size distribution information. Pairwise linkage analysis (LOD scores) using the program FASTLINK was performed assuming a recessive model with 95% penetrance. The admixture test of linkage in the presence of locus heterogeneity was performed using a simple program developed for this purpose (S. Diehl, unpublished).

## III. RESULTS AND DISCUSSION

Here we report results for a set of 17 microsatellite markers spanning all of chromosome 7. The average distance between microsatellite loci analyzed was 10 cM. The genetic map of the distal part of human chromo-

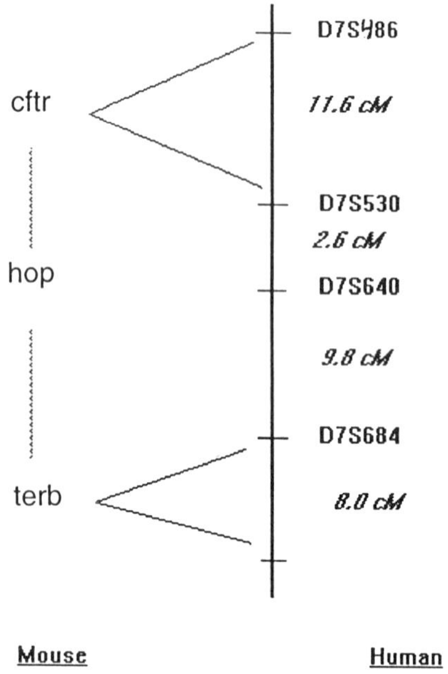


**Figure 1** Map of distal human chromosome 7q and the homologous portion of the mouse genome containing the hop gene.

some 7q is compared to its murine counterpart containing the hop mutation in Figure 1. Results of LOD score calculations assuming homogeneity for the recessive model tested are summarized in Table 1. No evidence of linkage was obtained after allowing for possible locus heterogeneity. Out of the 20 potentially informative families analyzed only eight provided substantial linkage information; LOD scores for the other families were on a level of 0.001–0.002. Although these results do not support the suggestion that the distal fragment of chromosome 7q may contain a gene involved in primary ciliary dyskinesia, the limited number of informative families contained in our study limits the statistical power of this exclusion. We are currently in the process of further evaluating this chromosomal region using a larger group of informative PCD families as part of our genome-wide scan for genes contributing to the susceptibility of this disorder.

**Table 1** LOD Scores for Microsatellite Markers on Chromosome 7 Assuming Locus Homogeneity

| | cM Distance | | | |
|---|---|---|---|---|
| Marker | 0 cM | 1 cM | 5 cM | 10 cM |
| D7S531 | −4.817 | −3.544 | −2.110 | −1.336 |
| D7S513 | −4.713 | −3.854 | −2.350 | −1.383 |
| D7S517 | −5.328 | −4.233 | −2.493 | −1.504 |
| D7S507 | −3.081 | −2.228 | −1.137 | −.0584 |
| D7S493 | −1.835 | −1.120 | −0.353 | −0.025 |
| D7S629 | −1.712 | −1.550 | −0.859 | −0.373 |
| D7S484 | −3.132 | −2.629 | −1.155 | −.0324 |
| D7S519 | −1.444 | −1.236 | −0.358 | 0.132 |
| D7S502 | −3.490 | −2.976 | −1.610 | −0.732 |
| D7S669 | −2.986 | −2.123 | −0.957 | −0.370 |
| D7S657 | −3.673 | −3.226 | −2.131 | −1.368 |
| D7S527 | −2.331 | −2.023 | −1.260 | −0.753 |
| D7S486 | −1.000 | −0.796 | −0.314 | −0.044 |
| D7S530 | −2.088 | −1.716 | −0.902 | −0.415 |
| D7S640 | −3.086 | −2.213 | −1.237 | −0.708 |
| D7S684 | −1.896 | −1.530 | −0.744 | −0.304 |
| D7S550 | −5.435 | −3.864 | −1.982 | −1.072 |

Markers are presented in the table corresponding to their order in the genetic map ranging from 7pter to 7qter.

## ACKNOWLEDGMENTS

We thank Nandita Chopra, Heena Beck, and Yi Jie Zhang for assistance in the laboratory and Yue-Fen Wang, Don Smith, Carla Bock, Mary Gregg, and Alejandro Schaffer for data management and statistical analyses.

## REFERENCES

1. Afzelius BA, Mossberg B. Immotile–cilia syndrome (primary ciliary dyskinesia), including Kartagener syndrome. In: CR Scriver et al. ed. The Metabolic Basis of Inherited Disease. New York: McGraw Hill, 1992; 2739–2750.

2. Kartagener syndrome (244400). In: McKusick V, ed. Mendelian Inheritance of Man. Baltimore: The Johns Hopkins University Press, 1992:1488–1490.
3. Schidlow DV. Primary ciliary dyskinesia (the immotile cilia syndrome). Ann Allergy 1994; 73:457–470.
4. Narayan D, Krishnan SN, Upender M et al. Unusual inheritance of primary ciliary dyskinesia (Kartagener syndrome). J Med Genet 1994; 31:493–496.
5. Luck DJL, Huang B, Piperno G. Genetic and biochemical analysis of the eukaryotic flagellum. Soc Exp Biol Symp 1982; 35:399–307.
6. Bryan JHD, Hughes RL, Bates TJ. Brain development in hydrocephalic-polydactyl, a recessive pleiotropic mutant in the mouse. Virchows Arch A Pathol Anat Histol 1977; 374:205–214.
7. Torikata C, Kijimoto C, Koto M. Ultrastructure of respiratory cilia of WIC-Hyd male rats. An animal model for human immotile cilia syndrome. Am J Pathol 1991; 138:341–347.
8. Edwards DF, Patton CS, Bemis DA et al. Immotile cilia syndrome in three dogs from a litter. J Am Vet Med Assoc 1983; 183:667–672.
9. Roperto F, Galati O, Rossacco P. Immotile cilia syndrome in pigs. A model for human disease. Am J Pathol 1993; 143:643–647.
10. Volz A, Weiss E, Trowsdale J, Ziegler A. Presence of an expressed β-tubulin gene (TUBB) in the HLA class I region may provide the genetic basis for HLA-linked microtubule dysfunction. J Hum Genet 1994; 93:42–46.
11. Gasparini P, Grifa A, Savasta S et al. The motilin gene: subregional localisation, tissue expression, DNA polymorphisms and exclusion as a candidate gene for the HLA-associated immotile cilia syndrome. Hum Genet 1994; 94: 671–674.
12. Handel MA. Allelism of hop and hpy. Mouse News Lett 1985; 72:124.
13. Bonneau D, Raymond F, Kremer C et al. Usher syndrome type 1 associated with bronchiectasis and immotile nasal cilia in two brothers. J Med Genet 1993; 30:253–254.
14. Narayan D, Desai T, Banks A et al. Localization of the human cytoplasmic dynein heavy chain (DNECL) to 14qter by fluorescence in situ hybridization. Genomics 1994; 22:660–661.
15. Weber JL, May PE. Abundant class of human DNA polymorphisms which can be typed using the polymerase chain reaction. Am J Hum Genet 1989; 44:388–396.
16. Gyapay G, Morissette J, Vignal A et al. The 1993–1994 Genethon human genetic linkage map. Nature Genet 1994; 7:246–339.
17. Reed PW, Davies JL, Copeman JB et al. Chromosome-specific microsatellite sets for fluorescence-based, semi-automated genome mapping., Nature Genet 1994; 7:390–395.
18. Ziegle JS, Su Y, Corcoran KP et al. Application of automated DNA sizing technology for genotyping microsatellite loci. Genomics 1992; 14:1026–1031.
19. Davies JL, Kawaguchi Y, Bennett ST et al. A genome-wide search for human type 1 diabetes susceptibility genes. Nature 1994; 371:130–136.

# 56
# Bend Shapes and Molecular Mechanisms of Flagella

**Michael Holwill**
*King's College, London, United Kingdom, and Albert Einstein College of Medicine, Bronx, New York*

## I. INTRODUCTION

Bends are formed and propagated along eukaryotic flagella by an internal mechanism that involves microtubule sliding induced by the molecular motor, axonemal dynein [6]. The shape of propagated bends provides information about the molecular interactions responsible for microtubule sliding. For example, if the nature and degree of axonemal twisting are known, the shape allows the extent of relative sliding between microtubules to be determined from simple geometric considerations [3]; the amount of sliding is related, inter alia, to the step size of the motor molecule. A detailed knowledge of sliding patterns gives important data from which to determine the relationship between the force-generating motor and the structures that resist sliding and convert the shear forces into bending moments. While the shape of a bend reflects the molecular mechanism responsible for generating it, the viscous and elastic loading of the system may also be determining factors. The hispid flagella of monads, the smooth flagellum of *Crithidia oncopelti* and the demembranated flagellum of *Crithidia*, provide significantly different resistances to be overcome by the molecular motors. A comparative analysis of bend shapes for these organelles suggests that in intact flagella the dynein arms exert forces that are large compared to the resistances offered, while in the demembranated organelles the two are of the same order of magnitude.

## II. MATERIAL AND METHODS

Images of swimming cells were recorded, under phase-contrast conditions, at 100 pictures per second using a Jackson high-speed video camera mounted on a Zeiss WL Research microscope. Illumination was provided by a Chadwick Helmuth point source stroboscope synchronised to the camera framing rate. The final magnification on the monitor screen was about 3000X. As a preliminary stage in digitizing prior to further processing, a flagellar image was traced onto a transparent film taped to the monitor screen. Tracings were digitized with a Summagraphics digitizing tablet, which features a 0.1 mm resolution, an order of magnitude better than the effective physical resolution of the recording system.

## III. ANALYSIS

Data analysis and assessment followed the procedure described in detail by Johnston et al. [4]. Specific analytical functions were selected, expressed in intrinsic coordinates (i.e., tangent angle $\phi$ at distance $s$ along the curve, with $\phi = 0$ at $s = 0$) and their sets of Fourier coefficients calculated. the Fourier series representing a waveform can be expressed as

$$\phi(s) = \sum_{n=1}^{\infty} b_{2n-1} \sin\left[(2n - 1)\frac{\pi s}{2S}\right]$$

since the coefficients of the cosine terms and of the even sine terms are all zero. In this equation $S$ is the total length of the curve under analysis. The coefficients are given by

$$b_{2n-1} = \frac{2}{S}\int_0^S \phi(s) \sin\left[(2n - 1)\frac{\pi s}{2S}\right] ds$$

### A. Analysis of Standard Waves

Specific analytical functions were selected, expressed in intrinsic coordinates, and their sets of Fourier coefficients calculated. Figure 1 shows the coefficients for arc-line waves in which the fraction of circular arc varied from 30% to 100%. Wave shapes with less than 30% arc are not a realistic representation of the flagellar wave. The second, third, and fourth coefficients ($b_3$, $b_5$, $b_7$), expressed as a percentage of $b_1$, for this range of shapes are plotted in Figure 1. Other test functions analysed were the

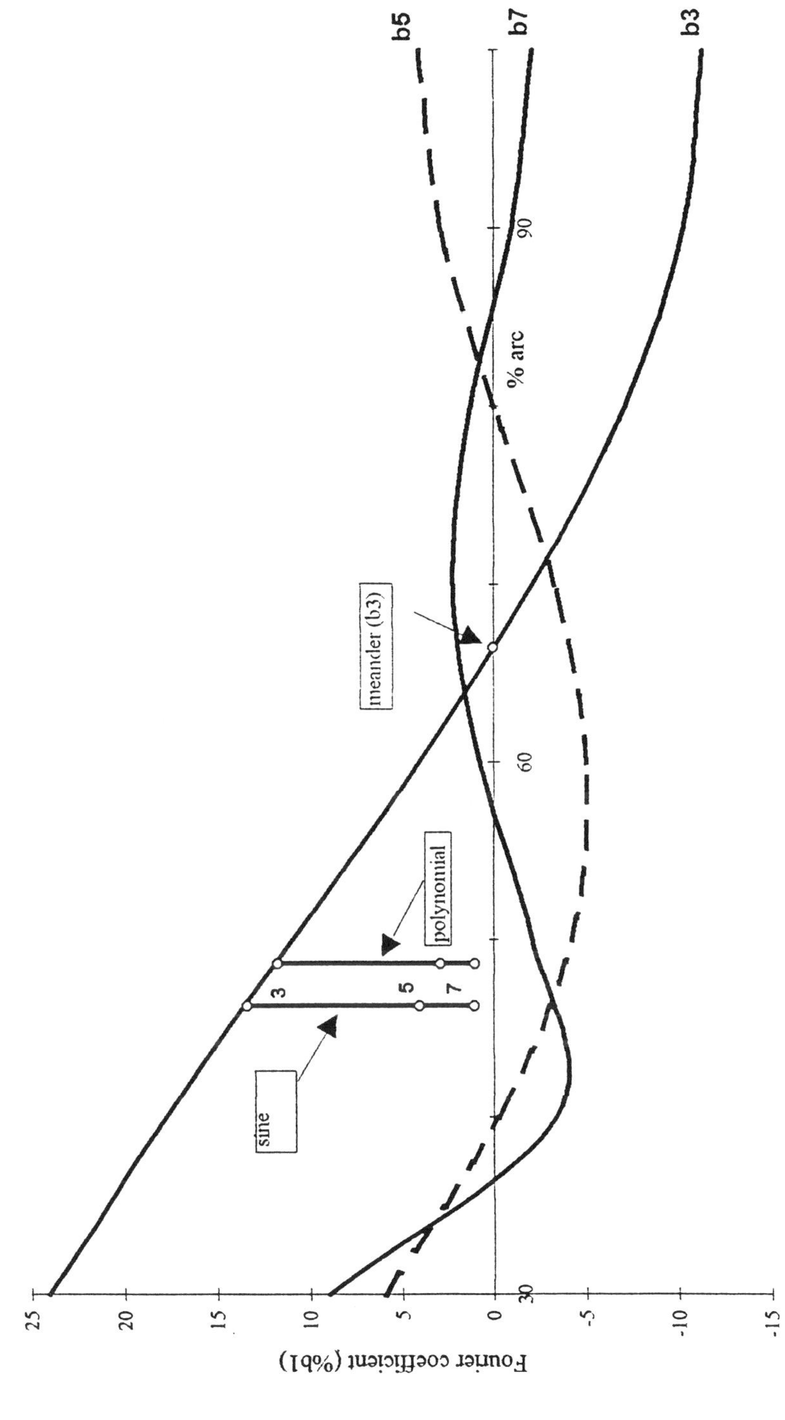


**Figure 1** Fourier coefficients for a number of standard waves.

meander, sine-generated wave, and a polynomial of third degree, all with wavelength/amplitude ratios similar to that for the arc-line series. A pure sine function was also included in the set of test functions for comparative purposes only, since it is usually apparent by visual inspection that the wave shapes are not sinusoidal.

### B. Analysis of Flagellar Waves

The shape assumed by a flagellum generally shows variations in amplitude, wavelength, and curvature as the wave propagates. Since the Fourier coefficients obtained for the reference functions relate to waves of constant amplitude, manipulation of the experimental data was necessary to provide comparable waves. Accordingly the waveforms were divided into segments corresponding to a quarter wavelength. Appropriate reflection and translation operations were performed on the quarter waves to generate complete, uniform waveforms, which were then transformed into intrinsic coordinates and the Fourier coefficients computed.

## IV. RESULTS

Comparison of the experimental and theoretical coefficients indicates that the wave shapes of the different flagella studied are as shown in the table below.

| Organism | Flagellum type | Bend shape | Reference |
|---|---|---|---|
| *Crithidia oncopelti* | smooth | arc-line | 4 |
| Monad species | hispid | arc-line | 2 |
| *Crithidia oncopelti* | demembranated (smooth) | meander | 5 |

## V. DISCUSSION

The mechanical loadings on smooth and hispid flagella are significantly different from each other, as illustrated by considering the situation at the wave crest of an undulating flagellum with no translational movement. For a smooth flagellum, the viscous forces are proportional to the local velocity of the organelle and, in the region of the wave crest, will be predominantly normal to the wave axis for a nontranslating organelle. Because the flagellum in this region is reversing direction, with its instantaneous (transverse) velocity at the crest becoming zero, the local forces here will be small. (Stresses generated by movements of other regions of the flagellum have, of course, to be sustained locally by the structure.) For a hispid flagellum, the mastigonemes at the wave crest sweep through an angle in the plane of the wave, with a lateral velocity that will be maximal at the wave crest, and elicit a thrust from the fluid such that there will be significant loading on the flagellar shaft in a direction parallel to the wave axis. The wave shapes on smooth and hispid flagella are of the same form, despite the different external viscous loading.

The meander shape adopted by the demembranated flagellum suggests that the elastic properties of the flagellum offer resistances to motion that are comparable to the forces generated by the dynein arms. Since the structural integrity of the axoneme has not been destroyed, this suggests that the forces generated by the dynein arms may have been reduced compared with those in the intact axoneme, perhaps through changes in the ionic environment.

## VI. CONCLUSIONS

Variations in external viscous forces have little effect on wave shape, which is determined therefore by interactions within the axoneme itself. The forces generated by the dynein arms, together with the elastic components that convert the shearing forces into bending moments operate in such a way that they are essentially unaffected by externally imposed forces. This result is consistent with the conclusion reached by Hamasaki et al. [1] in modeling the translocation of microtubules by a field of dynein molecules. Agreement between the computer simulation and the experimental results was achieved by assuming that the resistance offered by a translocating microtubule is negligible compared to the force generated by the dynein molecule.

The different wave shapes assumed by intact and demembranated flagella can be explained by assuming that the mechanism for bend propagation remains intact in the in vitro organelle, but the local environment reduces the magnitude of the forces that can be generated by the dynein arms.

## REFERENCES

1. Hamasaki T, Holwill MEJ, Barkalow K, Satir P. Mechanochemical aspects of axonemal dynein activity studied by *in vitro* microtubule translocation. Biophys J 1995; 69:2569–2579.
2. Holwill MEJ, Gaitzsch P. Bend shapes of hispid flagella. Mol Biol Cell 1996; 7:48a.
3. Holwill MEJ, Cohen HJ, Satir P. A sliding microtubule model incorporating axonemal twist and compatible with three-dimensional ciliary bending. J Exp Biol 1979; 78:265–280.
4. Johnston DN, Silvester NR, Holwill MEJ. An analysis of the shape and propagation of waves on the flagellum of *Crithidia oncopelti*. J Exp Biol 1979; 80: 299–315.
5. Marchese-Ragona SP, Glazzard AN, Holwill MEJ. A comparison of the wave parameters of flagellar axonemes with 9+2 and 9+1 microtubular configurations. J Submicroscop Cytol 1983; 15:55–59.
6. Satir P. Landmarks in cilia research from Leeuwenhoek to us. Cell Motil Cytoskeleton 1995; 32:90–94.

# Index